London Mathematical Society Lecture Note Series: 340



# Groups St Andrews 2005

## Volume 2

*Edited by*

C.M. Campbell
*University of St Andrews*

M.R. Quick
*University of St Andrews*

E.F. Robertson
*University of St Andrews*

G.C. Smith
*University of Bath*

CAMBRIDGE
UNIVERSITY PRESS

CAMBRIDGE UNIVERSITY PRESS
Cambridge, New York, Melbourne, Madrid, Cape Town, Singapore, São Paulo

Cambridge University Press
The Edinburgh Building, Cambridge CB2 2RU, UK

Published in the United States of America by Cambridge University Press, New York

www.cambridge.org
Information on this title: www.cambridge.org/9780521694698



Printed in the United Kingdom at the University Press, Cambridge

*A catalogue record for this publication is available from the British Library*

ISBN-13 978-0-521-69470-4 paperback
ISBN-10 0-521-69470-1 paperback

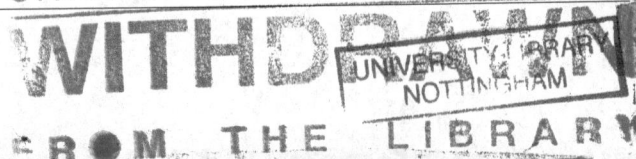

LONDON MATHEMATICAL SOCIE

Managing Editor: Professor N.J. Hitc St Giles, Oxford OXI 3LB, United Kingdom

The titles below are available from booksellers, or from Cambridge University Press at www.cambridge.org/mathematics

214 Generalised Euler-Jacobi inversion formula and asymptotics beyond all orders, V. KOWALENKO *et al*
215 Number theory 1992–93, S. DAVID (ed)
216 Stochastic partial differential equations, A. ETHERIDGE (ed)
217 Quadratic forms with applications to algebraic geometry and topology, A. PFISTER
218 Surveys in combinatorics, 1995, P. ROWLINSON (ed)
220 Algebraic set theory, A. JOYAL & I. MOERDIJK
221 Harmonic approximation., S.J. GARDINER
222 Advances in linear logic, J.-Y. GIRARD, Y. LAFONT & L. REGNIER (eds)
223 Analytic semigroups and semilinear initial boundary value problems, KAZUAKI TAIRA
224 Computability, enumerability, unsolvability, S.B. COOPER, T.A. SLAMAN & S.S. WAINER (eds)
225 A mathematical introduction to string theory, S. ALBEVERIO, *et al*
226 Novikov conjectures, index theorems and rigidity I, S. FERRY, A. RANICKI & J. ROSENBERG (eds)
227 Novikov conjectures, index theorems and rigidity II, S. FERRY, A. RANICKI & J. ROSENBERG (eds)
228 Ergodic theory of $\mathbf{Z}^d$ actions, M. POLLICOTT & K. SCHMIDT (eds)
229 Ergodicity for infinite dimensional systems, G. DA PRATO & J. ZABCZYK
230 Prolegomena to a middlebrow arithmetic of curves of genus 2, J.W.S. CASSELS & E.V. FLYNN
231 Semigroup theory and its applications, K.H. HOFMANN & M.W. MISLOVE (eds)
232 The descriptive set theory of Polish group actions, H. BECKER & A.S. KECHRIS
233 Finite fields and applications, S. COHEN & H. NIEDERREITER (eds)
234 Introduction to subfactors, V. JONES & V.S. SUNDER
235 Number theory 1993–94, S. DAVID (ed)
236 The James forest, H. FETTER & B. G. DE BUEN
237 Sieve methods, exponential sums, and their applications in number theory, G.R.H. GREAVES *et al*
238 Representation theory and algebraic geometry, A. MARTSINKOVSKY & G. TODOROV (eds)
240 Stable groups, F.O. WAGNER
241 Surveys in combinatorics, 1997, R.A. BAILEY (ed)
242 Geometric Galois actions I, L. SCHNEPS & P. LOCHAK (eds)
243 Geometric Galois actions II, L. SCHNEPS & P. LOCHAK (eds)
244 Model theory of groups and automorphism groups, D. EVANS (ed)
245 Geometry, combinatorial designs and related structures, J.W.P. HIRSCHFELD *et al*
246 $p$-Automorphisms of finite $p$-groups, E.I. KHUKHRO
247 Analytic number theory, Y. MOTOHASHI (ed)
248 Tame topology and o-minimal structures, L. VAN DEN DRIES
249 The atlas of finite groups: ten years on, R. CURTIS & R. WILSON (eds)
250 Characters and blocks of finite groups, G. NAVARRO
251 Gröbner bases and applications, B. BUCHBERGER & F. WINKLER (eds)
252 Geometry and cohomology in group theory, P. KROPHOLLER, G. NIBLO & R. STÖHR (eds)
253 The $q$-Schur algebra, S. DONKIN
254 Galois representations in arithmetic algebraic geometry, A.J. SCHOLL & R.L. TAYLOR (eds)
255 Symmetries and integrability of difference equations, P.A. CLARKSON & F.W. NIJHOFF (eds)
256 Aspects of Galois theory, H. VÖLKLEIN *et al*
257 An introduction to noncommutative differential geometry and its physical applications 2ed, J. MADORE
258 Sets and proofs, S.B. COOPER & J. TRUSS (eds)
259 Models and computability, S.B. COOPER & J. TRUSS (eds)
260 Groups St Andrews 1997 in Bath, I, C.M. CAMPBELL *et al*
261 Groups St Andrews 1997 in Bath, II, C.M. CAMPBELL *et al*
262 Analysis and logic, C.W. HENSON, J. IOVINO, A.S. KECHRIS & E. ODELL
263 Singularity theory, B. BRUCE & D. MOND (eds)
264 New trends in algebraic geometry, K. HULEK, F. CATANESE, C. PETERS & M. REID (eds)
265 Elliptic curves in cryptography, I. BLAKE, G. SEROUSSI & N. SMART
267 Surveys in combinatorics, 1999, J.D. LAMB & D.A. PREECE (eds)
268 Spectral asymptotics in the semi-classical limit, M. DIMASSI & J. SJÖSTRAND
269 Ergodic theory and topological dynamics, M.B. BEKKA & M. MAYER
270 Analysis on Lie groups, N.T. VAROPOULOS & S. MUSTAPHA
271 Singular perturbations of differential operators, S. ALBEVERIO & P. KURASOV
272 Character theory for the odd order theorem, T. PETERFALVI
273 Spectral theory and geometry, E.B. DAVIES & Y. SAFAROV (eds)

274 The Mandlebrot set, theme and variations, TAN LEI (ed)
275 Descriptive set theory and dynamical systems, M. FOREMAN *et al*
276 Singularities of plane curves, E. CASAS-ALVERO
277 Computational and geometric aspects of modern algebra, M.D. ATKINSON *et al*
278 Global attractors in abstract parabolic problems, J.W. CHOLEWA & T. DLOTKO
279 Topics in symbolic dynamics and applications, F. BLANCHARD, A. MAASS & A. NOGUEIRA (eds)
280 Characters and automorphism groups of compact Riemann surfaces, T. BREUER
281 Explicit birational geometry of 3-folds, A. CORTI & M. REID (eds)
282 Auslander-Buchweitz approximations of equivariant modules, M. HASHIMOTO
283 Nonlinear elasticity, Y. FU & R.W. OGDEN (eds)
284 Foundations of computational mathematics, R. DEVORE, A. ISERLES & E. SÜLI (eds)
285 Rational points on curves over finite, fields, H. NIEDERREITER & C. XING
286 Clifford algebras and spinors 2ed, P. LOUNESTO
287 Topics on Riemann surfaces and Fuchsian groups, E. BUJALANCE *et al*
288 Surveys in combinatorics, 2001, J. HIRSCHFELD (ed)
289 Aspects of Sobolev-type inequalities, L. SALOFF-COSTE
290 Quantum groups and Lie theory, A. PRESSLEY (ed)
291 Tits buildings and the model theory of groups, K. TENT (ed)
292 A quantum groups primer, S. MAJID
293 Second order partial differential equations in Hilbert spaces, G. DA PRATO & J. ZABCZYK
294 Introduction to the theory of operator spaces, G. PISIER
295 Geometry and Integrability, L. MASON & YAVUZ NUTKU (eds)
296 Lectures on invariant theory, I. DOLGACHEV
297 The homotopy category of simply connected 4-manifolds, H.-J. BAUES
298 Higher operads, higher categories, T. LEINSTER
299 Kleinian Groups and Hyperbolic 3-Manifolds Y. KOMORI, V. MARKOVIC & C. SERIES (eds)
300 Introduction to Möbius Differential Geometry, U. HERTRICH-JEROMIN
301 Stable Modules and the D(2)-Problem, F.E.A. JOHNSON
302 Discrete and Continuous Nonlinear Schrödinger Systems, M. J. ABLORWITZ, B. PRINARI & A. D. TRUBATCH
303 Number Theory and Algebraic Geometry, M. REID & A. SKOROBOGATOV (eds)
304 Groups St Andrews 2001 in Oxford Vol. 1, C.M. CAMPBELL, E.F. ROBERTSON & G.C. SMITH (eds)
305 Groups St Andrews 2001 in Oxford Vol. 2, C.M. CAMPBELL, E.F. ROBERTSON & G.C. SMITH (eds)
306 Peyresq lectures on geometric mechanics and symmetry, J. MONTALDI & T. RATIU (eds)
307 Surveys in Combinatorics 2003, C. D. WENSLEY (ed.)
308 Topology, geometry and quantum field theory, U. L. TILLMANN (ed)
309 Corings and Comodules, T. BRZEZINSKI & R. WISBAUER
310 Topics in Dynamics and Ergodic Theory, S. BEZUGLYI & S. KOLYADA (eds)
311 Groups: topological, combinatorial and arithmetic aspects, T. W. MÜLLER (ed)
312 Foundations of Computational Mathematics, Minneapolis 2002, FELIPE CUCKER *et al* (eds)
313 Transcendental aspects of algebraic cycles, S. MÜLLER-STACH & C. PETERS (eds)
314 Spectral generalizations of line graphs, D. CVETKOVIC, P. ROWLINSON & S. SIMIC
315 Structured ring spectra, A. BAKER & B. RICHTER (eds)
316 Linear Logic in Computer Science, T. EHRHARD *et al* (eds)
317 Advances in elliptic curve cryptography, I. F. BLAKE, G. SEROUSSI & N. SMART
318 Perturbation of the boundary in boundary-value problems of Partial Differential Equations, D. HENRY
319 Double Affine Hecke Algebras, I. CHEREDNIK
321 Surveys in Modern Mathematics, V. PRASOLOV & Y. ILYASHENKO (eds)
322 Recent perspectives in random matrix theory and number theory, F. MEZZADRI & N. C. SNAITH (eds)
323 Poisson geometry, deformation quantisation and group representations, S. GUTT *et al* (eds)
324 Singularities and Computer Algebra, C. LOSSEN & G. PFISTER (eds)
325 Lectures on the Ricci Flow, P. TOPPING
326 Modular Representations of Finite Groups of Lie Type, J. E. HUMPHREYS
328 Fundamentals of Hyperbolic Manifolds, R. D. CANARY, A. MARDEN & D. B. A. EPSTEIN (eds)
329 Spaces of Kleinian Groups, Y. MINSKY, M. SAKUMA & C. SERIES (eds)
330 Noncommutative Localization in Algebra and Topology, A. RANICKI (ed)
331 Foundations of Computational Mathematics, Santander 2005, L. PARDO, A. PINKUS, E. SULI & M. TODD (eds)
332 Handbook of Tilting Theory, L. ANGELERI HÜGEL, D. HAPPEL & H. KRAUSE (eds)
333 Synthetic Differential Geometry 2ed, A. KOCK
334 The Navier-Stokes Equations, P. G. DRAZIN & N. RILEY
335 Lectures on the Combinatorics of Free Probability, A. NICA & R. SPEICHER
336 Integral Closure of Ideals, Rings, and Modules, I. SWANSON & C. HUNEKE
337 Methods in Banach Space Theory, J. M. F. CASTILLO & W. B. JOHNSON (eds)
338 Surveys in Geometry and Number Theory N. YOUNG (ed)

# Contents of Volume 2

# Contents of Volume 1

# Introduction

Groups St Andrews 2005 was held in the University of St Andrews from 30 July to 6 August 2005. This was the seventh in the series of Groups St Andrews group theory conferences organised by Colin Campbell and Edmund Robertson of the University of St Andrews. The first three were held in St Andrews and subsequent conferences held in Galway, Bath and Oxford before returning to St Andrews in 2005. We are pleased to say that the conference was, we believe, a success having been attended by 230 participants from 37 countries. The lectures and talks were given in the Mathematical Institute and the School of Physics and Astronomy of the University of St Andrews. Accommodation was provided in New Hall and in Fife Park.

The Scientific Organising Committee of Groups St Andrews 2005 was: Colin Campbell (St Andrews), Nick Gilbert (Heriot-Watt), Steve Linton (St Andrews), John O'Connor (St Andrews), Edmund Robertson (St Andrews), Nik Ruskuc (St Andrews), Geoff Smith (Bath). The Committee received very valuable support from our Algebra colleagues at St Andrews, both staff and postgraduate students. Once again, we believe that the support of the two main British mathematics societies, the Edinburgh Mathematical Society and the London Mathematical Society has been an important factor in the success of these conferences.

The main speakers at the meeting were Peter J Cameron (Queen Mary, London), Rostislav I Grigorchuk (Texas A&M), John C Meakin (Nebraska-Lincoln) and Akos Seress (Ohio State). Additionally there were seven one-hour invited speakers together with an extensive programme of over a hundred seminars; a lot to fit into a week! As has become the tradition, all the main speakers have written substantial articles for these Proceedings. Each volume begins with two such articles. All papers have been subjected to a formal refereeing process comparable to that of a major international journal. Publishing constraints have forced the editors to exclude some very worthwhile papers, and this is of course a matter of regret.

It is hoped that the next conference in this series will be held in 2009 and will be Groups St Andrews in Bath. As Colin and Edmund will have retired before Groups St Andrews 2009, they look forward to this conference with special interest and anticipate a slightly different role. These Proceedings do, however, mark the first 25 years of Groups St Andrews conferences. In addition to the mathematics, illustrated by these Proceedings, the seven conferences have contained a wide selection of social events. "Groups St Andrews tourism" has taken us to a variety of interesting and scenic venues in Scotland, Ireland, England and Wales. Bus trips have included Kellie Castle, Loch Earn and Loch Tay, Falkland Palace and Hill of Tarvit, Crathes Castle and Deeside, Loch Katrine and the Trossachs, House of Dun, Connemara and Kylemore Abbey, the Burren and the Cliffs of Moher, Tintern Abbey and Welsh Valleys, the Roman Baths in Bath, Salisbury Cathedral, Rufus Stone and the New Forest, Stonehenge, Wells Cathedral and the Cheddar Gorge, Blenheim Palace, Glamis Castle. We have been on boats on Loch Katrine, the Thames, and Galway Bay to the Aran Islands. There have also been: musical events with participants as the musicians, Scottish Country Dance evenings, barn dances, piano recitals, organ recitals, theatre trips, whisky tasting, putting, chess, walks along the Fife Coast, walks round Bath, and walks round Oxford. All these have provided opportunities for relaxation, but also opportunities to continue mathematical discussions. We hope that the cricket match (and balloon trip) postponed from 1997 will take place only 12 years late at Groups St Andrews 2009 in Bath!

Thanks to those authors who have contributed articles to these Proceedings. We have edited their articles to produce some uniformity without, we hope, destroying individual styles. For any inconsistency in, and errors introduced by, our editing we take full responsibility. Our final thanks go to Roger Astley and the rest of the Cambridge University Press team for their assistance and friendly advice throughout the production of these Proceedings.

Colin Campbell
Martyn Quick
Edmund Robertson
Geoff Smith

# GROUPS AND SEMIGROUPS: CONNECTIONS AND CONTRASTS

JOHN MEAKIN

Department of Mathematics, University of Nebraska, Lincoln, NE 68588, USA
Email: jmeakin@math.unl.edu

## 1 Introduction

Group theory and semigroup theory have developed in somewhat different directions in the past several decades. While Cayley's theorem enables us to view groups as groups of permutations of some set, the analogous result in semigroup theory represents semigroups as semigroups of functions from a set to itself. Of course both group theory and semigroup theory have developed significantly beyond these early viewpoints, and both subjects are by now integrally woven into the fabric of modern mathematics, with connections and applications across a broad spectrum of areas.

Nevertheless, the early viewpoints of groups as groups of permutations, and semigroups as semigroups of functions, do permeate the modern literature: for example, when groups act on a set or a space, they act by permutations (or isometries, or automorphisms, etc.), whereas semigroup actions are by functions (or endomorphisms, or partial isometries, etc.). Finite dimensional linear representations of groups are representations by invertible matrices, while finite dimensional linear representations of semigroups are representations by arbitrary (not necessarily invertible) matrices. The basic structure theories for groups and semigroups are quite different — one uses the ideal structure of a semigroup to give information about the semigroup for example — and the study of homomorphisms between semigroups is complicated by the fact that a congruence on a semigroup is not in general determined by one congruence class, as is the case for groups.

Thus it is not surprising that the two subjects have developed in somewhat different directions. However, there are several areas of modern semigroup theory that are closely connected to group theory, sometimes in rather surprising ways. For example, central problems in finite semigroup theory (which is closely connected to automata theory and formal language theory) turn out to be equivalent or at least very closely related to problems about profinite groups. Linear algebraic monoids have a rich structure that is closely related to the subgroup structure of the group of units, and this has interesting connections with the well developed theory of (von Neumann) regular semigroups. The theory of inverse semigroups (i.e., semigroups of partial one-one functions) is closely tied to aspects of geometric and combinatorial group theory.

In the present paper, I will discuss some of these connections between group theory and semigroup theory, and I will also discuss some rather surprising contrasts between the theories. While I will briefly mention some aspects of finite semigroup theory, regular semigroup theory, and the theory of linear algebraic monoids, I will focus primarily on the theory of inverse semigroups and its connections with geometric group theory.

For most of what I will discuss, there is no loss of generality in assuming that the semigroups under consideration have an identity — one can always just adjoin an identity to a semigroup if necessary — so most semigroups under consideration will be monoids, and on occasions the group of units (i.e., the group of invertible elements of the semigroup) will be of considerable interest.

## 2 Submonoids of Groups

It is perhaps the case that group theorists encounter semigroups (or monoids) most naturally as submonoids of groups. For example, if $P$ is a submonoid of a group $G$ such that $P \cap P^{-1} = \{1\}$, then the relation $\leq_P$ on $G$ defined by $g \leq_P h$ iff $g^{-1}h \in P$ is a left invariant partial order on $G$. This relation is also right invariant iff $g^{-1}Pg \subseteq P$ for all $g \in G$, and it is a total order iff $P \cup P^{-1} = G$. Note that $g \in P$ iff $1 \leq_P g$. Every left invariant partial order on $G$ arises this way. One says that $(G, P)$ is a partially ordered group with positive cone $P$. One may note that the partial order has the property that for all $g \in G$ there exists some $p \in P$ such that $g \leq_P p$ iff $G = PP^{-1}$, i.e., iff $G$ is the group of (right) quotients of $P$. The study of ordered groups is well over a hundred years old, and I will not attempt to survey this theory here.

The question of embeddability of a semigroup (monoid) in a group is a classical question that has received a lot of attention in the literature. Clearly a semigroup must be cancellative if it is embeddable in a group. It is easy to see that commutative cancellative semigroups embed in abelian groups, in fact such a semigroup embeds in its group of quotients in much the same way as an integral domain embeds in a field. For non-commutative semigroups, the situation is far more complicated. One useful condition in addition to cancellativity that guarantees embeddability of a semigroup $S$ in a group is the Ore condition. A semigroup $S$ satisfies the *Ore condition* if any two principal right ideals intersect, i.e., $sS \cap tS \neq \emptyset$ for all $s, t \in S$. (In the language of many subsequent authors in the group theory literature, $s$ and $t$ have at least one *common multiple* for each $s, t \in S$: in the language of classical semigroup theory, one says that $S$ is *left reversible*). The following well known result was essentially proved by Ore in 1931: a detailed proof may be found in Volume 1, Chapter 1 of the book by Clifford and Preston [30], which is a standard reference for basic classical results and notation in semigroup theory. There is an obvious dual result involving right reversible semigroups and groups of left quotients.

**Theorem 1** *A cancellative semigroup satisfying the Ore condition can be embedded in a group. In fact a cancellative semigroup $P$ can be embedded in a group $G = PP^{-1}$ of (right) quotients of $P$ if and only if $P$ satisfies the Ore condition.*

As far as I am aware, the first example of a cancellative semigroup that is not embeddable in a group was provided by Mal'cev in 1937 [85]. Necessary and sufficient conditions for the embeddability of a semigroup in a group were provided by

Mal'cev in 1939 [86]. Mal'cev's conditions are countably infinite in number and no finite subset of them will suffice to ensure embeddability of a semigroup in a group. A similar set of conditions, with a somewhat more geometric interpretation, was provided by Lambek in 1951. Chapter 10 of Volume 2 of Clifford and Preston [30] provides an account of the work of Mal'cev and Lambek and a description of the relationship between the two sets of conditions.

The question of when a monoid with presentation $P = Mon\langle X : u_i = v_i\rangle$ embeds in a group has been studied by many authors, and has received attention in the contemporary literature in group theory. Clearly such a monoid embeds in a group if and only if it embeds in the group with presentation $G = Gp\langle X : u_i = v_i\rangle$. Here the $u_i, v_i$ are *positive* words, i.e., $u_i, v_i \in X^*$, where $X^*$ denotes the free monoid on $X$. We allow for the possibility that some of the words $u_i$ or $v_i$ may be empty, (i.e., the identity of $X^*$). Also, we use the notation $Mon\langle X : u_i = v_i\rangle$ for the monoid presented by the set $X$ of generators and relations of the form $u_i = v_i$ to distinguish it from the group $Gp\langle X : u_i = v_i\rangle$ or the semigroup $Sgp\langle X : u_i = v_i\rangle$ with the same set of generators and relations. From an algorithmic point of view, the embeddability question is undecidable, as are many such questions about semigroup presentations or group presentations, since the property of being embeddable in a group is a Markov property (see Markov's paper [94]).

It is perhaps worth observing that being embeddable in a group is equivalent to *being* a group for special presentations where all defining relations are of the form $u_i = 1$. Recall that the *group of units* of a monoid $P$ is the set

$$U(P) = \{a \in P : ab = ba = 1 \text{ for some } b \in P\}.$$

**Proposition 1** *Let $P$ be a monoid with presentation of the form $P = Mon\langle X : u_i = 1,\ i = 1, \ldots, n\rangle$, where each letter of $X$ is involved in at least one of the relators $u_i$. Then $P$ is embeddable in a group if and only if it is a group.*

**Proof** Suppose that $P$ is embeddable in a group $G$, and consider a relation $u_i = 1$ in the set of defining relations of $P$. If $u_i = x_1x_2 \ldots x_n$ with each $x_j \in X$, then clearly $x_1$ is the inverse of $x_2 \ldots x_n$ in $G$, so $x_1$ is in the group of units of $P$ and $x_2 \ldots x_n x_1 = 1$ in $P$ also. It follows that $x_2$ is in the group of units of $P$, and similarly each $x_j$ must be in the group of units of $P$. Since this holds for each relator $u_i$, and since each letter in $X$ is involved in some such relator, every letter of $X$ (i.e., every generator of $P$) must lie in the group of units of $P$, so $P$ is a group. □

**Remark** We remark at this point that the word problem for one-relator monoids with a presentation of the form $M = Mon\langle X : u = 1\rangle$ was solved by Adian [2]. However the word problem for semigroups with one defining relation of the form $S = Sgp\langle X : u = v\rangle$ where both $u$ and $v$ are non-empty words in $X^*$ remains open, as far as I am aware. There has been considerable work done on the one-relator semigroup problem in general (see for example, the papers by Adian and Oganessian [3], Guba [54], Lallement [75], Watier [144], and Zhang [148]). Later in this paper, I will indicate how this problem is related to the membership problem for certain submonoids of one-relator groups.

Despite the difficulties in deciding embeddability of a semigroup in a group in general, there are many significant results in the literature that show that monoids (semigroups) with particular presentations may be embedded in the corresponding groups. Perhaps the first such general result along these lines was obtained by Adian [1].

Let $P$ be a semigroup with presentation $P = Sgp\langle X : u_i = v_i,\ i = 1, \ldots, n\rangle$, where $u_i, v_i$ are strictly positive (i.e., non-empty) words. The *left graph* for this presentation is the graph with set $X$ of vertices and with an edge from $x$ to $y$ if there is a defining relation of the form $u_i = v_i$ where $x$ is the first letter of $u_i$ and $y$ is the first letter of $v_i$. The *right graph* is defined dually. The semigroup $P$ is called an *Adian semigroup* and the corresponding group $G = Gp\langle X : u_i = v_i\rangle$ is called an *Adian group* if both the left graph and the right graph are cycle-free (i.e., if both graphs are forests). Of course a presentation is regarded as cycle-free if it contains no defining relations.

**Theorem 2 (Adian [1])** *Any Adian semigroup embeds in the corresponding Adian group.*

Remmers [118] gave a geometric proof of this using semigroup diagrams, and Stallings [132] gave another proof using a graph theoretic lemma. Sarkisian [124] apparently gave a proof of the decidability of the membership problem for an Adian semigroup $P$ in the corresponding Adian group $G$, and used this to solve the word problem for Adian groups: unfortunately there appears to be a gap in the proof in [124]. Adian's results have been extended in different directions in the work of several authors (see, for example, the papers by Kashintsev [70], Guba [53], Krstic [74], and Kilgour [72], where various small cancellation conditions are used to study embeddability of semigroups in groups).

We remark that in general an Adian group $G$ is not the group of quotients of the corresponding Adian semigroup $P$. For example, if we consider the presentation $P = Sgp\langle a, b : ab = b^2a\rangle$, then $P$ is an Adian semigroup whose associated Adian group is the Baumslag–Solitar group $G = BS(1, 2)$. Not all elements of $G$ belong to $PP^{-1}$, for example $a^{-1}ba \notin PP^{-1}$. However, every element of $G$ can be written as a product of two elements of $PP^{-1}$ — see Stallings [133] for a discussion of this example. Stallings shows that if $P$ is an Adian semigroup, then $PP^{-1}$ is a quasi-pregroup for the corresponding Adian group $G$ (that is, if $q_1, q_2, \ldots, q_n \in PP^{-1}$, $n > 1$ and $q_iq_{i+1} \notin PP^{-1}$ for all $i$, then $q_1q_2 \ldots q_n \neq 1$ in $G$).

As a second large class of important examples of semigroups that are embeddable in groups, we turn to a brief discussion of braid groups and Artin groups. The braid monoid on $n + 1$ strings is the monoid $P_n$ with presentation

$$Mon\langle x_1, x_2, \ldots, x_n : x_ix_j = x_jx_i \text{ if } |i - j| > 1,$$
$$x_ix_{i+1}x_i = x_{i+1}x_ix_{i+1} \text{ if } i = 1, \ldots, n - 1\rangle.$$

The corresponding group with the same presentation as a group, is the braid group $B_n$ on $n + 1$ strands. Braid groups have been the object of intensive study in the literature (see for example the influential book of Birman [21], and many subsequent papers dealing with braids and their connection to other areas of mathematics). Braid monoids play a prominent role in the theory of braid groups. Garside [46] showed that such monoids satisfy the Ore condition, in fact the principal right ideals form a lattice: for each $a, b \in B_n$, there exists $c \in B_n$ such that $aB_n \cap bB_n = cB_n$, and also $B_n$ is cancellative. Thus, by Ore's theorem we have:

**Theorem 3 (Garside)** *For each $n$, the braid monoid $P_n$ embeds in the braid group $B_n$, and $B_n$ is the group of quotients of $P_n$. Furthermore, the principal right ideals of $P_n$ form a semilattice (in fact a lattice) under intersection.*

This result was simultaneously generalized by Brieskorn and Saito [23] and by Deligne [41] to Artin groups and monoids of finite type. Recall that a group $G$ is called an *Artin group* and the corresponding monoid is called an *Artin monoid* if it is presented by a set $X$ subject to relations of the form $prod(x, y; m_{x,y}) = prod(y, x; m_{x,y})$ if $m_{x,y} < \infty$. (Here $m_{x,x} = 1$ and $m_{x,y} = m_{y,x} \in \{2, 3, \ldots, \infty\}$ for $x, y \in X$, and $prod(x, y; m_{x,y})$ stands for the alternating word $xyxy\ldots$ of length $m_{x,y}$). An Artin group (monoid) is said to be *of finite type* if the corresponding Coxeter group is finite. These results were further generalized by Dehornoy and Paris [37] to a class of groups known as *Garside groups*, and were generalized further by Dehornoy [36] to a class of groups that admit a *thin* group of fractions, and to a group that arises in the study of left self distributivity and its connection to mathematical logic (see the book by Dehornoy [35] for full details about this). Many properties of braid groups, Artin groups of finite type, Garside groups and the more general groups considered by Dehornoy are proved by a deep study of the associated monoid of positive elements. We refer to the papers of Dehornoy cited above for further references and details. These groups admit a presentation where every relation is of the form $xu = yv$ for $x \neq y \in X$ and admit one such relation for each pair $x \neq y \in X$, so their left graphs are in fact cliques. Thus this class of groups is very different from the class of Adian groups. We also refer to the recent papers by Paris [109] and Godelle and Paris [50] where the authors solve Birman's conjecture [22] for braid groups and right angled Artin groups by studying the emebedding of singular braid monoids (Artin monoids) in the corresponding groups.

Several authors have studied the question of embeddability of general Artin monoids in Artin groups: for example special cases of this question have been considered by Charney [25] and Cho and Pride [28]. Much additional information about embeddability of semigroups in groups may be found in the paper by Cho and Pride. Paris [108] has established the following deep general result about Artin groups and Artin monoids.

**Theorem 4 (Paris)** *Every Artin monoid embeds in the corresponding Artin group.*

It is worth remarking that while Artin groups of finite type and the more general groups considered by Dehornoy et al. are groups of fractions of their corresponding monoids of positive elements, this is not the case for Artin groups in general. The fact that braid groups, Artin groups of finite type, thin groups of fractions, etc., are all groups of fractions of their positive monoids leads to a fast algorithm for solving the word problem for such groups — they have quadratic isoperimetric inequality and admit an automatic structure. However, the word problem for Artin groups in general remains open, as far as I am aware.

I will close this section with brief mention of another prominent example of a monoid that embeds in its group of fractions. Recall that the Thompson group $F$ can be defined by the presentation $F = Gp\langle x_0, x_1, \ldots : x_n x_k = x_k x_{n+1} \text{ for } k < n\rangle$.

This group has appeared in numerous settings, having been originally introduced by R. Thompson (see [101]) as a group that acts naturally on bracketed expressions by moving the brackets, i.e., by applying the associative law. We refer the reader to the monograph by Cannon, Floyd and Parry [24] for an introduction to the Thompson group $F$ and some of its many connections with other areas of mathematics. The Thompson monoid is the monoid defined by the same relations as those that define $F$ as a group. The following result appears to be well known.

**Theorem 5** *The Thompson monoid embeds in the Thompson group $F$. Furthermore, $F$ is the group of fractions of the Thompson monoid.*

A closely related group is the group $G_{LD}$ introduced by Dehornoy [34] to describe the geometry of the left self-distributive law $x(yz) = (xy)(xz)$. See Dehornoy's book [35] for further information and deep connections with mathematical logic. It is known that the group $G_{LD}$ is the group of quotients of an appropriate submonoid of this group, but a presentation for that submonoid seems to be unknown.

There are several ways to associate an **inverse monoid** with the situation when $P$ is a monoid that embeds in a group $G$ with the same presentation. We recall first some basic definitions and facts about regular and inverse monoids, and some of the rather extensive structure theory for such monoids.

## 3 Regular and Inverse Monoids

A monoid $M$ is called a *(von Neumann) regular* monoid, if for each $a \in M$ there exists some element $b \in M$ such that $a = aba$ and $b = bab$. Such an element $b$ is called *an inverse* of $a$ (it is not necessarily unique). Note that regular monoids have in general lots of *idempotents*: if $b$ is an inverse of $a$ in $M$ then $ab$ and $ba$ are both idempotents of $M$, i.e., $(ab)^2 = ab$ and $(ba)^2 = ba$ (and in general $ab \neq ba$). A monoid $M$ is called an *inverse monoid* if for each $a \in M$ there exists a unique inverse (denoted by $a^{-1}$) in $M$ such that

$$a = aa^{-1}a \quad \text{and} \quad a^{-1} = a^{-1}aa^{-1}.$$

Equivalently, $M$ is inverse iff it is regular and the idempotents of $M$ commute. Thus if $M$ is an *inverse* monoid then the idempotents of $M$ form a commutative idempotent semigroup with respect to the product in $M$. Since a commutative idempotent semigroup may be viewed as a lower semilattice (with meet operation equal to the product), we normally refer to such semigroups as *semilattices*. We will consistently denote the set of idempotents of a monoid $M$ by $E(M)$. Thus if $M$ is an inverse monoid, then $E(M)$ is a submonoid of $M$ that is a semilattice, referred to as the *semilattice of idempotents* of $M$. Every inverse monoid $M$ comes equipped with a *natural partial order* defined by

$$a \leq b \text{ if and only if } a = eb \text{ for some idempotent } e \in M.$$

If $e = e^2$ is an idempotent of a monoid $M$, then the set

$$H_e = \{a \in M : ae = ea = e \text{ and } \exists b \in M \text{ such that } ab = ba = e\}$$

is a *subgroup* of $M$ with identity $e$ (i.e., it forms a group with identity $e$ relative to the multiplication in $M$). Clearly $H_e$ is the largest subgroup of $M$ with identity $e$, and $H_e \cap H_f = \emptyset$ if $e \neq f$. It is also clear that $H_1 = U(M)$, the group of units of the monoid $M$. The subgroups $H_e$, $e \in E(M)$, are referred to as the *maximal subgroups* of $M$. The semilattice of idempotents and the maximal subgroups of an inverse monoid $M$ give us a good deal of information about $M$, but do not by any means determine the structure of $M$: in general, not all elements of an inverse monoid need belong to subgroups of the monoid.

A standard example of a regular monoid is the *full transformation monoid* on a set $X$, which consists of all functions from $X$ to itself with respect to composition of functions. The group of units of this monoid is of course the symmetric group on $X$. Idempotents in this monoid consist of functions that are identity maps on their ranges, and the maximal subgroup corresponding to such an idempotent is isomorphic to the symmetric group on the range of the map. Every semigroup can be embedded in an appropriate full transformation monoid (see Clifford and Preston [30], Volume 1).

Another standard example of a regular semigroup is the *full linear monoid* $M_n(k)$ of $n \times n$ matrices with entries in a field $k$, with respect to matrix multiplication. The group of units of $M_n(k)$ is the general linear group $GL_n(k)$. From elementary linear algebra we know that an idempotent matrix of rank $r$ is similar to the diagonal matrix with block diagonal identity matrix $I_r$ in the top left hand corner and zeroes elsewhere. The group $GL_n(k)$ acts by conjugation on the set of idempotent matrices, and the orbits of this action consist of idempotent matrices of a fixed rank. The idempotent matrices in $M_n(k)$ may be identified with pairs of opposite parabolic subgroups of $GL_n(k)$. The maximal subgroup corresponding to an idempotent matrix of rank $r$ is isomorphic to the general linear group $GL_r(k)$. Of course the idempotents of $M_n(k)$ do not form a subsemigroup if $n > 1$. We refer to Okninski's book [106] for a detailed description of this monoid, and to Putcha's book [114] for an introduction to the elegant theory of linear algebraic monoids. A linear algebraic monoid is regular if and only if its group of units is a reductive group: the subgroup structure of the group of units of a linear algebraic monoid provides very detailed information about the structure of the monoid (see [114]).

Clearly every group is an inverse monoid (in fact groups are just regular monoids with precisely one idempotent), and every semilattice $E$ is an inverse monoid with $e^{-1} = e$ and with $H_e = \{e\}$, for all $e \in E$. A more enlightening example of an inverse monoid is the *symmetric inverse monoid* on a set $X$, denoted by $SIM(X)$.

The monoid $SIM(X)$ is the monoid of all partial one-one maps (i.e., one-one maps from subsets of $X$ to subsets of $X$) with respect to multiplication of partial maps: if $\alpha$ and $\beta$ are partial one-one maps, then $\alpha\beta(x) = \alpha(\beta(x))$ whenever this makes sense, i.e., if $x \in dom(\beta)$ and $\beta(x) \in dom(\alpha)$. The group of units of $SIM(X)$ is obviously the symmetric group (the group of permutations) on $X$, and the idempotents of $SIM(X)$ are the identity maps on subsets of $X$, so the semilattice of idempotents of $SIM(X)$ is isomorphic to the lattice of subsets of $X$. The empty subset corresponds to the zero of $SIM(X)$: a product $\alpha\beta$ of two partial one-one maps on $X$ is zero (the empty map) if $range(\beta) \cap dom(\alpha) = \emptyset$. The maximal subgroup corresponding to the identity map on the subset $Y$ of $X$ is the symmetric group on $Y$. The natural partial order on $SIM(X)$ is defined by domain restriction of a partial one-one map, i.e., $\alpha \leq \beta$ iff $dom(\alpha) \subseteq dom(\beta)$ and $\alpha = \beta|_{dom(\alpha)}$. (I note that the definition of $SIM(X)$ given here is the dual of the usual definition found in many books on semigroup theory, where functions are traditionally written on the right rather than the left.)

Symmetric inverse monoids are in a sense generic inverse monoids.

**Theorem 6 (Vagner–Preston)** *Every inverse monoid embeds in a suitable symmetric inverse monoid.*

Thus inverse monoids may be viewed as monoids of partial one-one maps, in much the same way as groups may be viewed as groups of permutations. Inverse monoids arise naturally whenever one encounters partial one-one maps throughout mathematics. For example, the Vagner–Preston theorem has been extended by Barnes [10] and Duncan and Paterson [44] to show that every inverse monoid embeds as a monoid of partial isometries of some Hilbert space, and from this point of view, inverse monoids play an increasingly important role in the theory of operator algebras (see the book by Paterson [110] for an introduction to the role of inverse monoids in this theory). The book by Petrich [111] or the more recent book by Lawson [76] provide an account of the general theory of inverse monoids and some of their connections with other areas of mathematics.

Another natural class of examples of inverse monoids arises in connection with submonoids of groups. Note that any submonoid $P$ of a group must be a left and right cancellative monoid. Let $P$ be any left cancellative monoid. The left regular representation $a \to \lambda_a$, where $\lambda_a : x \to ax$ for all $a, x \in P$, defines an embedding of $P$ into the symmetric inverse monoid $SIM(P)$, since each map $\lambda_a$ is clearly a partial one-one map on $P$ with domain $P$ and range $aP$. The submonoid of $SIM(P)$ generated by the image of $P$ in this embedding into $SIM(P)$ is an inverse monoid, referred to as the *(left) inverse hull* $I_l(P)$ of $P$. Of course there is a dual inverse monoid $I_r(P)$ that arises from the right regular representation of a right cancellative monoid $P$. This is the most obvious way in which inverse monoids arise in connection with submonoids of groups. I will discuss some other ways of associating inverse monoids with submonoids of groups later in this paper.

The ideal structure of a monoid provides a basic tool for beginning to study the structure of the monoid. It will be convenient to introduce some standard terminology along these lines. There are five equivalence relations, known as the

*Green's relations* $\mathcal{R}$, $\mathcal{L}$, $\mathcal{J}$, $\mathcal{H}$ and $\mathcal{D}$ that play a prominent role in the theory. For a monoid $M$ we define

$$\begin{aligned}\mathcal{R} &= \{(a,b) \in M \times M : aM = bM\},\\ \mathcal{L} &= \{(a,b) \in M \times M : Ma = Mb\},\\ \mathcal{J} &= \{(a,b) \in M \times M : MaM = MbM\},\\ \mathcal{H} &= \mathcal{R} \cap \mathcal{L},\end{aligned}$$

and $\mathcal{D} = \mathcal{R} \vee \mathcal{L}$ (the join of $\mathcal{R}$ and $\mathcal{L}$ in the lattice of equivalence relations on $M$).

The corresponding equivalence classes containing $a \in M$ are denoted by $R_a$, $L_a$, $J_a$, $H_a$ and $D_a$ respectively. Clearly $\mathcal{H} \subseteq \mathcal{R}$, $\mathcal{L} \subseteq \mathcal{D} \subseteq \mathcal{J}$. It is a fortunate fact in semigroup theory that the equivalence relations $\mathcal{R}$ and $\mathcal{L}$ commute, i.e., $\mathcal{R} \circ \mathcal{L} = \mathcal{L} \circ \mathcal{R}$, and it follows that $\mathcal{D} = \mathcal{R} \circ \mathcal{L} = \mathcal{L} \circ \mathcal{R}$. Thus $a\mathcal{D}b$ in $M$ iff $\exists c \in M$ such that $a\,\mathcal{R}\,c\,\mathcal{L}\,b$ iff $\exists d \in M$ such that $a\,\mathcal{L}\,d\,\mathcal{R}\,b$. For an inverse monoid $M$ it is easy to see that $a\,\mathcal{R}\,b$ iff $aa^{-1} = bb^{-1}$ and $a\,\mathcal{L}\,b$ iff $a^{-1}a = b^{-1}b$. It is a well-known fact that if $P$ is a cancellative monoid, then $P$ embeds as the $\mathcal{R}$-class $R_1$ of 1 in its right inverse hull $I_r(P)$ and as the $\mathcal{L}$-class $L_1$ of 1 in its left inverse hull $I_l(P)$.

It is informative to provide an explicit description of the Green's relations in the full linear monoid $M_n(k)$. If $A$ and $B$ are two matrices in $M_n(k)$, then

$A\,\mathcal{R}\,B$ iff $A\,GL_n(k) = B\,GL_n(k)$ iff $Col(A) = Col(B)$,

$A\,\mathcal{L}\,B$ iff $GL_n(k)\,A = GL_n(k)\,B$ iff $Nul(A) = Nul(B)$,

$A\,\mathcal{J}\,B$ iff $GL_n(k)\,A\,GL_n(k) = GL_n(k)\,B\,GL_n(k)$ iff $rank(A) = rank(B)$, and

$\mathcal{J} = \mathcal{D}$.

Furthermore, for each fixed $r \leq n$, the group $GL_n(k)$ acts transitively by left multiplication [resp., right multiplication] on the set of $\mathcal{R}$-classes of $M_n(k)$ [resp., $\mathcal{L}$-classes of $M_n(k)$] within the $\mathcal{J}$-class $J_r$ consisting of the matrices of rank $r$. In addition, if $Y_r$ denotes the set of all matrices of rank $r$ that are in reduced row echelon form and if $X_r$ is the set of transposes of elements of $Y_r$, then the $\mathcal{R}$-classes of $M_n(k)$ in the $\mathcal{J}$-class $J_r$ are in one-one correspondence with the matrices in $X_r$, and the $\mathcal{L}$-classes of $M_n(k)$ in the $\mathcal{J}$-class $J_r$ are in one-one correspondence with the matrices in $Y_r$. Every matrix in $J_r$ has a unique decomposition of the form $XGY$ with $X \in X_r$, $Y \in Y_r$ and $G \in GL_n(k)$. Proofs of all of these facts and much additional interesting information about full linear monoids may be found in [106].

It is a well known fact in semigroup theory that if a $\mathcal{D}$-class contains a regular element, then every element of that $\mathcal{D}$-class is regular. The structure of regular $\mathcal{D}$-classes is very nice: for example, all $\mathcal{H}$-classes within the $\mathcal{D}$-class are of the same cardinality, an $\mathcal{H}$-class is a maximal subgroup iff it contains an idempotent, and two maximal subgroups contained in the same $\mathcal{D}$-class are isomorphic. A $\mathcal{D}$-class is regular iff it contains an idempotent, and if $a$ is a regular element of $M$, then every inverse of $a$ lies in the $\mathcal{D}$-class $D_a$. A product $ab$ lies in the $\mathcal{H}$-class $R_a \cap L_b$ iff $L_a \cap R_b$ contains an idempotent. Proofs of these facts may be found in any standard book on semigroup theory, for example [30].

A semigroup $S$ is called *simple* [resp., *bisimple*] if it contains just one $\mathcal{J}$-class [resp., $\mathcal{D}$-class]. A semigroup with just one $\mathcal{H}$-class is a group. A semigroup $S$ with a zero element 0 is called 0-*simple* if $S^2 \neq 0$ and $S$ has only two $\mathcal{J}$-classes ($\{0\}$ and $S - \{0\}$). Such a semigroup is called 0-*bisimple* if it has just two $\mathcal{D}$-classes ($\{0\}$ and $S - \{0\}$). The structure of finite simple and 0-simple semigroups was determined by Suschkewitsch in 1928 [142]. This was extended by Rees [117] in 1940 to a class of simple [resp., 0-simple] semigroups known as *completely simple* [resp., *completely 0-simple*] semigroups. We refer to [30] for an account of this important work.

Bisimple inverse monoids may be constructed from right cancellative monoids whose principal left ideals form a semilattice. The following theorem was proved by Clifford [29] in 1953.

**Theorem 7 (Clifford, 1953)** *Let $M$ be a bisimple inverse monoid with identity 1 and let $R = R_1$, the $\mathcal{R}$-class of 1. Then $R$ is a right cancellative monoid and the principal left ideals of $R$ form a semilattice under intersection, i.e., for each $a, b \in R$, there exists $c \in R$ such that $Ra \cap Rb = Rc$. Conversely, let $R$ be a right cancellative monoid in which the intersection of any two principal left ideals is a principal left ideal. Then the (right) inverse hull of $R$ is a bisimple inverse monoid and the $\mathcal{R}$-class of 1 in this monoid is a submonoid that is isomorphic to $R$.*

Again, there is an obvious dual construction of bisimple inverse monoids from left cancellative monoids whose principal right ideals form a semilattice. We thus have the following corollary of Garside's theorem (Theorem 3) and Clifford's theorem (Theorem 7): the result extends to Artin groups of finite type, Garside groups, thin groups of quotients, etc.

**Corollary 1** *The inverse hull of the braid monoid $B_n$ is a bisimple inverse monoid.*

If $S$ is an inverse semigroup, then the natural partial order induces a homomorphism from $S$ onto its maximal group homomorphic image. For $S$ an inverse semigroup and $a, b \in S$ we define an equivalence relation $\sigma$ on $S$ by $a \,\sigma\, b$ iff $\exists c \in S$ such that $c \leq a$ and $c \leq b$. It is easy to see that $\sigma$ is a *congruence* on $S$ (i.e., it is compatible with respect to multiplication on both sides), so the set of $\sigma$-classes of $S$ forms a semigroup $S/\sigma$ with respect to the obvious multiplication, and there is a natural map (which we denote again by $\sigma$) from $S$ onto $S/\sigma$. It is straightforward to see that $S/\sigma$ is a group, the maximal group homomorphic image of $S$.

The inverse semigroup $S$ is called $E$-*unitary* if the inverse image under $\sigma$ of the identity of the group $S/\sigma$ consists just of the semilattice $E(S)$ of idempotents of $S$. Equivalently, $S$ is $E$-unitary if $a \geq e$ and $e \in E(S)$ implies $a \in E(S)$. $E$-unitary inverse semigroups play an essential role in the theory of inverse semigroups. Their structure has been determined by McAlister [96] by means of a group acting by order automorphisms on a partially ordered set with an embedded semilattice. Furthermore, McAlister proved [97] that if $S$ is any inverse semigroup, then there is some $E$-unitary inverse semigroup $T$ and an *idempotent-separating* homomorphism $f$ from $T$ onto $S$ (a homomorphism $f : T \to S$ is called "idempotent-separating" if distinct idempotents of $T$ are mapped to distinct idempotents of $S$). In this situation, we refer to $T$ as an $E$-*unitary cover* of $S$ over the group $G$, where $G$ is the maximal group homomorphic image of $T$. We refer to Lawson [76] for a proof of McAlister's results and for an account of the importance of $E$-unitary inverse semigroups in the theory.

An inverse monoid $M$ is called an *$F$-inverse monoid* if each $\sigma$-class of $M$ contains a unique maximal element in the natural partial order on $M$. An $F$-inverse monoid is $E$-unitary. It is known that every inverse semigroup has an $F$-inverse cover, but it appears to be unknown at present whether every *finite* inverse semigroup has a *finite* $F$-inverse cover. One can define a multiplication on the maximal elements in $\sigma$-classes of an $F$-inverse monoid $M$ as follows: if $a$ and $b$ are maximal elements in $M$ (i.e., maximal elements in their respective $\sigma$-classes), then define $a.b$ to be the maximal element in the $\sigma$-class containing $ab$. With respect to this multiplication, the set of maximal elements of an $F$-inverse monoid forms a group that is isomorphic to the maximal group homomorphic image of $M$.

As an example of this situation, we recall some of the ideas developed by Birget [18] to study the complexity of the word problem for the Thompson group $V$. A subset $R$ of a monoid $M$ is called a right ideal if $RM \subseteq R$. A function $\theta : R \to R'$ is called a right ideal isomorphism if $R$ and $R'$ are right ideals of $M$ and $\theta(rm) = \theta(r)m$ for all $r \in R$, $m \in M$. The collection of right ideal isomorphisms is an inverse monoid. A right ideal $R$ of $M$ is called an *essential right ideal* if $R \cap R' \neq \emptyset$ for all right ideals $R'$. The collection of right ideal isomorphisms between essential right ideals is an inverse monoid. Every essential right ideal of $M$ is of the form $R = CM$ where $C$ is a maximal prefix code, and every finitely generated essential right ideal of $M$ is of the form $R = CM$ where $C$ is a maximal finite prefix code. The collection of right ideal isomorphisms between the essential finitely generated right ideals of the free monoid $\{a, b\}^*$ on 2 letters is an $F$-inverse monoid whose maximal group image is Thompson's group $V$: the multiplication that defines the group $V$ is just the multiplication of maximal elements in $\sigma$-classes defined above (see Birget's paper [18] for details). The tree representation of prefix codes connects this definition and the definition by action on finite trees used in [24].

There are many other examples of this situation in the literature. I will briefly mention the work of Lawson [77] on *Möbius inverse monoids.* Recall that a *Möbius transformation* is a linear fractional transformation $\alpha$ of the complex plane having the form $\alpha(z) = (az + b)/(cz + d)$ where $a, b, c, d$ are complex numbers and $ad - bc \neq 0$. Möbius transformations are either one-one functions (when $c = 0$) or partial one-one functions with restricted domain (when $c \neq 0$). Thus they are elements of the symmetric inverse monoid $SIM(\mathbf{C})$. The *Möbius inverse monoid* is the (inverse) submonoid of $SIM(\mathbf{C})$ generated by the Möbius transformations. It is an $F$-inverse monoid with maximal group image the *Möbius group.* One may refer to [77] or Lawson's book [76] for details.

An $E$-unitary inverse monoid with a zero must be essentially trivial — more precisely, it must be a semilattice with 0 and 1. An inverse semigroup $S$ with 0 is called $E^*$-unitary if, whenever $a \geq e = e^2$ and $e \neq 0$ in $S$, then $a = a^2$. Similarly, an inverse monoid $M$ is called $F^*$-inverse if each non-zero element of $M$ has a unique maximal element in its $\sigma$-class. The multiplication on the maximal elements of each $\sigma$-class that was defined above for $F$-inverse monoids essentially carries over to the case of $F^*$-inverse monoids. The product of two such maximal elements $a.b$

is defined as before if $ab \neq 0$ and it is undefined if $ab = 0$. It is of interest to know when the partial group of maximal elements of such a monoid embeds in a group: this happens, for example, if this partial group is a pregroup in the sense of Stallings [130]. $F$-inverse monoids and $F^*$-inverse monoids arise naturally in connection with submonoids of groups.

Let us return now to the situation where $P = Mon\langle X : u_i = v_i \rangle$ embeds as the positive submonoid in the group $G = Gp\langle X : u_i = v_i \rangle$ with the same presentation. For each $g \in G$ denote by $\beta_g$ the restriction of the left regular representation of $G$ to $P$, i.e., $dom(\beta_g) = \{t \in P : gt \in P\}$, $range(\beta_g) = \{s \in P : g^{-1}s \in P\}$, and $\beta_g$ acts by left translation by $g$ on elements of its domain. Clearly $\beta_g \in SIM(P)$ for each $g \in G$. Let $T(G, P)$ be the inverse submonoid of $SIM(P)$ generated by $\{\beta_g : g \in G\}$. $T(G, P)$ is referred to as the *Toeplitz inverse monoid of* $(G, P)$ because of its natural connection to the Toeplitz algebra associated with $G$ and $P$. It was introduced in the paper of Nica [104] who constructed a locally compact space on which the monoid $T(G, P)$ acts and who showed that this action leads very naturally to the *Wiener–Hopf groupoid* whose reduced $C^*$-algebra is the Toeplitz algebra associated with $(G, P)$. The locally compact space on which $T(G, P)$ acts may be viewed essentially as a compactifiction of the set of $\mathcal{R}$-classes in the $\mathcal{D}$-class of 1 in $T(G, P)$ (see the paper by Nica [104] for details about this).

For example, if $P = \mathbf{N}$ and $G = \mathbf{Z}$, then $T(G, P)$ is the *bicyclic monoid*, often defined by the monoid presentation $Mon\langle a, b : ab = 1 \rangle$. The bicyclic monoid is a bisimple $F$-inverse inverse monoid with an infinite descending chain of idempotents $1 > ba > b^2a^2 > b^3a^3 > \dots$. If $P = X^*$ and $G = FG(X)$ (the free group on $X$), then $T(G, P)$ is the *polycyclic monoid on* $X$, which is defined by the monoid presentation $Mon\langle X \cup X^{-1} : x^{-1}x = 1,\ x^{-1}y = 0 \text{ for } x \neq y \rangle$. The polycyclic monoid was first introduced by Nivat and Perrot [105] as the syntactic monoid of the language of correctly parenthesized expressions. It is $F^*$-inverse and 0-bisimple. It has also surfaced in operator algebras in connection with Cuntz $\mathbf{C}^*$-algebras, where it is referred to as the *Cuntz monoid* (see [110]) and in many other contexts, including Birget's work on Thompson's group $V$ [18]. For the polycyclic monoid, the space constructed by Nica on which this monoid acts is a subspace of the space of ends of the tree of the free monoid $X^*$.

The following result follows from the work of Nica [104].

**Theorem 8** *Suppose that $P = Mon\langle X : u_i = v_i \rangle$ embeds as the positive submonoid in the group $G = Gp\langle X : u_i = v_i \rangle$ with the same presentation. Then*

(1) *$T(G, P)$ contains a zero iff $PP^{-1} \neq G$.*

(2) *$T(G, P)$ is a simple $F$-inverse monoid if $G = PP^{-1}$ and a 0-simple $F^*$-inverse monoid if $G \neq PP^{-1}$.*

Additional information about the structure of $T(G, P)$ and its connection with the inverse hull $I_l(P)$ has been developed by Margolis and Lawson (private communication). For braid groups, Artin groups of finite type, Garside groups, etc., the corresponding Toeplitz inverse monoid is bisimple, and is equal to the (left) inverse hull of the monoid, but for general Artin groups, Adian groups, etc., the corresponding Toeplitz inverse monoid is $F^*$ inverse and 0-simple.

Inverse monoids arise in many other ways in the context of geometric group theory. In order to describe some of these connections, it will be convenient to recall some of the beautiful theory of free inverse monoids.

## 4 Free Inverse Monoids, Equations

Inverse monoids form a variety of algebras in the sense of universal algebra, defined by associativity and the identities:

$$a = aa^{-1}a,\ (a^{-1})^{-1} = a,\ (ab)^{-1} = b^{-1}a^{-1},\ aa^{-1}bb^{-1} = bb^{-1}aa^{-1}.$$

It follows that free inverse monoids exist. We will denote the free inverse monoid on a set $X$ by $FIM(X)$. Clearly $FIM(X)$ is the quotient of the free monoid $(X \cup X^{-1})^*$ by the smallest congruence that makes the identities above hold. An elegant solution to the word problem for $FIM(X)$ was obtained by Munn [103] in 1974.

Denote by $FG(X)$ the free group on $X$ and by $\Gamma(X)$ the Cayley graph of $(FG(X), \emptyset)$. For each word $w \in (X \cup X^{-1})^*$, denote by $MT(w)$ the finite subtree of the tree $\Gamma(X)$ obtained by reading the word $w$ as the label of a path in $\Gamma(X)$, starting at 1. Thus, for example, if $w = aa^{-1}bb^{-1}ba^{-1}abb^{-1}$, then $MT(w)$ is the tree pictured below.

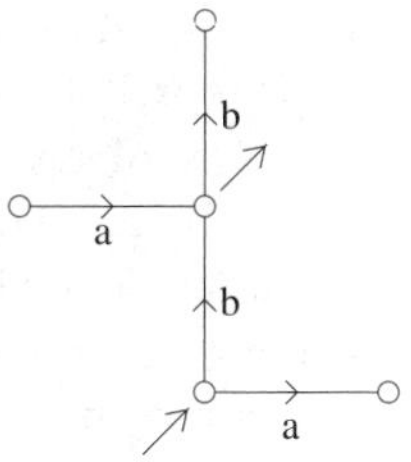


One may view $MT(w)$ as a birooted tree, with initial root 1 and terminal root $r(w)$, the reduced form of the word $w$ in the usual group-theoretic sense. Munn's solution to the word problem in $FIM(X)$ may be stated in the following form.

**Theorem 9 (Munn)** *If $u, v \in (X \cup X^{-1})^*$, then $u = v$ in $FIM(X)$ iff $MT(u) = MT(v)$ and $r(u) = r(v)$.*

Thus elements of $FIM(X)$ may be viewed as pairs $(MT(w), r(w))$ (or as birooted edge-labelled trees, which was the way that Munn described his results). An equivalent description of $FIM(X)$ in terms of Schreier subsets of $FG(X)$ was provided independently by Scheiblich [125]. Multiplication in $FIM(X)$ is performed as follows. If $u, v \in (X \cup X^{-1})^*$, then $MT(uv) = MT(u) \cup r(u).MT(v)$ (just translate $MT(v)$ so that its initial root coincides with the terminal root of $MT(u)$ and take the union of $MT(u)$ and the translated copy of $MT(v)$: the terminal root is of course $r(uv)$).

It is straightforward to see that $FIM(X)$ is an $F$-inverse monoid with maximal group image $FG(X)$: the natural map $\sigma : FIM(X) \to FG(X)$ takes $(MT(u), r(u))$

onto $r(u)$. The maximum element in a $\sigma$-class is just the the reduced word in that $\sigma$-class. The multiplication of maximal elements in the $F$-inverse monoid $FIM(X)$ is just the well-known multiplication in the free group: $r(u).r(v) = r(r(u)r(v))$. The idempotents of $FIM(X)$ correspond to the Dyck words in $(X \cup X^{-1})^*$, i.e., the words whose reduced form is 1. Two such words represent the same idempotent in $FIM(X)$ iff they span the same Munn tree. The maximal subgroups of $FIM(X)$ are all trivial (i.e., $FIM(X)$ is *combinatorial*). Free inverse monoids are residually finite and the word problem is decidable in linear time by Munn's theorem.

The construction of $FIM(X)$ from finite birooted subtrees of $\Gamma(X)$ discussed above can be extended to the Cayley graph of any group presentation. Let $\Gamma(G, X)$ be the Cayley graph of a group $G$ relative to a set $X$ of generators. Let $M(G, X) = \{(\Delta, g) : \Delta$ is a finite connected subgraph of $\Gamma(G, X)$ containing 1 and $g\}$, with multiplication $(\Delta_1, g_1).(\Delta_2, g_2) = (\Delta_1 \cup g_1.\Delta_2, g_1g_2)$. Then $M(G, X)$ is an $X$-generated $E$-unitary inverse monoid with maximal group image $G$, in fact it is the "universal" $X$-generated $E$-unitary inverse monoid with maximal group image $G$ (see the paper by Margolis and Meakin [87] for details about this construction).

There are some similarities and some striking differences between the theory of free groups and the theory of free inverse monoids. It is easy to see that inverse submonoids of free inverse monoids are not necessarily free: for example, the set of idempotents of a free inverse monoid is an inverse submonoid that is a semilattice, so it is certainly not a free inverse monoid. However, the study of *closed* inverse submonoids of free inverse monoids has some interesting relationships with the study of subgroups of free groups: this will be described in the next section of this paper. There is a developing theory of presentations of inverse monoids by generators and relations: some results about inverse monoid presentations will be discussed in the final section of this paper. The theory of equations in free inverse monoids is significantly different than the corresponding theory for free groups or free monoids. We describe some aspects of this situation in the remainder of this section.

Let $X$ be an alphabet that is disjoint from $A$. It is convenient to denote $X \cup X^{-1}$ by $\tilde{X}$ and $A \cup A^{-1}$ by $\tilde{A}$ throughout the remainder of this section. We will view letters of $\tilde{X}$ as *variables* and elements of $\tilde{A}^*$ as *constants*. The sets $A$ and $X$ will be assumed to be *finite and non-empty*. An *equation* in $FG(A)$ or in $FIM(A)$ with coefficients in $FG(A)$ (or in $FIM(A)$) is a pair $(u, v)$, where $u, v \in (\tilde{A} \cup \tilde{X})^*$. Usually we will denote such an equation by $u = v$: if necessary for emphasis we will denote $u$ and $v$ by $u(X, A)$ and $v(X, A)$ if there is any possibility of confusion about the sets of variables and constants in the equation. Similarly an equation in $A^*$ is a pair $(u, v)$ with $u, v \in (A \cup X)^*$, and again we will denote this by $u = v$. If needed to distinguish where equations are being viewed, we will denote an equation $u = v$ in $A^*$, [resp., $FG(A)$, $FIM(A)$] by $u =_M v$ [resp., $u =_G v$, $u =_I v$].

Any map $\phi : X \to \tilde{A}^*$ extends to a homomorphism (again denoted by $\phi$) from $(\tilde{A} \cup \tilde{X})^*$ to $\tilde{A}^*$ in such a way that $\phi$ fixes the letters of $A$. We say that $\phi$ is a *solution* to the equation $u =_G v$ in $FG(A)$ [resp., $u =_I v$ in $FIM(A)$ or $u =_M v$ in $A^*$] if $\phi(u) = \phi(v)$ in the appropriate setting. A solution to a set of equations $u_i = v_i$ for $i = 1, \ldots, n$ is a map $\phi$ that is a solution to each equation in the set. If a set of equations has at least one solution it is called *consistent*: otherwise it is called *inconsistent*. It is easy to give examples of equations that are inconsistent in any of the three possible settings where we are considering such equations, and it is easy to give examples of equations that are consistent in $FG(A)$ but not in $FIM(A)$ or in $A^*$. For example, if $A = \{a, b\}$, then the equation $ax = xb$ is inconsistent in all three settings, while the equation $ax = b$ is consistent in $FG(A)$ but inconsistent in $A^*$ and in $FIM(A)$ if $a \neq b$. [It is obvious that this equation is inconsistent in $A^*$ — no word in a free monoid can start with two distinct letters: one sees easily that this equation is also inconsistent in $FIM(a, b)$ — no matter what $x$ is, when one multiplies $MT(a)$ by $MT(x)$, there must be an edge labelled by $a$ in the resulting tree, so this cannot be the Munn tree of $b$.]

On the other hand, it is obvious that any set of equations that is consistent in $FIM(A)$ must be consistent in $FG(A)$: if $\psi$ is any solution to a set of equations in $FIM(A)$ and $\psi(x) = w_x \in \tilde{A}^*$ for each $x \in X$, then $\phi : X \to \tilde{A}^*$ defined by $\phi(x) = r(w_x)$ is a solution to the same set of equations, viewed as equations in $FG(A)$.

The *consistency problem* for systems of equations in $A^*$ [resp., $FG(A)$, $FIM(A)$] is the problem of determining whether there is an algorithm that, on input a finite set $\{u_i = v_i : i = 1, \ldots, n\}$ of equations in $A^*$ [resp., $FG(A)$, $FIM(A)$], produces an output of "Yes" if the system is consistent and "No" if it is inconsistent. Theorems of Makanin [84],[83] imply that the consistency problems for systems of equations in $A^*$ and in $FG(A)$ are decidable. Much work has been done on solutions to systems of equations in free monoids and free groups: we refer the reader to [80, 55, 113, 116, 32] for just some of the extensive literature on this subject. On the other hand, a theorem of Rozenblat [121] shows that while the consistency problem for systems of equations in $FIM(A)$ is decidable if $|A| = 1$, this problem is undecidable if $|A| > 1$. The consistency problem for equations in $FIM(A)$ of some restricted type (for example, single variable equations, or quadratic equations) is open as far as I am aware. Some work on special cases of this problem has been done by Deis [38]. For example, Deis [38] has shown that while the consistency problem for single *multilinear* equations in $FIM(A)$ is decidable, the consistency problem for finite *systems* of multilinear equations is undecidable.

Now consider an equation $u =_I v$ in $FIM(A)$, let $\psi$ be a solution to this in $FIM(A)$, and let $\phi$ be a solution to the corresponding equation in $FG(A)$, where $\phi(x)$ is a reduced word for each $x \in X$. We say that $\psi$ is an *extension* of $\phi$ (or that $\phi$ *extends to* $\psi$) if for each $x \in X$ there is some Dyck word $e_x$ such that $\psi(x) = e_x\phi(x)$. If $\psi$ is a solution to an equation $u = v$ in $FIM(A)$ and if $\phi(x) = r(\psi(x))$ for each $x \in X$, then of course $\phi$ is a solution to $u = v$ in $FG(A)$ and $\psi$ is an extension of $\phi$.

A given solution $\phi$ to an equation $u = v$ in $FG(A)$ may admit finitely many extended solutions, infinitely many extended solutions, or no extended solutions, to the same equation in $FIM(A)$. For example, the equation $bb^{-1}x = aa^{-1}bb^{-1}$ has trivial solution in $FG(a, b)$, and this has exactly two extensions $\psi_1(x) = aa^{-1}bb^{-1}$ and $\psi_2(x) = aa^{-1}$ in $FIM(a, b)$. The equation $bb^{-1}x = aa^{-1}x$ has trivial solution in $FG(a, b)$ that extends to infinitely many solutions $\psi_e(x) = e$ for any idempotent

$e \leq aa^{-1}bb^{-1}$ in the natural order on $FIM(a,b)$. The equation $a^{-1}ax = aa^{-1}$ has trivial solution in $FG(a,b)$ but no solution in $FIM(a,b)$. These facts are easy to check via the multiplication of Munn trees in the free inverse monoid, as described above.

A natural question arises here: when does a solution to an equation $u = v$ in $FG(A)$ extend to a solution to the same equation in $FIM(A)$? We refer to the corresponding algorithmic problem as the *extendibility problem* for equations in $FIM(A)$. More precisely, the extendibility problem for equations in $FIM(A)$ asks whether there is an algorithm that, on input a finite set $\{u_i = v_i : i = 1, \ldots, n\}$ of equations in $FIM(A)$ that is consistent in $FG(A)$ and a solution $\phi$ to this system in $FG(A)$, produces the output "Yes" if $\phi$ can be extended to a solution to the system of equations in $FIM(A)$ and "No" if $\phi$ cannot be extended to a solution to this system in $FIM(A)$. Some special cases of the extendibility problem were considered by Deis [38]. The main result of the paper by Deis, Meakin and Sénizergues [39] shows that the extendibility problem is decidable.

In [39] it is shown that the requirement that $\phi$ should be extendible to some solution $\psi(x) = e_x\phi(x)$ to the system in $FIM(A)$ translates as follows. Consider the system of equations

$$\sum_x \alpha_{i,x} \cdot T_x + \beta_i = \sum_x \alpha'_{i,x} \cdot T_x + \beta'_i : i = 1, \ldots, n \tag{1}$$

Here the $\alpha_{i,x}, \alpha'_{i,x}, \beta_i, \beta'_i$ are finite subsets of $FG(A)$ and the $T_x$ are unknowns. A solution of (1) is any collection of subsets $T_x$ $(x \in X)$ of $FG(A)$ that satisfies this system of equations. We would like to decide whether the system of equations (1) has at least one solution such that each $T_x$ is both finite and prefix closed. (A subset $T$ of $FG(A)$ is *prefix closed* if the corresponding set of reduced words is prefix closed.) In [39] it is shown that this problem is decidable by appealing to Rabin's tree theorem [115]. From the discussion above, this shows that the extendibility problem is decidable. Rabin's tree theorem is a theorem in second order monadic logic.

We assume some familiarity with basic definitions and ideas of (first order) logic. See, for example, Barwise [12]. In second order monadic logic, quantifiers refer to sets (i.e., unary or monadic predicates) as well as to individual members of a structure. The syntax and semantics of terms and well formed formulae are defined inductively in the usual way. Atomic formulae include those of the form $t \in Y$ where $t$ is a term and $Y$ is a set variable. A sentence of the form $\forall Y \nu(Y)$ where $Y$ is a set variable, in particular, is true in a structure $M$ iff $\nu(Y)$ is (inductively) true in $M$ for all subsets $Y$ of the universe of $M$. If a sentence $\theta$ is true in a structure $M$ we write $M \models \theta$ and we define $Th_2(M) = \{\theta : M \models \theta\}$. The (second order monadic) theory of $M$ is *decidable* if there is an algorithm that tests whether a given sentence $\theta$ of the language of $M$ is in $Th_2(M)$ or not.

Let $A$ be a countable set and consider the structure $T_A = (A^*, \{r_a : a \in A\}, \leq)$. Here $r_a : A^* \to A^*$ is right multiplication by $a$, $xr_a = xa, \forall x \in A^*$ and $\leq$ is the prefix order $x \leq y$ iff $\exists u \in A^*(xu = y)$. The theory $Th_2(T_A)$ is called the theory of $A$-successor functions. For $|A| = 2$ this is often denoted by $S2S$, and sentences in $Th_2(T_A)$ can be reformulated as sentences in $S2S$. Rabin's tree theorem [115] is that $Th_2(T_A)$ is decidable. It is one of the most powerful decidability results known in model theory: the decidability of many other results can be reduced to $Th_2(T_A)$ (see, for example, [12]). A rather delicate application of Rabin's tree theorem enables a proof of the main theorem of Deis, Meakin and Sénizergues [39].

**Theorem 10 (Deis, Meakin, Sénizergues)** *There is an algorithm that will decide, on input a system of equations of the form* (1), *whether this system of equations has at least one solution* $\{T_x : x \in X\}$ *such that each* $T_x$ *is a finite prefix-closed subset of* $FG(A)$. *Hence the extendability problem for equations in* $FIM(A)$ *is decidable.*

This result can be used to study the consistency problem for equations in $FIM(A)$ when one has good control over the set of solutions to the corresponding equations in $FG(A)$. For example, a result in [129] that is attributed to James Howie states that if $w(x) = 1$ is a single-variable equation in $FG(A)$ and the exponent sum of the single variable $x$ in $w(x)$ is not zero, then the equation $w(x) = 1$ can have at most one solution in $FG(A)$. This implies that the consistency problem for such an equation in $FIM(A)$ is decidable, by Theorem 10. It is an open problem whether the consistency problem for all single variable equations in $FIM(A)$ is decidable. Some cases where this problem is shown to be decidable are contained in the paper [39]. In view of established literature providing parametric solutions to single variable equations in free groups (see Lyndon [81], Appel [6], Lorents [79]), it is plausible that the consistency problem for single variable equations in $FIM(A)$ might be decidable.

## 5 Subgroups of Free Groups and Closed Inverse Submonoids of Free Inverse Monoids

In his paper [131], Stallings showed how immersions between finite graphs (i.e., locally injective graph morphisms) may be used to study finitely generated subgroups of free groups. In this section, we show that inverse monoids play the same role in the theory of immersions that groups play in the theory of coverings. This leads to the study of closed inverse submonoids of free inverse monoids and to the use of finite inverse monoids to classify finitely generated subgroups of free groups.

By a *graph* $\Gamma = (V(\Gamma), E(\Gamma))$ we will mean a graph in the sense of Serre [128]. Thus every edge $e : v \to w$ comes equipped with an inverse edge $e^{-1} : w \to v$ and $e^{-1} \neq e$: the initial vertex of $e$ is denoted by $\alpha(e)$ and the terminal vertex of $e$ is denoted by $\omega(e)$. There is an evident notion of morphism between graphs. If $v \in V(\Gamma)$, let $star(\Gamma, v) = \{e \in E(\Gamma) : \alpha(e) = v\}$: a morphism $f : \Gamma \to \Gamma'$ induces a map $f_v : star(\Gamma, v) \to star(\Gamma', f(v))$ between star sets. We say that $f$ is a *cover* if each $f_v$ is a bijection and that $f$ is an *immersion* if each $f_v$ is an injection. It is clear from the definition of graph that it is possible to label the edges of a graph with labels coming from a set $X \cup X^{-1}$ in such a way that if $x$ labels an edge $e$, then $x^{-1}$ labels $e^{-1}$, and so that the labelling is consistent with an immersion over $B_X$, the bouquet of $|X|$ circles: by this we mean that we choose a labelling set $X$ such that no two edges with the same initial or terminal vertex are assigned the same label. The associated natural immersion $f_\Gamma : \Gamma \to B_X$ preserves edge labelling. *All graphs that we will consider in this paper will be so labelled and all immersions will preserve edge labelling.* For example, the Cayley graph $\Gamma(G, X)$ of a group $G$ relative to a set $X$ of generators is obviously labelled over $X \cup X^{-1}$. Munn trees are finite trees with edges labelled over $X \cup X^{-1}$ for some appropriate set $X$.

A simple example illustrating these ideas is provided in the diagram below: the natural map from the graph $\Gamma_1$ to $B_{\{a,b\}}$ is a cover, while the natural map from $\Gamma_2$ to $B_{\{a,b\}}$ is an immersion that is not a cover.

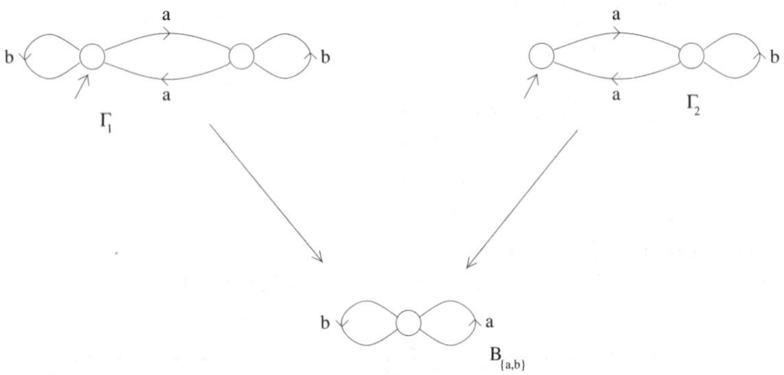


It is well known (see for example, Lyndon and Schupp [82], or Stallings [131] or Stillwell [140]) that covers of a connected graph $\Gamma$ may be classified via subgroups of the fundamental group $\pi_1(\Gamma, v)$ based at some base point $v \in V(\Gamma)$. It is convenient for us for later use to adopt the slightly more general point of view of Higgins [60] and consider the fundamental groupoid of the graph $\Gamma$. Here the notion of homotopy on the set $P(\Gamma)$ of paths of $\Gamma$ is defined from the equivalence relation $\sim$ on $P(\Gamma)$ induced by removing a pair of consecutive edges of the form $ee^{-1}$ from a path $p$ to get a path $q$. If we view $\sim$ as a congruence on the free category $C(\Gamma)$ over $\Gamma$, then the quotient category $\pi_1(\Gamma)$ is a *groupoid* (i.e., a small category in which each morphism is an isomorphism), called the *fundamental groupoid* of $\Gamma$. Denote the $\sim$-equivalence class containing the path $p$ by $[p]$ and for each vertex $v \in V(\Gamma)$ let $\pi_1(\Gamma, v) = \{[p] \in \pi_1(\Gamma) : \alpha(p) = \omega(p) = v\}$. Then $\pi_1(\Gamma, v)$ is a group, called the *fundamental group of* $\Gamma$ *based at* $v$. The fundamental groups are free groups, and fundamental groups at different base points in the same connected component of $\Gamma$ are isomorphic. If $\Gamma$ is connected and $T$ is a spanning tree of $\Gamma$, then the rank of $\pi_1(\Gamma, v)$ is the number of positively oriented edges in $\Gamma - T$. For example, $\pi_1(B_X) \cong FG(X)$. We refer to standard sources such as Higgins [60] or Lyndon and Schupp [82] or Cohen [31] for the basic theory of coverings of a connected graph via subgroups of the fundamental group of the graph.

Closed inverse submonoids of free inverse monoids provide the analogous tool for classifying immersions between graphs. An inverse submonoid $N$ of an inverse monoid $M$ is called a *closed inverse submonoid of* $M$ if, whenever $a \in N$, $b \in M$ and $b \geq a$ in the natural partial order on $M$, then $b \in N$. For example, by definition, an inverse monoid $M$ is $E$-unitary iff $E(M)$ is a closed inverse submonoid of $M$. Closed inverse submonoids of an inverse monoid $M$ arise naturally from transitive representations of $M$ by partial one-one maps, in essentially the same way as subgroups of a group arise in connection with transitive representations of the group by permutations. We briefly review Schein's theory [126] of representations of inverse monoids by injective maps.

An inverse monoid $M$ acts (on the left by injective functions) on a set $Q$ if there is a morphism from $M$ to $SIM(Q)$. If $m \in M$ and $q \in Q$ then we denote by $mq$ the image of $q$ under the action of $m$ if $q \in Dom(m)$. An action is transitive if for all $p, q \in Q$ there exists $m \in M$ such that $mq = p$. Notice that this implies that $m^{-1}p = q$. For every $q \in Q$, $Stab(q) = \{m \in M : mq = q\}$ is a closed inverse submonoid of $M$. Conversely, given a closed inverse submonoid $N$ of $M$, we can construct a transitive representation of $M$ as follows. Let $m$ be such that $m^{-1}m \in N$. A subset of $M$ of the form $(mN)^\omega = \{s : s \geq mn \text{ for some } n \in N\}$ is called a *left $\omega$-coset* of $N$. Notice that $N = (1N)^\omega$ is a left $\omega$-coset of itself. Let $X_N$ be the set of left $\omega$-cosets of $N$. If $m \in M$, define an action on $X_N$ by $m.Y = (mY)^\omega$ if $(mY)^\omega \in X_N$ and undefined otherwise. This defines a transitive action of $M$ on $X_N$. Conversely, if $M$ acts transitively on $Q$, then this action is equivalent to the action of $M$ on the left $\omega$-cosets of $Stab(q)$ for any $q \in Q$. For details we refer to Schein's paper [126] or Petrich's book [111]. Note that if $M$ is a group, this just reduces to the usual coset representation of $M$ on some subgroup $N$. Clearly there is a dual notion of right action and right $\omega$-cosets of a closed inverse submonoid of $M$.

Now let $\Gamma$ be a (connected) graph. In order to classify the immersions over $\Gamma$ we make use of the free inverse category over $\Gamma$. A category $C$ is called an *inverse category* (see [87]) if for each morphism $p$ of $C$ there is a unique morphism $p^{-1}$ of $C$ such that $p = pp^{-1}p$ and $p^{-1} = p^{-1}pp^{-1}$. Denote the loop monoid at a vertex (object) $v$ of $C$ by $Mor(v, v)$: that is, $Mor(v, v)$ is the set of morphisms $p$ from $v$ to $v$, with the multiplication induced by $C$. It is clear that each loop monoid is an inverse monoid. The *free inverse category* $FIC(\Gamma)$ over $\Gamma$ is the quotient of the free category on $\Gamma$ by the congruence $\sim_I$ induced by all relations of the form $p = pp^{-1}p$, $p^{-1} = p^{-1}pp^{-1}$ and $pp^{-1}qq^{-1} = qq^{-1}pp^{-1}$ when $\alpha(p) = \alpha(q)$ for paths $p, q$ in $\Gamma$. Then $FIC(\Gamma)$ is an inverse category. If $\Gamma = B_X$, then $FIC(\Gamma) = FIM(X)$. The word problem in $FIC(\Gamma)$ may be solved in a manner similar to the way in which Munn solved the word problem for $FIM(X)$ by passing to the universal cover of $\Gamma$. If $\Gamma$ is a connected graph labelled over $X \cup X^{-1}$ as described above, then each loop monoid of $FIC(\Gamma)$ is a closed inverse submonoid of $FIM(X)$. See the paper by Margolis and Meakin [88] for details and proofs of these facts.

Free inverse categories and their loop monoids serve the same role in classifying connected immersions as do fundamental groupoids and their corresponding fundamental groups in classifying connected covers of graphs. Let $\Gamma$ be a graph labelled over $X \cup X^{-1}$ as above. Denote the loop monoid of $FIC(\Gamma)$ at vertex $v$ by $L(\Gamma, v)$, a closed inverse submonoid of $FIM(X)$. If $v$ and $w$ are vertices in the same connected component of a graph $\Gamma$ then $L(\Gamma, v)$ is *conjugate* to $L(\Gamma, w)$, that is, there exists $m \in FIM(X)$ such that $m^{-1}L(\Gamma, v)m \subseteq L(\Gamma, w)$ and $mL(\Gamma, w)m^{-1} \subseteq L(\Gamma, v)$. (This does not imply that the submonoids $L(\Gamma, v)$ and $L(\Gamma, w)$ are isomorphic however — see [88].) The following result is proved in [88].

**Theorem 11 (Margolis and Meakin)** *Let $f : \Delta \to \Gamma$ be an immersion over $\Gamma$, where $\Delta$ and $\Gamma$ are connected graphs labelled over $X \cup X^{-1}$. If $v \in V(\Gamma)$ and $v' \in V(\Delta)$ such that $f(v') = v$, then $f$ induces an embedding of $L(\Delta, v')$ into $L(\Gamma, v)$. Conversely, let $\Gamma$ be a graph labelled over $X \cup X^{-1}$ and let $H$ be a closed inverse submonoid of $FIM(X)$ such that $H \subseteq L(\Gamma, v)$ for some vertex $v \in V(\Gamma)$. Then there is a graph $\Delta$, an immersion $f : \Delta \to \Gamma$ and a vertex $v' \in V(\Delta)$ such that $f(v') = v$ and $f(L(\Delta, v')) = H$. Furthermore, $\Delta$ is unique (up to graph isomorphism) and $f$ is unique (up to equivalence). If $H$ and $K$ are two closed inverse submonoids of $FIM(X)$ with $H, K \subseteq L(\Gamma, v)$ then the corresponding immersions $f : \Delta \to \Gamma$ and $g : \Delta' \to \Gamma$ are equivalent if and only if $H$ and $K$ are conjugate in $FIM(X)$.*

While closed inverse submonoids of free inverse monoids are not necessarily free, they share many properties in common with free inverse monoids and they may be built from free actions of groups on trees and they admit idempotent-pure morphisms onto free inverse monoids ([88]).

**Theorem 12 (Margolis and Meakin)** *Let $G$ be a group that acts freely on the left on a (simplicial) tree $T$ (so that $G$ is a free group). Fix a root $v_0 \in V(T)$. Let $M(T, G, v_0) = \{(t, g) : t$ is a finite subtree of $T$, $g \in G, v_0, g.v_0 \in V(t)\}$ with multiplication $(t_1, g_1).(t_2, g_2) = (t_1 \cup g_1.t_2, g_1 g_2)$. Then $M(T, G, v_0)$ is isomorphic to a closed inverse submonoid of $FIM(X)$ where $X$ may be identified with the positively oriented edges of the quotient graph of the action of $G$ on $T$. Conversely, every closed inverse submonoid of a free inverse monoid arises this way. Furthermore, $M(T, G, v_0)$ admits an idempotent-pure morphism onto some free inverse monoid $FIM(Y)$ (i.e., there is a surjective morphism $f : M(T, G, v_0) \to FIM(Y)$ such that the inverse image of each idempotent in $FIM(Y)$ consists only of idempotents in $M(T, G, v_0)$).*

Closed inverse submonoids of free inverse monoids admit remarkable finiteness properties. We say that a closed inverse submonoid $N$ of $M$ is *finitely generated* (in the closed sense) if there is a finite set $A$ of elements of $M$ such that $N = \langle A \rangle^{\omega}$, where $\langle A \rangle$ denotes the inverse submonoid of $M$ generated by $A$. We say that *$N$ is of finite index in $M$* if there are finitely many left $\omega$-cosets of $N$ in $M$. A subset $U$ of a monoid $M$ is called *rational* if it can be built from the one-element subsets of $M$ by finitely many applications of the Kleene operations of union, concatenation and submonoid generation: $U$ is called *recognizable* if there is a finite monoid $K$ and a subset $P \subseteq K$ and a morphism $f : M \to K$ such that $U = f^{-1}(P)$ (see Berstel [17]) for details). Rational and recognizable sets coincide in finitely generated free monoids by Klenne's theorem (see any book on formal language theory, e.g., Hopcroft and Ullman [61], for a proof of Kleene's theorem), but they do not coincide in general. A well known result of Anissimov and Seifert [5] shows that a subgroup $H$ of a finitely generated group $G$ is recognizable iff $H$ has finite index in $G$ and rational iff $H$ is finitely generated. By way of contrast, for closed inverse submonoids of free inverse monoids, we have the following surprising result that is proved in [88].

**Theorem 13 (Margolis and Meakin)** *Let $N$ be a closed inverse submonoid of $FIM(X)$ for $X$ a finite set. Then the following conditions are equivalent:*
(1) *$N$ has finite index in $FIM(X)$;*
(2) *$N$ is recognizable;*
(3) *$N$ is rational;*
(4) *$N$ is finitely generated (in the closed sense).*

In his paper [131], Stallings shows how finite $X \cup X^{-1}$-labelled graphs (i.e., finite immersions over $B_X$) may be used to study finitely generated subgroups of free groups. We recall the basic construction of these graphs. Let $H$ be a subgroup of $FG(X)$ generated by the reduced words $h_1, h_2, \ldots, h_n \in FG(X)$. Form a "rose" with a distinguished vertex $v_0$ and one "petal" labelled by each of the generators $h_i$. This is not necessarily a graph labelled over $X \cup X^{-1}$ in the sense described above, because there may be two edges with the same label starting at $v_0$: if there are two such edges, then form a new graph by identifying these edges (one says that the edges are "folded" together). This may result in additional vertices with two edges with the same label starting at the vertex. Continue folding edges until no further folding can occur. Clearly this process stops after a finite number of steps since we started with a finite graph. It can be proved that the folding process is confluent, and also that the resulting graph is independent of the choice of generators for $H$. Denote the resulting graph by $\Gamma(H)$. Clearly the labelling of this graph over $X \cup X^{-1}$ is consistent with an immersion over $B_X$ in the sense described above (we are assuming implicitly that an edge from $v$ to $w$ labelled by a letter $x \in X \cup X^{-1}$ has an inverse edge from $w$ to $v$ labelled by $x^{-1}$ of course). One may view the graph $\Gamma(X)$ as the "core" graph that is obtained from the coset graph of $H$ in $FG(X)$ by pruning trees off the coset graph.

These ideas are illustrated in the following simple examples, where $\Gamma_i = \Gamma(H_i)$ for $i = 1, 2, 3$.

**Example 1** $H_1 = \langle aba^{-1}b^{-1}, a^2, aba, ba^2 \rangle \leq FG(a, b)$.

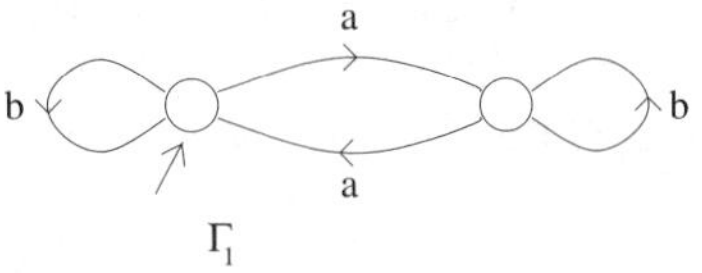


**Example 2** $H_2 = \langle aba^{-1}, a^2 \rangle \leq FG(a, b)$.

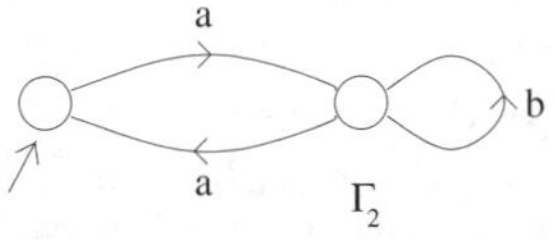

**Example 3** $H_3 = \langle aba^{-1} \rangle \leq FG(a,b)$.

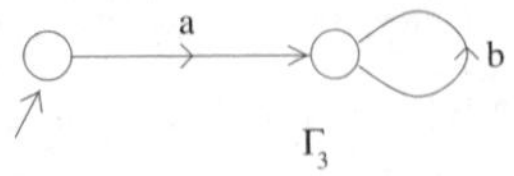


**Example 2 (continued)** Note that the graph $\Gamma(H_2)$ may be obtained from the right coset graph of $H_2$ in $FG(a,b)$ by pruning trees off this graph as illustrated below.

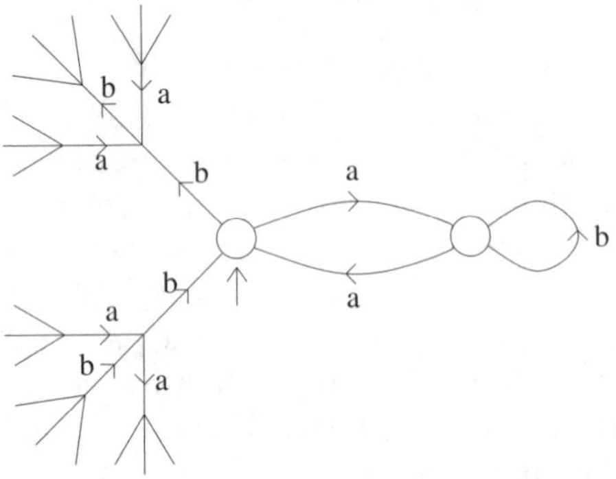


One may regard $\Gamma(H)$ as an automaton with initial and terminal state equal to the image of the distinguished vertex $v_0$ in the folded graph. The distinguished state (vertex) of each automaton $\Gamma(H_i)$ for the subgroups $H_i$ of examples 1, 2 and 3 are indicated by arrows at the corresponding vertices. From this point of view, we can study the *language* accepted by $\Gamma(H)$, namely $L(\Gamma(H)) = \{w \in (X \cup X^{-1})^* :$ $w$ labels a path in $\Gamma(H)$ starting and ending at $v_0\}$. The set of reduced words in this language coincides with the set of reduced words representing elements of $H$, i.e., the image of this language in $FG(X)$ is just $H$. The image of this language in $FIM(X)$ is a closed inverse submonoid of $FIM(X)$, which we will denote by $\bar{H}$. Conversely, if $\Gamma$ is any finite connected graph whose edges are labelled over $X \cup X^{-1}$ consistent with an immersion to $B_X$ and if $v_0$ is a fixed base point of $V(\Gamma)$, then by viewing $\Gamma$ as an automaton with initial and terminal state $v_0$, the language accepted by this automaton gives rise to a finitely generated subgroup $H$ of $FG(X)$ and a finitely generated closed inverse submonoid $\bar{H}$ of $FIM(X)$. If we change the base point, the associated subgroups (closed inverse submonoids) are conjugate. Much additional information about closed inverse submonoids of free inverse monoids is provided in the thesis of Ruyle [122].

It is natural from an automaton point of view to also consider the *syntactic monoid* of this automaton. Each letter $x \in X$ induces a partial one-one map (denoted again by $X$) on the set $Q = V(\Gamma(H))$ of vertices of $\Gamma(H)$. The letter $x$ maps vertex $v$ to vertex $w$ iff there is an edge labelled by $x$ from $v$ to $w$: clearly $x^{-1}$ is the inverse partial one-one map: $x^{-1}$ maps $w$ to $v$ iff $x$ maps $v$ to $w$. The submonoid of $SIM(Q)$ generated by these partial one-one maps induced by the letters in $X \cup X^{-1}$ is an inverse monoid, which we denote by $I(H)$ and refer to as the *syntactic monoid* of $H$. Clearly $I(H)$ is finite because $\Gamma(H)$ is finite.

**Examples 1, 2 and 3 (continued)** Consider the subgroups $H_1 = \langle aba^{-1}b^{-1}, a^2, aba, ba^2\rangle$, $H_2 = \langle aba^{-1}, a^2\rangle$, and $H_3 = \langle aba^{-1}\rangle$ of $FG(a,b)$. It is easy to check from the resulting automata $\Gamma(H_i)$ displayed above that $I(H_1) \cong \mathbf{Z}_2$ (the cyclic group of order 2), while $I(H_2) \cong SIM(a,b)$ and $I(H_3)$ is isomorphic to the combinatorial Brandt monoid of order 6 (i.e., the monoid consisting of the $2 \times 2$ matrix units together with the zero and identity matrix).

The graph $\Gamma(H)$ associated with a finitely generated subgroup of a free group $FG(X)$ has been used by many authors to study various properties of the subgroup $H$. For example, this construction enables one to quickly determine the rank of $H$ and to find a free basis for $H$, to solve the membership problem for $H$ in $FG(X)$, to solve the conjugacy problem for finitely generated subgroups of $FG(X)$, to give a quick proof of Marshall Hall's theorem that shows that finitely generated subgroups of free groups are closed in the profinite topology [57], to determine whether $H$ has finite index in $FG(X)$, to determine whether $H$ is normal in $FG(X)$, to determine whether $H$ is malnormal in $FG(X)$, to give a quick proof of Howson's theorem [62], and of Takashasi's theorem [143], to study algebraic extensions of subgroups of free groups, and to study the well-known conjecture of Hanna Neumann about the rank of the intersection of two finitely generated subgroups of a free group. We refer to the paper by Kapovich and Myasnikov [68] for many references to the literature and a survey of some of the results that have been obtained for subgroups of free groups via Stallings foldings.

While many results about finitely generated subgroups of free groups may be obtained just by studying the graphs $\Gamma(H)$ directly, there are other results that require an analysis of the structure of the syntactic monoid $I(H)$. The structure of this (finite) inverse monoid provides some information about how the subgroup $H$ sits inside the free group $FG(X)$, and may be used to analyze the complexity of some algorithmic problems about finitely generated subgroups of free groups. Recall that a subgroup $H$ of a group $G$ is called a *pure* subgroup if $g^n \in H$ for some integer $n > 1$ implies $g \in H$. The second statement in the following theorem is proved in the paper of Birget, Margolis, Meakin, and Weil [20]; the first statement is a slight restatement of a well known fact (see [88] or [68]).

**Theorem 14** *Let $H = \langle h_1, h_2, \ldots, h_n\rangle$ be a finitely generated subgroup of a free group $FG(X)$. Then*

(1) *$H$ has finite index in $FG(X)$ iff $I(H)$ is a group (iff $\Gamma(H)$ is a cover over $B_X$). One can decide whether $h_1, h_2, \ldots, h_n$ generate a finite index subgroup of $FG(X)$ in polynomial time.*

(2) *$H$ is a pure subgroup of $FG(X)$ iff $I(H)$ is combinatorial (i.e., all maximal subgroups of $I(H)$ are trivial). The problem of deciding whether $h_1, h_2, \ldots, h_n$ generate a pure subgroup of $FG(X)$ is PSPACE-complete.*

Thus for the subgroups $H_1, H_2, H_3$ of $FG(a,b)$ considered in Examples 1, 2 and 3 above, we see that $H_1$ is of finite index in $FG(a,b)$ but is not a pure subgroup, $H_2$ is not of finite index and is not a pure subgroup, and $H_3$ is a pure subgroup but is not of finite index.

The complexity analysis involved in the proof of part (2) of the theorem above involves a series of reductions eventually to a theorem of C. Bennett [13] about the space complexity of injective Turing machines, and shows that several other natural problems about finite immersions are PSPACE complete (see [20] for details). In general terms, if an algorithmic problem about finitely generated subgroups of free groups can be decided by just examining properties of the graphs $\Gamma(H)$ directly, then these problems are usually "easy" from a complexity point of view, while they tend to be "hard" from this point of view if computation of the syntactic monoid $I(H)$ is required.

The classification of connected immersions via closed inverse submonoids of free inverse monoids and several of the associated ideas developed in this section have been extended by Delgado, Margolis and Steinberg [40] to study subgroups of arbitrary finitely presented groups: in particular, they use inverse-monoid theoretic methods to study quasiconvexity and separability properties of subgroups of groups. Similar techniques are used by Gitik [48],[49] to study separability properties and quasiconvexity in hyperbolic groups and by Steinberg [134],[136],[135] to study monoid presentations and profinite topologies on free groups. Methods related to Stallings foldings, making use of 2-complexes instead of graphs, have been employed by McCammond and Wise [100] to study coherence in groups and by Schupp [127] to study surface groups and Coxeter groups of extra large type. Kapovich, Weidmann and Myasnikov [69] have extended the methods of Stallings foldings to study subgroups of fundamental groups of graphs of groups. Recently, Luda Markus-Epstein [95] has used an extension of the methods of Stallings foldings to study algorithmic problems for finitely generated subgroups of amalgamated free products of finite groups. Her construction is related to a construction used by Cherubini, Meakin and Piochi [27] to prove that the word problem for amalgamated free products of finite inverse semigroups in the category of inverse semigroups is decidable: this is in contrast to a result of Sapir [123] that shows that the word problem for amalgamated free products of finite semigroups (in the category of semigroups) is undecidable in general. It seems plausible that further investigation of these techniques might prove fruitful in the study of algorithmic problems for subgroups of finitely presented groups and for other algorithmic problems about groups and semigroups. Indeed, a construction that employs Stallings foldings will be employed in the final section of this paper to study presentations of inverse monoids.

## 6 Finite Inverse Monoids and Infinite Groups

The theory of finite monoids is very closely related to the theory of finite automata and the theory of regular languages. Indeed Kleene's theorem shows that if $A$ is a finite alphabet, then a language $L \subset A^*$ is regular (equivalently rational) if and only if its syntactic monoid is finite. Eilenberg's variety theorem [112], sets up a one-one correspondence between pseudovarieties of finite monoids and varieties of regular languages. This has led to an intensive study of pseudovarieties of finite semigroups (monoids): this is one of the central areas of research in finite semigroup theory.

Inspired by the connection between finite semigroups and finite automata, Krohn and Rhodes [73] proved the "prime decomposition theorem" for finite semigroups. This can be formulated to show that every finite semigroup $S$ is a homomorphic image of a subsemigroup of some iterated wreath product $A_1 \circ G_1 \circ A_2 \circ G_2 \ldots \circ G_n \circ A_{n+1}$, where each $A_i$ is a semigroup with only trivial subgroups and each $G_i$ is a group. The smallest such integer $n$ for which $S$ can be so expressed is called the *group complexity* of $S$, and the problem of algorithmically deciding the group complexity of a finite semigroup is a problem of central importance in the theory. There are several books and survey articles devoted to the theory of finite semigroups and connections between this theory and the theory of profinite groups and profinite monoids (for example, Almeida [4], Pin [112], Eilenberg [45], Henckell, Margolis, Pin and Rhodes [59], Straubing [141], Weil [145], ...). In the present section, I will discuss briefly some results of Ruyle [122] linking pseudovarieties of finite inverse monoids and the theory of equations in free groups, and I will discuss (again briefly) some connections between finite inverse monoids, finite immersions, and profinite topologies on free groups.

A *pseudovariety* of finite [inverse] monoids is a class $\mathbf{V}$ of finite [inverse] monoids that is closed under taking [inverse] submonoids, homomorphic images and finite direct products. It is well-known that if $K$ is a set of finite [inverse] monoids, then the pseudovariety of finite [inverse] monoids generated by $K$ (i.e., the smallest pseudovariety of [inverse] monoids containing $K$) is $\langle K \rangle = HSP(K)$, the class of [inverse] monoids that are homomorphic images of [inverse] submonoids of finite direct products of monoids in $K$. A question of interest in finite semigroup theory is the question of decidability of membership in a pseudovariety $\mathbf{V}$: i.e., given a pseudovariety $\mathbf{V}$ or a set $K$ of finite [inverse] monoids, and a finite [inverse] monoid $M$, is there an algorithm to decide whether or not $M \in \mathbf{V}$ or $M \in \langle K \rangle$.

In his thesis [122], Ruyle developed an Eilenberg-type variety theorem linking pseudovarieties of finite inverse monoids and "varieties" of edge-labelled graphs. A *variety* of (edge-labelled) graphs is a function $\nu$ that associates with each set $X$ a set $\nu(X)$ of finite connected $X \cup X^{-1}$-labelled graphs such that:

(1) If $\Gamma, \Lambda \in \nu(X)$, then each connected component of $\Gamma \times \Lambda$ is in $\nu(X)$.

(2) If $\Gamma \in \nu(X)$ and $\Gamma \to \Lambda$ is a covering of graphs, then $\Lambda \in \nu(X)$.

(3) If $\Gamma \in \nu(Y)$ and $\psi : FIM(X) \to FIM(Y)$ is a morphism between free inverse monoids, then each component of $\Gamma\psi^{-1}$ is in $\nu(X)$.

Here the *product* $\Gamma \times \Lambda$ of two $X \cup X^{-1}$-labelled graphs is the graph with set $V(\Gamma \times \Lambda) = V(\Gamma) \times V(\Lambda)$ of vertices and with and an edge labelled by $x \in (X \cup X^{-1})$ from $(u_1, u_2)$ to $(v_1, v_2)$ iff there is an edge labelled by $x$ from $u_1$ to $v_1$ in $\Gamma$ and an edge labelled by $x$ from $u_2$ to $v_2$ in $\Lambda$. Also, if $\psi : FIM(X) \to FIM(Y)$ is a morphism between free inverse monoids and if $\Gamma$ is a $Y \cup Y^{-1}$-labelled graph, then $\psi^{-1}(\Gamma)$ is the $X \cup X^{-1}$-labelled graph with the same set of vertices as $\Gamma$ and with an edge labelled by $x \in X \cup X^{-1}$ from $v_1$ to $v_2$ iff $\psi(x)$ labels a path from $v_1$ to $v_2$ in $\Gamma$.

For example, if $X = \{x, y, z\}$ and $Y = \{a, b\}$ and $\psi : FIM(X) \to FIM(Y)$ is defined by $x \to a, y \to a^2, z \to bb^{-1}$, then the result of applying inverse edge substitution $\psi^{-1}$ to a graph $\Gamma$ is shown in the diagram below.

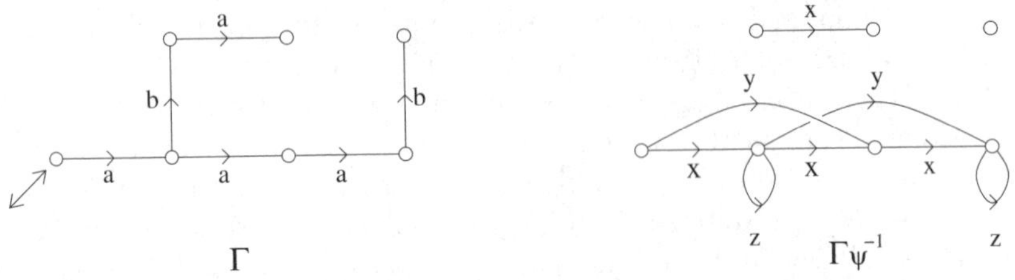


For $\Gamma$ a finite connected $X \cup X^{-1}$-labelled graph, denote by $I(\Gamma)$ the syntactic inverse monoid of $\Gamma$ (i.e., the inverse monoid of partial one-one maps on the set of vertices of $\Gamma$ induced by the letters of $X$). The following theorem was proved by Ruyle [122] in 1997.

**Theorem 15 (Ruyle)** *There is a one-one correspondence between pseudovarieties of finite inverse monoids and varieties of finite connected edge-labelled graphs, given by*

$\mathbf{V} \to \nu$ *(where $\nu(X)$ is the set of $X \cup X^{-1}$-labelled graphs $\Gamma$ such that $I(\Gamma) \in \mathbf{V}$),*

*and*

$\nu \to \mathbf{V} = \langle I(\Gamma) : \Gamma \in \nu(X),$ *some* $X\rangle$.

Now denote by $\mathcal{T}(X)$ the class of all $X \cup X^{-1}$-labelled Munn trees, and by $\langle \mathcal{T} \rangle$ the variety of all graphs generated by all Munn trees. Denote by $\mathbf{T}$ the corresponding pseudovariety of finite inverse monoids. In his thesis [122], Ruyle showed that $\langle \mathcal{T} \rangle = CPIn(\mathcal{T})$, i.e., every graph in $\langle \mathcal{T} \rangle$ is covered by some finite product of inverse images of Munn trees. He also proved the following interesting connection between membership in $\langle \mathcal{T} \rangle$ (or equivalently in $\mathbf{T}$) and the theory of equations in free groups.

**Theorem 16 (Ruyle)** *The membership problem in $\langle \mathcal{T} \rangle$ is algorithmically equivalent to deciding consistency of coefficient-free systems of equations and inequations ($u \neq v$) in free groups.*

An immediate corollary of this and Makanin's work [84] on decidability of the universal theory of free groups is the following:

**Corollary 2** *There is an algorithm to decide membership in $\langle \mathcal{T} \rangle$.*

An important result in finite semigroup theory is a theorem of Ash [7] that proves that the pseudovariety of finite semigroups generated by the class of finite inverse semigroups is precisely the class of finite semigroups whose idempotents commute: thus membership in the pseudovariety of semigroups generated by the finite inverse semigroups is decidable.

Ash also provided the solution to another celebrated problem in finite semigroup theory with his solution [8] to the Rhodes "Type II" conjecture. An alternative proof of this conjecture was provided by Ribes and Zalesskiĭ [119], using methods of profinite groups. It is a well-known result of Marshall Hall [57] that a finitely generated subgroup of a free group is closed in the profinite topology: Ribes and Zalesskiĭ showed that the product $H_1H_2 \dots H_n$ of finitely many finitely generated subgroups of a free group is closed in the profinite topology, and this implies that Rhodes' "Type II" conjecture is true (see the survey paper by Henckell, Margolis, Pin and Rhodes [59] for connections between these problems).

Let $\mathbf{V}$ be a pseudovariety of finite groups. Recall [57] that in the pro-$\mathbf{V}$ topology on a group $G$, a basis of clopen neighborhoods of 1 is given by the normal subgroups $N$ of $G$ such that $N$ has finite index in $G$ and $G/N \in \mathbf{V}$. (The profinite topology on $G$ corresponds to the case where the pseudovariety $\mathbf{V}$ consists of all finite groups of course.) A pseudovariety $\mathbf{V}$ of groups is said to be *extension-closed* if every finite extension of a group in $\mathbf{V}$ is also in $\mathbf{V}$.

Recall that if $H$ is a finitely generated subgroup of a free group $F(X)$, then the syntactic monoid $I(H)$ of $H$ is the inverse monoid of partial one-one maps on the vertices of the associated immersion $\Gamma(H)$ induced by the letters in $X$. We say that $H$ *is* $\mathbf{V}$*-extendible* if the immersion $\Gamma(H)$ can be embedded in a finite cover over $B_Y$ for some finite set $Y$ containing $X$ in such a way that the syntactic monoid of this cover is a group in $\mathbf{V}$. That is, the set of partial one-one maps in $I(H)$ can be extended to a set of permutations on a possibly larger set of states generating a group in $\mathbf{V}$. This is equivalent to saying that $I(H)$ has an $E$-unitary cover over some finite group in $\mathbf{V}$. The following result was proved by Ribes and Zalesskiĭ [120] and by Margolis, Sapir and Weil [93].

**Theorem 17** *Let $H$ be a finitely generated subgroup of a free group $FG(X)$. If $\mathbf{V}$ is extension-closed, then the following are equivalent*

(1) *$H$ is closed in the pro-$\mathbf{V}$ topology.*

(2) *$H$ is a free factor of a clopen subgroup of $FG(X)$.*

(3) *$H$ is $\mathbf{V}$-extendible.*

In [93], Margolis, Sapir and Weil provided an algorithm for computing the pro-$p$ closure and the pro-nil closure of a finitely generated subgroup of a free group $FG(X)$. The question of providing an algorithm to compute the pro-solvable closure of a finitely generated subgroup of a free group remains open as far as I am aware. For additional information about the extension problem for finite inverse monoids of partial one-one maps, the reader is referred to the paper by Steinberg [135] and the paper by Auinger and Steinberg [9].

## 7 Presentations of Inverse Monoids

The inverse monoid $M$ presented by a set $X$ of generators and relations of the form $u_i = v_i$ will be denoted by $M = Inv\langle X : u_i = v_i\rangle$. Here $u_i, v_i \in (X \cup X^{-1})^*$. Clearly $M$ is the image of the free inverse monoid $FIM(X)$ by the congruence generated by the defining relations, or equivalently, it is the quotient of $(X \cup X^{-1})^*$ obtained by applying the relations $u_i = v_i$ and the identities that define the variety of inverse monoids. It is easy to see that $G = Gp\langle X : u_i = v_i\rangle$ is the maximal group homomorphic image of the inverse monoid $M = Inv\langle X : u_i = v_i\rangle$. For example, the bicyclic monoid has a presentation $Inv\langle a : a^{-1}a = 1\rangle$ as an inverse monoid (and its maximal group image is $\mathbf{Z}$), a polycyclic monoid may be defined by $Inv\langle X : x^{-1}x = 1,\ x^{-1}y = 0,\ x \neq y\rangle$ (and it has trivial maximal group image), and the free group $FG(X)$ may be defined by $FG(X) = Inv\langle X : xx^{-1} = x^{-1}x = 1 \text{ for all } x \in X\rangle$. Of course $FIM(X) = Inv\langle X : \emptyset\rangle$ and its maximal group image is $FG(X)$.

If $P = Mon\langle X : u_i = v_i\rangle$ embeds in $G = Gp\langle X : u_i = v_i\rangle$, then we have associated two inverse semigroups with this situation: the inverse hull of $P$, and the Toeplitz inverse monoid $T(G,P)$ (see section 3 of this paper to recall the definitions). We clearly have $\beta_{x^{-1}}\beta_x = 1$ in the Toeplitz inverse monoid $T(G,P)$. This expresses the fact that each $\beta_x, x \in X$ has domain $P$. Clearly if $\beta_g \in T(G,P)$ has domain $P$ then $g \in P$. This simply says that $P$ embeds as the $\mathcal{L}$-class of 1 in $T(G,P)$. If we identify each $x \in X$ with $\beta_x \in T(G,P)$, then $u_i = v_i$ is a relation in $T(G,P)$ as well. However, even if $T(G,P)$ is bisimple (as is the case for example for the bicyclic monoid, or if $G$ is a Garside group, etc. — see sections 2 and 3), then $T(G,P)$ has in general other relations that are not obviously consequences of the defining relations for $P$.

For example, if $P = Mon\langle a,b : ab = ba\rangle$ (the free commutative monoid on 2 generators), then $P$ embeds as the positive cone of the free abelian group of rank 2, $PP^{-1} = G$, and principal left ideals of $P$ form a lattice, so $T(G,P)$ is bisimple and coincides with the left inverse hull $I_l(P)$. However, a presentation for $T(G,P)$ in this case is given by $T(G,P) = Inv\langle a,b : a^{-1}a = b^{-1}b = 1,\ ab = ba,\ ab^{-1} = b^{-1}a\rangle$, i.e., $T(G,P)$ is a direct product of two bicyclic monoids. It will be apparent with the results to be described in this section that the relation $ab^{-1} = b^{-1}a$ is not a consequence of the other defining relations for this inverse monoid. More generally, one can provide a presentation for the Toeplitz inverse monoid associated with the embedding of a partially commutative monoid in its corresponding right angled Artin group. Recall that a *Coxeter matrix* is a matrix $(m_{x,y}),\ x,y \in X$, such that $m_{x,y} = m_{y,x},\ m_{x,x} = 1$ and $m_{x,y} \in \{2,\infty\}$. The associated *right-angled Artin group* is the group $G = Gp\langle X : xy = yx \text{ if } m_{x,y} = 2\rangle$ and the corresponding partially commutative monoid $P$ with the same presentation as a monoid embeds in $G$. This is well-known, and follows from the theorem of Paris (Theorem 4 above) of course. The corresponding Toeplitz inverse monoid $T(G,P)$ has presentation

$$T(G,P) = Inv\langle X : x^{-1}x = 1,\ \ xy = yx,\ \ x^{-1}y = yx^{-1} \text{ if } m_{x,y} = 2,$$
$$x^{-1}y = 0 \text{ if } m_{x,y} = \infty,\ \ x,y \in X\rangle.$$

This presentation is implicit in the paper of Crisp and Laca [33]. Note that $PP^{-1} \neq G$ in general for right-angled Artin groups, in fact from Nica's theorem (Theorem 8 above), a right-angled Artin group $G$ is the group of (right) quotients of its associated partially commutative monoid $P$ if and only if it is a free abelian group (i.e., $m_{x,y} = 2$ for all $x \neq y$).

It is of interest to study presentations for Toeplitz inverse monoids in general. If $PP^{-1} \neq G$ then by Theorem 8, $T(G,P)$ has a zero and there will be relations similar to the relations in the polycyclic monoid above. If $PP^{-1} = G$, then $T(G,P)$

is $F$-inverse with maximal group image $G$, so the relations in $T(G,P)$ must be consequences of the defining relations in the group $G$.

There are several other natural ways to associate inverse monoids with monoid presentations $P = Mon\langle X : u_i = v_i \rangle$ that embed in the associated group with the same presentation. For example, the inverse monoids $S = Inv\langle X : u_i = v_i \rangle$ and $T = Inv\langle X : u_i^{-1}v_i = 1 \rangle$ (or its dual) and $M = Inv\langle X : x^{-1}x = 1,\ u_i^{-1}v_i = 1\ \ \forall x \in X \rangle$ (or its dual) are clearly all inverse monoids with maximal group image $G$. Clearly there are surjective morphisms $S \to T \to M$. As will be apparent from some of the results discussed later in this section, these inverse monoids are not isomorphic in general. In good cases, one or more of these monoids may be $E$-unitary, and it may be possible to exploit this fact to study the group $G$, although little work has been done along these lines as far as I am aware.

Many other interesting examples of presentations of inverse monoids arise in the theory of $C^*$-algebras. For example, Hancock and Raeburn [58] provide a presentation for the *Cuntz–Krieger* inverse monoid associated with the well-known Cuntz–Krieger algebra $\mathcal{O}_A$ associated with an $n \times n$ zero-one matrix $A = (A_{i,j})$ with at least one non-zero entry in each row and column. The corresponding Cuntz–Krieger inverse monoid has presentation $C_A = Inv\langle s_1, \ldots, s_n : s_i^{-1}s_j = 0,\ i \neq j,\ s_i^{-1}s_is_js_j^{-1} = A_{i,j}s_js_j^{-1}\ \ \forall i,j \rangle$. We refer to [58] for details and to the book of Paterson [110] or the extensive literature on graph $C^*$-algebras for indications of many more examples of inverse monoids arising in connection with $C^*$-algebras. Other interesting examples of inverse semigroups arise in connection with the study of tilings (see for example the work of Kellendonk and Lawson [71]) and in the study of self-similar actions of inverse monoids (see the paper by Bartholdi, Grigorchuk, and Nekrasevych [11] for an introduction to this subject). It would be of interest to study such semigroups from the point of view of presentations.

The main tool that has been exploited to study the word problem for presentations of inverse monoids is the construction of the associated Schützenberger graphs, initiated by Stephen [138]. If $M = Inv\langle X : u_i = v_i \rangle$, then one may consider the corresponding Cayley graph $\Gamma(M, X)$. This graph has the elements of the inverse monoid $M$ as vertices and has an edge labelled by $x \in X \cup X^{-1}$ from $m$ to $mx$ for each $m \in M$. One may view this edge as starting at $m$, ending at $mx$, and labelled by $x$. However, the Cayley graph $\Gamma(M, X)$ suffers from the deficiency that it is not strongly connected in general (unless $M$ happens to be a group) — there is not necessarily an edge labelled by $x^{-1}$ starting at $mx$ and ending at $m$: as an extreme example, if the monoid $M$ happens to have a zero, then there may be many edges in this graph ending at zero, but all edges that start at zero must obviously end at zero. Thus the Cayley graph is not an $X$-labelled graph in the sense defined in section 5 of this paper. Unlike the situation for Cayley graphs of groups, the Cayley graph of an inverse monoid is not naturally equipped with a word metric that turns it into a geodesic metric space.

To overcome this difficulty, it is convenient to study the *strongly connected components* of $\Gamma(M, X)$. A strongly connected component of the Cayley graph $\Gamma(M, X)$ corresponds exactly to the restriction of $\Gamma(M, X)$ to an $\mathcal{R}$-class of $M$. Such graphs are referred to in the literature as the *Schützenberger graphs* associated with $M$ by virtue of their connection with the Schützenberger representation of $M$. In more detail, for each word $u \in (X \cup X^{-1})^*$, we denote by $S\Gamma(M, X, u)$ the graph with set $R_u = \{m \in M : mm^{-1} = uu^{-1} \text{ in } M\}$ of vertices, and with an edge labelled by $x \in X \cup X^{-1}$ from $m$ to $mx$ if $m, mx \in R_u$. It is not difficult to see that if $x$ labels an edge from $m$ to $mx$ in $S\Gamma(M, X, u)$, then $x^{-1}$ labels an edge from $mx$ to $m$ in this graph. Thus the Schützenberger graphs $S\Gamma(M, X, u)$ are strongly connected subgraphs of the Cayley graph $\Gamma(M, x)$: in fact it is not difficult to see that these graphs are geodesic metric spaces with respect to the word metric, similar to the situation for Cayley graphs of groups.

There is a Schützenberger graph corresponding to each idempotent of the inverse monoid $M$ and there is an obvious natural graph morphism (not necessarily an injection) from each Schützenberger graph to the Cayley graph $\Gamma(G, X)$ where $G = Gp\langle X : u_i = v_i\rangle$ is the maximal group image of $M$. It is not very difficult to show that the inverse monoid $M$ is $E$-unitary if and only if each Schützenberger graph of $M$ naturally *embeds* in the Cayley graph $\Gamma(G, X)$. This provides an appropriate geometric interpretation of the notion of an $E$-unitary inverse monoid that has will be exploited later in this section.

It is useful to consider the *Schützenberger automaton* $\mathcal{A}(M, X, u) = (uu^{-1}, S\Gamma(M, X, u), u)$ with initial state (vertex) $uu^{-1} \in M$, terminal state $u \in M$ and set $S\Gamma(M, X, u)$ of states. From this point of view, the *language* accepted by this automaton is the set

$$L(u) = \{v \in M : v \text{ labels a path in } \mathcal{A}(M, X, u) \text{ from } uu^{-1} \text{ to } u \text{ in } S\Gamma(M, X, u)\}.$$

(Here we are interpreting $u$ and $v$ both as words in $(X \cup X^{-1})^*$ and as elements of $M$, and we are regarding the language $L(u)$ as a subset of $M$. Strictly speaking, if $v \in (X \cup X^{-1})^*$, then the corresponding element of $M$ is $\tau(v)$ where $\tau$ is the natural homomorphism from $(X \cup X^{-1})^*$ to $M$, but I will suppress this notation for ease of exposition: the context makes it clear whether $v$ is being considered as a word in $(X \cup X^{-1})^*$ or as an element of $M$.)

The following result of Stephen [138] is central to the theory.

**Theorem 18 (Stephen)** *Let $M = Inv\langle X : u_i = v_i\rangle$ and let $u, v \in (X \cup X^{-1})^*$ (also interpreted as elements of $M$ as above). Then*

(1) *$L(u) = \{v \in M : v \geq u$ in the natural partial order on $M\}$.*

(2) *$u = v$ in $M$ iff $L(u) = L(v)$ iff $u \in L(v)$ and $v \in L(u)$ iff $\mathcal{A}(u)$ and $\mathcal{A}(v)$ are isomorphic as birooted edge-labelled graphs.*

Thus we are able to solve the word problem for $M$ if we can effectively construct the automata $\mathcal{A}(M, X, u)$ for each $u \in (X \cup X^{-1})^*$. This is of course analogous to solving the word problem for a group presentation by effectively constructing the corresponding Cayley graph. In his paper [138], Stephen provided an iterative procedure for constructing the Schützenberger automaton of a word $u \in (X \cup X^{-1})^*$, analogous to the Todd–Coxeter procedure for iteratively constructing the Cayley graph of a group presentation: I will briefly review his construction. Start with the "linear automaton" of the word $u$, i.e., the $X \cup X^{-1}$-labelled graph with the word $u$

labelling a linear path from an initial vertex $\alpha$ to a terminal vertex $\omega$. Build intermediate automata by successively applying one of the following two operations: (1) Stallings foldings, and (2) "expansions". Here an expansion is described as follows. If in an intermediate automaton there is a path from some vertex $\gamma$ to some vertex $\delta$ labelled by one side (say $u_i$) of one of the defining relations in $M$, then sew on (add) a new path from $\gamma$ to $\delta$ labelled by the other side ($v_i$) of the defining relation in $M$. Stephen shows in [138] that these operations are confluent, that there is a well-defined process of closing with respect to these two operations, and that the resulting process limits (in a sense made precise in his paper) to the Schützenberger automaton $\mathcal{A}(M, X, u)$.

For example, if $M = Inv\langle X : \emptyset\rangle$, the free inverse monoid on $X$, then no expansions apply: it is easy to see that if one starts with the linear automaton of a word $u$, then successive applications of Stallings foldings produces the Munn tree $MT(u)$ (which coincides with the Schützenberger automaton in this case), so Munn's theorem (Theorem 9) follows as a special case of Stephen's results, as would be expected.

As another example, consider the inverse monoid presentation $M_1 = Inv\langle a, b : a^{-1}a = b^{-1}b = 1,\ ab = ba\rangle$. If we start with a single vertex (call it 1), then at this vertex it is possible to read the empty path (i.e., the path labelled by 1), so an expansion can be applied and we sew on a path labelled by $a^{-1}a$ starting and ending at the vertex 1. An application of a Stallings folding collapses this to a single edge labelled by the letter $a$ ending at the vertex 1 and starting at some other vertex (let's call it $a^{-1}$): similarly we attach an edge labelled by the letter $b$ ending at the vertex 1 and starting at some other vertex (let's call it $b^{-1}$). Repeat the process at the vertex $b^{-1}$: we sew on an edge labelled by $a^{-1}$ ending at the vertex $b^{-1}$. Let's call the initial vertex of this edge $a^{-1}b^{-1}$. This creates a path labelled by $ab$ from $a^{-1}b^{-1}$ to 1, so we must sew in a path labelled by $ba$ from the vertex $a^{-1}b^{-1}$ to the vertex 1. But then there are two edges labelled by the letter $a$ ending at the vertex 1, so we identify these by a Stallings folding, to obtain a square as indicated in the diagram below. Repetition of this argument at any vertex shows that the Schützenberger graph of 1 consists of the third quadrant of the $a - b$ plane as shown in the diagram below.

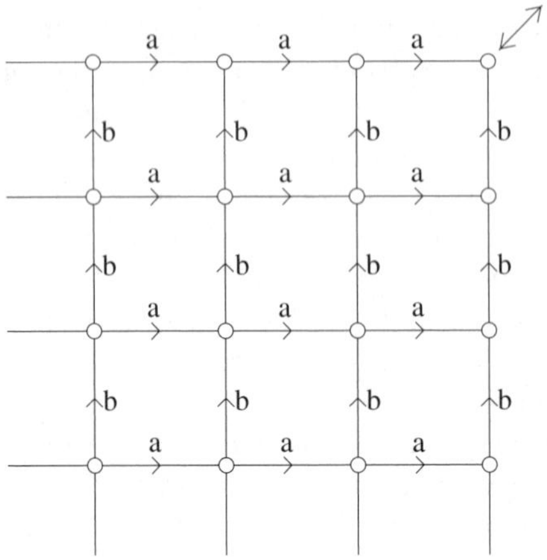


Similarly, if $u$ is any word, the Schützenberger graph of $u$ is closed under adding all squares of the Cayley complex of the free abelian group on $\{a, b\}$ to the "south-west" of any vertex constructed. One sees that all Schützenberger graphs of this monoid embed in the Cayley graph of the free abelian group on $\{a, b\}$, so $M_1$ is $E$-unitary. For example, the Schützenberger graph of the word $b^{-1}a$ is depicted in the diagram below.

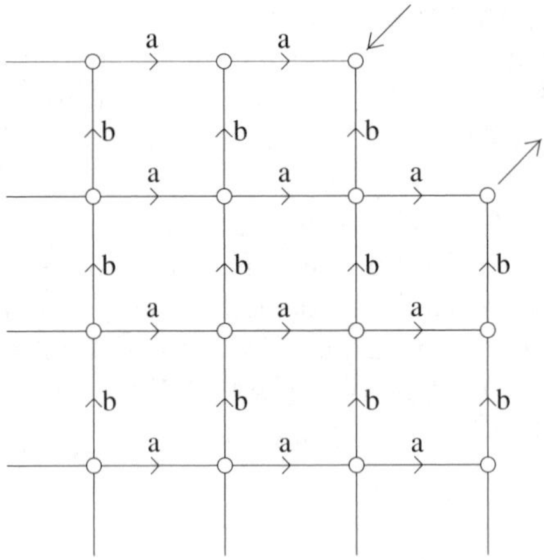


Note that $ab^{-1} \notin L(b^{-1}a)$ so $ab^{-1} \neq b^{-1}a$ in the inverse monoid $M_1$. On the other hand, if one considers the Toeplitz inverse monoid $M_2 = Inv\langle a, b : a^{-1}a = b^{-1}b = 1,\ ab = ba,\ a^{-1}b = ba^{-1}\rangle$, then the Schützenberger graphs of $M_2$ are subgraphs of the Cayley graph of the free abelian group on $\{a, b\}$ that are closed under addition of squares to the "south-west" of any vertex as above, but also are closed under completing all squares of the Cayley complex whenever two consecutive edges of a square are present. Thus there is no word whose Schützenberger graph in $M_2$ is equal to the Schützenberger graph of $b^{-1}a$ in $M_1$. It follows that the inverse monoids $M_1$ and $M_2$ are not isomorphic.

Schützenberger automata have played a pivotal role in much of the theory of presentations of inverse monoids. In his paper [138], Stephen used these automata to solve the word problem for the free inverse monoid on $n$ commuting generators (this monoid is not $E$-unitary if $n > 2$, but it is $E$-unitary if $n = 2$ [98]). Margolis and Meakin [89] made use of these automata, together with Rabin's tree theorem, to solve the word problem for inverse monoids of the form $Inv\langle X : e_i = f_i,\ i = 1, \ldots, n\rangle$ where $e_i$ and $f_i$ are Dyck words (i.e., idempotents of $FIM(X)$). The automata were used by several authors to study free products of inverse semigroups [67], and various classes of amalgamated free products and HNN extensions in the category of inverse semigroups (e.g., [14], [65], [26], [27].) Related work on the structure of amalgams and HNN extensions of inverse semigroups has been done by Haataja, Margolis and Meakin [56], Bennett [15], Yamamura [146], [147], and Gilbert [47].

The notion of the *Schützenberger complex* of a presentation of an inverse monoid has been introduced by Steinberg [137] by analogy with the corresponding notion of the Cayley complex of a group presentation: this notion has been used by Lindblad [78] in his work on the prefix membership problem for one-relator groups. Margolis, Meakin and Stephen [91] used Schützenberger automata to solve the word problem for finitely generated free Burnside inverse semigroups in the variety of inverse semigroups defined by an identity of the form $w^a = w^{a+b}$ where $b \leq a$: the word problem for free Burnside inverse semigroups in the corresponding variety where $b > a$ remains unknown as far as I am aware. [As an aside, it is interesting to note that the word problem for finitely generated free Burnside semigroups in the variety of *semigroups* defined by the identities $w^a = w^{a+b}$ is known for all values of $a, b > 0$ except $a = 2$ and $b = 1$ — see the papers of McCammond [99], Guba [52], de Luca and Varricchio [42], and do Lago [43] for much information about Burnside varieties of semigroups.]

In [63], Ivanov, Margolis and Meakin made heavy use of Schützenberger graphs as well as van Kampen diagram arguments to study one relator inverse monoids. In view of the difficulty of solving the word problem for one-relation semigroups of the form $Sgp\langle X : u = v\rangle$ where $u, v \in X^+$, I will restrict attention to one-relator inverse monoids of the form $Inv\langle X : w = 1\rangle$. If $w$ is a Dyck word (i.e., an idempotent in $FIM(X)$), then there is a fast algorithm provided in the paper of Birget, Margolis and Meakin [19] to solve the word problem. However, in general this problem remains unsolved. In fact, the following result in [63] shows that this problem is at least as difficult as the one-relation semigroup problem, even if we restrict to the case where $w$ is a reduced word.

**Theorem 19 (Ivanov, Margolis, Meakin)** *If the word problem is decidable for every one-relator inverse monoid of the form $Inv\langle X : w = 1\rangle$, for $w$ a reduced word, then the word problem for every one relation semigroup $Sgp\langle X : u = v\rangle$ (for $u, v \in X^+$) is decidable.*

**Outline of Proof** A result of Adian and Oganessian [3] reduces the word problem for one relation semigroups to the study of the word problem for semigroups with a presentation of the form $S = Sgp\langle X : aub = avc\rangle$ where $a, b, c \in X,\ \ b \neq c$

and $u, v \in X^*$. Consider the associated one relator inverse monoid $M = Inv\langle X : aubc^{-1}v^{-1}a^{-1} = 1\rangle$. Note that $aubc^{-1}v^{-1}a^{-1}$ is a reduced word (but not a cyclically reduced word) since $b \neq c$. In [63], Ivanov, Margolis and Meakin show, by using some results of Adian and Oganessian [3] and by using the inverse hull of the semigroup $S$, that $S$ embeds in $M$. It follows that if the word problem is decidable for $M$, then it must also be decidable for $S$. □

In view of this result, I will restrict attention further to the consideration of inverse monoids of the form $Inv\langle X : w = 1\rangle$ where $w$ is a *cyclically reduced* word. In this case, Ivanov, Margolis and Meakin [63] proved the following structural result about such monoids.

**Theorem 20 (Ivanov, Margolis, Meakin)** *Every inverse monoid of the form $Inv\langle X : w = 1\rangle$, where $w$ is a cyclically reduced word, is $E$-unitary.*

I will outline some of the arguments used in the proof of this theorem, since they are of independent interest and provide a nice blend of techniques from geometric group theory and inverse semigroup theory. First note that if $w$ can be factored in the free monoid $(X \cup X^{-1})^*$ as $w \equiv uv$, then it does not necessarily follow that the cyclic conjugate $vu$ of $w$ is also 1 in the inverse monoid $M = Inv\langle X : w = 1\rangle$. (For example, we know from looking at the bicyclic monoid that $ab = 1$ does not imply that $ba = 1$.) The cyclic conjugate $vu$ of the word $w \equiv uv$ is called a *unit cyclic conjugate* of $w$ if $vu = 1$ in $M$.

In order to prove that the inverse monoid $M$ is $E$-unitary, it is necessary to prove that every word that is 1 in the corresponding maximal group homomorphic image $G = Gp\langle X : w = 1\rangle$ must be an idempotent of $M$. Now words that are 1 in $G$ are precisely the words that label the boundary of some van Kampen diagram over $G$, by van Kampen's lemma (see Lyndon and Schupp [82] or Ol'shanskii [107]). Consequently, we need to show that every word labelling the boundary of every van Kampen diagram over $G$ must be an idempotent in $M$. A lemma in [63] shows that this happens if, for every van Kampen diagram $\Delta$ over $G$, there is some cell $\Pi$ of $\Delta$ and some vertex $\alpha \in \partial\Delta \cap \partial\Pi$ such that the cyclic conjugate of $w^{\pm 1}$ obtained by reading around the boundary of $\Pi$, starting at $\alpha$, is a *unit* cyclic conjugate of $w^{\pm 1}$. In this situation, we say that *a unit cyclic conjugate of $w$ starts on the boundary of* $\Delta$. Thus the problem reduces to showing that there is a unit cyclic conjugate of $w$ starting on the boundary of every van Kampen diagram over $G$. Unfortunately, we do not know of any general procedure for testing whether a cyclic conjugate of a cyclically reduced word $w$ is a unit cyclic conjugate. In fact, the situation can be quite complicated, as the following example in a paper by Margolis, Meakin, and Stephen [90] shows.

**Example 4** Let $w = abcdacdadabbcdacd$, a 17-letter word in $\{a, b, c, d\}^*$. In [90] it is shown that there is a van Kampen diagram $\Delta$ with four cells $\Pi_i$, $i = 1, \ldots, 4$, over $w$ such that for every cell $\Pi_i$ of $\Delta$, the vertex on $\partial(\Pi_i)$ at which one reads $w^{\pm 1}$ around $\partial(\Pi_i)$ is an interior vertex of the diagram $\Delta$. The situation is represented in the picture below, where the arrows depict vertices at which the word $w$ is read around the corresponding cell.

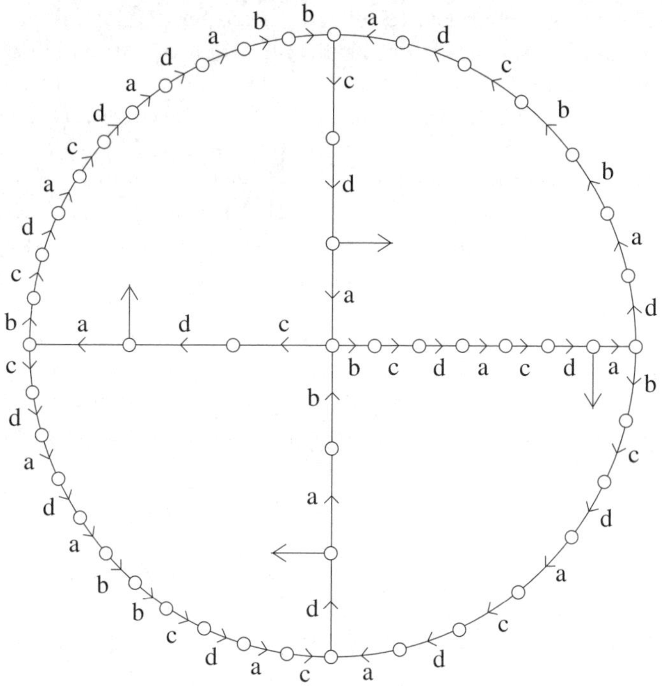


Note that one may construct this diagram $\Delta$ by starting with any of the vertices labelled by arrows and sewing on a copy of a cell whose boundary is labelled by $w$ at that vertex, then successively sewing on three more cells with boundaries labelled by $w$ at appropriate vertices, and applying Stallings foldings. Thus the 1-skeleton of this diagram is an intermediate graph along the way to constructing the Schützenberger graph of 1 for the corresponding inverse monoid $M = Inv\langle a, b, c, d : w = 1\rangle$. It follows by Stephen's theorem (Theorem 18 above), that any word that labels a loop based at any of the interior vertices labelled by arrows must be 1 in $M$. In particular, the cyclic conjugates of $w$ obtained by starting at a vertex marked by an arrow and reading around the boundary of the adjacent cell, must be a unit cyclic conjugate of $w$. This provides us with a method of constructing new unit cyclic conjugates of $w$. In particular, an analysis of this example shows that every cyclic conjugate of this word $w$ that starts with the letter $a$ must be a unit cyclic conjugate of $w$ (see [90] for details). Now the Freiheitssatz tells us that the letter $a$ must occur on the boundary of *every* van Kampen diagram over $w$, so there is a unit cyclic conjugate of $w$ starting on the boundary of every van Kampen diagram over $w$, and this particular inverse monoid $M$ must be $E$-unitary. □

The argument outlined in this example is indicative of some of the general arguments used in [63] to prove Theorem 20. Essential use is also made of a theorem of Ivanov and Meakin [64] about diagrammatic asphericity of a class of two-relator groups. The reader is encouraged to refer to the paper [63] for full details.

Theorem 20 implies that the Schützenberger graph $S\Gamma(M, X, 1)$ of 1 for a one-relator inverse monoid $M = Inv\langle X : w = 1\rangle$ with $w$ cyclically reduced must embed as a subgraph of the Cayley graph $\Gamma(G, X)$ of the corresponding one-relator group $G$. From results of Stephen [139] it follows that $S\Gamma(M, X, 1)$ is a *full* subgraph of $\Gamma(G, X)$, and also that membership in $L(u)$ is decidable for each $u \in (X \cup X^{-1})^*$ iff membership in $L(1)$ is decidable. It is clear that the set of vertices of $S\Gamma(M, X, 1)$ is precisely the set of elements in the submonoid $P_w$ of $G$ generated by the set of prefixes of the word $w$. Hence we have the following corollary to Theorem 20.

**Corollary 3 (Ivanov, Margolis, Meakin)** *The word problem for the inverse monoid $M = Inv\langle X : w = 1\rangle$ (for $w$ a cyclically reduced word) is decidable if the membership problem for the submonoid $P_w$ of the one-relator group $G = Gp\langle X : w = 1\rangle$ is decidable, i.e., if there is an algorithm that will decide, on input an element of $G$, whether or not this element belongs to $P_w$.*

We refer to this membership problem as the *prefix membership problem* for $w$. Note that the prefix membership problem depends on the specific word $w$: a cyclic conjugate of $w$ gives rise to the same group $G$ but a different inverse monoid $M$ and a different submonoid $P_w$ of $G$. In general, the prefix membership problem for cyclically reduced words remains open, but several special cases of the problem have been solved. The following theorem collects some of the results about this problem that are proved in the papers by Ivanov, Margolis and Meakin [63], and the paper by Margolis, Meakin and Šunik [92].

**Theorem 21** *Let $G = Gp\langle X : w = 1\rangle$ where $w$ is a cyclically reduced word and let $P_w$ be the prefix monoid of $w$ in $G$. Then the membership problem for $P_w$ is decidable in the following cases:*

(1) *Some letter of $a^{\pm 1}$ occurs only once in $w$ ([63]);*

(2) *$w$ is a cyclic permutation of $[a_1, b_1][a_2, b_2] \ldots [a_n, b_n]$ ([63] and also [92]);*

(3) *there is a morphism $f$ from $G$ onto $\mathbf{Z}$ such that $f(v) > 0$ for every proper prefix $v$ of $w$ ([63]);*

(4) *$w \equiv a^{n_0}b^{m_1} \ldots a^{n_{k-1}}b^{m_k}a^{n_k}$ where $exp_a(w) = 0$ and the polynomial $m_1x^{n_0} + m_2x^{n_0+n_1} + \ldots + m_kx^{n_0+n_1+\ldots+n_{k-1}}$ has a positive real root ([92]). (This includes Baumslag–Solitar groups $w = ab^ma^{-1}b^k$ and most Adian groups $w = uv^{-1}$ where $u, v \in X^+$ and $uv^{-1}$ is cyclically reduced, for example.)*

**Example 5** As an example of what is involved in solving this problem, consider the cyclically reduced word $w = aba^{-1}b^{-1}cdc^{-1}d^{-1}$ (so the corresponding one-relator group $G = \langle a, b, c, d : w = 1\rangle$ is the fundamental group of the orientable surface of genus 2). I am aware of at least four distinct ways of solving the prefix membership problem for this word $w$. I will outline the ideas involved in one of the solutions that I find most geometrically appealing. Consider the iterative process of constructing the Schützenberger complex of 1 inside the Cayley complex of the surface group $G$. Starting at the vertex 1, we sew on a 2-cell based at 1 whose boundary is labelled by the word $w$, then sew on 2-cells based at all 0-cells (vertices) of this 2-cell, inside the Cayley complex of $G$, and continue this process, iteratively constructing 2-cells based at all resulting vertices of the Cayley graph of $G$ that are reached by this process. We are interested in knowing which vertices of this Cayley graph are eventually reached via this iterative process. The diagram below shows a portion of the Schützenberger complex of 1 corresponding to sewing on 2-cells at the vertices 1, $ab$, and $dcd^{-1}$ of the Cayley graph of $G$.

One can show that there are 19 different cone types of vertices in this Schützenberger complex: representatives of the 19 different cone types of these vertices are marked on the diagram below: the vertex corresponding to the element $aba^{-1}b^{-1}$ of $G$ has the same cone type as the vertex 1: all other vertices of the Schützenberger graph of 1 have the same cone type as one of these 19 vertices. This was first proved by Haataja (unpublished, circa 1990), who essentially constructed an automatic structure on this Schützenberger graph in order to solve the prefix membership problem for this word.

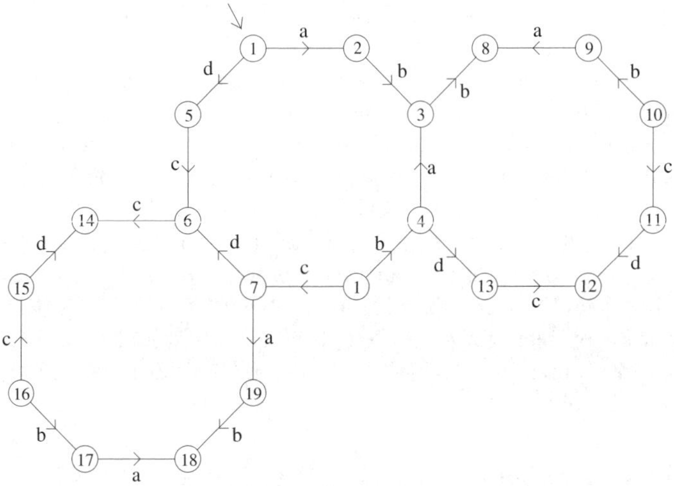


A detailed discussion of this example is contained in the thesis of Lindblad [78], who shows that the dual graph of this Schützenberger complex of 1 has a tree-like structure, and extends these ideas to solve the prefix membership problem for a class of cyclically reduced words that have essentially no overlaps between pieces. □

**Example 6 (Haataja)** This example, recently constructed by Haataja, shows that while the Schützenberger graph of 1 for a one-relator inverse monoid $M = Inv\langle X : w = 1\rangle$ corresponding to a cyclically reduced word $w$ embeds in the Cayley graph of the corresponding one-relator group, this embedding is not in general an isometry. Consider the word $w = abcabaccaacdda^{-1}c^{-1}$ and the corresponding inverse monoid $M = Inv\langle a, b, c : w = 1\rangle$ and group $G = Gp\langle a, b, c : w = 1\rangle$. By some results contained in the paper of Ivanov, Margolis, and Meakin [63], the monoid $M$ has trivial group of units. Also, the group $G$ is a small cancellation group (in fact a 1/6th group), so $G$ is word hyperbolic. Now the word $u = abaccaac$ labels a path in $S\Gamma(M, \{a, b, c\}, 1)$ but its inverse, $c^{-1}b^{-1}a^{-1}cad^{-1}d^{-1}$ in $G$ is shorter than $u$ and does not label a path in $S\Gamma(M, \{a, b, c\}, 1)$. Thus the embedding of $S\Gamma(M, \{a, b, c\}, 1)$ in the Cayley graph $\Gamma(G, \{a, b, c\})$ is not an isometry.

It would be of interest to determine conditions under which the embedding of the Schützenberger graph of 1 for a one-relator inverse monoid embeds isometrically into the Cayley graph of the corresponding group. This does happen for the word $w = aba^{-1}b^{-1}cdc^{-1}d^{-1}$ of Example 5, for example. □

The prefix membership problem is a special case of the more general problem of deciding membership in finitely generated submonoids of a group. There seems to be little known about this problem in general, even for groups that are known to have decidable generalized word problem. It is well known that membership in rational subsets of free groups is decidable — the original proof is due to Benois [16], and alternative proofs have been provided by other authors. Thus the membership problem for finitely generated submonoids of free groups is decidable: an alternate proof of this fact was provided in the paper by Ivanov, Margolis and Meakin [63]. Membership in rational subsets of finitely generated free abelian groups is decidable, and hence membership in finitely generated submonoids of such groups is decidable (see Grunschlag [51]). Grunschlag showed that if membership in rational subsets of some class of groups is decidable, then membership in rational subsets of groups that are virtually in that class is also decidable, so membership in finitely generated submonoids of virtually free groups or virtually abelian groups is decidable. Recently, James [66] has shown that if membership in rational subsets of two groups $G_1$ and $G_2$ is decidable, then membership in finitely generated submonoids of the free product $G_1 * G_2$ is decidable: his proof is an adaptation of the proof of Mihailova [102] that the generalized word problem for $G_1 * G_2$ is decidable if the generalized word problem for each $G_i$ is decidable. It would be of interest to find other classes of groups for which the membership problem for finitely generated submonoids is decidable.

## 8 Acknowledgements

I would like to express my gratitude to the organizers of the Groups St Andrews 2005 conference for the invitation to speak at the conference. This paper is an outgrowth of the lectures that I presented at that conference. I would also like to thank Stuart Margolis and Steven Haataja for several discussions concerning the content of this paper. Thanks also go to Tim Deis, Luda Markus-Epstein, Geraud Sénizergues, Steven Haataja, and Justin James, for their permission to cite some of their as yet unpublished work.

### References

[1] S. I. Adian, On the embeddability of semigroups in groups, English translation: *Soviet Mathematics* **1** (1960), Amer. Math. Soc.

[2] S. I Adian, Defining relations and algorithmic problems for groups and semigroups, *Trudy Math. Inst. Steklov* **85** (1966).

[3] S. I Adian and J. Oganessian, Problems of equality and divisibility in semigroups with a single defining relation, *Mat. Zametki* **41** (1987), 412–421.

[4] J. Almeida, *Finite Semigroups and Universal Algebra*, World Scientific, 1994.

[5] A. W. Anissimov and F. D. Seifert, Zur algebraischen Charakteristik der durch kontext-freie Sprachen definierten Gruppen, *Elektron. Inform. Verarb. u. Kybern* **11** (1975) 695–702.

[6] K. I. Appel, One-variable equations in free groups, *Proc. Amer. Math. Soc.* **19** (1968), 912–918.

[7] C. J. Ash, Finite semigroups with commuting idempotents, *J. Austral. Math. Soc. (Series A)* **43** (1987), 81–90.

[8] C. J. Ash, Inevitable graphs: a proof of the type II conjecture and some related decision procedures, *Internat. J. Algebra Comput.* **1** (1991), 411–436.

[9] K. Auinger and B. Steinberg, On the extension problem for partial permutations, *Proc. Amer. Math. Soc.* **131** (2003), no. 9, 2693–2703 (electronic).

[10] B. A. Barnes, Representation of the $l_1$-algebra of an inverse semigroup, *Trans. Amer. Math. Soc.* **218** (1976), 361–396.

[11] L. Bartholdi, R. Grigorchuk and V. Nekrashevych, From fractal groups to fractal sets, in *Fractals in Graz (2001)*, 25–118, Trends Math. Birkhäuser, Basel (2003).

[12] J. Barwise (ed.) *Handbook of Mathematical Logic*, North Holland, 1978.

[13] C. Bennett, Time/space tradeoffs for reversible computation, *SIAM J. Computing* **18** (1989), 766–776.

[14] P. Bennett, Amalgamated free products of inverse semigroups, *J. Algebra* **198** (1997), 499–537.

[15] P. Bennett, On the structure of inverse semigroup amalgams, *Internat. J. Algebra Comput.* **7** (1997), no. 5, 577–604.

[16] M. Benois, Parties rationnelles du groupe libre, *C. R. Acad. Sci. Paris Sér. A–B* **269** (1969), A1188–A1190.

[17] J. Berstel, *Transductions and Context-free Languages*, Teubner Studienbücher, 1979.

[18] J.-C. Birget, The groups of Richard Thompson and complexity, *Internat. J. Algebra Comput.* **14** (2004), nos. 5 & 6, 569–626.

[19] J.-C. Birget, S. Margolis and J. Meakin, The word problem for inverse monoids presented by one idempotent relator, *Theoretical Computer Science* **123** (1994), 273–289.

[20] J.-C. Birget, S. Margolis, J. Meakin, and P. Weil, PSPACE-complete problems for subgroups of free groups and inverse finite automata, *Theoretical Computer Science* **242** (2000), no. 1–2, 247–281.

[21] J. S. Birman, *Braids, Links, and Mapping Class Groups*, Annals of Math. Studies **82**, Princeton Univ. Press, 1975.

[22] J. S. Birman, New points of view in knot theory, *Bull. Amer. Math. Soc. (N.S.)* **28** (1993), 253–287.

[23] E. Brieskorn and K. Saito, Artin-Gruppen und Coxeter-Gruppen, *Invent. Math.* **17** (1972), 245–271.

[24] J. W. Cannon, W. J. Floyd and W. R Parry, Introductory notes on Richard Thompson's groups, *L'Enseignement Mathématique* **42** (1996), 215–256.

[25] R. Charney, Injectivity of the positive monoid for some infinite type Artin groups, in *Geometric Group Theory Down Under, Proceedings of a special year in geometric group theory, Canberra, Australia, 1966*, J. Cossey et al. (eds.), W. de Gruyter, Berlin, 1999.

[26] A. Cherubini, J. Meakin and B. Piochi, Amalgams of free inverse semigroups, *Semigroup Forum* **54** (1997), 199–220.

[27] A. Cherubini, J. Meakin and B. Piochi, Amalgams of finite inverse semigroups, *J. Algebra* **285** (2005) 706–725.

[28] J. R. Cho and S. J. Pride, Embedding semigroups into groups, and the asphericity

of semigroups, *Internat. J. Algebra Comput.* **3** (1993), no. 1, 1–13.
[29] A. H. Clifford, A class of $d$-simple semigroups, *Amer. J. Math.* **75** (1953), 547–556.
[30] A. H. Clifford and G. B. Preston, *The Algebraic Theory of Semigroups, Vols. 1 and 2*, Amer. Math. Soc., Math. Surveys **7**, 1961/1967.
[31] D. E. Cohen, *Combinatorial Group Theory: A Topological Approach*, Cambridge University Press, 1989.
[32] L. P. Comerford and C. C. Edmunds, Products of commutators and products of squares in a free group, *Internat. J. Algebra Comput.* **4** (1994), no. 3, 469–480.
[33] J. Crisp and M. Laca, On the Toeplitz algebras of right-angled and finite-type Artin groups, *J. Austral. Math. Soc.* **72** (2002), no. 2, 223–245.
[34] P. Dehornoy, Braid groups and left distributive operations, *Trans. Amer. Math. Soc.* **345** (1994), no. 1, 115–151.
[35] P. Dehornoy, *Braids and Self-distributivity*, Progress in Math. **192**, Birkhäuser, 2000.
[36] P. Dehornoy, Thin groups of fractions, in *Combinatorial and geometric group theory (New York, 2000/Hoboken, NJ 2001)*, 95–128, Contemp. Math. **296**, Amer. Math. Soc. 2002.
[37] P. Dehornoy and L. Paris, Gaussian groups and Garside groups: two generalizations of Artin groups, *Proc. London Math. Soc. (3)* **79** (1999), 569–604.
[38] T. Deis, *Equations in Free Inverse Monoids*, Ph.D. Thesis, Univ. of Nebraska (1999).
[39] T. Deis, J. Meakin and G. Sénizergues, Equations in free inverse monoids, preprint.
[40] M. Delgado, S. Margolis and B. Steinberg, Combinatorial group theory, inverse monoids, automata, and global semigroup theory, *Internat. J. Algebra Comput.* **12** (2002), nos. 1 & 2, 179–211.
[41] P. Deligne, Les immeubles des groupes de tresses généralisés, *Invent. Math.* **17** (1972), 273–302.
[42] A. de Luca and S. Varricchio, On non-counting regular classes, *Theoretical Computer Science* **100** (1992), 67–104.
[43] A. P. do Lago, On the Burnside semigroups $x^m = x^{n+m}$, *Internat. J. Algebra Comput.* **6** (1996), no. 2, 179–227.
[44] J. Duncan and A. L. T. Paterson, $C^*$-algebras of inverse semigroups, *Proc. Edinburgh Math. Soc.* **28** (1985), 41–58.
[45] S. Eilenberg, *Automata, Languages and Machines, Vol. B*, Academic Press, New York, 1976.
[46] F. A. Garside, The braid group and other groups, *Quart. J. Math. Oxford Ser. (2)* **20** (1969), 235–254.
[47] N. D. Gilbert, $HNN$-extensions of inverse semigrops and groupoids, *J. Algebra* **272** (2004), no. 1. 27–45.
[48] R. Gitik, Graphs and separability properties of groups, *J. Algeba* **188** (1997), 125–143.
[49] R. Gitik, On quasiconvex subgroups of negatively curved groups, *J. Pure Appl. Algebra* **119** (1998), 155–169.
[50] E. Godelle and L. Paris, On singular Artin monoids, *Contemporary Math.* **372** (2005), Amer. Math. Soc., 43–57.
[51] Z. Grunschlag, *Algorithms in Geometric Group Theory*, Ph.D. Thesis, University of California—Berkeley (1999).
[52] V. S. Guba, The word problem for the relatively free semigroup satisfying $t^m = t^{m+n}$ with $m \geq 3$, *Internat. J. Algebra Comput.* **3** (1993), no. 3, 335–348.
[53] V. Guba, On the embeddability of semigroups in groups, *Mat. Zametki* **54** (1994), 3–14 (in Russian).
[54] V. Guba, On the relationship between the word problem and the divisibility problems in one-relator semigroups, *Russian Math. Izv.* **61** (1997).

[55] C. Gutiérrez, Satisfiability of equations in free groups is in PSPACE, *32nd Ann. ACM Symp. Theory Comput. (STOC'2000)*, ACM Press 2000.

[56] S. Haataja, S. Margolis and J. Meakin, Bass–Serre theory for groupoids and the structure of full regular semigroup amalgams, *J. Algebra* **183** (1996), 38–54.

[57] M. Hall Jr., A toplogy for free groups and related groups, *Ann. of Math. (2)* **52** (1950), 127–139.

[58] R. Hancock and I. Raeburn, The $C^*$-algebras of some inverse semigroups, *Bull. Austral. Math. Soc.* **42** (1990), 335–348.

[59] K. Henckell, S. Margolis, J.-E. Pin and J. Rhodes, Ash's type-II theorem, profinite topology and Malcev products, *Internat. J. Algebra Comput.* **1** (1991), 411–436.

[60] P. Higgins, *Notes on Categories and Groupoids*, Mathematical Studies Vol. 32, Van Nostrand Reinhold Company, London, 1971.

[61] J. E. Hopcroft and J. D. Ullman, *Introduction to Automata Theory, Languages and Computation*, Addison-Wesley, New York, 1979.

[62] A. G. Howson, On the intersection of finitely generated free groups, *J. London Math. Soc.* **29** (1954), 428–434.

[63] S. V. Ivanov, S. W. Margolis and J. C. Meakin, On one-relator inverse monoids and one-relator groups, *J. Pure Appl. Algebra* **159** (2001), 83–111.

[64] S. V. Ivanov and J. C. Meakin, On asphericity and the Freiheitssatz for certain finitely presented groups, *J. Pure Appl. Algebra* **159** (2001), 113–121.

[65] T. Jajcayova, *HNN extensions of inverse semigroups*, Ph.D. Thesis, University of Nebraska—Lincoln (1997).

[66] J. James, (private communication, paper in preparation).

[67] P. R. Jones, S. W. Margolis, J. Meakin and J. B. Stephen, Free products of inverse semigroups II, *Glasgow Math. J.* **33** (1991), 373–387.

[68] I. Kapovich and A. Myasnikov, Stallings foldings and subgroups of free groups, *J. Algebra* **248** (2002), 608–668.

[69] I. Kapovich, R. Weidman and A. Myasnikov, Foldings, graphs of groups and the membership problem, *Internat. J. Algebra Comput.* **15** (2005), no. 1, 95–128.

[70] E. V. Kashintsev, Small cancellation conditions and embeddability of semigroups in groups, *Internat. J. Algebra Comput.* **2** (1992), 433–441.

[71] J. Kellendonk and M. V. Lawson, Tiling semigroups, *J. Algebra* **224** (2000), 140–150.

[72] C. Kilgour, Embedding monoids in groups, preprint.

[73] K. Krohn and J. Rhodes, Algebraic theory of machines I: Prime decomposition theorem for finite semigroups and machines, *Trans. Amer. Math. Soc.* **116** (1965), 450–464.

[74] S. Krstic, Embedding semigroups in groups — a geometric approach, *Publ. Inst. Math. (Beograd) (N.S.)* **38** (52) (1995), 69–82.

[75] G. Lallement, On monoids presented by a single relation, *J. Algebra* **32** (1977), 370–388.

[76] M. V. Lawson, *Inverse Semigroups: the Theory of Partial Symmetries*, World Scientific, 1998.

[77] M. V. Lawson, The Möbius inverse monoid, *J. Algebra* **200** (1998), 428–438.

[78] S. Lindblad, *Inverse monoids presented by a single sparse relator*, Ph.D. Thesis, University of Nebraska (2003).

[79] A. A. Lorents, Representations of sets of solutions of systems of equations with one unknown in a free group, *Dokl. Akad. Nauk. SSSR* **178** (1968), 290–292 (Russian).

[80] M. Lothaire, *Algebraic Combinatorics on Words*, Cambridge University Press, 2001.

[81] R. C. Lyndon, Equations in free groups, *Trans. Amer. Math. Soc.* **96** (1960), 445–457.

[82] R. C. Lyndon and P. E. Schupp, *Combinatorial Group Theory*, Ergebnisse der Mathematik und ihrer Grenzgebiete, Band **89**, Springer-Verlag, Berlin, 1977.

[83] G. S. Makanin, Problem of Solvability of Equations in Free Semigroup, *Math. Sbornik* **103** (1977), 147–236; English transl. in *Math. USSR Sbornik* **32** (1977).
[84] G. S. Makanin, Equations in a Free Group, *Izv. Akad. Nauk. SSR, Ser. Math* **46** (1983), 1199–1273; English transl. in *Math. USSR Izv.* **21** (1983).
[85] A. I. Mal'cev, On the immersion of an algebraic ring into a field, *Math. Ann.* **133** (1937), 686–691.
[86] A.I. Mal'cev, On the immersion of associative systems in groups, *Mat. Sbornik (N.S.)* **6** (1939), 331–336.
[87] S. Margolis and J. Meakin, $E$-unitary inverse monoids and the Cayley graph of a group presentation, *J. Pure Appl. Algebra* **58** (1989), 45–76.
[88] S. Margolis and J. Meakin, Free inverse monoids and graph immersions, *Internat. J. Algebra Comput.* **3** (1993), no. 1, 79–99.
[89] S. Margolis and J. Meakin, Inverse monoids, trees and context-free languages, *Trans. Amer. Math. Soc.* **335** (1993), no. 1, 259–276.
[90] S. Margolis, J. Meakin and J. Stephen, Some decision problems for inverse monoid presentations, in *Semigroups and their applications* (S. M. Goberstein and P. M. Higgins, eds.), Reidel, Dordrecht, 1987, 99–110.
[91] S. Margolis, J. Meakin and J. Stephen, Free objects in certain varieties of inverse semigroups, *Canadian J. Math.* **42** (1990), 1084–1097.
[92] S. Margolis, J. Meakin and Z. Šunik, Distortion functions and the membership problem for submonoids of groups and monoids, *Contemporary Mathematics* **372** (2005), 109–129.
[93] S. Margolis, M. Sapir and P. Weil, Closed subgroups in pro-**V** topologies and the extension problem for inverse automata, *Internat. J. Algebra Comput.* **11** (2001), no. 4, 405–445.
[94] A. Markov, Impossibility of algorithms for recognizing some properties of associative systems, *Dokl. Acad. Nauk SSSR* **77** (1951), 953–956.
[95] L. Markus-Epstein, *Algorithmic problems in subgroups of some finitely presented groups*, Ph.D. Thesis, Bar-Ilan University (2005).
[96] D. B. McAlister, Groups, semilattices and inverse semigroups, *Trans. Amer. Math. Soc.* **192** (1974), 227–244.
[97] D. B. McAlister, Groups, semilattices and inverse semigroups II, *Trans. Amer. Math. Soc.* **196** (1974), 251–270.
[98] D. B. McAlister and R. McFadden, The free inverse monoid on two commuting generators, *J. Algebra* **32** (1974), 215–233.
[99] J. McCammond, The solution to the word problem for the relatively free semigroups satisfying $t^a = t^{a+b}$ with $a \geq 6$, *Internat. J. Algebra Comput.* **1** (1991), 1–32.
[100] J. McCammond and D. Wise, Coherence, local quasiconvexity, and the perimeter of 2-complexes, preprint.
[101] R. McKenzie and R. J. Thompson, An elementary construction of unsolvable word problems in group theory, in *Word Problems*, Z. Boone et al. (eds.), Studies in Logic **71**, North Holland, Amsterdam, 1973.
[102] K. A. Mihailova, The occurrence problem for free products of groups, *Mat. Sb.* **75 (117)** (1968), 199–210.
[103] W. D. Munn, Free inverse semigroups, *Proc. London Math. Soc. (3)* **29** (1974), 385–404.
[104] A. Nica, On a groupoid construction for actions of certain inverse semigroups, *Internat. J. Math.* **5** (1994), no. 3, 349–372.
[105] M. Nivat and J-F. Perrot, Une généralisation du monoïde bicyclique, *C. R. Acad. Sci. Paris* **271** (1970), 824–827.
[106] J. Okninski, *Semigroups of matrices*, World Scientific, 1998.

[107] A. Yu. Ol'shanskii, Geometry of defining relations in groups, *Nauka, Moscow* (1989); English translation in: *Math. and its Applications (Soviet series)* **70**, Kluwer Dordrecht, 1991.
[108] L. Paris, Artin monoids inject in their groups, *Comment. Math. Helv.* **77** (2002), no. 3, 609–637.
[109] L. Paris, The proof of Birman's conjecture on singular braid monoids, *Geom. Topol.* **8** (2004), 1281–1300 (electronic).
[110] Alan L. T. Paterson, *Groupoids, Inverse Semigroups, and their Operator Algebras*, Birkhäuser, 1998.
[111] M. Petrich, *Inverse Semigroups*, Wiley, 1984.
[112] J.-E. Pin, *Varieties of Formal Languages*, Plenum, London, 1986.
[113] W. Plandowski, Satisfiability of word equations with constants is in PSPACE, in *Proc. 40th Ann. Symp. Found. Comput. Sci. (FOCS'99), IEEE Computer Society Press*, (1999), 495–500.
[114] M. Putcha, *Linear Algebraic Monoids*, London Math. Soc. Lecture Note Series **133**, Cambridge Univ Press, Cambridge, UK, 1988.
[115] M. O. Rabin, Decidability of second order theories and automata on infinite trees, *Trans. Amer. Math. Soc.* **141** (1969), 1–35.
[116] A. A. Razborov, On Systems of Equations in a Free Group, *Math. USSR Izv.* **25** (1985), 115–162.
[117] D. Rees, On semigroups, *Proc. Cambridge Philos. Soc.* **36** (1940), 387–400.
[118] J. H. Remmers, On the geometry of semigroup presentations, *Adv. in Math.* **36** (1980), no. 3, 283–296.
[119] L. Ribes and P.A. Zalesskiĭ, On the profinite topology on a free group, *Bull. London Math. Soc.* **25** (1993), 37–43.
[120] L. Ribes and P.A. Zalesskiĭ, The pro-$p$ topology of a free group and algorithmic problems in semigroups, *Internat. J. Algebra Comput.* **4** (1994), 359–374.
[121] B. V. Rozenblat, Diophantine theories of free inverse semigroups, *Sibirsk. Mat. Zh.* **26** (1986), no. 6, 101–107 (Russian); English transl. in pp. 860–865.
[122] R. L. Ruyle, *Pseudovarieties of Inverse Monoids*, Ph.D. Thesis, University of Nebraska — Lincoln (1997).
[123] M. V. Sapir, Algorithmic problems for amalgams of finite semigroups, *J. Algebra* **229** (2000), no. 2, 514–531.
[124] O. Sarkesian, On identity and divisibility problems in semigroups and groups without cycles, *Izv. Akad. Nauk SSSR, Ser. Mat.* **45**, 6 (1981), 1424–1441.
[125] H. E. Scheiblich, Free inverse semigroups, *Proc. Amer. Math. Soc.* **38** (1973), 1–7.
[126] B. M. Schein, Representations of generalized groups, *Izv. Vyss. Ucebn. Zav. Matem.* **3** (1962), 164–176 (Russian).
[127] P. Schupp, Coxeter groups, 2-completion, perimeter reduction and subgroup separability, *Geom. Dedicata* **96** (2003), 179–198.
[128] J.P. Serre, *Trees*, Springer-Verlag, New York, 1980.
[129] P. V. Silva, Word equations and inverse monoid presentations, in *Semigroups and Applications, Including Semigroup Rings*, S. Kublanovsky, A. Mikhalev, P. Higgins and J. Ponizovskii (eds.), "Severny Ochag", St. Petersburg (1999).
[130] J. R. Stallings, *Group Theory and Three-Dimensional Manifolds*, Yale Math. Monograph **4**, Yale Univ. Press, 1971.
[131] J. R. Stallings, Topology of finite graphs, *Invent. Math.* **71** (1983), no. 3, 551–565.
[132] J. R. Stallings, A graph-theoretic lemma and group embeddings, in *Combinatorial Group Theory and Topology*, S. Gersten and J. Stallings (eds.), Princeton University Press (1987), 145–155.
[133] J. Stallings, Adian groups and pregroups, in *Essays in Group Theory*, Math. Sci.

Res. Inst. Publ. **8**, Springer, New York (1987), 321–342.

[134] B. Steinberg, Fundamental groups, inverse Schützenberger automata, and monoid presentations, *Comm. Algebra* **28** (2000), 5235–5253.

[135] B. Steinberg, Inevitable graphs and profinite topologies: some solutions to algorithmic problems in monoid and automata theory stemming from group theory, *Internat. J. Algebra Comput.* **11** (2001), no. 1, 25–71.

[136] B. Steinberg, Inverse automata and profinite topologies on a free group, *J. Pure Appl. Algebra* **167** (2002), no. 2–3, 341–359.

[137] B. Steinberg, A topological approach to inverse and regular semigroups, *Pacific J. Math.* **208** (2003), no. 2, 367–396.

[138] J. B. Stephen, Presentations of inverse monoids, *J. Pure Appl. Algebra* **63** (1990), 81–112.

[139] J. B. Stephen, Inverse monoids and rational subsets of related groups, *Semigroup Forum* **46** (1993), no. 1, 98–108.

[140] J. Stillwell, *Classical Topology and Combinatorial Group Theory*, Springer-Verlag, New York, 1980.

[141] H. Straubing, *Finite Automata, Formal Logic, and Circuit Complexity*, Birkhäuser Boston Inc., Boston, MA., 1994.

[142] A. Suschkewitsch, Über die endlichen Gruppen ohne das Gesetz der eindeutigen Umkehrbarkeit, *Math. Ann.* **99** (1928), 30–50.

[143] M. Takahasi, Note on chain conditions in free groups, *Osaka Math. J.* **3** (1951), 221–225.

[144] G. Watier, On the word problem for single relation monoids with an unbordered relator, *Internat. J. Algebra Comput.* **7** (1997), no. 6, 749–770.

[145] P. Weil, Profinite methods in semigroup theory, *Internat. J. Algebra Comput.* **12** (2002), nos. 1 & 2, 137–178.

[146] A. Yamamura, $HNN$ extensions of inverse semigroups and applications, *Internat. J. Algebra Comput.* **7** (1997), no. 5, 605–624.

[147] A. Yamamura, $HNN$-extensions of semilattices, *Internat. J. Algebra Comput.* **9** (1999), 555–596.

[148] L. Zhang, A short proof of a theorem of Adian, *Proc. Amer. Math. Soc.* **116** (1992), no. 1, 1–3.

# TOWARD THE CLASSIFICATION OF $s$-ARC TRANSITIVE GRAPHS

ÁKOS SERESS[1]

Department of Mathematics, The Ohio State University, Columbus, Ohio 43210, USA

## Abstract

We survey the efforts toward the classification of $s$-arc transitive graphs for $s \geq 2$.

*AMS Classification:* 05C25, 20B25

## 1 Introduction

Throughout this paper, $\Gamma(V, E)$ denotes a finite, connected, simple graph (that is, no loops or multiple edges), and we suppose that the minimal valency of vertices of $\Gamma$ is at least two. An *$s$-arc* in $\Gamma$ is a sequence $(\alpha_0, \alpha_1, \ldots, \alpha_s)$ of vertices such that $\{\alpha_i, \alpha_{i+1}\} \in E(\Gamma)$ for $0 \leq i < s$ and $\alpha_i \neq \alpha_{i+2}$ for $0 \leq i < s - 1$. For a subgroup $G \leq \mathsf{Aut}(\Gamma)$, we say that $\Gamma$ is *$(G, s)$-arc transitive* if $G$ acts transitively on the $s$-arcs of $\Gamma$; that is, for any two $s$-arcs $(\alpha_0, \alpha_1, \ldots, \alpha_s)$, $(\beta_0, \beta_1, \ldots, \beta_s)$ of $\Gamma$, there exists $g \in G$ such that $\alpha_i^g = \beta_i$ for $0 \leq i \leq s$. If $G = \mathsf{Aut}(\Gamma)$ then we simply say that $\Gamma$ is $s$-arc transitive. We say that $\Gamma$ is *$(G, s)$-transitive* if it is $(G, s)$-arc transitive but it is not $(G, s+1)$-arc transitive. We also denote by $s(G)$ the value $s$ for which $\Gamma$ is $(G, s)$-transitive.

Because of the restriction on the minimal degree of $\Gamma$, any $s_1$-arc can be extended to an $s_2$-arc for $s_2 > s_1$. This means that $(G, s_2)$-arc transitivity implies $(G, s_1)$-arc transitivity; in particular, $(G, s)$-arc transitivity for some $s$ implies $(G, 0)$-arc transitivity, and the latter property says that $G$ acts transitively on $V(\Gamma)$. Therefore, $\Gamma$ is regular of valency $d$ for some $d \geq 2$.

Cycles are $s$-arc transitive for any $s$, but for graphs of valency $d \geq 3$, $s$-arc transitivity is a quite severe restriction. For example, the complete graph $K_n$ and complete bipartite graph $K_{n,n}$ have very large automorphism groups, but they are only 2-arc transitive and 3-arc transitive, respectively. The reason for that is that their *girth* (the length of the shortest cycle) is small.

**Lemma 1.1 ([39, 7.61])** *Let $\Gamma$ be an $s$-arc transitive graph of girth $g$ and minimal valency at least 3. Then $g \geq 2s - 2$ or, equivalently, $s \leq (g + 2)/2$.*

**Proof** By the definition of girth, $\Gamma$ contains a cycle $[\alpha_0, \alpha_1, \ldots, \alpha_{g-1}]$ of length $g$. Let $\gamma$ be a neighbor of $\alpha_{g-1}$, different from $\alpha_{g-2}$ and $\alpha_0$. Then the $g$-arc $(\alpha_0, \alpha_1, \ldots, \alpha_{g-1}, \alpha_0)$ cannot be mapped to the $g$-arc $(\alpha_0, \alpha_1, \ldots, \alpha_{g-1}, \gamma)$ by a graph automorphism, and so $s < g$. Let $\beta$ be a neighbor of $\alpha_{s-1}$, different from $\alpha_{s-2}$ and $\alpha_s$. Since $\Gamma$ is $s$-arc transitive, there is an automorphism $h$ of $\Gamma$ mapping

[1]Supported in part by the NSF and the NSA

the $s$-arc $(\alpha_0, \alpha_1, \ldots, \alpha_{s-1}, \alpha_s)$ to the $s$-arc $(\alpha_0, \alpha_1, \ldots, \alpha_{s-1}, \beta)$. Then $h$ maps the $(g-s+1)$-arc $(\alpha_{s-1}, \alpha_s, \ldots, \alpha_{g-1}, \alpha_0)$ to another $(g-s+1)$-arc which starts at $\alpha_{s-1}$ and ends at $\alpha_0$. These two $(g-s+1)$-arcs form a closed walk, and hence there exists a cycle of length at most $2(g-s+1)$. Therefore $g \leq 2(g-s+1)$, and so $g \geq 2s-2$. □

The girth of $d$-regular graphs can be arbitrarily large, so Lemma 1.1 does not exclude the existence of $s$-transitive graphs for arbitrarily large $s$. However, as we shall see, $s$ must be bounded. The first result in this direction was the following beautiful theorem of Tutte.

**Theorem 1.1 ([38])** *Let $\Gamma$ be an $s$-transitive graph of valency 3. Then $s \leq 5$.*

The bound 5 in this theorem is smallest possible, as the following example shows.

**Example 1.2** Let $\Gamma(V_1 \cup V_2, E)$ be a bipartite graph, where $V_1$ consists of all 2-element subsets of $X = \{1, 2, 3, 4, 5, 6\}$ and $V_2$ consists of all partitions of $X$ into three subsets of size 2. Some $\alpha \in V_1$ is adjacent to some $\beta \in V_2$ if and only if $\alpha$ is a subset in the partition $\beta$. Then $\Gamma$ is a 3-regular, 5-transitive graph on 30 vertices.

Since Tutte's seminal paper [38], a lot of effort has been expended toward finding a bound of the transitivity of graphs of valency $d > 3$, and then toward the classification of all $s$-arc transitive graphs for $s \geq 2$. The values $s \geq 2$ are chosen here because there are too many 1-arc transitive graphs: under very mild conditions on a finite group $G$, there are $(G, 1)$-arc transitive graphs as the following construction shows.

Suppose that a group $G$ has a core-free subgroup $H$ and an element $g \in (G \setminus H)$ satisfying $g^2 \in H$. Then we define the *coset graph* $\mathsf{Cos}(G, H, HgH)$ to have vertex set $[G : H]$, the right cosets of $H$ in $G$. Two cosets $Hx$, $Hy$ are adjacent if and only if $yx^{-1} \in HgH$.

**Lemma 1.2** *The coset graphs $\mathsf{Cos}(G, H, HgH)$ are undirected and $(G, 1)$-arc transitive. Conversely, any $(G, 1)$-arc transitive graph $\Gamma$ is isomorphic to a coset graph $\mathsf{Cos}(G, H, HgH)$ for some $H \leq G$ and $g \in (G \setminus H)$ satisfying $g^2 \in H$.*

**Proof** The condition $g^2 \in H$ implies that $HgH = Hg^{-1}H$. Therefore, for any $x, y \in G$, $yx^{-1} \in HgH$ if and only if $xy^{-1} \in HgH$ and so $\Delta := \mathsf{Cos}(G, H, HgH)$ is undirected. Moreover, if $\{Hx, Hy\} \in E(\Delta)$ then $(yz)(xz)^{-1} \in HgH$ for any $z \in G$, so $G \leq \mathsf{Aut}(\Delta)$. Finally, we show that $\Delta$ is $(G, 1)$-arc transitive. Let $\{Hx, Hy\} \in E(\Delta)$ arbitrary; then $yx^{-1} = h_1 g h_2$ for some $h_1, h_2 \in H$. So $Hx = H(h_2 x)$ and $Hy = (Hh_1 g)(h_2 x)$ and therefore $(Hx, Hy)$ is in the $G$-orbit of the 1-arc $(H, Hg)$.

Conversely, if $\Gamma(V, E)$ is $(G, 1)$-arc transitive then, for some fixed $\alpha \in V(\Gamma)$ and $\{\alpha, \beta\} \in E(\Gamma)$, let $H$ be the stabilizer of $\alpha$ in $G$ and let $g \in G$ map the 1-arc $(\alpha, \beta)$ to $(\beta, \alpha)$. Then $\Gamma \cong \mathsf{Cos}(G, H, HgH)$. □

The two major methods attempting the classification of 2-arc transitive graphs are called *local analysis* and *global analysis*, and our discussion is also divided into these two major categories. Previous surveys of the local analysis are in [41] and in [17].

We finish this introduction by some further definitions and notation. For a permutation group $G \le \mathsf{Sym}(V)$, the *orbitals* of $G$ are the $G$-orbits $(\alpha,\beta)^G$ for the action of $G$ on the ordered pairs of $V$. The *pair* of an orbital $(\alpha,\beta)^G$ is $(\beta,\alpha)^G$, and $(\alpha,\beta)^G$ is called *self-paired* if it is equal to its pair. Self-paired orbitals of $G$ are exactly the edge sets of $(G,1)$-arc transitive graphs on $V$ (and so, by Lemma 1.2, they correspond to coset graphs of $G$).

For a graph $\Gamma(V,E)$ and $\alpha \in V(\Gamma)$, let $\Gamma(\alpha)$ denote the set of neighbors of $\alpha$ in $\Gamma$. For $G \le \mathsf{Aut}(\Gamma)$, $G_\alpha$ denotes the stabilizer of $\alpha$ and $G_\alpha^{\Gamma(\alpha)}$ is the restriction of the permutation action of $G_\alpha$ to $\Gamma(\alpha)$. The kernel of this action is the pointwise stabilizer of $\{\alpha\} \cup \Gamma(\alpha)$ in $G_\alpha$, and it is denoted by $G_\alpha^{[1]}$. More generally, $\Gamma_i(\alpha)$ is the set of vertices of distance at most $i$ from $\alpha$, and $G_\alpha^{[i]}$ is the pointwise stabilizer of $\Gamma_i(\alpha)$ in $G_\alpha$.

## 2 Local analysis

Let $\Gamma(V,E)$ be a $(G,2)$-arc transitive graph. Local analysis refers to methods which attempt to find the structure of the vertex stabilizers $G_\alpha$. The two major problems we consider are the following.

**Problem 2.1** *Let $\Gamma$ be a $(G,2)$-arc transitive graph of valency $d$.*
(i) *Give an absolute bound $M$ such that $s(G) \le M$.*
(ii) *Prove that there exists a function $f : \mathbb{N} \to \mathbb{N}$ such that $|G_\alpha| \le f(d)$.*

It is easy to see that $G_\alpha^{\Gamma(\alpha)}$ is 2-transitive (just look at the 2-arcs with middle point $\alpha$). The 2-transitive permutation groups are known, and there are not too many of them, so one could try to characterize the possibilities for $G_\alpha$ by considering $G_\alpha^{\Gamma(\alpha)}$ and the kernel $G_\alpha^{[1]}$ of this action. It turns out that it is better to consider $G_{\alpha\beta}^{[1]} := G_\alpha^{[1]} \cap G_\beta^{[1]}$ for adjacent vertices $\alpha,\beta$.

**Theorem 2.2 ([10])** *Let $\Gamma(V,E)$ be finite, and let $G \le Aut(\Gamma)$ such that $G$ is transitive on $V$ and $G_\alpha^{\Gamma(\alpha)}$ is primitive for $\alpha \in V$. Then for any two adjacent vertices $\alpha,\beta$, $G_{\alpha\beta}^{[1]}$ is a $p$-group for some prime $p$.*

The proof is based on the so-called Thompson–Wielandt theorem [29], [47, Thm. 6.6]. Essentially the same statement was proven by Knapp [18] (see Remark 2.7). Of course, $(G,2)$-arc transitive graphs satisfy the hypothesis of Theorem 2.2.

Starting from Theorem 2.2, R. Weiss solved Problem 2.1(i) in a series of papers, culminating in [46].

**Theorem 2.3 ([46])** *Let $\Gamma$ be a $(G,2)$-arc transitive graph of valency $d$. Then the following holds.*
(i) *If $G_{\alpha\beta}^{[1]} = 1$ then $s(G) \le 3$.*

(ii) *If* $s(G) \geq 4$ *then* $d = p^e + 1$ *for some prime* $p$ *and* $G_\alpha^{\Gamma(\alpha)} \rhd \mathsf{PSL}(2, p^e)$. *Moreover,* (a) $s(G) = 4$*; or* (b) $s(G) = 5$ *and* $p = 2$*; or* (c) $s(G) = 7$ *and* $p = 3$.

A precursor of Theorem 2.3 is the work of Gardiner [11], [12], who proved that if $G_\alpha^{\Gamma(\alpha)}$ is doubly primitive then $s(G) \leq 5$ or $s(G) = 7$. (Recall that a permutation group $G \leq \mathsf{Sym}(\Omega)$ is called *doubly primitive* if $G$ is transitive and the point stabilizer $G_\omega$ is primitive on $\Omega \setminus \{\omega\}$.)

The key step in the proof of Theorem 2.3 is the following lemma. We sketch its proof because it illustrates nicely the mixture of group-theoretic and graph-theoretic arguments characteristic in the topic.

**Lemma 2.1** *Let* $\Gamma$ *be a* $(G, 2)$*-arc transitive graph of valency* $d$*, and suppose that* $G_{\alpha\beta}^{[1]} \neq 1$. *Then* $d = (q^n - 1)/(q - 1)$ *for some* $n \geq 2$ *and prime power* $q = p^e$*, where* $p$ *is the unique prime dividing the order of* $G_{\alpha\beta}^{[1]}$. *Moreover,* $G_\alpha^{\Gamma(\alpha)} \rhd \mathsf{PSL}(n, p^e)$.

**Proof** Let $(\alpha, \beta, \gamma)$ be a 2-arc in $\Gamma$. We first claim that

$$G_{\alpha\beta}^{[1]} \not\leq G_\gamma^{[1]}. \tag{1}$$

Suppose, on the contrary, that $G_{\alpha\beta}^{[1]} \leq G_\gamma^{[1]}$. Then, since $G_{\alpha\beta}^{[1]}$ is obviously a subgroup of $G_\beta^{[1]}$, we have $G_{\alpha\beta}^{[1]} \leq G_\beta^{[1]} \cap G_\gamma^{[1]} = G_{\beta\gamma}^{[1]}$. However, $G_{\alpha\beta}^{[1]}$ and $G_{\beta\gamma}^{[1]}$ are conjugate in $G$, therefore have the same order, and so we can conclude that $G_{\alpha\beta}^{[1]} = G_{\beta\gamma}^{[1]}$. We obviously have $G_{\alpha\beta}^{[1]} \lhd G_{\alpha\beta}$, and the previous sentence implies that $G_{\alpha\beta}^{[1]} \lhd G_{\beta\gamma}$. Thus $G_{\alpha\beta}^{[1]} \lhd \langle G_{\alpha\beta}, G_{\beta\gamma} \rangle$. The next observation is that

$$\langle G_{\alpha\beta}, G_{\beta\gamma} \rangle = G_\beta, \tag{2}$$

because the 2-transitivity of $G_\beta^{\Gamma(\beta)}$ implies that $G_{\alpha\beta}$ and $G_{\beta\gamma}$ act transitively on $\Gamma(\beta) \setminus \{\alpha\}$ and $\Gamma(\beta) \setminus \{\gamma\}$, respectively. Hence $\langle G_{\alpha\beta}, G_{\beta\gamma} \rangle^{\Gamma(\beta)}$ acts transitively on $\Gamma(\beta)$ and $\langle G_{\alpha\beta}, G_{\beta\gamma} \rangle_\alpha = (G_\beta)_\alpha$, implying (2).

So far, we have shown that $G_{\alpha\beta}^{[1]} \lhd G_\beta$. Since $\Gamma$ is $(G, 1)$-arc transitive, there exists $g \in G$ exchanging $\alpha$ and $\beta$. Obviously $g$ normalizes $G_{\alpha\beta}^{[1]}$, therefore $G_{\alpha\beta}^{[1]} \lhd H := \langle G_\beta, g \rangle$. The group $H$ is an edge-transitive group of automorphisms of $\Gamma$, because $G_\beta$ maps the edge $\{\alpha, \beta\}$ to all edges incident to $\beta$, and it conjugates the subgroup $G_\beta^g = G_\alpha \leq H$ to the subgroups $G_\delta \leq H$, for all $\delta \in \Gamma(\beta)$. Then $G_\delta$, $\delta \in \Gamma(\beta)$, maps the edge $\{\delta, \beta\}$ to all edges incident to $\delta$ and conjugates $G_\beta$ to all subgroups $G_\varepsilon$ for $\{\delta, \varepsilon\} \in E(\Gamma)$, etc. Since $\Gamma$ is connected, eventually $\{\alpha, \beta\}$ is mapped to all edges. The final conclusion is that for all edges $\{\delta, \varepsilon\} \in E(\Gamma)$, $H$ conjugates $G_{\alpha\beta}^{[1]}$ to $G_{\delta\varepsilon}^{[1]}$; but $G_{\alpha\beta}^{[1]} \lhd H$ and so $G_{\alpha\beta}^{[1]}$ stabilizes all vertices of $\Gamma$. This contradicts the fact $G_{\alpha\beta}^{[1]} \neq 1$, eventually proving (1).

We have $G_{\alpha\beta}^{[1]} \lhd G_\beta^{[1]} \lhd G_{\beta\gamma}$, because each of the first two groups in this subgroup chain is the kernel of an action of the next group in the chain. So, by Theorem 2.2,

$G_{\alpha\beta}^{[1]}$ is a nontrivial subnormal $p$-subgroup of $G_{\beta\gamma}$ for some prime $p$, implying that $G_{\beta\gamma}$ contains the nontrivial normal $p$-subgroup $K := \langle (G_{\alpha\beta}^{[1]})^{G_{\beta\gamma}} \rangle$. This group $K$ fixes $\gamma$ and, by (1), acts nontrivially on $\Gamma(\gamma)$. Hence the 2-transitive group $G_\gamma^{\Gamma(\gamma)}$ has the property that the stabilizer of a point contains a nontrivial normal $p$-subgroup. Inspecting the list of 2-transitive groups, this implies that either $G_\gamma^{\Gamma(\gamma)}$ is of affine type (that is, it has an elementary abelian normal subgroup), or $G_\gamma^{\Gamma(\gamma)}$ is almost simple, containing a normal subgroup isomorphic to $\mathsf{PSU}(3,q)$, ${}^2B_2(q)$, ${}^2G_2(q)$, or $\mathsf{PSL}(n,q)$, for some power $q = p^e$ and for some $n \geq 2$. However, if a $(G,2)$-arc transitive group $G_\gamma^{\Gamma(\gamma)}$ is of affine type then by [43] $G_{\alpha\beta}^{[1]} = 1$, contradicting our assumption. Similarly, if $G_\gamma^{\Gamma(\gamma)}$ contains the normal subgroup $\mathsf{PSU}(3,q)$, ${}^2B_2(q)$, or ${}^2G_2(q)$ then by [45] we have $G_{\alpha\beta}^{[1]} = 1$ (this latter result also follows from [5], [6], and [3]). □

The proof of Theorem 2.3(b) can be finished by appealing to [42], where it is shown that if $G_\alpha^{\Gamma(\alpha)}$ contains $\mathsf{PSL}(n,q)$ as a normal subgroup for some $n \geq 3$ then $s(G) \leq 3$, and by appealing to [44], where the structure of $G_\alpha$ is studied in the case $G_\alpha^{\Gamma(\alpha)} \rhd \mathsf{PSL}(2,q)$.

The bound on the transitivity of graphs in Theorem 2.3(b) is best possible. For the case $p = 2$, we have seen a 5-transitive graph in Example 1.2. For the case $p = 3$, we have

**Example 2.4** Let $T = G_2(3^m)$, and let $H$ be a maximal parabolic subgroup of $T$. In $T$, there are two conjugacy classes of maximal parabolic subgroups, and there exists an involution $g \in \mathsf{Aut}(T)$ fusing these conjugacy classes. Let $G = \langle T, g \rangle$. Then $\mathsf{Cos}(G, H, HgH)$ is $(G,7)$-transitive of valency $3^m$.

Examples 1.2 and 2.4 are generalized polygons. Recall that a *generalized $n$-gon* [40] is a bipartite graph of diameter $n$ and girth $2n$ such that any two vertices are contained in a cycle of length $2n$. Our examples are generalized 4-gons and 6-gons, respectively. It is well-known that finite generalized $n$-gons exist for $n \in \{2,3,4,6,8\}$; then why cannot we use generalized 8-gons to construct 9-transitive graphs? The problem is that in generalized 8-gons $\Gamma(A \cup B, E)$, the valencies of vertices in $A$ are not equal to the valencies of vertices in $B$ and so $\Gamma$ is not even 1-arc transitive. Nevertheless, there are lots of symmetries: $\mathsf{Aut}(\Gamma)$ acts transitively on 9-arcs starting at the same vertex (or more generally, starting in the same class of the bipartition). This motivates the following definition.

Let $\Gamma$ be a graph and let $G \leq \mathsf{Aut}(\Gamma)$. We say that $\Gamma$ is *locally $(G,s)$-arc transitive* if for all vertices $\alpha \in V(\Gamma)$, $G$ acts transitively on the $s$-arcs starting at $\alpha$.

Some of the results about arc-transitive graphs (e.g., Lemma 1.1) are valid for locally arc-transitive graphs as well, but in certain aspects the behavior of locally arc-transitive graphs is markedly different. It is easy to see that locally $(G,1)$-arc transitive, but not $(G,0)$-arc transitive graphs are bipartite, and $G$ acts transitively on the two classes of the bipartition. In this survey, we do not deal with locally arc-transitive graphs. The only theorem we mention here is an unpublished result of

Stellmacher: if $\Gamma$ has minimal valency at least 3 and is locally $(G, s)$-arc transitive then $s \leq 9$.

We now turn to Problem 2.1(ii). If $G_{\alpha\beta}^{[1]} = 1$ then clearly $|G_\alpha| \leq d! \cdot (d-1)!$ and, as the graph $K_{d,d}$ shows, this bound is best possible. By Lemma 2.1, the only possibility for $G_{\alpha\beta}^{[1]} \neq 1$ is when $G_\alpha^{\Gamma(\alpha)} \rhd \mathsf{PSL}(n, q)$, in the natural 2-transitive permutation representation. In the case $n = 2$, [44] contains enough information on the structure of $G_\alpha$ to establish a bound on $|G_\alpha|$. The case $n \geq 3$ is much more complicated and was eventually resolved in a series of papers by Trofimov [31], [32], [33], [34], [35], [36], [37].

**Theorem 2.5 ([30])** *Let $\Gamma$ be $(G, 2)$-arc transitive, and suppose that $G_\alpha^{\Gamma(\alpha)}$ contains a normal subgroup that is permutationally isomorphic to $\mathsf{PSL}(n, p^e)$ in its natural 2-transitive permutation representation for some integers $n \geq 3$ and $e \geq 1$, and prime $p$. Moreover, let $\{\alpha, \beta\} \in E(\Gamma)$. Then $G_\alpha^{[2]} = 1$ for $p > 3$, $G_{\alpha\beta}^{[3]} := G_\alpha^{[3]} \cap G_\beta^{[3]} = 1$ for $p = 3$, and $G_{\alpha\beta}^{[5]} = 1$ for $p = 2$. Consequently, $G_\alpha^{[6]} = 1$ for all $p$ and $|G_\alpha|$ is bounded from above by a function of the valency of $\Gamma$.*

**Example 2.6** Let $T = F_4(q)$ with $q = 2^m$, and let $H$ be a maximal parabolic subgroup of $T$ of structure $[q^{20}].(\mathsf{GL}(3, q) \times \mathsf{SL}(2, q))$ (that is, a parabolic subgroup corresponding to an inner node of the Dynkin diagram). In $T$, there are two conjugacy classes of such parabolic subgroups, and there exists an involution $g \in \mathsf{Aut}(T)$ fusing these conjugacy classes. Let $G = \langle T, g \rangle$. Then $\Gamma := \mathsf{Cos}(G, H, HgH)$ is of valency $q^2 + q + 1$, it is $(G, 3)$-transitive, and we have $G_\alpha^{\Gamma(\alpha)} \cong \mathsf{PGL}(3, q)$ and $|G_\alpha^{[5]}| = q^2$.

**Remark 2.7** The interest in Problem 2.1(ii) partially stems from its similiarity to *Sims's conjecture* [27]. This conjecture states that there exists a function $f : \mathbb{N} \to \mathbb{N}$ such that if $G \leq \mathsf{Sym}(\Omega)$ is a finite primitive permutation group, $\alpha \in \Omega$, and $G_\alpha$ has an orbit of length $d$ on $\Omega \setminus \{\alpha\}$ then $|G_\alpha| \leq f(d)$. Orbits of $G_\alpha$ are in one-to-one correspondence with the orbital graphs of $G$, and recall that self-paired orbital graphs are just the $(G, 1)$-arc transitive graphs. The orbital graphs of primitive groups are connected, and some of the results quoted earlier [3], [5], [6], [18] are formulated in terms of orbital graphs of primitive groups. These results are applicable in our setting because their proof does not use the primitivity of $G$, just the fact that the graph they study is connected and $(G, 1)$-arc transitive for some group $G$. Sims's conjecture was eventually resolved by Cameron et al. [4].

Given $\Gamma = \mathsf{Cos}(G, H, HgH)$ of valency $d$, let $\alpha$ and $\beta$ denote the vertices $H$ and $Hg$, respectively, and let $K$ be the stabilizer of the edge $\{\alpha, \beta\}$ in $G$. Then the subgroups $H, K$ satisfy the properties

$$G = \langle H, K \rangle, \quad |H : H \cap K| = d, \quad \text{and} \quad |K : H \cap K| = 2.$$

If $\Gamma$ is $(G, 2)$-arc transitive then the proofs of Theorems 2.3 and 2.5 also provide structural information about $H$ and $K$, and in a lot of cases the exact isomorphism types of $H$ and $K$ are known. For example, if $d = 3$ then there are only six possibilities for the pair $(H, K)$ by [7] and [14], and Theorem 2.5 implies that for any $d$ there are only finitely many possibilities. The *amalgam method* [17] tries to reverse this process: given $H, K$ with a common subgroup $L$ of index $d$ and 2, respectively, what are the possibilities for $\langle H, K\rangle$? The largest such group is the free amalgamated product $\mathcal{A}(H, K)$ of $H$ and $K$. This is a finitely presented group $\mathcal{A}(H, K) = \langle E \mid \mathcal{R}\rangle$, where the generator set is $E = H \cup K$, and the relators are all equations $ab = c$ from the multiplication tables of $H$ and $K$, together with identifications $a = b$ from the common subgroup $L$ of $H$ and $K$. Finding all 2-arc transitive graphs is equivalent to describing all finite factor groups of $\mathcal{A}(H, K)$, for all possible $H, K$. Unfortunately, although the amalgam method provided some interesting examples of graphs, there seems to be little hope for finding all finite factor groups (see [17] for a more detailed discussion).

## 3 Global analysis

Global analysis refers to a program initiated by Praeger [24] for the classification of all 2-arc transitive graphs. Instead of concentrating on the point stabilizer of the automorphism group, it tries to describe the permutation theoretic structure of the entire group.

Praeger's method is based on the following Lemma 3.1. To state this lemma, we need some definitions.

Suppose that $G \leq \mathsf{Aut}(\Gamma)$ and $N \lhd G$. If $N$ is intransitive then we define the *quotient graph* $\Gamma_N$ to have vertex set the orbits of $N$, with two orbits $\Delta_1, \Delta_2$ connected if and only if there exists $\delta_1 \in \Delta_1$ and $\delta_2 \in \Delta_2$ adjacent in $\Gamma$.

We say that $\Gamma$ is a *cover* of a graph $\Sigma$ if there is a surjective homomorphism $\varphi : V(\Gamma) \to V(\Sigma)$ that maps edges of $\Gamma$ to edges of $\Sigma$, and for all vertices $\alpha \in V(\Gamma)$, the valency of $\alpha$ is the same as the valency of $\alpha^\varphi \in V(\Sigma)$. Equivalently, $\Gamma$ can be obtained from $\Sigma$ by replacing each vertex $\alpha \in V(\Sigma)$ by a set $S(\alpha)$, $V(\Gamma) = \bigcup_{\alpha \in V(\Sigma)} S(\alpha)$, and adding a perfect matching between $S(\alpha)$ and $S(\beta)$ to $E(\Gamma)$, whenever $\{\alpha, \beta\} \in E(\Sigma)$.

A cover $\Gamma$ is called a *normal cover* of $\Sigma$ if there exists $G \leq \mathsf{Aut}(\Gamma)$ and $N \lhd G$ such that $\Sigma$ is the quotient graph $\Gamma_N$.

**Lemma 3.1 ([24])** *Suppose that $\Gamma$ is $(G, 2)$-arc transitive, $N \lhd G$, and $N$ is intransitive on $V(\Gamma)$. Then*

(i) *If $N$ has two orbits then $\Gamma$ is bipartite, the orbits of $N$ are the classes of the bipartition, and $\Gamma_N \cong K_2$.*

(ii) *If $N$ has at least three orbits then $\Gamma_N$ is $(G, s)$-arc transitive for the permutation action of $G$ induced on the orbits of $N$, and $\Gamma$ is a normal cover of $\Gamma_N$. Moreover, $N_\alpha = 1$ for all $\alpha \in V(\Gamma)$.*

**Proof** Since $G$ permutes the orbits of $N$, there is no $g \in G$ mapping two vertices in the same orbit of $N$ to two vertices in different orbits of $N$. Therefore, because $\Gamma$ is $(G, 1)$-arc transitive, either all edges of $\Gamma$ are within orbits of $N$, or the restrictions of $\Gamma$ to the orbits of $N$ are empty graphs. Since $\Gamma$ is connected, the latter of these possibilities must occur.

The argument in the previous paragraph proves (i). For (ii), consider a 2-arc $(\alpha_0, \alpha_1, \alpha_2)$ of $\Gamma$, with the $\alpha_i$ in three different $N$-orbits $\Delta_i$, $i = 0, 1, 2$. Such 2-arcs exist, because $\Gamma$ is connected. Using again that $G$ permutes the orbits of $N$, we conclude that $(\alpha_0, \alpha_1, \alpha_2)$ cannot be mapped by $G$ to a 2-arc $(\beta_0, \beta_1, \beta_2)$ with $\beta_0, \beta_2$ in the same $N$-orbit. Since $\Gamma$ is $(G, 2)$-arc transitive, this means that we cannot have any 2-arcs in $\Gamma$ starting and ending in the same $N$-orbit. In other words, if $\Delta_1, \Delta_2$ are $N$-orbits and $\beta_1 \in \Delta_1$, then $\beta_1$ is connected to at most one vertex in $\Delta_2$. Moreover, if $\beta_1$ is adjacent to some $\beta_2 \in \Delta_2$ then, using the transitivity of $N$ on $\Delta_1$, we see that $\{\beta_1, \beta_2\}^N$ is a set of edges connecting each $\beta \in \Delta_1$ with exactly one vertex in $\Delta_2$. Hence $\Gamma$ is a cover of $\Gamma_N$.

The assertion that $\Gamma_N$ is $(G, s)$-arc transitive is obvious, since for any $s$-arc $A = (\Delta_0, \Delta_1, \ldots, \Delta_s)$ in $\Gamma_N$, it is clear that there is an $s$-arc $A^* = (\alpha_0, \alpha_1, \ldots, \alpha_s)$ in $\Gamma$ with $\alpha_i \in \Delta_i$. Hence for any two $s$-arcs $A_1, A_2$ of $\Gamma_N$, the element of $G$ mapping $A_1^*$ to $A_2^*$ in $\Gamma$ also maps $A_1$ to $A_2$.

Finally, we show that $N_\alpha = 1$ for all $\alpha \in V(\Gamma)$. If, on the contrary, there exists $\alpha \in V(\Gamma)$ and $g \in N_\alpha$, $g \neq 1$, then the connectedness of $\Gamma$ implies that there exists $\beta \in V(\Gamma)$ and $\gamma, \delta \in \Gamma(\beta)$ such that $\beta^g = \beta$ and $\gamma^g = \delta \neq \gamma$. This would mean, however, that $\beta$ is connected to two different vertices in the $N$-orbit $\gamma^N$, a contradiction. □

Permutation groups $G \leq \mathsf{Sym}(\Omega)$ with the property that all nontrivial normal subgroups of $G$ act transitively on $\Omega$ are called *quasiprimitive*. All primitive groups are quasiprimitive. Based on Lemma 3.1, the following approach was proposed in [24] for the classification of all nonbipartite 2-arc transitive graphs.

**Problem 3.1** (i) *Determine all graphs admitting a quasiprimitive $s$-arc transitive group of automorphisms.*

(ii) *Find all $s$-arc transitive normal covers of these graphs.*

Work so far has concentrated on Problem 3.1(i). First of all, we need a description of quasiprimitive groups. This was done in [24], along the lines of the O'Nan–Scott theorem for primitive permutation groups. Our description does not use the notation of [24]; rather we follow [25], where quasiprimitive groups were divided into eight categories. Four of these categories are the following. (Recall that the *socle* $\mathsf{Soc}(G)$ of a group $G$ is the subgroup generated by all minimal normal subgroups of $G$.)

HA (Holomorph Affine) This class consists of all quasiprimitive groups with an elementary abelian minimal normal subgroup. Such groups are subgroups of $\mathsf{AGL}(d, p)$, for some prime $p$ and positive integer $d$, acting on the vectors in $\mathsf{GF}(p)^d$. The stabilizer of the vector 0 is an irreducible subgroup of $\mathsf{GL}(d, p)$. In fact, all groups of type HA are primitive.

AS (Almost Simple) These are all the almost simple groups $G$ with transitive socle $T$, that is $T \leq G \leq \mathsf{Aut}(T)$. The group $G$ is primitive if and only if a point stabilizer is a maximal subgroup of $G$ not containing $T$.

TW (Twisted Wreath) We follow the description in [28, p. 269]. Let $T$ be a finite nonabelian simple group, $P$ a group with a proper subgroup $Q$ and let

$\phi : Q \to \mathsf{Aut}(T)$ be a homomorphism. We define the *complete base group* $\mathcal{B}$ to be the set of maps $f : P \to T$ under pointwise multiplication and so $\mathcal{B} \cong T^{|P|}$. The group $P$ acts on $\mathcal{B}$ by

$$f^p(x) = f(px) \quad \text{for all } x, p \in P, \quad f \in \mathcal{B}. \tag{3}$$

We define $\mathcal{X}$ to be the semidirect product $\mathcal{B} \rtimes P$ with respect to this action. Define the *$\phi$–base group*

$$B_\phi = \{f : P \to T \mid f(pq) = f(p)^{\phi(q)} \text{ for all } p \in P, q \in Q\}.$$

This group is isomorphic to $T^k$ where $k = |P : Q|$. Also $B_\phi$ is normalized by $P$, and so $B_\phi$ and $P$ generate the subgroup $X_\phi = B_\phi \rtimes P$ of $\mathcal{X}$ which we call the *twisted wreath product* $T \operatorname{twr}_\phi P$ of $T$ by $P$ with respect to $\phi$. The action of $X_\phi$ on its base group is quasiprimitive if and only if $\phi^{-1}(\mathsf{Inn}(T))$ is a core-free subgroup of $P$. The socle of $X_\phi$ is regular, and it is the unique minimal normal subgroup of $X_\phi$. The action of $X_\phi$ is primitive if and only if $\phi$ extends to no larger subgroup of $P$ and $\mathsf{Inn}(T) \leq \phi(Q)$ (see [1]).

PA (Product Action) This is the class of quasiprimitive groups that differs most from the corresponding class of primitive groups. A group $G$ of type PA preserves some partition $\mathcal{P}$ (possibly trivial in the sense of having parts of size 1) of the permutation domain $\Omega$ upon which $G$ acts faithfully preserving a product structure $\Delta^k$. Furthermore, $N = T^k \leq G \leq H \wr S_k$ where $H$ acts quasiprimitively on $\Delta$ of type AS with nonregular socle $T$ and $G$ acts transitively by conjugation on the set of simple direct factors of $N$. Let $\delta \in \Delta$ and $B = (\delta, \ldots, \delta) \in \mathcal{P}$. Then $N_B = T_\delta^k$ and for $\omega \in B$, the point stabilizer $N_\omega$ is a subdirect subgroup of $N_B$, that is, $N_\omega$ projects onto $T_\delta$ in every coordinate. The action of $G$ is primitive if and only if $\mathcal{P}$ is trivial and the action of $H$ on $\Delta$ is primitive.

We singled out these four classes of quasiprimitive groups because of the following fact.

**Theorem 3.2 ([24])** *If $\Gamma$ is $(G,2)$-arc transitive and $G$ acts quasiprimitively on $V(\Gamma)$ then $G$ has type* HA, AS, TW, *or* PA.

The situation is quite satisfactory in the HA case.

**Theorem 3.3 ([16])** *All $(G,2)$-arc transitive graphs with a quasiprimitive group $G$ of automorphisms of type* HA *are known. There are three infinite families and two sporadic examples. In all graphs, the number of vertices is a power of two.*

One infinite family is the set of complete graphs $K_n$ with $n = 2^m$ for some integer $m \geq 2$. Another family is given in the following example.

**Example 3.4** Let $m$ be even and let $\Gamma = Q_{m+1}$ be the $(m+1)$-cube, that is, the vertex set is $\mathbb{Z}_2^{m+1}$, with two vertices connected if and only if they differ in exactly one coordinate. Then $G := \mathsf{Aut}(\Gamma) \cong \mathbb{Z}_2 \wr S_{m+1}$. Let $N \cong \mathbb{Z}_2$ be the normal subgroup exchanging antipodal vertices of $\Gamma$ (that is, $N = \langle g \rangle$, with $\alpha^g = (1, 1, \ldots, 1) - \alpha$ for all $\alpha \in V(\Gamma)$). Then the quotient graph $\Gamma_N$ is called the *folded* $(m+1)$*-cube*, and it is $(G/N, 2)$-arc transitive of type HA with $2^m$ vertices.

Much less is known in the AS case. There are numerous examples, but the complete analysis was done only when $\Gamma$ is $(G, 2)$-arc transitive with $\mathsf{Soc}(G) = {}^2B_2(2^{2m+1})$ [8], $\mathsf{Soc}(G) = {}^2G_2(3^{2m+1})$ [9], or $\mathsf{Soc}(G) = \mathsf{PSL}(2, q)$ [15]. The reason is that in these rank 1 groups the complete subgroup structure of $G$ is known and it is possible to check all candidates $H, g$ for an appropriate coset graph $\mathsf{Cos}(G, H, HgH)$. In all three of these papers previously unknown 2-arc transitive graphs were found, so the complete analysis is a worthwhile exercise. Here we give only a classical example.

**Example 3.5 ([48])** Let $T = \mathsf{PSL}(2, p)$, where $p$ is a prime and $p = \pm 1 \pmod{16}$. Let $H \cong S_4 < T$ and let $D \cong D_8$ be a Sylow 2-subgroup of $H$. Let $g_1 \in N_T(D) \setminus D$ such that $g_1^2 \in D$. Then the coset graph $\mathsf{Cos}(T, H, Hg_1H)$ is a connected $(T, 4)$-arc transitive graph of valency 3.

In a sense, the $(G, 2)$-arc transitive graphs with $G$ of type TW are characterized [2], although no explicit list of the possible pairs $(\Gamma, G)$ as in the case HA and rank 1 AS is available. Let $\Gamma$ be a $(G, 2)$-arc transitive graph, with $G = T \operatorname{twr}_\phi P$. Then $\mathsf{Soc}(G)$ acts regularly on $V(\Gamma)$, so we can identify $V(\Gamma)$ with the $\phi$-base group $B_\phi$ and $\Gamma$ with a Cayley graph $\mathsf{Cay}(B_\phi, Y)$, where $Y$ is the set of functions in $B_\phi$ connected to the identity element of $B_\phi$. The elements of $Y$ are involutions, because $\mathsf{Cay}(B_\phi, Y)$ is an undirected graph. Also, the range of the function $F : Y \to T$, defined as $F(f_\omega) = f_\omega(1_P)$ for $f_\omega \in Y$ is a set of elements in $T$, each of order 1 or 2, which generate $T$. Moreover, the set $Y$ is invariant for the conjugation action of $P$ on $B_\phi$, as defined in (3), and the action of $P$ on $Y$ is a 2-transitive permutation group.

Conversely, given a nonabelian simple group $T$, a group $P$ acting 2-transitively on a set $\Omega$, and a function $F : \Omega \to T$ whose range consists of elements of order at most 2, such that the range generates $T$, it is shown in [2] how to construct a twisting automorphism $\phi$ such that $G = T \operatorname{twr}_\phi P$ acts 2-arc transitively on the Cayley graph $\mathsf{Cay}(B_\phi, Y)$, where $Y$ consists of all functions $f \in B_\phi$ such that $f(1_P) = F(\omega)$ for some $\omega \in \Omega$, and further values of $f$ are defined by $f(p) := F(\omega^p)$. This group $G$ is quasiprimitive if and only if it satisfies the criterion that $\phi^{-1}(\mathsf{Inn}(T))$ is a core-free subgroup of $P$.

The fourth possibility allowed by Theorem 3.2 is that a graph is $(G, 2)$-arc transitive, where $G$ is quasiprimitive of type PA. It was only recently established [21] that such graphs exist. We use the notation introduced in the definition of types of quasiprimitive groups. There are two, quite different, types of examples, depending on whether $N_\omega$ is a diagonal subgroup of $N_B$ or not. A partial classification of all 2-arc transitive graphs with a quasiprimitive group of automorphisms is under way [20]; in the nondiagonal case also a complete classification seems feasible. Interestingly, although quasiprimitive groups of type PA belong to case (i) of Problem 3.1, the classification in the diagonal subcase requires the study of covers of certain graphs of type AS, which is similar to case (ii) of that Problem. Here we give only an example.

**Example 3.6 ([21])** Let Let $T = \mathsf{PSL}(2,p)$, where $p$ is a prime and $p = \pm 1 \pmod{16}$, and let $\Sigma = \mathsf{Cos}(T, H, Hg_1H)$ be the connected $(T,4)$-arc transitive graph of valency 3 described in Example 3.5. Recall that $H \cong S_4 < T$ and $D \cong D_8$ is a Sylow 2-subgroup of $H$, and $g_1 \in N_T(D) \setminus D$.

We construct a quasiprimitive group $G$ with socle $N := T_1 \times T_2$, $T_i \cong T$ for $i = 1,2$. Let $g_2 = g_1c$, where $\langle c \rangle = Z(D)$. Moreover, let $N_\alpha = \{(h,h) \mid h \in H\}$ (a diagonal subgroup of $H \times H$) and $g = (g_1, g_2)$. Then $\Gamma = \mathsf{Cos}(N, N_\alpha, N_\alpha g N_\alpha)$ is a connected $(N,4)$-transitive graph of valency 3.

Let $\tau = (1_T, 1_T)(1,2) \in T \wr \mathbb{Z}_2$ and $G = \langle N, \tau \rangle$. Then $G \le \mathsf{Aut}(\Gamma)$, the group $G$ acts on $V(\Gamma)$ as a quasiprimitive group of product action type, and $\Gamma$ is $(G,5)$-transitive.

In the study of permutation groups, the natural combinatorial reduction (action on orbits, and then action on blocks of imprimitivity) leads to primitive permutation groups. Since the theory of primitive groups is so well-developed, it would be desirable to have an extension of Lemma 3.1 that further reduces the study of 2-arc transitive graphs to $(G,2)$-arc transitive graphs with a *primitive* group of automorphisms $G$. Unfortunately, such reduction is not possible in general. As pointed out in [22], $\mathsf{PGL}(2,7)$ has a valency 3 coset graph $\Gamma$ (on the cosets of a subgroup $D_{12}$) with 28 vertices. The group $G = \mathsf{PSL}(2,7)$ is quasiprimitive of type AS on $V(\Gamma)$, and $\Gamma$ is $(G,2)$-arc transitive. The group $G$ has two block systems of imprimitivity, but $G$ is only 1-arc transitive on the quotient graphs. Hence we cannot obtain $\Gamma$ as the normal cover of a 2-arc transitive graph with a primitive group of automorphisms.

Nevertheless, highly arc-transitive graphs with a primitive group of automorphisms are interesting. For example, Li [19] determined all vertex-primitive 4-arc transitive graphs. An exciting new construction in this paper is a 4-transitive, valency 14 coset graph of the Monster on the cosets of a 13-local subgroup.

We finish this survey with a remark on the bipartite case. Lemma 3.1(ii) gives a reduction to the quasiprimitive case only if there is a normal subgroup with at least three orbits. In the bipartite case, a reduction theorem was established in [23]. The base cases are the *bi-quasiprimitive* groups: these are transitive groups $G$ with a subgroup $G^+$ of index 2 such that $G^+$ has two orbits and it acts quasiprimitively on them (and every normal subgroup of $G$ has at most two orbits). An O'Nan–Scott type characterization of bi-quasiprimitive groups was obtained in [26]. The papers [16] and [19] also classify all bi-primitive affine, and at least 4-arc transitive graphs, respectively. The study of bipartite $s$-arc transitive graphs is also related to the study of locally $s$-arc transitive graphs, but in the local case there are further difficulties. In [13], examples of locally 3-arc transitive graphs are given, such that the action on both classes of the bipartition is quasiprimitive, but of different type.

**Acknowledgement** I am indebted to Cai Heng Li and Cheryl Praeger for their comments on the manuscript.

## References


[1] R. W. Baddeley, Primitive permutation groups with a regular nonabelian normal subgroup, *Proc. London Math. Soc. (3)* **67** (1993), no. 3, 547–595.

[2] Robert W. Baddeley, Two-arc transitive graphs and twisted wreath products, *J. Algebraic Combin.* **2** (1993), no. 3, 215–237.

[3] Michael Bürker and Wolfgang Knapp, Zur Vermutung von Sims über primitive Permutationsgruppen. II, *Arch. Math. (Basel)* **27** (1976), no. 4, 352–359.

[4] P. J. Cameron, C. E. Praeger, J. Saxl and G. M. Seitz, On the Sims conjecture and distance transitive graphs, *Bull. London Math. Soc.* **15** (1983), no. 5, 499–506.

[5] Ulrich Dempwolff, A factorization lemma and an application, *Arch. Math. (Basel)* **27** (1976) no. 1, 18–21.

[6] Ulrich Dempwolff, A factorization lemma and an application. II, *Arch. Math. (Basel)* **27** (1976), no. 5, 476–479.

[7] Dragomir Ž. Djoković and Gary L. Miller, Regular groups of automorphisms of cubic graphs, *J. Combin. Theory Ser. B* **29** (1980), no. 2, 195–230.

[8] Xin Gui Fang and Cheryl E. Praeger, Finite two-arc transitive graphs admitting a Suzuki simple group, *Comm. Algebra* **27** (1999), no. 8, 3727–3754.

[9] Xin Gui Fang and Cheryl E. Praeger, Finite two-arc transitive graphs admitting a Ree simple group, *Comm. Algebra* **27** (1999), no. 8, 3755–3769.

[10] A. Gardiner, Arc transitivity in graphs, *Quart. J. Math. Oxford Ser. (2)* **24** (1973), 399–407.

[11] A. Gardiner, Arc transitivity in graphs. II, *Quart. J. Math. Oxford Ser. (2)* **25** (1974), 163–167.

[12] Anthony Gardiner, Doubly primitive vertex stabilisers in graphs, *Math. Z.* **135** (1973/74), 257–266.

[13] Michael Giudici, Cai Heng Li and Cheryl E. Praeger, Analysing finite locally $s$-arc transitive graphs, *Trans. Amer. Math. Soc.* **356** (2004), no. 1, 291–317 (electronic).

[14] David M. Goldschmidt, Automorphisms of trivalent graphs, *Ann. of Math. (2)* **111** (1980), no. 2, 377–406.

[15] Akbar Hassani, Luz R. Nochefranca and Cheryl E. Praeger, Two-arc transitive graphs admitting a two-dimensional projective linear group, *J. Group Theory* **2** (1999), no. 4, 335–353.

[16] A. A. Ivanov and Cheryl E. Praeger, On finite affine 2-arc transitive graphs, *European J. Combin.* **14** (1993), no. 5, 421–444.

[17] A. A. Ivanov and S. V. Shpectorov, Applications of group amalgams to algebraic graph theory, in *Investigations in algebraic theory of combinatorial objects*, volume 84 of *Math. Appl. (Soviet Ser.)*, pages 417–441, Kluwer Acad. Publ., Dordrecht, 1994.

[18] Wolfgang Knapp, On the point stabilizer in a primitive permutation group, *Math. Z.* **133** (1973), 137–168.

[19] Cai Heng Li, The finite vertex-primitive and vertex-biprimitive $s$-transitive graphs for $s \geq 4$, *Trans. Amer. Math. Soc.* **353** (2001), no. 9, 3511–3529 (electronic).

[20] Cai Heng Li and Ákos Seress, Finite quasiprimitive two-arc transitive graphs of product action type, in preparation.

[21] Cai Heng Li and Ákos Seress, Constructions of quasiprimitive two-arc transitive graphs of product action type, in *Finite Geometries, Groups, and Computation*, pages 115–123. de Gruyter, Berlin–New York, 2006.

[22] Cheryl E. Praeger, Finite vertex transitive graphs and primitive permutation groups, in *Coding theory, design theory, group theory (Burlington, VT, 1990)*, Wiley-Intersci. Publ., pages 51–65. Wiley, New York, 1993.

[23] Cheryl E. Praeger, On a reduction theorem for finite, bipartite 2-arc-transitive graphs, *Australas. J. Combin.* **7** (1993), 21–36.

[24] Cheryl E. Praeger, An O'Nan-Scott theorem for finite quasiprimitive permutation groups and an application to 2-arc transitive graphs, *J. London Math. Soc. (2)* **47** (1993), no. 2, 227–239.
[25] Cheryl E. Praeger, Finite quasiprimitive graphs, in *Surveys in combinatorics, 1997 (London)*, London Math. Soc. Lecture Note Ser. **241**, pages 65–85. Cambridge Univ. Press, Cambridge, 1997.
[26] Cheryl E. Praeger, Finite transitive permutation groups and bipartite vertex-transitive graphs, *Illinois J. Math.* **47** (2003), no. 1–2, 461–475.
[27] Charles C. Sims, Graphs and finite permutation groups, *Math. Z.* **95** (1967), 76–86.
[28] Michio Suzuki, *Group theory I*, Grundlehren der Mathematischen Wissenschaften [Fundamental Principles of Mathematical Sciences] **247**, Springer-Verlag, Berlin, 1982.
[29] John G. Thompson, Bounds for orders of maximal subgroups, *J. Algebra* **14** (1970), 135–138.
[30] V. I. Trofimov, Vertex stabilizers of graphs with projective suborbits, *Dokl. Akad. Nauk SSSR* **315** (1990), no. 3, 544–546.
[31] V. I. Trofimov, Graphs with projective suborbits, *Izv. Akad. Nauk SSSR Ser. Mat.* **55** (1991), no. 4, 890.
[32] V. I. Trofimov, Graphs with projective suborbits. Cases of small characteristics. I, *Izv. Ross. Akad. Nauk Ser. Mat.* **58** (1994), no. 5, 124–171.
[33] V. I. Trofimov, Graphs with projective suborbits. Cases of small characteristics. II, *Izv. Ross. Akad. Nauk Ser. Mat.* **58** (1994), no. 6, 137–156.
[34] V. I. Trofimov, Graphs with projective suborbits. Exceptional cases of characteristic 2. I, *Izv. Ross. Akad. Nauk Ser. Mat.* **62** (1998), no. 6, 159–222.
[35] V. I. Trofimov, Graphs with projective suborbits. Exceptional cases of characteristic 2. II, *Izv. Ross. Akad. Nauk Ser. Mat.* **64** (2000), no. 1, 175–196.
[36] V. I. Trofimov, Graphs with projective suborbits. Exceptional cases of characteristic 2. III, *Izv. Ross. Akad. Nauk Ser. Mat.* **65** (2001), no. 4, 151–190.
[37] V. I. Trofimov, Graphs with projective suborbits. Exceptional cases of characteristic 2. IV, *Izv. Ross. Akad. Nauk Ser. Mat.* **67** (2003), no. 6, 193–222.
[38] W. T. Tutte, A family of cubical graphs, *Proc. Cambridge Philos. Soc.* **43** (1947), 459–474.
[39] W. T. Tutte, *Connectivity in graphs*, Mathematical Expositions, No. 15, University of Toronto Press, Toronto, Ont., 1966.
[40] Hendrik van Maldeghem, *Generalized polygons*, Monographs in Mathematics **93**, Birkhäuser Verlag, Basel, 1998.
[41] R. Weiss, $s$-transitive graphs, in *Algebraic methods in graph theory, Vol. I, II (Szeged, 1978)*, Colloq. Math. Soc. János Bolyai **25**, pages 827–847, North-Holland, Amsterdam, 1981.
[42] Richard Weiss, Über symmetrische Graphen und die projektiven Gruppen, *Arch. Math. (Basel)* **28** (1977), no. 1, 110–112.
[43] Richard Weiss, An application of $p$-factorization methods to symmetric graphs, *Math. Proc. Cambridge Philos. Soc.* **85** (1979), no. 1, 43–48.
[44] Richard Weiss, Groups with a $(B, N)$-pair and locally transitive graphs, *Nagoya Math. J.* **74** (1979), 1–21.
[45] Richard Weiss, Permutation groups with projective unitary subconstituents, *Proc. Amer. Math. Soc.* **78** (1980), no. 2, 157–161.
[46] Richard Weiss, The nonexistence of 8-transitive graphs, *Combinatorica* **1** (1981), no. 3, 309–311.
[47] H. Wielandt, *Subnormal subgroups and permutation groups*, Ohio State University Lecture Notes, Columbus, OH, 1971.

[48] Warren J. Wong, Determination of a class of primitive permutation groups, *Math. Z.* **99** (1967), 235–246.

# NON-CANCELLATION GROUP COMPUTATION FOR SOME FINITELY GENERATED NILPOTENT GROUPS

HABTAY GHEBREWOLD

University of Asmara (University of the Western Cape), P.O. Box 1220, Asmara, Eritrea
Email: habtay@uoa.edu.er

## Abstract

For any group $G$, the non-cancellation set of $G$, denoted by $\chi(G)$, is the set of all isomorphism classes of groups $H$ such that $H \times \mathbb{Z} \simeq G \times \mathbb{Z}$ (see [5], [16] and [17]). The Mislin genus (see [10], [6] and [1]) of a finitely generated nilpotent group $N$ denoted by $\mathcal{G}(N)$ is the set of all isomorphism classes of a finitely generated nilpotent groups $M$ such that for every prime $p$, the groups $N$ and $M$ have isomorphic $p$-localizations. The Mislin genus ($\mathcal{G}(N)$) and the non-cancellation set ($\chi(N)$) for a finitely generated nilpotent group $N$ are the same [14]. In this paper we discuss some properties and calculate the non-cancellation set (Mislin genus) for finitely generated nilpotent groups of the type

$$G_{m,h,k,l} = \langle\, z, x, y, w \mid z^m = 1,\ z \text{ central},\ [x,w] = z^h,\ [x,y] = z^k,\ [y,w] = z^l \,\rangle$$

where $m$, $h$, $k$, and $l$ are non-negative integers.

## 1 Introduction

The *cancellation* (*non-cancellation*) phenomenon of groups naturally arises from the assumption that given groups $F$, $G$ and $H$, such that $F \times G \simeq F \times H$, then is $G \simeq H$? Or more generally, what can be said about the relationship between the groups $G$ and $H$? It is known that ([9, Lemma 1]) if $F$, $G$ and $H$ are finitely generated abelian groups such that $F \times G \simeq F \times H$, then $G \simeq H$. But if $G$ and $H$ are any two groups or if $F$ is not a finitely generated group then the relationship cannot be easily described. For a group $F$ not finitely generated one can see the example given by Walker [13].

However, for finitely generated groups with a finite commutator subgroup and $F \simeq \mathbb{Z}$, the relationship between $G$ and $H$ is found to be very interesting. Walker in his paper [13], considered the example constructed by Scott and presented it in the form of a theorem as follows:

**Theorem 1.1 ([13, Theorem 13])** *There exist groups $G$ and $H$ such that*

$$G \not\simeq H, \quad \textit{yet} \quad G \times \mathbb{Z} \simeq H \times \mathbb{Z}.$$

In his proof he constructed $G$ and $H$ to be infinite metacyclic groups with isomorphic torsion subgroups, but having non-isomorphic torsion centers. Later on, this phenomenon is well discussed and discovered by a number of authors. Among some of the major contributors in this area are Hilton [5] and Witbooi [16, 17]. In particular, P. Witbooi formulated a group structure on the non-cancellation set ([17]) for finitely generated groups having a finite commutator subgroup (he denoted the above class of groups by $\mathcal{X}_0$) and introduced the notation $\chi(G)$ for the set of all isomorphism classes of groups $[H]$ such that $G \times \mathbb{Z} \simeq H \times \mathbb{Z}$.

On the other hand, the Mislin genus $\mathcal{G}(N)$ (see [10], [6] and [1]) of a finitely generated nilpotent group $N$, which is defined to be the set of all isomorphism classes of a finitely generated nilpotent groups $M$ such that for every prime $p$ the groups $M$ and $N$ have isomorphic $p$-localizations, has a close relationship to the non-cancellation phenomena of groups. The relationship was early formulated by Warfield [14] and recently developed by P. Witbooi [16, 17].

The following is a theorem of Warfield.

**Theorem 1.2 ([14, Theorem 3.6])** *Let $N$ and $M$ be finitely generated nilpotent groups having finite commutator subgroups. The following conditions are equivalent.*

(a) $N \times \mathbb{Z} \simeq M \times \mathbb{Z}$

(b) *For every prime $p$, the $p$-localizations of $N$ and $M$ are isomorphic.*

(c) *$\mathcal{F}(N) = \mathcal{F}(M)$, where $\mathcal{F}(N)$ is the set of all isomorphism classes of finite quotient groups of $N$ (see [11]).*

One of the applications of the above theorem is that one can use the methods of finding the non-cancellation group in order to calculate the Mislin genus of a finitely generated nilpotent group having a finite commutator subgroup.

## 2 Some properties of the groups

In the paper by Ghebrewold [4], the non-cancellation group and some properties of the groups

$$G = G_{m,k} = \langle\, z, x, y \mid z^m = 1,\ z \in \mathbf{Z}(G),\ [x, y] = z^k \,\rangle,$$

were discussed. In fact, it was indicated that two groups of the form $G_{m,h}$ and $G_{m,k}$ are isomorphic if and only if $h \equiv rk \mod m$ for some $r$ relatively prime to $m$ ([4, Proposition 5.2]). Moreover, it was shown that the non-cancellation group $(\chi(G))$ of the group $G = G_{m,k}$ is trivial ([4, Proposition 5.4]).

In this paper we consider the group $G$ given by the following presentation,

$$G = G_{m,h,k,l} = \langle\, z, x, y, w \mid z^m = 1,\ z \in \mathbf{Z}(G),\ [x, w] = z^h,\ [x, y] = z^k,\ [y, w] = z^l \,\rangle,$$

where $m$, $h$, $k$ and $l$ are non-negative integers. In this section we discuss some of the properties of the group $G$. In the subsequent section we compute the non-cancellation group of the above group by first considering the special case where $l \equiv 0 \mod m$ and later we consider where $l \not\equiv 0 \mod m$.

The next two propositions describe some of the properties of the above group $G$.

**Proposition 2.1** *For the group $G = G_{m,h,k,l}$, the following are true:*

(i) *For any element $g$ in $G$, $g$ can be uniquely expressed as $g = z^r x^s y^t w^u$ for some $r$, $s$, $t$ and $u$ integers and $0 \leq r \leq m-1$.*

(ii) *Given $g_1 = z^{r_1} x^{s_1} y^{t_1} w^{u_1}$ and $g_2 = z^{r_2} x^{s_2} y^{t_2} w^{u_2}$ in $G$,*

$$g_1 g_2 = z^{r_1+r_2-s_1u_1h-s_1t_1k-t_1u_1l} x^{s_1+s_2} y^{t_1+t_2} w^{u_1+u_2}.$$

(iii) *For $g$ in $G$, $g = z^r x^s y^t w^u$, its inverse is $g^{-1} = z^{-r-suh-stk-tul} x^{-s} y^{-t} w^{-u}$.*

**Proof** (i) From the assumptions of the group $G$ we have $xwx^{-1}w^{-1} = z^h$ which is the same as $xw = z^h wx$ or $z^{-h}xw = wx$. Similarly, we get $z^{-k}xy = yx$ and $z^{-l}yw = wy$. Furthermore, by repeated applications of the $z^{-h}xw = wx$, $z^{-k}xy = yx$ and $z^{-l}yw = wy$ we get $w^u x^s = z^{-suh} x^s w^u$, $y^t x^s = z^{-stk} x^s y^t$ and $w^u y^t = z^{-tul} y^t w^u$ for any integers $s$, $t$ and $u$.

Hence, using the above assumptions and the fact that $z$ is in the centre it follows that for any $g \in G$, $g = z^r x^s y^t w^u$ for some $r$, $s$, $t$ and $u$ integers and $0 \leq r \leq m-1$.

The proofs of (ii) and (iii) follow from (i). □

**Proposition 2.2** *The group $G = G_{m,h,k,l}$ is a nilpotent group of class two.*

**Proof** To show $G$ is nilpotent group of class two, we need to show that $[G,G] \neq 1$ and $[G', G] = 1$. But these follows from the fact that for any $g_1$ and $g_2$ in $G$, $g_1 g_2 g_1^{-1} g_2^{-1} = z^r$ and the assumption that $z$ is in the centre. □

Note that the group $G = G_{m,h,k,l}$ is a central extension of a cyclic group of order $m$ by a free abelian group of rank three.

$$\mathbb{Z}_m \hookrightarrow G_{m,h,k,l} \longrightarrow \mathbb{Z}^3$$

## 3 Non-cancellation group (Mislin genus)

In this section we will discuss the non-cancellation group (Mislin genus) for the groups given by the presentation:

$$G = G_{m,h,k,l} = \langle\, z, x, y, w \mid z^m = 1,\ z \in \mathbf{Z}(G),\ [x, w] = z^h,\ [x, y] = z^k,\ [y, w] = z^l \,\rangle.$$

First we consider a special case where $l \equiv 0 \mod m$. In this case the group will be of the form:

$$G = G_{m,h,k,0} = \langle\, z, x, y, w \mid z^m = 1,\ z \in \mathbf{Z}(G),\ [x, w] = z^h,\ [x, y] = z^k,\ wy = yw \,\rangle.$$

**Definition 3.1** Let $G$ be an infinite group such that $G$ is finitely generated having finite commutator subgroup. Denote the torsion subgroup of $G$ by $T_G$. Let $a$ be the exponent of $T_G$, let $b$ be the exponent of $\mathrm{Aut}(T_G)$ and let $c$ be the exponent of the torsion subgroup of the centre of $G$. Then $n(G)$ is defined to be the product of the integers $a$, $b$ and $c$, i.e., $n(G) = abc$ (see [17]).

Note that for the group $G = G_{m,h,k,l}$, $n(G) = m^2\phi(m)$ ($\phi$ is the Euler phi function).

**Theorem 3.2 ([17, Theorem 4.2])** *Let $G$ be any group in $\mathcal{X}_0$ and let $H$ be any group such that $H \times \mathbb{Z} \simeq G \times \mathbb{Z}$. Then $H$ is isomorphic to a subgroup $L$ of $G$ of finite index in $G$ such that $[G : L]$ is relatively prime to $n = n(G)$ (where $n(G)$ is as defined in 3.1).*

**Remark 3.3** For the group $G = G_{m,h,k,l}$, we know that $n = n(G) = m^2\phi(m)$. Let $X$ be the set of all positive integers that are relatively prime to $n$. Using [17] we have a well defined surjective function $\mu : X \to \chi(G)$, defined by the rule $\mu(x) = [H]$ where $H$ is a subgroup of $G$ of index $x$ and $[\cdot]$ denotes isomorphism class. Let $\eta : X \to \mathbb{Z}_n^*/\{\pm 1\}$ be the "reduction mod $n$"-function. Thus, we get the unique function $\theta : \mathbb{Z}_n^*/\{\pm 1\} \to \chi(G)$ satisfying $\theta \circ \eta = \mu$ ([17]). However, one can consider the "reduction mod $n$"-function $\eta : X \to \mathbb{Z}_n^*$ and the corresponding unique function $\theta :\ Z_n^* \to \chi(G)$ satisfying $\theta \circ \eta = \mu$. Then, using the same method used in the proof of [17, Theorem 5.1], it is possible to show that the fiber $\theta^{-1}[G]$ is a subgroup of $\mathbb{Z}_n^*$ and for any $[H] \in \chi(G)$, $\theta^{-1}[H]$ is a coset of $\theta^{-1}[G]$. Thus, we have a bijection

$$\Theta : \mathbb{Z}_n^*/(\theta^{-1}[G]) \to \chi(G).$$

Hence in order to calculate the non-cancellation groups for finitely generated groups having a finite commutator subgroup, it is sufficient to get hold of the kernel of $\theta$. This is the same as identifying the set $S$ of subgroups of $G$ such that

$$S = \{H \leq G \mid H \simeq G \text{ and } [G : H] = t\},$$

where $t$ is relatively prime to $n = n(G)$. Note that the function $\theta : \mathbb{Z}_n^* \to \chi(G)$ induces a group structure on $\chi(G)$, which coincides with the Hilton–Mislin genus if $G$ is a nilpotent group [17, Theorem 5.2].

The following proposition characterizes the non-cancellation group for the group $G_{m,h,k,0}$.

**Proposition 3.4** *The non-cancellation group of the group $G_{m,h,k,0}$ is trivial, i.e.,*

$$\chi(G_{m,h,k,0}) = 1.$$

**Proof** In order to prove that the non-cancellation group of $G$ $(G = G_{m,h,k,0})$ is trivial, we have to show that, for any group $H$, if $H \times \mathbb{Z} \simeq G \times \mathbb{Z}$, then $H \simeq G$. Moreover, using Theorem 3.2, it is sufficient to show that any subgroup $H$ of $G$ with $[G : H] = d$ where $d$ is relatively prime to $m^2\phi(m)$ is isomorphic to $G$ (i.e., $\theta^{-1}(G) \simeq \mathbb{Z}_n^*$, Theorem 3.2. and Remark 3.3).

Furthermore, one can choose the subgroup $H = \langle z, x^d, y, w \rangle$ of $G$ where $d$ is relatively prime to $m^2\phi(m)$ ([17, Theorem 4.3.]).

Define the function

$$\alpha : G \to H,\ \ z^r x^s y^t w^u \mapsto z^{dr} x^{ds} y^t w^u.$$

(i) To show that $\alpha$ is a homomorphism:

Let $z^{r_1}x^{s_1}y^{t_1}w^{u_1}$ and $z^{r_2}x^{s_2}y^{t_2}w^{u_2}$ be any two elements of $G$, then using Proposition 2.1. we have

$$(z^{r_1}x^{s_1}y^{t_1}w^{u_1})(z^{r_2}x^{s_2}y^{t_2}w^{u_2}) = z^{r_1+r_2-s_2u_1h-s_2t_1k}x^{s_1+s_2}y^{t_1+t_2}w^{u_1+u_2},$$

and

$$\alpha(z^{r_1+r_2-s_2u_1h-s_2t_1k}x^{s_1+s_2}y^{t_1+t_2}w^{u_1+u_2}) \\ = z^{d(r_1+r_2-s_2u_1h-s_2t_1k)}x^{d(s_1+s_2)}y^{t_1+t_2}w^{u_1+u_2},$$

which is the same as

$$\alpha(z^{r_1}x^{s_1}y^{t_1}w^{u_1})\alpha(z^{r_2}x^{s_2}y^{t_2}w^{u_2}).$$

Thus $\alpha$ is a homomorphism.

(ii) To show that $\alpha$ is one to one, let $\alpha(z^r x^s y^t w^u) = 1$, then $z^{dr}x^{ds}y^t w^u = 1$. Thus, we get

$$dr \equiv 0 \mod m, \;\; ds = 0, \;\; t = 0 \text{ and } u = 0.$$

Since $d$ is relatively prime to $m^2\phi(m)$, we have, $r \equiv 0 \mod m$ and $s = 0$. Hence $\alpha$ is one to one.

(iii) The homomorphism $\alpha$ is onto, since for $z^r x^{ds} y^t w^u$ in $H$ there is $z^{ar}x^s y^t w^u$ in $G$ such that $\alpha(z^{ar}x^s y^t w^u) = z^r x^{ds} y^t w^u$ where $ad \equiv 1 \mod m$.

Hence, the subgroup $H$ is isomorphic to $G$, which shows that the non-cancellation group of $G$ is trivial. □

For the rest of this section we will consider the general case (where $h$, $k$ and $l$ are not all congruent to 0 modulo $m$).

**Remark 3.5** ([2]) The following property of $3 \times 3$ matrices will be used in the proof of the next theorem.

If the matrix

$$\begin{pmatrix} s_1 & t_1 & u_1 \\ s_2 & t_2 & u_2 \\ s_3 & t_3 & u_3 \end{pmatrix}$$

is unimodular, then

$$\begin{pmatrix} t_2u_3 - t_3u_2 & s_2u_3 - s_3u_2 & s_2t_3 - s_3t_2 \\ t_1u_3 - t_3u_1 & s_1u_3 - s_3u_1 & s_1t_3 - s_3t_1 \\ t_1u_2 - t_2u_1 & s_1u_2 - s_2u_1 & s_1t_2 - s_2t_1 \end{pmatrix}$$

is a unimodular matrix.

**Theorem 3.6** *Let the group $G = G_{m,h,k,l}$, and suppose $H$ and $K$ are subgroups of $G$ such that $[G : H] = d$ and $[G : K] = e$, where $d$ and $e$ are both relatively prime to $m^2\phi(m)$. If $H \simeq K$, then $e^2 \equiv \pm d^2 r^3 \mod m$ for some $r$ relatively prime to $m$.*

**Proof** Without loss of generality, one can choose (see [17, Theorem 4.3]) the subgroups $H$ and $K$ of $G$ to be as follows:

$$H = \langle z, x^d, y, w\rangle \qquad \text{and} \qquad K = \langle z, x^e, y, w\rangle.$$

Suppose $H \simeq K$, and let $\beta$ be the isomorphism from $H$ into $K$. Then $\beta(z) = z^r$ for some $r$ relatively prime to $m$. Furthermore, there exists $r_i, s_i, t_i$ and $u_i$ (for $i = 1, 2, 3$) in $\mathbb{Z}$ such that

$$\begin{aligned}\beta(x^d) &= z^{r_1}(x^e)^{s_1}y^{t_1}w^{u_1},\\ \beta(y) &= z^{r_2}(x^e)^{s_2}y^{t_2}w^{u_2},\\ \beta(w) &= z^{r_3}(x^e)^{s_3}y^{t_3}w^{u_3}.\end{aligned}$$

From the assumption that $\beta$ is an isomorphism we have

$$H/\langle z\rangle \simeq \mathbb{Z}^3 \simeq K/\langle z\rangle.$$

Thus, we have the unimodular matrix $\begin{pmatrix} s_1 & t_1 & u_1\\ s_2 & t_2 & u_2\\ s_3 & t_3 & u_3\end{pmatrix}$. Since $[x, w] = z^h$, we have $[x^d, w] = z^{dh}$. By substituting into the formula

$$z^{rdh} = \beta(z^{dh}) = \beta([x^d, w]) = \beta(x^d)\beta(w)\beta(x^{-d})\beta(w^{-1}),$$

the reader may calculate that

$$z^{es_1u_3h - es_3u_1h + es_1t_3k - es_3t_1k + t_1u_3l - t_3u_1l} = z^{rdh}.$$

Which is the same as:

$$he(s_1u_3 - s_3u_1) + ke(s_1t_3 - s_3t_1) + l(t_1u_3 - t_3u_1) \equiv rdh \mod m.$$

In a similar way and using the assumptions $[x^d, y] = z^{dk}$ and $[y, w] = z^l$ we get the following

$$he(s_1u_2 - s_2u_1) + ke(s_1t_2 - s_2t_1) + l(t_1u_2 - t_2u_1) \equiv rdk \mod m$$

and

$$he(s_2u_3 - s_3u_2) + ke(s_2t_3 - s_3t_2) + l(t_2u_3 - t_3u_2) \equiv rl \mod m.$$

Collecting the above equations, we have the following system of linear congruencies:

$$\begin{aligned} l(t_2u_3 - t_3u_2) + he(s_2u_3 - s_3u_2) + ke(s_2t_3 - s_3t_2) &\equiv rl \mod m\\ l(t_1u_3 - t_3u_1) + he(s_1u_3 - s_3u_1) + ke(s_1t_3 - s_3t_1) &\equiv rdh \mod m\\ l(t_1u_2 - t_2u_1) + he(s_1u_2 - s_2u_1) + ke(s_1t_2 - s_2t_1) &\equiv rdk \mod m\end{aligned}$$

As the result we get the following:

$$\begin{pmatrix} t_2u_3 - t_3u_2 & e(s_2u_3 - s_3u_2) & e(s_2t_3 - s_3t_2) \\ t_1u_3 - t_3u_1 & e(s_1u_3 - s_3u_1) & e(s_1t_3 - s_3t_1) \\ t_1u_2 - t_2u_1 & e(s_1u_2 - s_2u_1) & e(s_1t_2 - s_2t_1) \end{pmatrix} \begin{pmatrix} l \\ h \\ k \end{pmatrix}$$
$$\equiv \begin{pmatrix} r & 0 & 0 \\ 0 & rd & 0 \\ 0 & 0 & rd \end{pmatrix} \begin{pmatrix} l \\ h \\ k \end{pmatrix} \quad \bmod m$$

Note that the $3 \times 3$ matrix in the left hand side as well as the $3 \times 3$ matrix in the right hand side do not depend on the particular choice of $h, k$ and $l$. Thus, the determinant of the $3 \times 3$ matrix in the left hand side and the determinant of the $3 \times 3$ matrix in the right hand side must be congruent modulo $m$. Furthermore, the determinant of the left hand side matrix is $e^2$ (Remark 3.5). Hence, we have

$$e^2 \equiv \pm d^2 r^3 \mod m,$$

for some $r$ relatively prime to $m$. □

In order to determine the non-cancellation group of the group $G = G_{m,h,k,l}$, we use the following corollary to describe the subgroups of $G$ that are isomorphic to the group $G = G_{m,h,k,l}$. That is, to find the non-cancellation group of the group $G = G_{m,h,k,l}$, it is sufficient to find $S$ (the kernel of the surjective map $\theta : Z_n^* \to \chi(G)$ as given in Remark 3.3.), where

$$S = \{H \leq G \mid H \simeq G \text{ and } [G : H] = t\},$$

for $t$ relatively prime to $n = n(G)$ and $G = G_{m,h,k,l}$.

**Corollary 3.7** *Let $G = G_{m,h,k,l}$ and let $H$ be a subgroup of $G$ such that $[G : H] = d$, where $d$ is relatively prime to $m^2\phi(m)$. If $H$ is isomorphic to $G$, then $d^2r^3 \equiv \pm 1 \bmod m$ for some $r$ relatively prime to $m$.*

**Lemma 3.8** *Let*

$$M = \{x \in \mathbb{Z}^*_{m^2\phi(m)} \mid x^2 r^3 \equiv \pm 1 mod m \text{ for some } r \text{ relatively prime to } m\}.$$

*Then $M$ is a subgroup of $\mathbb{Z}^*_{m^2\phi(m)}$.*

**Proof** The proof follows from the fact that, if $d_1^2 r_1^3 \equiv \pm 1 \mod m$ and $d_2^2 r_2^3 \equiv \pm 1 \mod m$, then $(d_1d_2)^2(r_1r_2)^3 \equiv \pm 1 \mod m$ and $(d_1^{-1})^2(r_1^{-1})^3 \equiv \pm 1 \mod m$ (where $d_1^{-1}d_1 \equiv 1 \mod m$ and $r_1^{-1}r_1 \equiv 1 \mod m$). □

Using Corollary 3.7 and Lemma 3.8 we now prove the main theorem of this section.

**Theorem 3.9** *The group $G$ denoted by $G_{m,h,k,l}$, given by the presentation*

$$G_{m,h,k,l} = \langle z, x, y, w \mid z^m = 1,\ z \text{ is central},\ [x, w] = z^h,\ [x, y] = z^k,\ [y, w] = z^l \rangle$$

*has a non-cancellation group $\chi(G)$ isomorphic to*

$$\mathbb{Z}^*_{m^2\phi(m)}/M,$$

*where $M$ is the subgroup of $\mathbb{Z}^*_{m^2\phi(m)}$ as defined in Lemma 3.8.*

**Proof** The proof follows from the assumptions of Corollary 3.7 and Lemma 3.8, and the discussions given in Remark 3.3. □

Another assertion that can be easily deduced from the above discussion is the following theorem.

**Theorem 3.10** *Given the groups $G_1 = G_{m,h_1,k_1,l_1}$ and $G_2 = G_{m,h_2,k_2,l_2}$, then $\chi(G_1) \simeq \chi(G_2)$ provided that the $h_i$'s, $k_i$'s and the $l_i$'s, for $i = 1, 2$, are not all congruent to 0 modulo $m$.*

**Corollary 3.11** *The non-cancellation group of the group $G = G_{m,h,k,l}$, for $m \leq 6$ is trivial, i.e., $\chi(G) = 1$ .*

**Proof** Using direct computation we have $\mathbb{Z}^*_{m^2\phi(m)} \simeq M$ for $m \leq 6$, where $M$ is the subgroup defined as in Lemma 3.8. □

Note that the non-cancellation group (Mislin genus) of the group $G = G_{m,h,k,l}$, for $m > 6$, is not necessarily trivial. This assertion is easily shown in Example 3.12.

**Example 3.12** The non-cancellation group of the group $G = G_{m,h,k,l}$ for $m = 7$, is non-trivial.

For $m = 7$, the group $G$ is given by

$$G = G_{7,h,k,l} = \langle z, x, y, w \mid z^7 = 1,\ z \text{ is central},\ [x, w] = z^h,\ [x, y] = z^k,\ [y, w] = z^l \rangle,$$

where $h$, $k$ and $l$ are all not congruent to 0 modulo 7. Thus, $\mathbb{Z}^*_{m^2\phi(m)} = \mathbb{Z}^*_{294} \simeq \mathbb{Z}_{42} \times \mathbb{Z}_2$, and $M = \{x \in \mathbb{Z}^*_{m^2\phi(m)} \mid x^2 r^3 \equiv \pm 1 \mod 7 \text{ for some } r \text{ relatively prime to } 7\}$. Thus, we get $M \simeq \mathbb{Z}_{14} \times \mathbb{Z}_2$. Hence,

$$\chi(G) \simeq \mathbb{Z}^*_{m^2\phi(m)}/M \simeq \mathbb{Z}_{42} \times \mathbb{Z}_2/(\mathbb{Z}_{14} \times \mathbb{Z}_2) \simeq \mathbb{Z}_3.$$

Moreover, the representatives of the isomorphism classes of the non-cancellation group of $G$ can be chosen to be: $G_0 = G$, while $G_1$ and $G_2$ are subgroups of $G$, defined as, $G_1 = \langle z, x^5, y, w \rangle$ and $G_2 = \langle z, x^{5^2}, y, w \rangle$.

Furthermore, the group operation can be defined as follows:

$$G_i + G_j = \langle z, x^{5^{i+j}}, y, w \rangle,$$

for $i, j = 0, 1, 2$.

Note that any subgroup $H$ of $G$ of finite index such that $[G : H]$ relatively prime to 294 is isomorphic to one of the above groups (Theorem 3.6). □

## References

[1] C. Casacuberta and P. Hilton, Calculating the Mislin genus for a certain family of nilpotent groups, *Comm. Algebra* **19** (1991), no. 7, 2051–2069.

[2] P. J. Davis, *The mathematics of matrices*, John Wiley and Sons Inc., New York, 1973.

[3] A. Fransman and P. Witbooi, Non-cancellation sets of direct powers of certain metacyclic groups, *Kyungpook Math. J.* **41** (2001), 191–197.

[4] H. Ghebrewold, Non-cancellation phenomena in the class of finitely generated groups with finite commutator subgroups, *Quaest. Math.* **25** (2002), no. 3, 333–339.

[5] P. Hilton, Non-cancellation properties for certain finitely presented groups, *Quaest. Math.* **9** (1986), 281–292.

[6] P. Hilton and G. Mislin, On the genus of a nilpotent group with finite commutator subgroup, *Math. Z.* **146** (1976), 201–211

[7] P. Hilton and D. Scevenels, Calculating the genus of a direct product of certain nilpotent groups, *Publ. Mat.* **39** (1995), 263–269.

[8] P. Hilton and C. Schuck, Non-cancellation phenomena in a class of finitely generated nilpotent groups, in Y. Ershove et al. (Eds.), *Algebra proceedings of the Third International Conference on Algebra, Krasnoyarsk, Russia, August 23–28 1993*, Walter de Gruyter, Berlin 1996, 93–101.

[9] D. L. Johnson, Non-cancellation and non-abelian tensor squares, in K. N. Cheng and Y. K. Leong (Eds.), *Group Theory (Proceedings of the Singapore Group Theory Conference, National University of Singapore, June 1987)*, Walter de Gruyter, Berlin 1989, 405–408.

[10] G. Mislin, Nilpotent groups with finite commutator subgroups, in P. Hilton (Ed.), *Localization in Group Theory and Homotopy Theory*, Lecture Notes in Mathematics **418**, Springer-Verlag Berlin 1974, 103–120.

[11] P. F. Pickel, Finitely generated nilpotent groups with isomorphic finite quotients, *Trans. Amer. Math. Soc.* **160** (1971) 327–341.

[12] D. Scevenels and P. Witbooi, Non-cancellation and Mislin genus of certain groups and $H_0$-spaces, *J. Pure Appl. Algebra* **170** (2002), 309–320.

[13] E. A. Walker, Cancellation in direct sums of groups, *Proc. Amer. Math. Soc.* **7** (1956), 898–902.

[14] R. Warfield, Genus and cancellation for groups with finite commutator subgroup, *J. Pure Appl. Algebra* **6** (1975), 125–132.

[15] P. Witbooi, Non-cancellation for certain classes of groups, *Comm. Algebra* **27** (1999), no. 8, 3639–3646.

[16] P. Witbooi, Non-unique direct product decomposition of direct powers of certain metacyclic groups, *Comm. Algebra* **28** (2000), no. 5, 2565–2576.

[17] P. Witbooi, Generalizing the Hilton–Mislin genus group, *J. Algebra* **239** (2001), 327–339.

# PERMUTATION AND QUASI-PERMUTATION REPRESENTATIONS OF THE CHEVALLEY GROUPS

MARYAM GHORBANY

Department of Mathematics, Iran University of Science and Technology, Narmak, Tehran 16844, Iran
Email: m-ghorbani@iust.ac.ir

## Abstract

A square matrix over the complex field with non-negative integral trace is called a quasi-permutation matrix. For a finite group $G$ the minimal degree of a faithful permutation representation of $G$ is denoted by $p(G)$. The minimal degree of a faithful representation of $G$ by quasi-permutation matrices over the rationals and the complex numbers are denoted by $q(G)$ and $c(G)$ respectively. Finally $r(G)$ denotes the minimal degree of a faithful rational valued complex character of $G$. In this paper $p(G)$, $q(G)$, $c(G)$ and $r(G)$ are calculated for the group $G_2(q^n)$, $q \neq 3$.

*AMS Classification:* 20C15
*Keywords:* General linear group, Quasi-permutation.

## 1 Introduction

Let $G$ be a finite linear group of degree $n$, that is, a finite group of automorphisms of an $n$-dimensional complex vector space. We shall say that $G$ is a quasi-permutation group if the trace of every element of $G$ is a non-negative rational integer. The reason for this terminology is that, if $G$ is a permutation group of degree $n$, its elements, considered as acting on the elements of a basis of an $n$-dimensional complex vector space $V$, induce automorphisms of $V$ forming a group isomorphic to $G$. The trace of the automorphism corresponding to an element $x$ of $G$ is equal to the number of letters left fixed by $x$ and so is a non-negative integer. Thus, a permutation group of degree $n$ has a representation as a quasi-permutation group of degree $n$ (see [13, 14]). In [1] the author investigated further the analogy between permutation groups and quasi-permutation groups. They also worked over the rational field and found some interesting results.

By a quasi-permutation matrix we mean a square matrix over the complex field $C$ with non-negative integral trace. Thus every permutation matrix over $C$ is a quasi-permutation matrix. For a given finite group $G$, let $p(G)$ denote the minimal degree of a faithful permutation representation of $G$, let $q(G)$ denote the minimal degree of a faithful representation of $G$ by quasi-permutation matrices over the rational field $Q$, and let $c(G)$ be the minimal degree of a faithful representation of $G$ by complex quasi-permutation matrices. By a rational valued character we mean a complex character $\chi$ of $G$ such that $\chi(g) \in Q$ for all $g \in G$. As the values of the character of a complex representation are algebraic numbers, a rational valued character is in fact integer valued. A quasi-permutation representation of $G$ is then simply a complex representation of $G$ whose character values are rational and non-negative. The module of such a representation will be called a quasi-permutation module. We will call a homomorphism from $G$ to $GL(n, Q)$ a rational representation of $G$ and its corresponding character will be called a rational character of $G$. Let $r(G)$ denote the minimal degree of a faithful rational valued character of $G$. It is easy to see that for a finite group $G$ we have

$$r(G) < c(G) \leq q(G) \leq p(G).$$

Finding the above quantities have been carried out in some papers, for example in [6] and [5] we found these for the groups $GL(2, q)$ and $SU(3, q^2)$ and $PSU(3, q^2)$ respectively. In [2] we found the rational character table and the values of $r(G)$, $c(G)$, $q(G)$ and $p(G)$ for the group $PGL(2, q)$. In this paper we will apply the algorithms in [1] to the group $G_2(q^n)$, $q \neq 3$. In [7] I found some above quantities for the group $G_2(q)$ and in this paper I have completed the calculations. We will prove

**Theorem** *Let $G = G_2(q)$ where $q = p^n$, $p \neq 3$ and $\epsilon = 1$ or $-1$ according as $q \equiv 1$ or $-1 \pmod 3$. Then*
(1) $r(G) = q^3 + \epsilon$,
(2) $c(G) = q(G) = q^3 + \epsilon + |\min\{\chi_{32}(g) : g \in G\}|$,
(3) $p(G) = (q^3 + 1)(q^2 + q + 1)$.

Let $G$ be a finite group and $\chi$ be an irreducible complex character of $G$. Let $m_Q(\chi)$ denote the Schur index of $\chi$ over $Q$ and $\Gamma(\chi)$ be the Galois group $Q(\chi)$ over $Q$. It is known that

$$\sum_{\alpha \in \Gamma(\chi)} m_Q(\chi)\chi^\alpha \qquad (1)$$

is a character of an irreducible $QG$-module [10, Corollary 10.2(b)]. So by knowing the character table of a group and Schur indices of each of the irreducible characters of $G$, we can find the irreducible rational characters of $G$.

## 2 Chevalley groups

We treat here the Chevalley groups in general. Let $\Sigma$ be a simple root system for a Lie algebra $L$. We fix an order in $\Sigma$ so that we can talk about positive, negative and fundamental roots. Let $P$ be the additive group generated by $\Sigma$. For $r, s$ in $\Sigma$ define the integer $s(r)$ by $s(r) = p - q$, where $q$ (resp. $-p$) is maximum (resp. minimum) integer $i$ such that $s + ir$ is a root. Then for each $r \in \Sigma$, the map $s \longrightarrow s(r)$ can be extended to a linear map on $P$. The symmetry $w_r$ with respect to $r \in \Sigma$ is the permutation of $\Sigma$ defined by $w_r(s) = s - s(r)r$. The Weyl group $W$ of $\Sigma$ is the group generated by all the $w_r$, $r \in \Sigma$. Now we fix an arbitrary field $K$, and denote by $K^*$ the multiplicative group of $K$. For $r \in \Sigma$ and $z \in K^*$, define the homomorphism $\chi_{r,z} : P \longrightarrow K^*$ by $\chi_{r,s}(u) = z^{u(r)}$. Denote by $X$ the group generated by all the $\chi_{r,s}$, where $r \in \Sigma$, $z \in K^*$. It can be shown that

$w(s)(w(r)) = s(r)$ for any $r, s \in \Sigma$ and $w \in W$. From this it follows that $\chi o w \in X$ for any $\chi \in X$ and $w \in W$. It is shown in [11] that, for any pair of $\Sigma$ and $K$, there exists a group $G$ (denoted by $G'$ in [11]), which we shall call the Chevalley group of type $\Sigma$ over $K$.

In [3] the character table of the groups $G_2(q)$, $q \neq 2, 3$, is given. Each of the irreducible characters is expressed as a linear combination of induced characters with integral coefficients. In this paper we use the same notations as used in [3] for the names of conjugacy classes and irreducible characters of these groups.

## 3 Algorithms for $r(G)$, $c(G)$ and $q(G)$

Let $\chi$ be a character of $G$ such that, for all $g \in G$, $\chi(g) \in Q$ and $\chi(g) \geq 0$. Then we say that $\chi$ is a non-negative rational valued character.

**Notation:** Let $\Gamma(\chi)$ be the Galois group of $Q(\chi)$ over $Q$.

Let $G$ be a finite group and $\chi$ be an irreducible complex character of $G$. Then we define

(1) $d(\chi) = |\Gamma(\chi)|\chi(1)$

(2) $$m(\chi) = \begin{cases} 0 & \chi = 1_G \\ \left|\min\left\{ \sum_{\alpha \in \Gamma(\chi)} \chi^\alpha(g) : g \in G \right\}\right| & \text{otherwise} \end{cases}$$

(3) $$c(\chi) = \sum_{\alpha \in \Gamma(\chi)} \chi^\alpha + m(\chi)1_G.$$

Now by using Lemma 3.5 and Corollary 3.7 of [1] if $\chi \in \mathrm{Irr}(G)$, then $\sum_{\alpha \in \Gamma(\chi)} \chi^\alpha$ is a rational valued character of $G$. Moreover $c(\chi)$ is a non-negative rational valued character of $G$ and $c(\chi)(1) = d(\chi) + m(\chi)$. By definition of $c(\chi)(1)$ and $d(\chi)$ we have $c(\chi)(1) \geq d(\chi) \geq \chi(1)$. And if $\chi \in \mathrm{Irr}(G)$, then $\ker \chi = \ker \sum_{\alpha \in \Gamma(\chi)} \chi^\alpha$. Moreover $\chi$ is faithful if and only if $\sum_{\alpha \in \Gamma(\chi)} \chi^\alpha$ is faithful.

Now according to Corollary 3.11 of [1] and above statements the following result is useful for calculation of $r(G)$, $c(G)$ and $q(G)$.

**Result 3.1** *Let $G$ be a finite group with a unique minimal normal subgroup. Then*

(1) $r(G) = \min\{d(\chi) : \chi$ *is a faithful irreducible complex character of* $G\}$

(2) $c(G) = \min\{c(\chi)(1) : \chi$ *is a faithful irreducible complex character of* $G\}$

(3) $q(G) = \min\{m_Q(\chi)c(\chi)(1) : \chi$ *is a faithful irreducible complex character of* $G\}$.

If the Schur index of each non-principal irreducible character of $G$ over $Q$ is equal to $m$, then from [1] Corollary 3.15 we have $q(G) = mc(G)$.

For the group $G = G_2(2^n)$ we proved that:

(1) $r(G) = q^3 + (-1)^n$

(2) $q(G) = c(G) = q^3 + q$

(3) $p(G) = (q^3 + 1)(q^2 + q + 1)$.
Now we have

**Theorem 3.2** *Let* $G = G_2(q)$ *where* $q = p^n$, $p \neq 3$ *and* $\epsilon = 1$ *or* $-1$ *according as* $q \equiv 1$ *or* $-1 \pmod 3$. *Then*
(1) $r(G) = q^3 + \epsilon$
(2) $c(G) = q(G) = q^3 + \epsilon + |\min\{\chi_{32}(g) : g \in G\}|$.

**Proof** By [3] we have the irreducible characters of $G_2(q)$, now by definition of $d(\chi)$ and character table $G_2(q)$ we obtain the Table (I) as follows:

**Table (I)**

| $\chi$ | $d(\chi)$ |
|---|---|
| $\chi_1$ | $\geq (q+1)^2(q^4+q^2+1)$ |
| $\chi_2$ | $\geq (q-1)^2(q^4+q^2+1)$ |
| $\chi_a$ | $\geq q^6 - 1$ |
| $\chi_b$ | $\geq q^6 - 1$ |
| $\chi_3$ | $\geq (q^2-1)^2(q^2-q+1)$ |
| $\chi_6$ | $\geq (q^2-1)^2(q^2+q+1)$ |
| $\chi_{1a}$ | $\geq q(q+1)(q^4+q^2+1)$ |
| $\chi'_{1a}$ | $\geq (q+1)(q^4+q^2+1)$ |
| $\chi_{1b}$ | $\geq q(q+1)(q^4+q^2+1)$ |
| $\chi'_{1b}$ | $\geq (q+1)(q^4+q^2+1)$ |
| $\chi_{2a}$ | $\geq q(q-1)(q^4+q^2+1)$ |
| $\chi'_{2a}$ | $\geq (q-1)(q^4+q^2+1)$ |
| $\chi_{2b}$ | $\geq q(q-1)(q^4+q^2+1)$ |
| $\chi'_{1b}$ | $\geq (q-1)(q^4+q^2+1)$ |
| $\chi_{21}$ | $q^2(q^4+q^2+1)$ |
| $\chi_{22}$ | $q^4+q^2+1$ |
| $\chi_{23}$ | $q(q^4+q^2+1)$ |
| $\chi_{24}$ | $q(q^4+q^2+1)$ |
| $\chi_{31}$ | $q^3(q^3+\epsilon)$ |
| $\chi_{32}$ | $q^3+\epsilon$ |
| $\chi_{33}$ | $q(q+\epsilon)(q^3+\epsilon)$ |
| $\chi_{12}$ | $q^6$ |
| $\chi_{13}$ | $q(q^4+q^2+1)/3$ |
| $\chi_{14}$ | $q(q^4+q^2+1)/3$ |
| $\chi_{15}$ | $q(q+1)^2(q^2-q+1)/2$ |
| $\chi_{16}$ | $q(q+1)^2(q^2+q+1)/6$ |
| $\chi_{17}$ | $q(q-1)^2(q^2+q+1)/2$ |
| $\chi_{18}$ | $q(q-1)^2(q^2-q+1)/6$ |
| $\chi_{19}$ | $q(q^2-1)^2/3$ |
| $\overline{\chi}_{19}$ | $q(q^2-1)^2/3$ |

Now by Result 3.1 and Table (I), we have

$$\min\{d(\chi) : \chi \text{ is a faithful irreducible complex character of } G\} = q^3 + \epsilon.$$

By [9] we know that the Schur index $m_Q(\chi)$ of any complex irreducible character $\chi$ of $G_2(q)$ with respect to $Q$ is equal to 1. Therefore $q(G) = c(G)$ for the Chevalley group $G_2(q)$, $q$ odd. Now by Result 3.1, character table $G_2(q)$ (see [3]), Table (I) and definition of $c(\chi)$ we have

$$\begin{aligned} c(G) &= \min\{c(\chi)(1) : \chi \text{ is a faithful irreducible complex character of } G\} \\ &= q^3 + \epsilon + |\min\{\chi_{32}(g) : g \in G\}|. \end{aligned}$$

□

For example let $G = G_2(5)$ we have $|\min\{\chi_{32}(g) : g \in G\}| = 20$ and therefore $c(G) = 124 + 20 = 144$.

## 4 Permutation representation

Using definition of $p(G)$ it is proved in [1] that

$$p(G) = \min\left\{\sum_{i=1}^{n}[G : H_i] : H_i \leq G, \bigcap_{i=1}^{n}\bigcap_{x\in G} H_i^x = 1\right\}$$

By [1] Corollary 2.4 we know that if $G$ is a finite group with a minimal normal subgroup then $p(G)$ is the smallest index of a subgroup with trivial core. Therefore for the simple group $G_2(q)$ we introduce the maximal subgroups of $G_2(q)$, $q \neq 3$, and then calculate $p(G)$ for this groups. So we regard $G_2(q)$ as a subgroup of the simple orthogonal group $G_0 = P\Omega_8^+(q)$, and our working definition of $G_2(q)$ is the centralizer in $G_0$ of the triality automorphism $\tau \in \mathrm{Aut}(G_0)$ (see 9.1.3.d of [8]). Thus $G_2(q)$ acts reducibly on the ambient space, which is the natural 8-dimensional (projective) module for $G_0$. This setup allows us to exploit certain aspects of the so-called geometry of triality.

Let $V$ be an 8-dimensional vector space over $F = GF(q)$, where $q = p^n$ and $p$ is an odd prime.

It is well known that $D \cong G_0.2^2$, where $D$ is the group of inner and diagonal automorphisms of $G_0$. We may choose a group $\eta \in \mathrm{Aut}(G_0)$ of graph automorphisms such that $\eta \cong S_3$. The group $\eta$ acts naturally on the Dynkin diagram of $G_0$ and

$$\Theta = D.\eta \cong G_0.S_4.$$

Also we can find a group $\Phi$ of field automorphisms of $G_0$ satisfying $\Phi = \mathrm{Aut}(F) \cong Z_n$ and $[\Phi, \eta] = 1$. The group $\Phi$ is central modulo $G_0$, and we have

$$A = \mathrm{Aut}(G_0) = \Theta.\Phi \cong G_0.(S_4 \times Z_n).$$

Let $\tau$ be an element of order 3 in $\eta$. Then $\tau$ is a triality automorphism (that is, $\tau$ induces a symmetry of order 3 on the Dynkin diagram of $G_0$), and

$$H_0 = C_{G_0}(\tau) \cong G_2(q).$$

The order of $H_0$ is $q^6(q^6-1)(q^2-1)$, and many of its basic properties are outlined in [4], for example.

The group $\Phi$ centralizes $\tau$, and so $\Phi$ normalizes $H_0$; we write

$$H_1 = H_0.\Phi.$$

Suppose for the moment that there exists $x \in C_{H_1}(H_0)$ of prime order $r$. By 7.2 and 9.1.1 of [8], $x$ induces a field automorphism on $G_0$, and $H_0 \leq O^{p'}(C_{G_0}(x)) \cong P\Omega_8^+(q^{1/r})$, which is impossible by Lagrange's theorem. Thus $C_{H_1}(H_0) = 1$ and we may regard $H_1$ as a subgroup of $\mathrm{Aut}(H_0)$. Indeed, $H_1$ is the group of inner and field automorphisms of $H_0$. When $p \geq 5$, we have

$$H_1 = \mathrm{Aut}(H_0) \qquad (p \geq 5).$$

When $p = 3$, however, then $H_0$ admits a graph automorphism (by graph automorphism we mean an automorphism inducing a non-trivial symmetry on the Dynkin diagram of $H_0$) and

$$|\mathrm{Aut}(H_0) : H_1| = 2 \qquad (p = 3).$$

In all cases, $\mathrm{Out}(H_0)$ is cyclic.

Throughout this paper, $H$ denotes a group satisfying

$$H_0 \leq H \leq \mathrm{Aut}(H_0)$$

and $M$ denotes a maximal subgroup of $H$ not containing $H_0$. We also define

$$M_0 = M \cap H_0,$$

and

$$S = \mathrm{soc}(M_0).$$

Now by the above statements we can introduce the maximal subgroups of the group $G_2(q)$ $(q \neq 2, 3)$ (see [11] Theorem A).

**Theorem 4.1** *Assume that $H_0 \leq H \leq H_1$, where $H_0 \cong G_2(q)$ $(q = p^n$ is odd$)$ and $H_1$ are as above. Let $M$ be a maximal subgroup of $H$ not containing $H_0$. Then $M_0 = M \cap H_0$ is $H_0$-conjugate to one of the following groups*

**Table (II)**

| *Groups* | *Structure* | *Remarks* |
|---|---|---|
| $P_a$ | $[q^5]:GL_2(q)$ | *Parabolic* |
| $P_b$ | $[q^5]:GL_2(q)$ | *Parabolic* |
| $C_{H_0}(s_2)$ | $(SL_2(q)\circ SL_2(q)).2$ | *involution centralizer* |
| $I$ | $2^3L_3(2)$ | $q=p$ |
| $K_+$ | $SL_3(q):2$ | *reducible* |
| $K_+^\gamma$ | $SL_3(q):2$ | $p=3, irreducible$ |
| $K_-$ | $SU_3(q):2$ | *reducible* |
| $K_-^\gamma$ | $SU_3(q):2$ | $p=3, irreducible$ |
| $C_{H_0}(\varphi_\alpha)$ | $G_2(q_0))$ | $q=q_0^\alpha, \alpha\ prime$ |
| $C_{H_0}(\varphi_2)$ | $Re(q)={}^2G_2(q)$ | $p=3, n\ odd$ |
| $PGL_2(q)$ | | $p\geq 7, q\geq 11$ |
| $L_2(8)$ | | $p\geq 5,\ F=F_p[\omega],\ \omega^3-3\omega+1=0$ |
| $L_2(13)$ | | $p\neq 13,\ F=F_p[\sqrt{}13]$ |
| $G_2(2)$ | | $q=p\geq 5$ |
| $J_1$ | | $q=11$ |

*Conversely, if $K\leq H_0$ is $H_0$-conjugate to one of these groups, then $N_H(K)$ is maximal in $H$.*

Note that $H_0$ contains a unique class of irreducible groups $PGL_2(q)$, $L_2(8)$, $L_2(13)$, $G_2(2)$, and $J_1$, for the appropriate values of $q$ indicated above.

**Theorem 4.2** *Let $G=G_2(q),\ \ q\neq 3$, then $p(G)=(q^3+1)(q^2+q+1)$.*

**Proof** We know that $G_2(q),\ \ q\neq 2$, is a simple group, so by definition of $p(G)$ we have $p(G)$ is equal to the smallest index of a maximal subgroup. Now by above theorem and table (II) we can calculate our result

$$p(G)=(q^3+1)(q^2+q+1)).$$

□

## References


[1] H. Behravesh, Quasi-permutation representations of $p$-groups of class 2, *J. London Math. Soc. (2)* **55** (1997), 251–260.

[2] H. Behravesh, A. Daneshkhah, M. R. Darafsheh and M. Ghorbany, The rational character table and quasi-permutation representations of the group $PGL(2,q)$, *Ital. J. Pure Appl. Math.* **11** (2001), 9–18.

[3] B. Chang and R. Ree, The characters of $G_2(q)$, *Sympos. Math. Vol. XIII* (1972), 395–413.

[4] R. W. Carter, *Simple groups of Lie type*, Wiley, London (1972).

[5] M. R. Darafsheh and M.Ghorbany, Quasi-permutation representations of the groups $SU(3,q^2)$ and $PSU(3,q^2)$, *Southeast Asian Bull. Math.* **26** (2002), 395–406.

[6] M. R. Darafsheh, M. Ghorbany, A. Daneshkhah and H. Behravesh, Quasi-permutation representation of the group $GL(2, q)$, *J. Algebra* **243** (2001), 142–167.
[7] M. Ghorbany, Special representations of the group $G_2(q)$, in *Proceedings of 34th Iranian Mathematics Conference*, to appear.
[8] D. Gorenstein and R. Lyons, The local structure of finite groups of characteristic 2 type, *Mem. Amer. Math. Soc.* **42** (1983), no. 276, Amer. Math. Soc., Providence, RI.
[9] R. Gow, Schur indices of some groups of Lie type, *J. Algebra* **42** (1976), 102–120.
[10] I. M. Isaacs, Character theory of finite groups, Academic Press, New York (1976).
[11] P. B. Kleidman, The maximal subgroups of the Chevalley groups $G_2(q)$ with $q$ odd, the Ree groups ${}^2G_2(q)$, and their automorphism groups, *J. Algebra* **117** (1988), 30–71.
[12] R. Ree, A family of simple groups associated with the simple Lie algebra of type $G_2$, *Amer. J. Math.* **83** (1961), 432–462.
[13] W. J. Wong, Linear groups analogous to permutation groups, *J. Austral. Math. Soc. Ser. A* **3** (1963), 180–184.
[14] W. J. Wong, On linear $p$-groups, *J. Austral. Math. Soc. Ser. A* **4** (1964), 174–178.

# THE SHAPE OF SOLVABLE GROUPS WITH ODD ORDER

S. P. GLASBY

Department of Mathematics, Central Washington University, WA 98926-7424, USA
Email: Stephen.Glasby@cwu.EDU, http://www.cwu.edu/~glasbys/

## Abstract

The minimal composition length, $c$, of a solvable group with solvable length $d$ satisfies $9^{(d-3)/9} < c < 9^{(d+1)/5}$. The minimal composition length, $c^o$, of a group with odd order and solvable length $d$ satisfies $7^{(d-2)/5} < c^o < 2^d$.

*AMS classification:* 20F16, 20F14, 20E34

## 1 Introduction

Let $c(G)$ and $d(G)$ denote the composition length and solvable (or derived length) of a finite and solvable $G$. All groups in this paper are finite and solvable unless otherwise stated. If $|G| = p_1 \cdots p_r$ where the $p_i$ are primes, then $G$ has a composition factor of order $p_i$ for $i = 1, \ldots, r$ and $c(G) = r$. The derived series for $G$ is defined recursively: $G^{(0)} := G$ and $G^{(i+1)} := [G^{(i)}, G^{(i)}]$ for $i \geq 0$. By definition $d(G)$ is the smallest $d \geq 0$ such that $G^{(d)} = 1$.

We seek to understand solvable and nilpotent groups that have solvable length $d$, and smallest possible composition length. Set

$$c_N(d) := \min\{c(G) \mid G \text{ is nilpotent and } d(G) = d\}$$
$$c_S(d) := \min\{c(G) \mid G \text{ is solvable and } d(G) = d\}.$$

Let $c_N^o(d)$ (resp. $c_S^o(d)$) be defined similarly except that $G$ ranges over nilpotent (resp. solvable) groups of *odd* order. Section 2 is devoted to a proof of the result in the abstract.

A simple way to construct groups with large solvable length is via permutational wreath products. Let $S_m$ denote the symmetric group of degree $m$. The group $S_m \operatorname{wr} S_n$ can be viewed as an imprimitive subgroup of $S_{mn}$. (We shall not view $S_m \operatorname{wr} S_n$ as a subgroup of $S_{m^n}$ with product action.) If $H \leq S_m$ and $K \leq S_n$ are both transitive, then $H \operatorname{wr} K \leq S_{mn}$ is transitive and $d(H \operatorname{wr} K) = d(H) + d(K)$, see [10, Corollary 1]. The wreath product $G_r = S_2 \operatorname{wr} \cdots \operatorname{wr} S_2$ with $r$ copies of $S_2$ has $d(G_r) = r$ and $c(G_r) = 2^r - 1$. P. Hall [6, Satz III.2.12] showed that $G^{(i)} \leq \gamma_{2^i}(G)$. If $G$ is a $p$-group and $d(G) = d > 1$, then $2^{d-1} + 1 \leq c(G)$ holds as $2^{d-1} \leq c(G/\gamma_{2^{d-1}}(G)) \leq c(G/G^{(d-1)})$. The following bounds hold for $d \geq 1$:

$$2^{d-1} \leq c_N(d) \leq 2^d - 1.$$

Therefore $d = \lfloor \log_2 c_N(d) \rfloor + 1$. In Section 3 some general remarks concerning the sharpening bounds for $c_N(d)$ and $c_S(d)$ are made, and the difficulty of constructing groups with solvable length $d$ and minimal composition length is considered.

## 2 Solvable groups

Given a solvable group with solvable length $d$ and minimal composition length $c$, we bound $d$ in terms of $c$, and $c$ in terms of $d$.

**Theorem 1** *Denote by $c$ (resp. $c_0$) the minimal composition length of a solvable group (resp. odd-order group) with solvable length $d>0$. Then*
(a) $\gamma \log_2 c - \frac{2}{3} < d < (\gamma+1)\log_2 c + 3$ *where* $\gamma = 5\log_9 2 \approx 1.58$,
(b) $\log_2 c_0 < d < (\gamma_0+1)\log_2 c_0 + 2$ *where* $\gamma_0 = 2\log_7 2 \approx 0.71$.
*Hence* $9^{(d-3)/9} < c_S(d) < 9^{(d+1)/5}$ *and* $7^{(d-2)/5} < c_S^o(d) < 2^d$ *where* $9^{1/9} \approx 1.27\cdots$, $9^{1/5} \approx 1.55\cdots$, *and* $7^{1/5} \approx 1.47\cdots$.

**Proof** (a) Suppose that $G$ is a group with solvable length $d > 0$, and minimal composition length. Then $c(G)$ equals $c := c_S(d)$. If $N$ is a nontrivial normal subgroup of $G$, then $c(G/N) < c(G)$ and so $d(G/N) < d(G)$. It follows that $G$ cannot have distinct minimal normal subgroups $N_1, N_2$ otherwise $G$ embeds in $G/N_1 \times G/N_2$ and $d(G) > \max\{d(G/N_1), d(G/N_2)\}$. Let $N$ be *the* unique minimal normal subgroup of $G$. As $N$ is characteristically simple, it is an elementary abelian $p$-group for some prime $p$. Let $P$ be the maximal normal $p$-subgroup of $G$. Then $P \neq 1$, and $G$ has no normal subgroups with order coprime to $p$. By a theorem of Hall and Higman [6, Hilfssatz VI.6.5] $G/P$ acts faithfully and completely reducibly on the vector space $P/\Phi(P)$. We shall view $P/\Phi(P)$ as an $r$-dimensional vector space over the field $\mathbb{F}_p$ with $p$ elements. As $P/\Phi(P)$ is abelian $d(G) \leq d(G/P) + 1 + d(\Phi(P))$ holds. Suppose $|\Phi(P)| = p^s$. Then $s := c(\Phi(P))$ and $d(\Phi(P)) \leq \log_2 s + 1$. Because $G/P \leq \mathrm{GL}(r, \mathbb{F}_p)$ is completely reducible, [10, Theorem C] gives

$$d(G/P) \leq 5\log_9(r/8) + 8 = \gamma \log_2(r/8) + 8$$

where $\gamma = 5\log_9 2$. Since $c(G) = c(G/P) + c(P/\Phi(P)) + c(\Phi(P))$, we see that $c = c(G/P) + r + s$ and $s \leq c - r$. Therefore

$$\begin{aligned} d = d(G) &\leq (\gamma \log_2(r/8) + 8) + 1 + (\log_2 s + 1) \\ &\leq \gamma \log_2(r/8) + \log_2(c - r) + 10 \end{aligned}$$

Using calculus, the maximum of $\gamma \log_2(r/8) + \log_2(c-r) + 10$, with $0 < r < c$, occurs when $r = \gamma c/(\gamma+1)$. Thus $d \leq (\gamma+1)\log_2 c + \delta$ where

$$\delta = \gamma \log_2(\gamma/8) - (\gamma+1)\log_2(\gamma+1) + 10 \approx 2.78 < 3.$$

This establishes the upper bound for (a).

The lower bound for (a) is obtained by constructing five families of transitive permutation groups $G_n \leq S_n$. Our exposition is influenced by [10]. Let $G_9$ denote the maximal solvable primitive permutation subgroup $\mathrm{GL}(2,3) \ltimes 3^2 \leq S_9$, and set $G_8 = \mathrm{GL}(2,3) \leq S_8$. Let $m = 9^r$, and define $G_m \leq S_m$ to be $G_9 \,\mathrm{wr} \cdots \mathrm{wr}\, G_9$ with $r$ copies of $G_9$. Set $G_{2m} = S_2 \,\mathrm{wr}\, G_m$, $G_{3m} = S_3 \,\mathrm{wr}\, G_m$, $G_{4m} = S_4 \,\mathrm{wr}\, G_m$, and $G_{8m} = G_8 \,\mathrm{wr}\, G_m$. Note that $d(G_m) = 5r$ and $c(G_m) = 7(m-1)/8$ because $|G_m| = |G_9|^{(m-1)/8}$. Thus $c_S(5r) \le 7(m-1)/8 < 7m/8$, and

$$5\log_9 c_S(5r) - 2/3 < 5\log_9(7m/8) - 2/3 < 5r$$

because $\log_9 m = r$ and $5\log_9(7/8) - 2/3 \approx -0.97$. Similar calculations are summarized below. In each case $c(G_n)$ equals $(k_n m - 7)/8$ for some $k_n \in \mathbb{Z}$, and we abbreviate $5\log_9(k_n/8) - 2/3$ by $x_n$.

| $G_n$ | $d(G_n)$ | $\lvert G_n\rvert$ | $c(G_n)$ | $x_n$ | $\log_9 c(G_n) - \frac{2}{3}$ |
|---|---|---|---|---|---|
| $G_m$ | $5r$ | $\lvert G_9\rvert^{(m-1)/8}$ | $(7m-7)/8$ | $\approx -0.97$ | $< 5r$ |
| $G_{2m}$ | $5r+1$ | $\lvert S_2\rvert^m \lvert G_m\rvert$ | $(15m-7)/8$ | $\approx 0.8$ | $< 5r+1$ |
| $G_{3m}$ | $5r+2$ | $\lvert S_3\rvert^m \lvert G_m\rvert$ | $(23m-7)/8$ | $\approx 1.7$ | $< 5r+2$ |
| $G_{4m}$ | $5r+3$ | $\lvert S_4\rvert^m \lvert G_m\rvert$ | $(39m-7)/8$ | $\approx 2.9$ | $< 5r+3$ |
| $G_{8m}$ | $5r+4$ | $\lvert G_8\rvert^m \lvert G_m\rvert$ | $(47m-7)/8$ | $\approx 3.4$ | $< 5r+4$ |

In all five cases $5\log_9 c(G_n) - 2/3 < d(G_n)$ holds. Indeed, if $\lceil x\rceil$ denotes the least integer $\ge x$, then $\lceil 5\log_9 c(G_n) - 2/3\rceil$ equals $d(G_n)$. Since $c_S(d(G_n)) \le c(G_n)$, the lower bound in part (a) holds for $d \ge 1$.

(b) The proof of this part follows the same pattern as part (a). The upper bound is proved similarly except instead of using [10] we use the following result [11, Theorem 4b]: If $G/P \le \mathrm{GL}(r, \mathbb{F})$ and $|G/P|$ is odd, then

$$d(G/P) \le 2\log_7(r/5) + 3 = \gamma_0 \log_2(r/5) + 3$$

where $\gamma_0 = 2\log_7 2$. Thus $d \le (\gamma_0+1)\log_2 c_0 + \delta_0$ where

$$\delta_0 = \gamma_0\log_2(\gamma_0/5) - (\gamma_0+1)\log_2(\gamma_0+1) + 5 \approx 1.7 < 2.$$

This establishes the upper bound for (b).

The lower bound for (b) follows from the trivial observation that $c_S^o(d) \le c_N^o(d)$, and the fact that P. Hall [6, Satz III.17.7] constructed groups of order $p^{2^d-1}$ and solvable length $d$ for each odd prime $p$. For aesthetic reasons, we outline an argument that mirrors that in part (a) even though it gives a poorer bound $\gamma_0\log_2 c_0 + \frac{2}{3} < d$ than $\log_2 c_0 < d$. We define as in [11] transitive subgroups $H_n \le S_n$ of degree $m = 7^r$, and $3m$. Set $H_m = H_7 \,\mathrm{wr} \cdots \mathrm{wr}\, H_7$ where there are $r$ copies of $H_7$ and $H_7 = \langle (2,3,5)(4,6,7), (1,2,3,4,5,6,7)\rangle \le S_7$, and set $H_{3m} = A_3 \,\mathrm{wr}\, H_m$ where $A_3$ has order 3 and degree 3. The bound $\gamma_0\log_2 c_S^0(d) + \frac{2}{3} < d$ is a consequence of

$$d(H_m) = 2r, \quad d(H_{3m}) = 2r+1, \quad c(H_m) = \frac{m-1}{3}, \quad c(H_{3m}) = \frac{4m-1}{3}.$$

It follows from $\gamma\log_2 c - \frac{2}{3} < d$ and $\log_2 c_0 < d$ that $c < 9^{(d+1)/5}$ and $c_0 < 2^d$. Similarly, $d < (\gamma+1)\log_2 c + 3$ and $d < (\gamma_0+1)\log_2 c_0 + 2$ together with

$$\frac{1}{9} < \frac{\gamma}{5(\gamma+1)} \approx 0.122 \quad \text{and} \quad \frac{1}{5} < \frac{\gamma_0}{2(\gamma_0+1)} \approx 0.208,$$

imply $9^{(d-3)/9} < c$ and $7^{(d-2)/5} < c_0$. This completes the proof. □

## 3 Examples

The bounds for $c_N(d)$ given in the introduction can be improved. P. Hall [6, Satz III.7.10] proved $2^{d-1} + d - 1 \leq c_N(d)$ by proving that a $p$-group $G$ with $G^{(i+1)} \neq 1$ satisfies $c(G^{(i)}/G^{(i+1)}) \geq 2^i + 1$. No such bound exists for solvable groups. We give an example below of an infinite residually solvable group $G$ for which $c(G^{(2i-1)}/G^{(2i)}) = 1$ for $i \geq 1$.

Let $Q$ denote the quaternion group of order 8, and $E$ the extraspecial group of order 27 and exponent 3. Denote by $Q_n$ (resp. $E_n$) the central product of $n$ copies of $Q$ (resp. $E$) amalgamating the centers. Then $|Q_n| = 2^{2n+1}$ and $|E_n| = 3^{2n+1}$. When $n = 0$ the groups $Q_0$ and $E_0$ are cyclic of order 2 and 3 respectively, and when $n > 0$ both $Q_n$ and $E_n$ are extraspecial groups. In [5, Section 7] an iterated split extension

$$G = Q_0 \ltimes E_0 \ltimes Q_1 \ltimes E_1 \ltimes Q_3 \ltimes E_4 \ltimes \cdots \ltimes Q_{a_n} \ltimes E_{b_n} \ltimes \cdots .$$

is constructed where $a_0 = b_0 = 0$ and $a_n = 3^{b_{n-1}}$, $b_n = 2^{-1+a_n}$ for $n \geq 1$. The orders of the derived quotients $G^{(i)}/G^{(i+1)}$ of $G$ are:

$$2, 3, 2^2, 2, 3^2, 3, 2^6, 2, 3^8, 3, 2^{162}, 2, 3^{2 \cdot 3^{81}}, 3, \ldots .$$

In particular, $c(G^{(2i-1)}/G^{(2i)}) = 1$ for $i \geq 1$. If $d \leq 10$ and $d \neq 7$, then $G/G^{(d)}$ has minimal composition length amongst all solvable groups of solvable length $d$, see [4]. This is surprising given that $c(G/G^{(d)})$ grows faster than any exponential function of $d$. It is shown in [4] that $c_S(d)$ equals $1, 2, 4, 5, 7, 8, 13, 15$ when $d = 1, 2, 3, 4, 5, 6, 7, 8$.

The notation $\beta_p(d)$ is used in [2] to denote the minimal composition length of a $p$-group with solvable length $d$. Clearly, $c_N(d) = \min_p \beta_p(d)$ where $p$ ranges over the prime numbers. The following values of $\beta_p(d)$ were known at the time of Burnside:

$$\beta_p(1) = 1, \;\; \beta_p(2) = 3, \;\; \beta_p(3) = 6 \text{ for } p \geq 5, \text{ and } \beta_2(3) = \beta_3(3) = 7.$$

It is shown in [1] that $\beta_p(4) = 14$ for $p \geq 5$, and in [2] that $\beta_p(d) \leq 2^d - 2$ for $p \geq 5$. The best known bounds for $c_N(d)$ are presently

$$2^{d-1} + 3d - 10 \leq c_N(d) \leq 2^d - 2.$$

The upper bound holds when $d \geq 3$, and the lower bound [12] which holds for $p \geq 5$ improves the bound of Mann [9] when $d \geq 7$. Although $c_N(d) \leq c_N^o(d)$, both are $\mathrm{O}(2^d)$. If it were the case that $c_S(d)$ and $c_S^o(d)$ are both $\mathrm{O}(k^d)$ for some constant $k$, then Theorem 1 shows that $7^{1/5} < k < 9^{1/5}$.

Mann [8] investigates subgroups of $S_n$ that have maximal order. These subgroups, which are wreath (and direct) products of $S_2$, $S_3$ and $S_4$, do not improve the bound $c_S(d) = \mathrm{O}(9^{d/5})$ of Theorem 1. The iterated wreath product $H \operatorname{wr} \cdots \operatorname{wr} H$ of $d$ copies of $H = S_2, S_3, S_4$ gives the bounds $c_S(d) = \mathrm{O}(2^d)$, $c_S(d) = \mathrm{O}(3^{d/2})$ and $c_S(d) = \mathrm{O}(4^{d/3})$. These bounds are less sharp because

$$9^{1/5} < 4^{1/3} < 3^{1/2} < 2 \quad \text{as} \quad 1.55\cdots < 1.58\cdots < 1.73\cdots < 2.$$

In the proof of Theorem 1(a), certain groups $G_n$ were used to establish an upper bound for $c_S(d)$. It is natural to ask whether proper subgroups $K_n$ of $G_n$ can produce sharper upper bounds which narrow the gap in Theorem 1(a). The permutation representation $S_9 \to \mathrm{GL}(9, \mathbb{F}_2)$ fixes the 8-dimensional subspace $\{(x_1, \dots, x_9) \mid x_1 + \cdots + x_9 = 0\}$. This observation may be used to construct a subgroup $K_{18}$ of $G_{18} = S_2 \,\mathrm{wr}\, G_9$ of index 2 satisfying $d(G_{18}) = d(K_{18})$. We define subgroups $K_{2m} \le G_{2m}$ where $m = 9^r \ge 9$ via $K_{2(9m)} = K_{2m} \,\mathrm{wr}\, G_9$. Although $d(K_{2m}) = d(G_{2m})$ and $|G_{2m} : K_{2m}| = 2^{m/18}$ is large, $c(K_{2m}) = \mathrm{O}(m) = c(G_{2m})$ and we obtain the same bound $c_S(d) = \mathrm{O}(9^{d/5})$ as before.

It is unclear whether or not the gap $2^{d-1} + 3d - 10 \le c_N(d) \le 2^d - 2$ can be closed appreciably by examples. The group $U_n$ of $n \times n$ unipotent upper-triangular matrices over $\mathbb{F}_p$ has $c(U_n) = n(n-1)/2$ and $d(U_n) = \lfloor \log_2(n-1) \rfloor + 1$. Although $U_n$ is generated by $n-1$ elements, it has subgroups with 2 or 3 generators and maximal solvable length [3]. Estimating the order of these subgroups appears difficult.

The wreath product $S_2 \,\mathrm{wr} \cdots \mathrm{wr}\, S_2$, with $d$ copies of the symmetric group $S_2$, gave rise to the bound $c_N(d) \le 2^d - 1$. It is natural to ask whether this group contains proper subgroups with derived length $d$. The answer is negative by Lemma 2 below. I am grateful to Csaba Schneider for showing me a proof of this lemma.

**Lemma 2** *Let $G_d = S_2 \,\mathrm{wr} \cdots \mathrm{wr}\, S_2$ with $d$ copies of $S_2$. Then every proper subgroup $G_d$ has solvable length less than $d$.*

**Proof** We use induction on $d$. The result is true when $d = 1$, and when $d = 2$ because proper subgroups of $S_2 \,\mathrm{wr}\, S_2$ are abelian. Assume that $d > 2$. It suffices to prove that each maximal subgroup $M$ of $G_d$ satisfies $M^{(d-1)} = 1$. Write $G_d = (H_1 \times H_2) \rtimes S_2$ where $H_i \cong G_{d-1}$. The result is true if $M = H_1 \times H_2$. Suppose that $M \ne H_1 \times H_2$. Set $N_i = M \cap H_i$. Since $|G_d : M| = 2$, we see $|H_i : N_i|$ equals 1 or 2. The former is impossible as $M \lhd G_d$ and $M \ne H_1 \times H_2$. Thus $|H_1 : N_1| = |H_2 : N_2| = 2$ and $N = N_1 \times N_2 \lhd G_d$. Since $M/N$ is a proper subgroup of $G/N \cong S_2 \,\mathrm{wr}\, S_2$, and $N^{(d-2)} = 1$ by induction, it follows that $M^{(d-1)} = 1$. □

## References


[1] S. Evans-Riley, On the derived length of finite, graded Lie rings with prime-power order, and groups with prime-power order, Ph.D. Thesis, The University of Sydney (2000).

[2] S. Evans-Riley, M. F. Newman and Csaba Schneider, On the soluble length of groups with prime-power order, *Bull. Austral. Math. Soc.* **59** (1999), 343–346.

[3] S. P. Glasby, Subgroups of the upper-triangular matrix group with maximal derived length and a minimal number of generators, in *Groups St Andrews 1997 in Bath, I,* edited by C. M. Campbell et al., London Mathematical Society Lecture Notes Series **260**, 275–281, Cambridge Univ. Press, 1999.

[4] S. P. Glasby, Solvable groups with a given solvable length, and minimal composition length, *J. Group Theory* (to appear).

[5] S. P. Glasby and R. B. Howlett, Writing representations over proper subfields, *Comm. Algebra* **25** (1997), no. 6, 1703–1712.
[6] B. Huppert, *Endliche Gruppen I*, Springer-Verlag, 1967.
[7] B. Huppert and N. Blackburn, *Finite Groups II*, Springer-Verlag, 1982.
[8] A. Mann, Soluble subgroups of symmetric and linear groups, *Israel J. Math.* **55** (1986), 162–172.
[9] A. Mann, The derived length of $p$-groups, *J. Algebra* **224** (2000), 263–267.
[10] M. F. Newman, The soluble length of soluble linear groups, *Math. Z.* **126** (1972), 59–70.
[11] P. P. Pálfy, Bounds for linear groups of odd order, in *Proceedings of the Second International Group Theory Conference (Bressanone, 1989)*, Rend. Circ. Mat. Palermo (2) Suppl. No. 23 (1990), 253–263.
[12] Csaba Schneider, On the derived subgroup of a finite $p$-group, *Gazette Austral. Math. Soc.* **26** (1999), no. 5, 232–237.

# EMBEDDING IN FINITELY PRESENTED LATTICE-ORDERED GROUPS: EXPLICIT PRESENTATIONS FOR CONSTRUCTIONS

A. M. W. GLASS*, VINCENZO MARRA† and DANIELE MUNDICI§

*Department of Pure Mathematics and Mathematical Statistics, Centre for Mathematical Sciences, Wilberforce Rd., Cambridge CB3 0WB, England
Email: amwg@dpmms.cam.ac.uk

†D.I.Co., Università degli Studi di Milano, via Comelico 39/49, I-20135 Milano, Italy
Email: marra@dico.unimi.it

§Dipartimento di Matematica, "Ulisse Dini", Università di Firenze, Viale Morgagni, 67/A, I-50134 Firenze, Italy
Email: mundici@math.unifi.it

## Abstract

[At the conference, I stated the analogue for lattice-ordered groups of Graham Higman's famous embedding theorem [7], *viz*:

**Theorem ([4])** *A finitely generated lattice-ordered group can be embedded in a finitely presented lattice-ordered group if and only if it can be defined by a recursively enumerable set of relations.*

In my talk, I gave an outline of the ideas needed in the constructive proof. I was asked if, given a finitely generated lattice-ordered group defined by a recursively enumerable set of defining relations, one can explicitly write down a finitely presented lattice-ordered group in which it can be embedded. My answer was "Yes, theoretically but no in practice". (A. M. W. Glass)]

In this article we show that, for certain constructions, an explicit answer can be given. Specifically, *inter alia*,

**Theorem A** *If $G$ is a lattice-ordered group that can be embedded explicitly in a finitely presented lattice-ordered group, then so can $G \operatorname{wr} (\mathbb{Z}, \mathbb{Z})$.*

**Theorem B** *If $G$ is a lattice-ordered group that can be embedded explicitly in a finitely presented lattice-ordered group and $H$ is any o-group that can be embedded explicitly in an o-group that is finitely presented as a lattice-ordered group and explicitly has a minimal strictly positive Archimedean class, then $H \overrightarrow{\otimes} G$ can be embedded explicitly in a finitely presented lattice-ordered group.*

*AMS Classification:* 06F15, 06F20, 20B27, 20F60.
*Keywords:* lattice-ordered group, presentation, lexicographic product, wreath product, recursive function.

## 1 Introduction

A finitely presented lattice-ordered group with insoluble group word problem was constructed in [5]. In [6] this construction was coupled with direct limits of simplicial groups to prove that every Abelian lattice-ordered group of finite rank that is defined by a recursively enumerable set of relations can be embedded in a finitely presented lattice-ordered group. The full analogue of Higman's Theorem was obtained in [4]. The purpose of this note is to give explicit embeddings under standard algebraic constructions when the constituent pieces have explicit embeddings into finitely presented lattice-ordered groups.

## 2 Background and notation

Throughout we will use $\mathbb{N}$ for the set of non-negative integers, $\mathbb{Z}_+$ for the set of positive integers, and $\mathbb{R}$ for the set of real numbers. The only order on $\mathbb{R}$ that we will consider will be the usual one.

As is standard, in any group $G$ we write $f^g$ for $g^{-1}fg$, and $[f,g]$ for $f^{-1}g^{-1}fg = f^{-1}f^g$.

A *lattice-ordered group* is a group which is also a lattice that satisfies the identities $x(y \wedge z)t = xyt \wedge xzt$ and $x(y \vee z)t = xyt \vee xzt$. Throughout we write $x \leq y$ as a shorthand for $x \vee y = y$ or $x \wedge y = x$, *ℓ-group* as a shorthand for lattice-ordered group, and *o-group* for a totally ordered group (i.e., if the ℓ-group is totally ordered). A sublattice subgroup of an ℓ-group is called an *ℓ-subgroup.*

Lattice-ordered groups are torsion-free and $f \vee g = (f^{-1} \wedge g^{-1})^{-1}$; moreover each element of $G$ can be written in the form $fg^{-1}$ where $f, g \in G^+ = \{h \in G : h \geq 1\}$ — see, e.g., [3], Corollary 2.1.3, Lemma 2.3.2 and Lemma 2.1.8. For each $g \in G$, let $|g| = g \vee g^{-1}$. Then $|g| \in G_+$ iff $g \neq 1$, where $G_+ = G^+ \backslash \{1\}$. Therefore, $(w_1 = 1 \;\&\; \ldots \;\&\; w_n = 1)$ iff $|w_1| \vee \cdots \vee |w_n| = 1$ [*ibid.*, Lemma 2.3.8 and Corollary 2.3.9]. Consequently, in the language of lattice-ordered groups (and in sharp contrast to group theory) any finite number of equalities can be replaced by a single equality.

We will write $f \perp g$ as a shorthand for $|f| \wedge |g| = 1$.

A *homomorphism* from one ℓ-group to another is a group and a lattice homomorphism. Kernels are precisely the normal ℓ-subgroups that are convex (if $k_1, k_2$ belong to the kernel and $k_1 \leq g \leq k_2$, then $g$ belongs to the kernel). They are called *ℓ-ideals.*

Free ℓ-groups on finite sets of generators exist by universal algebra. Finitely generated ℓ-groups are the homomorphic images of free ℓ-groups on that finite number of generators. As is standard, if the kernel is finitely generated as an ℓ-ideal, then the homomorphic image is said to be *finitely presented*; if the kernel is generated by a recursively enumerable set of elements (as an ℓ-ideal), then we say that the finitely generated homomorphic image has a *recursively enumerable set of defining relations* or is *recursively presented* (*sic*).

We will write

$$\langle Y : w_i(Y) = 1 \ (i \in I)\rangle$$

for the quotient $F_Y/K$ where $F_Y$ is the free $\ell$-group on the generating set $Y$ and $K$ is the $\ell$-ideal generated by $\{w_i(Y) : i \in I\}$.

The free $\ell$-group on a single generator is $\mathbb{Z} \oplus \mathbb{Z}$ ordered by: $(m_1, m_2) \geq (0,0)$ iff $m_1, m_2 \geq 0$; $(1,-1)$ is a generator since $(1,-1) \vee (0,0) = (1,0)$.

If $G_1, G_2$ are $\ell$-groups, then their *cardinal product* $G_1 \oplus G_2$ is their group product lattice ordered by: $(g_1, g_2) \geq (1,1)$ iff $g_j \in G_j^+$ $(j = 1,2)$.

There is another way to partially order the direct product of partially ordered groups $G_1$ and $G_2$, namely, $G_1 \overrightarrow{\otimes} G_2$:

$$(g_1, g_2) \geq (1,1) \quad \text{iff} \quad (g_1 \in (G_1)_+ \text{ or both } g_1 = 1 \text{ and } g_2 \in G_2^+).$$

This is an $\ell$-group if $G_1$ is an o-group and $G_2$ is an $\ell$-group.

If $G_1$, $G_2$ are partially ordered groups with $G_1 \subseteq \mathrm{Aut}(G_2, \cdot, \leq)$, then we can partially order the splitting extension by: $(g_2, g_1) \geq (1,1)$ iff $(g_1 \in (G_1)_+$ or both $g_1 = 1$ and $g_2 \in G_2^+)$. We write the result as $G_2 \overleftarrow{\rtimes} G_1$; it is an $\ell$-group whenever $G_1$ an o-group and $G_2$ an $\ell$-group.

The amalgamation property fails miserably for $\ell$-groups: there are $\ell$-groups $G, H_1, H_2$ with $\ell$-embeddings $\sigma_j : G \to H_j$ $(j = 1,2)$ such that there is no $\ell$-group $L$ such that $H_j$ can be $\ell$-embedded in $L$ $(j = 1,2)$ so that the resulting diagram commutes (see [9] or [3], Theorem 7.C). So HNN-extension tricks cannot be used in general (see [1]). Instead we use permutation group techniques.

Let $(\Omega, \leq)$ be a totally ordered set. Then $A(\Omega) := \mathrm{Aut}(\Omega, \leq)$ is an $\ell$-group when the group operation is composition and the lattice operations are just the pointwise supremum and infimum ($\alpha(f \vee g) = \max\{\alpha f, \alpha g\}$, *etc.*). There is an analogue of Cayley's Theorem for groups, namely the Cayley–Holland Theorem ([3], Theorem 7.A):

**Theorem (Holland [8])** *Every lattice-ordered group can be embedded in $A(\Omega)$ for some totally ordered set $(\Omega, \leq)$; every countable lattice-ordered group can be be embedded in $A(\mathbb{R})$.*

If $h \in A(\Omega)$, then the *support* of $h$, $\mathrm{supp}(h)$, is the set $\{\beta \in \Omega : \beta h \neq \beta\}$.

Since each real interval $(\alpha, \beta)$ is order-isomorphic to $(\mathbb{R}, \leq)$ we obtain:

**Corollary 2.1** *Let $\alpha, \beta \in \mathbb{R}$ with $\alpha < \beta$. Then every finitely generated $\ell$-group $G$ can be embedded in $A(\mathbb{R})$ so that* $\mathrm{supp}(g) \subseteq (\alpha, \beta)$ *for all $g \in G$.*

If $h \in A(\Omega)$ and $\alpha \in \mathrm{supp}(h)$, then the convexification of the $h$-orbit of $\alpha$ is called the *interval of support of $h$ containing* $\alpha$; i.e., the supporting interval of $h$ containing $\alpha$ is $\{\beta \in \Omega \colon (\exists m, n \in \mathbb{Z})(\alpha h^n \leq \beta \leq \alpha h^m)\}$. So the support of an element is the disjoint union of its supporting intervals. Supporting intervals are also called *bumps*.

By considering intervals of support, it is easy to establish the well-known fact

**Proposition 2.2** *For all $f, g \in A(\Omega)$, $\operatorname{supp}(f^g) = \operatorname{supp}(f)g$. Hence if $f^g \perp f$ and $g \geq 1$, then $|f|^m \leq g$ for all $m \in \mathbb{N}$.*

## 3 Proof of Theorem A

**Theorem A** *If $G$ is a lattice-ordered group that can be embedded explicitly in a finitely presented lattice-ordered group, then so can $G \operatorname{wr} (\mathbb{Z}, \mathbb{Z})$.*

**Proof** (c.f., [2]) It is enough to prove the theorem when $G$ is a finitely presented $\ell$-group, since $G \operatorname{wr} (\mathbb{Z}, \mathbb{Z})$ can be embedded in $G_0 \operatorname{wr} (\mathbb{Z}, \mathbb{Z})$ whenever $G$ can be embedded in $G_0$. So let

$$G = \langle g_1, \ldots, g_n : w(g_1, \ldots, g_n) = 1 \rangle$$

be a finitely presented $\ell$-group. Adjoin $g_0 = \bigvee_{i=1}^{n} |g_i|$ as a generator. Thus $g_0$ is a strong order unit in $G$. Let

$$H = \langle a, g_0, \ldots, g_n, h_0 : g_0 = \bigvee_{i=1}^{n} |g_i|, \;\; w(g_1, \ldots, g_n) = 1, \;\; h_0 \leq a, \\ g_0 \perp h_0, \;\; h_0 g_0^{-a} \perp g_0^a, \;\; h_0 h_0^{-a} \perp h_0^a \,\rangle.$$

Then $H$ is a finitely presented $\ell$-group.

Since $h_0 h_0^{-a} \geq 1$, an easy induction shows that $h_0 \geq h_0^{a^m}$ for all $m \in \mathbb{N}$. Since $h_0 g_0^{-a} \geq 1$, we have $h_0^{a^m} \geq g_0^{a^{m+1}}$ for all $m \in \mathbb{N}$. Hence $h_0 \geq g_0^{a^m}$ for all $m \in \mathbb{Z}_+$. Since $g_0 \perp h_0$, we get $g_0 \perp g_0^{a^m}$ for all $m \in \mathbb{Z}_+$. Thus $g_0^{a^m} \perp g_0^{a^n}$ for all distinct $m, n \in \mathbb{Z}$. By Proposition 2.2, the $\ell$-subgroup of $H$ generated by $G \cup \{a\}$ is a homomorphic image of $G \operatorname{wr} (\mathbb{Z}, \mathbb{Z})$

By Corollary 2.1, we may consider $G$ as an $\ell$-subgroup of $A(\mathbb{R})$ with each element of $G$ having support contained in $(0, 1)$. Let $\bar{g}_0 \in A(\mathbb{R})$ be $g_0$ on $(0, 1)$ and the identity off $(0, 1)$. Let $\bar{a}_0 \in A(\mathbb{R})$ be translation by 1 $(\alpha \bar{a}_0 = \alpha + 1)$. Define $\bar{h}_0 \in A(\mathbb{R})$ to have support contained in $\bigcup_{m \in \mathbb{Z}_+} (m, m+1)$ and $\bar{h}_0$ restricted to $(m, m+1)$ be $\bar{g}_0^{\bar{a}^m}$ $(m \in \mathbb{Z})$. Then all the relations of $H$ hold with $h_0, a_0$ replaced by $\bar{h}_0, \bar{a}_0$, respectively. Hence the $\ell$-subgroup $\bar{H}$ of $A(\mathbb{R})$ generated by $G \cup \{\bar{h}_0, \bar{a}_0\}$ is a homomorphic image of $H$. Since $G$ is embedded in $\bar{H}$, it is embedded in $H$. Consequently, the $\ell$-subgroup of $H$ generated by $g_0, g_1, \ldots, g_n, a$ is $G \operatorname{wr} (\mathbb{Z}, \mathbb{Z})$, and the $\ell$-subgroup generated by the conjugates of $g_0, g_1, \ldots, g_n$ by powers of $a$ is isomorphic to $\sum_{n \in \mathbb{Z}} G$. □

## 4 Proof of Theorem B

**Theorem B** *If $G$ is a lattice-ordered group that can be embedded explicitly in a finitely presented lattice-ordered group and $H$ is any o-group that can be embedded explicitly in an o-group that is finitely presented (explicitly) as a lattice-ordered group and explicitly has a minimal strictly positive Archimedean class, then $H \overrightarrow{\otimes} G$ can be embedded explicitly in a finitely presented lattice-ordered group.*

**Proof** If $H$ can be embedded in $H_0$ and $G$ in $G_0$, then $H \overrightarrow{\otimes} G$ can be embedded in $H_0 \overrightarrow{\otimes} G_0$. So again it suffices to assume that $H$ and $G$ are themselves finitely presented, $H$ is an o-group and that $H_+$ has a minimal Archimedean class.

Let

$$G = \langle g_0, \dots, g_m : w_0(g_0, \dots, g_m) = 1\rangle$$

and

$$H = \langle h_1, \dots, h_n : w_1(h_1, \dots, h_n) = 1\rangle,$$

where $g_0 = \bigvee_{i=1}^m |g_i|$.

Let $w_2(h_1, \dots, h_n) \in H_+$ have minimal Archimedean class in $H_+$. Let

$$\begin{aligned} L = \langle a, g_0, \dots, g_m, h_1, \dots, h_n :\ g_0 = \bigvee_{i=1}^{m} |g_i|,\ \ w_0(g_0, \dots, g_m) = 1, \\ w_1(h_1, \dots, h_n) = 1, [g_i, h_j] = 1,\ \ w_2(h_1, \dots, h_n) \geq g_0, \\ h_j^a = h_j,\ \ g_0^a = g_0^2\ \ (i = 0, \dots, m;\ j = 1, \dots, n)\rangle. \end{aligned}$$

Then $L$ is a finitely presented $\ell$-group and for all $k \in \mathbb{Z}_+$,

$$g_0^k \leq g_0^{2^k} = g_0^{a^k} \leq w_2(h_1, \dots, h_n)^{a^k} = w_2(h_1^{a^k}, \dots, h_n^{a^k}) = w_2(h_1, \dots, h_n);$$

so $g_0 \ll H_+$. Thus we have an explicit homomorphism $\psi$ from $H \overrightarrow{\otimes} G$ into $L$ (the natural identification). We must show that this homomorphism is injective.

Let $(W, \mathbb{R} \overleftarrow{\times} H) = (A(\mathbb{R}), \mathbb{R}) \operatorname{Wr} (H, H)$ and regard $G$ as an $\ell$-subgroup of $A(\mathbb{R})$. We embed $G$ in $W$ diagonally: $g \mapsto \hat{g}$ where $(r, \alpha)\hat{g} = (rg, \alpha)$ and embed $H$ in $W$ via: $h \mapsto \bar{h}$ where $(r, \alpha)\bar{h} = (r, \alpha h)$ $(r \in \mathbb{R};\ \alpha \in H)$. Note that $[\bar{h}, \hat{g}] = 1$ for all $g \in G,\ h \in H$; so $H \overrightarrow{\otimes} G$ is embedded in $W$ via: $hg \mapsto \bar{h}\hat{g}$.

Let $a_0 \in A(\mathbb{R})$ be such that $g_0^{a_0} = g_0^2$ (see [3], Lemma 8.3.3) and extend $a_0$ to $\hat{a} \in W$ diagonally: $(r, \alpha)\hat{a} = (ra_0, \alpha)$ $(r \in \mathbb{R};\ \alpha \in H)$. Then $[\bar{h}, \hat{a}] = 1$ for all $h \in H$ and $\hat{g}_0^{\hat{a}} = \hat{g}_0^2$. Hence $W$ contains the homomorphic image $L\phi$ of $L$ (the $\ell$-subgroup generated by $\hat{a}, \hat{g}_0, \dots, \hat{g}_m, \bar{h}_1, \dots, \bar{h}_n$) and $W$ contains $H \overrightarrow{\otimes} G$ (to within isomorphism $\theta$). Moreover, the diagram is commutative; i.e., $\theta = \psi\phi$. Therefore the homomorphism $\psi$ of $H \overrightarrow{\otimes} G$ into $L$ is injective. □

Every finitely generated *Abelian* o-group has a minimal strictly positive Archimedean class. We do not know if every finitely presented $\ell$-group that is a non-Abelian o-group has a minimal strictly positive Archimedean class. If so, then this extra hypothesis in the theorem would be unnecessary. The difficulty is the paucity of examples.

## 5 Theorem C

In Theorem B, the elements from $H$ acted as identity automorphisms on $G$. The purpose of the next result is to relax this restriction somewhat to $H$ being a subgroup of the automorphism group of $G$ (with a total order on $H$). We consider non-commutative extensions instead of the lexicographic sums. For this we must restrict to actual finitely presented $\ell$-groups as the embedding may not be feasible within $\mathrm{Aut}(G, \cdot, \vee)$.

**Theorem C** *If $G$ is a finitely presented lattice-ordered group, let $H$ be a subgroup of the group of automorphisms of $G$. If $H$ is an o-group that is finitely presented explicitly as a lattice-ordered group and explicitly has a minimal strictly positive Archimedean class, then $G \,\overline{\rtimes}\, H$ can be embedded explicitly in a finitely presented lattice-ordered group.*

**Proof** The proof is very similar to that of Theorem B.

Let

$$G = \langle g_0, \ldots, g_m : w_0(g_0, \ldots, g_m) = 1\rangle$$

and

$$H = \langle h_1, \ldots, h_n : w_1(h_1, \ldots, h_n) = 1\rangle,$$

where $g_0 = \bigvee_{i=1}^m |g_i|$.

For each $g_i, h_j$, we have $g_i^{h_j}, g_i^{h_j^{-1}} \in G$; say, $g_i^{h_j} = w_{i,j}(g_0, \ldots, g_m)$ and $g_i^{h_j^{-1}} = u_{i,j}(g_0, \ldots, g_m)$ $(i = 0, \ldots, m;\ j = 1, \ldots, n)$.

Let $w_2(h_1, \ldots, h_n) \in H_+$ have minimal Archimedean class in $H_+$. Let

$$\begin{aligned} L = \langle a, g_0, \ldots, g_m, h_1, \ldots, h_n : {} & w_0(g_0, \ldots, g_m) = 1,\ w_1(h_1, \ldots, h_n) = 1, \\ & g_0 = \bigvee_{i=1}^m |g_i|,\ g_i^{h_j} = w_{i,j}(g_0, \ldots, g_m),\ g_i^{h_j^{-1}} = u_{i,j}(g_0, \ldots, g_m), \\ & w_2(h_1, \ldots, h_n) \geq g_0,\ h_j^a = h_j,\ g_0^a = g_0^2\ (i = 0, \ldots, m;\ j = 1, \ldots, n)\rangle. \end{aligned}$$

Then $L$ is a finitely presented $\ell$-group and for all $k \in \mathbb{Z}_+$,

$$g_0^k \leq g_0^{2^k} = g_0^{a^k} \leq w_2(h_1, \ldots, h_n)^{a^k} = w_2(h_1^{a^k}, \ldots, h_n^{a^k}) = w_2(h_1, \ldots, h_n);$$

so $g_0 \ll H_+$. Also, $H$ acts on $G$ appropriately in $L$. Thus the natural map defines an explicit homomorphism $\psi$ from $G \,\overline{\rtimes}\, H$ into $L$. We must show that this homomorphism is injective.

Let $(W, \mathbb{R} \overleftarrow{\times} H) = (A(\mathbb{R}), \mathbb{R}) \operatorname{Wr} (H, H)$ where we regard $G$ as an $\ell$-subgroup of $A(\mathbb{R})$. We embed $H$ into $W$ as before: $h \mapsto \bar{h}$ where $(r, \alpha)\bar{h} = (r, \alpha h)$ $(r \in \mathbb{R};\ \alpha \in H)$. We embed $G$ into $W$ diagonally modulo $H$: $g \mapsto \hat{g}$ where $(r, \alpha)\hat{g} = (rg^{\alpha^{-1}}, \alpha)$ $(r \in \mathbb{R};\ \alpha \in H)$, $g^\beta$ being the image of $g$ under $\beta \in H \subseteq \mathrm{Aut}(G, \cdot, \vee)$ (see [3], proof of Theorem 7.G). These provide an embedding of $G \,\overline{\rtimes}\, H$ into $W$, for if $\alpha, f \in H$, $g \in G$ and $r \in \mathbb{R}$, then

$$(r, \alpha)\bar{f}^{-1}\hat{g}\bar{f} = (r, \alpha f^{-1})\hat{g}\bar{f} = (rg^{f\alpha^{-1}}, \alpha f^{-1})\bar{f} = (rg^{f\alpha^{-1}}, \alpha) = (r, \alpha)\widehat{g^f};$$

so $G \,\overline{\rtimes}\, H$ is embedded in $W$ via: $hg \mapsto \bar{h}\hat{g}$.

Let $a_0 \in A(\mathbb{R})$ be such that $g_0^{a_0} = g_0^2$ and extend $a_0$ to $\hat{a} \in W$ diagonally: $(r, \alpha)\hat{a} = (ra_0, \alpha)$ $(r \in \mathbb{R};\ \alpha \in H)$. Then $[\bar{h}, \hat{a}] = 1$ for all $h \in H$ and so

$$(r, \alpha)\hat{g}_0^{\hat{a}} = (ra_0^{-1}g_0^{\alpha^{-1}}a_0, \alpha) = (r(a_0^{-1}g_0a_0)^{\alpha^{-1}}, \alpha) = (rg_0^{2\alpha^{-1}}, \alpha) = (r, \alpha)\hat{g}_0^2,$$

$r \in \mathbb{R}$, $\alpha \in H$. Thus $\hat{a}^{-1}\hat{g}_0\hat{a} = \hat{g}_0^2$, and so all the relations of $L$ hold in $W$. Hence $W$ contains a homomorphic image $L\phi$ of $L$ (the $\ell$-subgroup generated by $\hat{a}, \hat{g}_0, \ldots, \hat{g}_m, \bar{h}_1, \ldots, \bar{h}_n$) and $W$ contains $G \,\bar{\wr}\, H$ (to within isomorphism $\theta$). Moreover, the diagram is commutative; i.e., $\theta = \psi\phi$. Therefore the homomorphism $\psi$ of $G \,\bar{\wr}\, H$ into $L$ is injective. □

**Acknowledgements:** This research was begun when the first author was visiting the Universities of Milano and Firenze in December 2002, and completed when he visited the latter in January and February 2004 (when the discussion and ideas begun in the hills of Tuscany eventually coalesced). It was made possible by funds from those universities in 2002, and from the Istituto Nazionale di Alta Matematica Francesco Severi, Italy, and Queens' College, Cambridge in 2004. We are extremely grateful to them for making these visits possible and providing us with the opportunity to develop this material together.

## References

[1] A. M. W. Glass, Results in partially ordered groups, *Comm. Algebra* **3** (1975), 749–761.

[2] A. M. W. Glass, Generating varieties of lattice-ordered groups: approximating Wreath products, *Illinois J. Math.* **30** (1986), 214–221.

[3] A. M. W. Glass, *Partially Ordered Groups*, Series in Algebra **7**, World Scientific Pub. Co., Singapore, 1999.

[4] A. M. W. Glass, Sublattice subgroups of finitely presented lattice-ordered groups (submitted).

[5] A. M. W. Glass and Y. Gurevich, The word problem for lattice-ordered groups, *Trans. Amer. Math. Soc.* **280** (1983), 127–138.

[6] A. M. W. Glass and V. Marra, Embedding finitely generated Abelian lattice-ordered groups: Higman's Theorem and a realisation of $\pi$, *J. London Math. Soc.* **68** (2003), 545–562.

[7] G. Higman, Subgroups of finitely presented groups, *Proc. Royal Soc. London Series A* **262** (1961), 455–475.

[8] W. C. Holland, The lattice-ordered group of automorphisms of an ordered set, *Michigan Math. J.* **10** (1963), 399–408.

[9] K. R. Pierce, *Amalgamations of lattice-ordered groups*, *Trans. Amer. Math. Soc.* **172** (1972), 249–260.

# A NOTE ON ABELIAN SUBGROUPS OF $p$-GROUPS

GEORGE GLAUBERMAN

Department of Mathematics, University of Chicago, 5734 University Avenue, Chicago, IL 60637-1514, USA
Email: gg@math.uchicago.edu

## 1 Introduction

Suppose $p$ is a prime and $A$ is an abelian subgroup of a finite $p$-group $S$. We ask:

*Does $S$ possess a normal abelian subgroup of the same order as $A$?*

It is easy to obtain an affirmative answer in some small cases, e.g., when $A$ has order at most $p^2$. However, a family of examples of J. Alperin ([8], p. 349; [4], pp. 324–325) shows that the answer is negative in general. Further work by Alperin and the author [2], [6] gives an affirmative answer when $A$ has a sufficiently small order (depending on $p$), namely, $n < (p+7)/4$, and proves an analogue for elementary abelian subgroups.

Alperin conjectured ([1], pp. 19–20) that there might be a "dual" result for $A$ of sufficiently small index in $S$. Recently the author has proved this [6]:

**Theorem 1** *Suppose $S$ is a finite $p$-group, $A$ is an abelian subgroup of $S$, $p^n = |S : A|$, and*

$$p > (n^2/2) + 2n - 3.$$

*Then $S$ possesses a normal abelian subgroup $A^*$ of the same order as $A$.*

The method of proof yields a surprising additional result:

**Theorem 2** *Suppose $S$ is a finite $p$-group of nilpotence class at most $p$. Then there exists a finite $p$-group $T$ and a bijection $\phi$ of $S/Z(S)$ onto $T/Z(T)$ such that*

(a) *$T$ has nilpotence class at most 2, and*

(b) *$\phi$ induces a bijection from the set of all abelian subgroups $A$ of $S$ that contain $Z(S)$ to the set of all abelian subgroups $B$ of $T$ that contain $Z(T)$, given by*

$$\phi(A/Z(S)) = B/Z(T).$$

In this paper, we discuss the ideas of the proofs and some open questions.

## 2 Ideas in proofs

We illustrate the main ideas in the proofs. First, we summarize the proof of Theorem 1.

Let $\widehat{A}$ be the normal closure of $A$ in $S$, i.e., the (normal) subgroup of $S$ generated by all the conjugates $A^x$ as $x$ ranges over $S$. By simple reductions, we may assume that

$$p \text{ is odd}, \quad n \geq 2, \quad \text{and} \quad Z(\widehat{A}) \leq A < \widehat{A} < S. \tag{1}$$

By calculations with joins and intersections of conjugates of $A$, we obtain

**Step 1:** $|\widehat{A}/Z(\widehat{A})| \leq p^{(p+1)/2}$ and the nilpotence class of $\widehat{A}$ is at most $p-1$.

For the next step, it would be convenient to have $Z(\widehat{A}) \leq \widehat{A}' = [\widehat{A}, \widehat{A}]$, but this need not be true. Instead, let $Z = Z(\widehat{A})$. Since $\widehat{A}$ has nilpotence class at most $p-1$, a beautiful theorem of M. Lazard ([9], pp. 121–124) says that there exist operations $+$ and $[\,,\,]$ on $\widehat{A}$ under which $\widehat{A}$ becomes a nilpotent Lie ring. Thus, we may regard $\widehat{A}/Z$ and $\widehat{A}'$ as additive groups and use the bracket multiplication $[\,,\,]$ on $\widehat{A}$ to induce a bi-additive mapping

$$\widehat{A}/Z \times \widehat{A}/Z \longrightarrow \widehat{A}'.$$

We use this mapping to construct a nilpotent Lie ring $L$ of class two: its additive group is the external direct sum $\widehat{A}/Z \oplus \widehat{A}'$ and its bracket multiplication is given by

$$[uZ \oplus v, u'Z \oplus v'] = 0 \oplus [u, u'].$$

Then

$$Z(L) = 0 \oplus \widehat{A}' = [L, L], \tag{2}$$

which turns out to be an adequate substitute for the desired condition on $\widehat{A}$.

For each $x$ in $S$, let $\alpha(x)$ be the automorphism of $L$ induced by conjugation of $\widehat{A}$ under $x$. Thus, for $x$ in $S$ and $\alpha = \alpha(x)$, we may regard $\alpha$ and $\alpha - 1$ as endomorphisms of the additive group of $L$. By using (2) and Step 1, we obtain

**Step 2:** $L(\alpha - 1)^p = 0$ whenever $x \in S$ and $\alpha = \alpha(x)$.

Now let $V$ be the set of Lie subrings of $L$ of the form

$$B/Z \oplus \widehat{A}$$

where $B$ ranges over all the abelian subgroups of $\widehat{A}$ that contain $Z$ and have the same order as $A$. Because of (2), some analogous conditions, and Step 2, the action of $S$ on $V$ is somewhat similar to the action of a nilpotent algebraic group on a projective variety. By an analogue of Borel's Fixed Point Theorem ([6], Theorem 2.4), $S$ fixes some member $A^*/Z \oplus \widehat{A}$ of $V$. Then $A^*$ satisfies the conclusion of Theorem 1.

We prove Theorem 2 by using the method of construction of the Lie ring $L$ and an extension of the proof of Lazard's Theorem.

## 3 Open questions

Yu. Berkovich [3] gave counterexamples to the conclusion of Theorem 1 in which $p \geq 5$, $|S| = p^{2p+1}$, $|A| = p^{(3p-1)/2}$, and $|S : A| = p^{(p+3)/2}$. In view of the linear bound of $n < (p+7)/4$ in the earlier work of Alperin and the author, we ask:

**Question 1** Can we reduce the bound of

$$p > (n^2/2) + 2n - 3,$$

in Theorem 1, perhaps to a linear bound?

Since the earlier results on abelian subgroups of small order have been extended [6] to subgroups of small class that may not be abelian, we ask:

**Question 2** Is there an analogue of Theorem 1 for $A$ of small nilpotence class relative to $p$?

After the conference, Dr. R. W. van der Waall kindly informed me of some related research. The results on stem groups in [7] might lead to a shorter proof of Step 2 above, and the results of [10] might help with Question 2. We thank Dr. van der Waall warmly for this information. We also thank the organizers for a very enjoyable conference, based solely on mathematical considerations, and the National Security Agency (USA) for its support through a grant.

## References

[1] J. L. Alperin, Large abelian subgroups of $p$-groups, *Trans. Amer. Math. Soc.* **117** (1965), 10–20.

[2] J. L. Alperin and G. Glauberman, Limits of abelian subgroups in finite $p$-groups, *J. Algebra* **203** (1998), 533–566.

[3] Yu. Berkovich, A certain nonregular $p$-group, *Siberian Math. J.* **12** (1971), 654–657.

[4] G. Glauberman, Large abelian subgroups of finite $p$-groups, *J. Algebra* **196** (1997), 301–338.

[5] G. Glauberman, Large subgroups of small class in finite $p$-groups, *J. Algebra* **272** (2004), 128–153.

[6] G. Glauberman, Abelian subgroups of small index in finite $p$-groups, *J. Group Theory* **8** (2005), 539–560.

[7] P. Hall, The classification of prime-power groups, *J. Reine Angew. Math.* **182** (1940), 130–141.

[8] B. Huppert, *Endliche Gruppen I*, Springer-Verlag, Berlin, 1967.

[9] E. I. Khukhro, *p-Automorphisms of finite p-groups*, London Math. Soc. Lecture Note Ser. **246**, Cambridge University Press, Cambridge, 1998.

[10] R. W. van der Waall, On $n$-isoclinic embeddings of $n$-isoclinic groups, *Indag. Math. (N.S.)* **15** (2004), no. 4, 595–600.

# ON KERNEL FLATNESS

AKBAR GOLCHIN

Department of Mathematics,University of Sistan and Baluchestan, Zahedan, Iran
E-mail: agdm@hamoon.usb.ac.ir

## Abstract

In this paper we will try to show that principal weak kernel flatness, weak kernel flatness and translation kernel flatness properties can be transferred from acts over monoids to their coproduct and vice versa.

## 1 Introduction

In [5], Valdis Laan introduced a mapping called $\phi$ corresponding to the pullback diagram $P(M, N, f, g, Q)$ and in accordance with surjectivity, injectivity or both of $\phi$ for some particular pullback diagrams, he gave equivalents of most flatness properties. In [2] three additional flatness properties called principal weak kernel flatness, weak kernel flatness and translation kernel flatness were introduced. Now if $A = \dot{\bigcup}_{i \in I} A_i$ is the coproduct of right $S$-acts $A_i$, $i \in I$, by proving that surjectivity and injectivity of the mapping $\phi$ corresponding to the pullback diagram $P(M, N, f, g, Q)$ for $A$ implies the surjectivity and injectivity of the mapping $\phi_i$ corresponding to the pullback diagram $P(M, N, f, g, Q)$ for $A_i$ and vice versa, we show that principal weak kernel flatness, weak kernel flatness and translation kernel flatness of $A$ implies these properties of $A_i$, $i \in I$, and vice versa.

Throughout this paper $S$ will denote a monoid. We refer the reader to [4] for basic definitions and terminology relating to semigroups and acts over monoids.

## 2 Preliminaries

Fix a right $S$-act $A_S$. Let us show that this gives rise to a functor $A_S \otimes_S -$ from the category of all left $S$-act to the category of sets. For objects ${}_SM$ of the category of left $S$-acts let $A_S \otimes_S -$ be defined by

$$ {}_SM \mapsto A_S \otimes {}_SM $$

and for morphisms $f : {}_SM \to {}_SN$ in the category of left $S$-acts (that is for homomorphisms of left $S$-acts) by

$$ f \mapsto id_A \otimes f $$

where

$$ id_A \otimes f : A_S \otimes {}_SM \to A_S \otimes {}_SN $$

is defined by

$$ (id_A \otimes f)(a \otimes m) = a \otimes f(m) $$

for all $a \in A_S,\ m \in {}_SM$. Then

$$(A_S \otimes_S -)(id_{{}_SM}) = id_{A_S} \otimes id_{{}_SM} = id_{A_S \otimes_S M} = id_{(A_S \otimes_S -)({}_SM)}$$

because

$$(id_{A_S} \otimes id_{{}_SM})(a \otimes m) = a \otimes id_{{}_SM}(m) = (a \otimes m) = id_{A_S \otimes_S M}(a \otimes m)$$

for all $a \in A_S,\ m \in {}_SM$. Take two homomorphisms $f : {}_SM \to {}_SN,\ g : {}_SN \to {}_SQ$ of left $S$-acts. Then

$$\begin{aligned}(id_{A_S} \otimes gf)(a \otimes m) &= a \otimes (gf)(m) = a \otimes g(f(m)) = (id_{A_S} \otimes g)(a \otimes f(m)) \\ &= (id_{A_S} \otimes g)((id_A \otimes f)(a \otimes m)) = ((id_A \otimes g)(id_A \otimes f))(a \otimes m)\end{aligned}$$

for all $a \in A_S,\ m \in {}_SM$. This means that

$$id_{A_S} \otimes gf = (id_{A_S} \otimes g)(id_{A_S} \otimes f),$$

or

$$(A_S \otimes_S -)(gf) = (A_S \otimes_S -)(g)(A_S \otimes_S -)(f).$$

Hence $A_S \otimes_S -$ is indeed a covariant functor. This functor is called the *functor of tensoring* by $A_S$.

**Lemma 2.1 ([3])** *Let $S$ be a monoid, $A \in$ Ens-$S$, $a, a' \in A$, $B \in S$-Ens, and $b, b' \in B$. Then $a \otimes b = a' \otimes b'$ if and only if there exist $a_1, \ldots, a_n \in A$, $b_2, \ldots, b_n \in B$, $s_1, \ldots, s_n \in S$ and $t_1, \ldots, t_n \in S$ such that*

$$\begin{array}{llcll} a & = a_1 s_1 & & s_1 b & = t_1 b_2 \\ a_1 t_1 & = a_2 s_2 & & s_2 b_2 & = t_2 b_3 \\ a_2 t_2 & = a_3 s_3 & & \vdots & \\ \vdots & & & s_n b_n & = t_n b'. \\ a_n t_n & = a' & & & \end{array}$$

A diagram

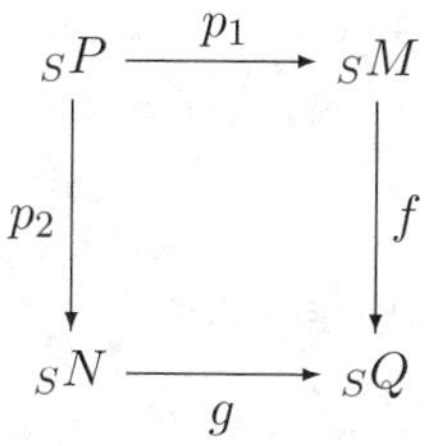


$(P_1)$

where ${}_SP$, ${}_SM$, ${}_SN$ and ${}_SQ$ are left $S$-acts and $f, g, p_1$ and $p_2$ are homomorphisms of left $S$-acts, is called a *pullback diagram*, or a *pullback square*, if $fp_1 = gp_2$ and for every left $S$-act ${}_S\bar{P}$, all homomorphisms $\bar{p_1} : {}_S\bar{P} \to {}_SM$ and $\bar{p_2} : {}_S\bar{P} \to {}_SN$

such that $fp_1 = gp_2$, there exists a unique homomorphism $h : {}_S\bar{P} \to {}_SP$ such that $p_1h = \bar{p_1}$ and $p_2h = \bar{p_2}$.

Every nonempty set can be considered as a left act over a trivial monoid. Homomorphisms of such left acts are just mappings of sets. Thus by the above definition, pullback diagrams in the category of sets are defined too.

Consider the diagram $(P_1)$ in the category of left $S$-acts.

**Lemma 2.2 ([6])** *If $(P_1)$ is a pullback diagram, then ${}_SP$ is isomorphic to the left $S$-act $\{(m,n) \in {}_SM \times {}_SN \mid f(m) = g(n)\}$ where $s(m,n) = (sm, sn)$ for all $s \in S$, $m \in {}_SM$ and $n \in {}_SN$*

So, if $(P_1)$ is a pullback of homomorphisms $f, g$ then ${}_SP$ is determined up to isomorphism and we may assume that it is

$${}_SP = \{(m,n) \in {}_SM \times {}_SN \mid f(m) = g(n)\}$$

and $p_1, p_2$ are the restriction of the projections, that is $p_1((m,n)) = m$ and $p_2((m,n)) = n$ for every $(m,n) \in {}_SP$. With this convention we denote the pullback diagram $(P_1)$ in the category of left $S$-acts by $P(M,N,f,g,Q)$.

Tensoring the pullback diagram $P(M,N,f,g,Q)$ by any right $S$-act $A_S$ one gets the commutative diagram

$$\begin{array}{ccc} A_S \otimes {}_SP & \xrightarrow{id_A \otimes p_1} & A_S \otimes {}_SM \\ {\scriptstyle id_A \otimes p_2}\downarrow & & \downarrow{\scriptstyle id_A \otimes f} \\ A_S \otimes {}_SN & \xrightarrow[id_A \otimes g]{} & A_S \otimes {}_SQ \end{array}$$

in the category of sets. For the pullback of mappings $id_A \otimes f$ and $id_A \otimes g$ in the category of sets we may take by our convention

$$P' = \{(a \otimes m, a' \otimes n) \in (A_S \otimes {}_SM) \times (A_S \otimes {}_SN) \mid a \otimes f(m) = a' \otimes g(n)\}$$

with $p'_1$ and $p'_2$ the restrictions of the projections. (Note that the existence of the pullback diagram $(P_1)$ implies the existence of the pullback diagram of $id_A \otimes f$ and $id_A \otimes g$.)

Now it follows from the definition of pullbacks that there exists a unique mapping $\phi : A_S \otimes {}_SP \to P'$ such that the diagram

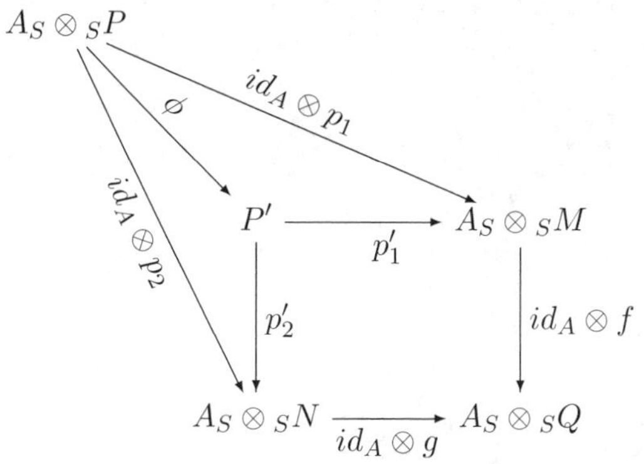


$(P_2)$

is commutative. We shall call this mapping *the $\phi$ corresponding to the pullback diagram* $P(M, N, f, g, Q)$. It is stated in [1] that the mapping $\phi$ in diagram $(P_2)$ is given by

$$\phi(a \otimes (m, n)) = (a \otimes m, a \otimes n)$$

for all $a \in A_S$ and $(m, n) \in {}_S P$.

Note that surjectivity of $\phi$ means that

$$\begin{aligned}(\forall a, a' \in A_S)(\forall m \in {}_S M)(\forall n \in {}_S N)[a \otimes f(m) = a' \otimes g(n) \Rightarrow \\ (\exists a'' \in A_S)(\exists m' \in {}_S M)(n' \in {}_S N) \\ (f(m') = g(n') \wedge a \otimes m = a'' \otimes m' \wedge a' \otimes n = a'' \otimes n')]\end{aligned}$$

and injectivity of $\phi$ means that

$$\begin{aligned}(\forall a, a' \in A_S)(\forall m, m' \in {}_S M)(\forall n, n' \in {}_S N) \\ [(f(m) = g(n) \wedge f(m') = g(n') \wedge a \otimes m = a' \otimes m' \wedge a \otimes n = a' \otimes n') \Rightarrow \\ a \otimes (m, n) = a' \otimes (m', n') \text{ in } A_S \otimes {}_S P]\end{aligned}$$

**Remark 2.3** Let $A = \dot{\bigcup}_{i \in I} A_i$ where $A_i$, $i \in I$, are right $S$-acts. Then by the mapping corresponding to the pullback diagram $P(M, N, f, g, Q)$ for $A_i$, $i \in I$, we mean the unique mapping $\phi_i$ which makes the diagram

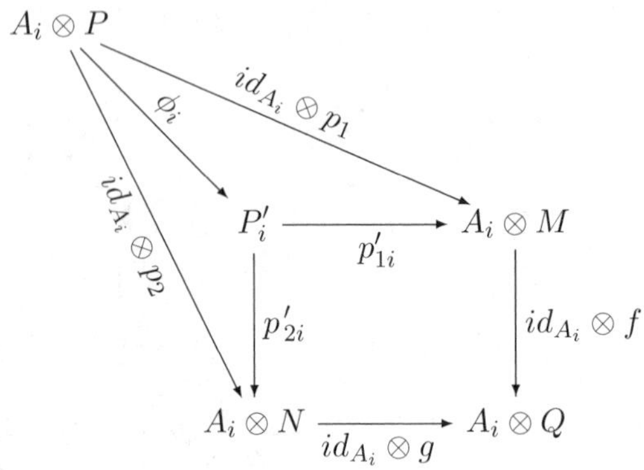


$(P_{2i})$

commutative, where

$$P'_i = \{(a \otimes m, a' \otimes n) \in (A_i \otimes M) \times (A_i \otimes N) \mid a \otimes f(m) = a' \otimes g(n)\}$$

and $p'_{1i}, p'_{2i}$ are restrictions of projections to $P'_i$.

**Definition 2.4** A right $S$-act $A$ is called

1. *principally weakly kernel flat* if the corresponding $\phi$ is bijective for every pullback diagram $P(Ss, Ss, f, f, S)$, $s \in S$.
2. *weakly kernel flat* if the corresponding $\phi$ is bijective for every pullback diagram $P(I, I, f, f, S)$, where $I$ is a left ideal of $S$.
3. *translation kernel flat* if the corresponding $\phi$ is bijective for every pullback diagram $P(S, S, f, f, S)$.

## 3 Results

**Lemma 3.1** *Let $A = \dot{\bigcup}_{i \in I} A_i$ where $A_i$, $i \in I$, are right $S$-acts. Let $B$ be a left $S$-act. If $a \otimes b = a' \otimes b'$ in $A_i \otimes B$, then $a \otimes b = a' \otimes b'$ in $A \otimes B$.*

**Proof** Since $a \otimes b = a' \otimes b'$ in $A_i \otimes B$, then by Lemma 2.1, there exist $a_1, \ldots, a_n \in A_i$, $b_2, \ldots, b_n \in B$, $s_1, \ldots, s_n \in S$ and $t_1, \ldots, t_n \in S$ such that

$$\begin{aligned} a &= a_1 s_1 \\ a_1 t_1 &= a_2 s_2 \\ a_2 t_2 &= a_3 s_3 \\ &\vdots \\ a_n t_n &= a' \end{aligned} \qquad \begin{aligned} s_1 b &= t_1 b_2 \\ s_2 b_2 &= t_2 b_3 \\ &\vdots \\ s_n b_n &= t_n b'. \end{aligned}$$

Since $A_i \subseteq A$, then $a_1, \ldots, a_n \in A$, and so $a \otimes b = a' \otimes b'$ in $A \otimes B$ as required. □

**Lemma 3.2** *Let $A = \dot{\bigcup}_{i \in I} A_i$ where $A_i$, $i \in I$, are right $S$-acts. Let $B$ be a left $S$-act and suppose that $a \otimes b = a' \otimes b'$ in $A \otimes B$. If $a \in A_i$ for some $i \in I$, then $a' \in A_i$.*

**Proof** Since $a\otimes b = a'\otimes b'$ in $A\otimes B$, then by Lemma 2.1, there exist $a_1, \ldots, a_n \in A$, $b_2, \ldots, b_n \in B$, $s_1, \ldots, s_n \in S$ and $t_1, \ldots, t_n \in S$ such that

$$\begin{aligned} a &= a_1 s_1 \\ a_1 t_1 &= a_2 s_2 \\ a_2 t_2 &= a_3 s_3 \\ &\vdots \\ a_n t_n &= a' \end{aligned} \qquad \begin{aligned} s_1 b &= t_1 b_2 \\ s_2 b_2 &= t_2 b_3 \\ &\vdots \\ s_n b_n &= t_n b'. \end{aligned}$$

Since $a \in A_i$, then $a_1 \in A_i$. Otherwise $a_1 \in A_j$ for some $j \neq i \in I$. Then $a_1 s_1 \in A_j$ and so $a = a_1 s_1 \in A_j$, which is a contradiction. Thus $a_1 \in A_i$ and so $a_1 t_1 \in A_i$. Since $a_1 t_1 = a_2 s_2$, then $a_2 \in A_i$. Otherwise there exists $j \neq i \in I$ such that $a_2 \in A_j$. Then $a_2 s_2 \in A_j$ and $a_1 t_1 = a_2 s_2$ implies that $a_1 t_1 \in A_i \cap A_j = \emptyset$ which is again a contradiction. By continuing this process we get $a_n \in A_i$ and therefore, $a' = a_n t_n \in A_i$ as required. □

**Corollary 3.3** *Let $A = \dot{\bigcup}_{i\in I} A_i$ where $A_i$, $i \in I$, are right $S$-acts. Let $B$ be a left $S$-act. If $a \otimes b = a' \otimes b'$ in $A \otimes B$ and $a \in A_i$, then $a \otimes b = a' \otimes b'$ in $A_i \otimes B$.*

**Proof** Since $a\otimes b = a'\otimes b'$ in $A\otimes B$, then by Lemma 2.1, there exist $a_1, \ldots, a_n \in A$, $b_2, \ldots, b_n \in B$, $s_1, \ldots, s_n \in S$ and $t_1, \ldots, t_n \in S$ such that

$$\begin{aligned} a &= a_1 s_1 \\ a_1 t_1 &= a_2 s_2 \\ a_2 t_2 &= a_3 s_3 \\ &\vdots \\ a_n t_n &= a' \end{aligned} \qquad \begin{aligned} s_1 b &= t_1 b_2 \\ s_2 b_2 &= t_2 b_3 \\ &\vdots \\ s_n b_n &= t_n b'. \end{aligned}$$

As we saw in Lemma 3.2, $a', a_1, \ldots, a_n \in A_i$ and so $a \otimes b = a' \otimes b'$ in $A_i \otimes B$ as required. □

**Lemma 3.4** *Let $A = \dot{\bigcup}_{i\in I} A_i$, where $A_i$, $i \in I$, are right $S$-acts and let $a \in A$, $b \in B$. If $a \in A_i$, then $a \otimes b \in A_i \otimes B$.*

**Proof** If there exists $j \neq i$ such that $a \otimes b \in A_j \otimes B$, then $a \otimes b = a' \otimes b'$ for some $a' \in A_j$ and $b' \in B$. But by Lemma 3.2, $a' \in A_i$ which is a contradiction. Therefore, $a \otimes b \in A_i \otimes B$ as required. □

**Corollary 3.5** *Let $A = \dot{\bigcup}_{i\in I} A_i$, where $A_i$, $i \in I$, are right $S$-acts. Let $\phi : A\otimes P \to P'$ be the mapping corresponding to the pullback diagram $P(M, N, f, g, Q)$ for $A$. If $\phi_i = \phi\,|_{A_i\otimes P}$, then $\phi_i : A_i \otimes P \to P'_i$.*

**Proof** Since $\phi(a \otimes (m, n)) = (a \otimes m, a \otimes n)$ for all $a \in A_S$ and $(m, n) \in {}_SP$, then it is sufficient to show that for $a \in A_i$, $m \in M$ and $n \in N$, $a \otimes m \in A_i \otimes M$ and $a \otimes n \in A_i \otimes N$ and this is true by Lemma 3.4. □

**Lemma 3.6** *Let $A = \dot{\bigcup}_{i \in I} A_i$, where $A_i$, $i \in I$, are right $S$-acts. Let $\phi : A \otimes P \to P'$ be a mapping and suppose that $\phi_i = \phi \mid_{A_i \otimes P}: A_i \otimes P \to P'_i$. Then $\phi$ is the mapping corresponding to the pullback diagram $P(M, N, f, g, Q)$ for $A$ if and only if $\phi_i$ is the mapping corresponding to the pullback diagram $P(M, N, f, g, Q)$ for $A_i$, $i \in I$.*

**Proof** Let $\phi$ be the the mapping corresponding to the pullback diagram $P(M, N, f, g, Q)$ for $A$. We show that the diagram $(P_{2i})$ is commutative. As it was mentioned, the diagram

$$\begin{array}{ccc} P'_i & \xrightarrow{\;p'_{1i}\;} & A_i \otimes M \\ {\scriptstyle p'_{2i}}\downarrow & & \downarrow{\scriptstyle id_{A_i} \otimes f} \\ A_i \otimes N & \xrightarrow[\;id_{A_i} \otimes g\;]{} & A_i \otimes Q \end{array}$$

is commutative. Now we show that $p'_{1i}\phi_i = id_{A_i} \otimes p_1$. Let $(a_i \otimes (m, n)) \in A_i \otimes P$. Then

$$\begin{aligned} p'_{1i}\phi_i(a_i \otimes (m, n)) &= p'_{1i}(a_i \otimes m, a_i \otimes n) = (a_i \otimes m) \\ &= id_{A_i}(a_i) \otimes p_1((m, n)) = (id_{A_i} \otimes p_1)(a_i \otimes (m, n)) \end{aligned}$$

It can also be seen that $p'_{2i}\phi_i = id_{A_i} \otimes p_2$. Since $\phi_i$ makes the diagram $(P_{2i})$ commutative, then by uniqueness, $\phi_i$ is the mapping corresponding to the pullback diagram $P(M, N, f, g, Q)$ for $A_i$.

Now let $\phi_i$ be the mapping corresponding to the pullback diagram $P(M, N, f, g, Q)$ for $A_i$, $i \in I$. Then the diagram $(P_{2i})$ is commutative. Since the diagram

$$\begin{array}{ccc} P' & \xrightarrow{\;p'_{1}\;} & A \otimes M \\ {\scriptstyle p'_{2}}\downarrow & & \downarrow{\scriptstyle id_{A} \otimes f} \\ A \otimes N & \xrightarrow[\;id_{A} \otimes g\;]{} & A \otimes Q \end{array}$$

is commutative, it is sufficient to show that $p'_1\phi = id_A \otimes p_1$ and $p'_2\phi = id_A \otimes p_2$. Let $a \otimes (m, n) \in A \otimes P$. Then there exists $i \in I$ such that $a \in A_i$. Then

$$\begin{aligned} p'_1\phi(a \otimes (m, n)) &= p'_1\phi_i(a \otimes (m, n)) = p'_1(a \otimes m, a \otimes n) = a \otimes m \\ &= id_{A_i}(a) \otimes p_1((m, n)) = (id_A(a) \otimes p_1((m, n)) = (id_A \otimes p_1)(a \otimes (m, n). \end{aligned}$$

The same argument shows that $p'_2\phi = id_A \otimes p_2$. □

**Theorem 3.7** *Let $\phi$ be the mapping corresponding to the pullback diagram $(P_1)$ for $A$ and let $\phi_i$, $i \in I$, be as in Lemma 3.6. Then $\phi$ is surjective if and only if $\phi_i$, $i \in I$, is surjective.*

**Proof** Let $\phi$ be surjective. We have to show that

$$(\forall a_i, a'_i \in A_i)(\forall m \in {}_SM)(\forall n \in {}_SN)[a_i \otimes f(m) = a'_i \otimes g(n) \Rightarrow \\ (\exists a''_i \in A_i)(\exists m' \in {}_SM)(n' \in {}_SN)$$

such that

$$f(m') = g(n') \wedge a_i \otimes m = a''_i \otimes m' \wedge a'_i \otimes n = a''_i \otimes n'].$$

Since $a_i, a'_i \in A_i \subseteq A$, then $a_i, a'_i \in A$. By Lemma 3.1, $a_i \otimes f(m) = a'_i \otimes g(n)$ in $A_i \otimes Q$ implies $a_i \otimes f(m) = a'_i \otimes g(n)$ in $A \otimes Q$. Since $\phi$ is surjective then

$$(\exists a''_i \in A_S)(\exists m' \in {}_SM)(n' \in {}_SN)$$

such that

$$f(m') = g(n') \wedge a_i \otimes m = a'' \otimes m' \wedge a'_i \otimes n = a'' \otimes n'$$

Since $a_i \otimes m = a'' \otimes m'$ in $A \otimes M$ and $a_i \in A_i$, then by Lemma 3.2, $a'' \in A_i$ and hence $\phi_i$ is surjective.

Now let $\phi_i$ be surjective for every $i \in I$ and suppose that

$$(\forall a, a' \in A_S)(\forall m \in {}_SM)(\forall n \in {}_SN)(a \otimes f(m) = a' \otimes g(n)).$$

Since $a \in A$, then there exists $i \in I$ such that $a \in A_i$. Then by Corollary 3.3, $a \otimes f(m) = a' \otimes g(n)$ in $A_i \otimes Q$. Since $\phi_i$ is surjective, then $a \otimes f(m) = a' \otimes g(n)$ implies that

$$(\exists a''_i \in A_i)(\exists m' \in {}_SM)(\exists n' \in {}_SN)$$

such that

$$(f(m') = g(n') \wedge a \otimes m = a'' \otimes m' \wedge a' \otimes n = a'' \otimes n'$$

Then by Lemma 3.1, $a \otimes m = a'' \otimes m' \wedge a' \otimes n = a'' \otimes n'$ in $A_i \otimes M$ and $A_i \otimes N$ implies $a \otimes m = a'' \otimes m' \wedge a' \otimes n = a'' \otimes n'$ in $A \otimes M$ and $A \otimes N$ respectively. Therefore, $\phi$ is surjective. □

**Theorem 3.8** *Let $\phi$ be the mapping corresponding to the pullback diagram $(P_1)$ for $A$ and let $\phi_i$, $i \in I$, be as in Lemma 3.6. Then $\phi$ is injective if and only if $\phi_i$, $i \in I$, is injective.*

**Proof** Let $\phi$ be injective and suppose that

$$(f(m) = g(n)) \wedge (f(m') = g(n')) \wedge (a \otimes m = a' \otimes m') \wedge (a \otimes n = a' \otimes n')$$

where $(a \otimes m = a' \otimes m') \wedge (a \otimes n = a' \otimes n')$ in $A_i \otimes M$ and $A_i \otimes N$ respectively for $i \in I$. Since $a, a' \in A_i$, then by Lemma 3.1, $a \otimes m = a' \otimes m'$ in $A \otimes M$ and $a \otimes n = a' \otimes n'$ in $A \otimes N$ respectively. Since $\phi$ is injective, then $a \otimes (m, n) = a' \otimes (m', n')$ in $A \otimes P$.

But $a \in A_i$ and so by Corollary 3.3, $a \otimes (m,n) = a' \otimes (m',n')$ in $A_i \otimes P$, that is $\phi_i$ is injective.

Conversely, Suppose that $\phi_i$ is injective for every $i \in I$ and let

$$(f(m) = g(n)) \wedge (f(m') = g(n')) \wedge (a \otimes m = a' \otimes m') \wedge (a \otimes n = a' \otimes n')$$

where $(a \otimes m = a' \otimes m') \wedge (a \otimes n = a' \otimes n')$ are respectively in $A \otimes M$ and $A \otimes N$. Since $a \in A$, then there exists $i \in I$ such that $a \in A_i$. Since $a \otimes m = a' \otimes m'$, in $A \otimes M$ and $a \otimes n = a' \otimes n'$ in $A \otimes N$, then by Corollary 3.3, $a \otimes m = a' \otimes m'$ and $a \otimes n = a' \otimes n'$ respectively in $A_i \otimes M$ and $A_i \otimes N$. By injectivity of $\phi_i$, we have $a \otimes (m,n) = a' \otimes (m',n')$ in $A_i \otimes P$ and hence, by Lemma 3.1, $a \otimes (m,n) = a' \otimes (m',n')$ in $A \otimes P$, that is $\phi$ is injective. □

**Theorem 3.9** *Let $S$ be a monoid and $A = \dot{\bigcup}_{i \in I} A_i$ where $A_i$, $i \in I$, are right $S$-acts. Then $A$ is principally weakly kernel flat, weakly kernel flat and translation kernel flat if and only if $A_i$ has these properties for every $i \in I$.*

**Proof** Let $A$ be principally weakly kernel flat. Then by Definition 2.3, the mapping $\phi$, corresponding to the pullback diagram $P(Ss, Ss, f, f, S)$, $s \in S$, for $A$ is bijective. Then by Theorems 3.7 and 3.8, the mapping $\phi_i$, corresponding to the pullback diagram $P(Ss, Ss, f, f, S)$, $s \in S$, for $A_i$ is also bijective and again by Definition 2.3, $A_i$ is principally weakly kernel flat.

Conversely, if $A_i$ is principally weakly kernel flat for every $i \in I$, then by Definition 2.3, the mapping $\phi_i$, corresponding to the pullback diagram $P(Ss, Ss, f, f, S)$, $s \in S$, for $A_i$ is bijective. Then by Theorems 3.7 and 3.8, the mapping $\phi$, corresponding to the pullback diagram $P(Ss, Ss, f, f, S)$, $s \in S$, for $A$ is also bijective and again by Definition 2.3, $A$ is principally weakly kernel flat.

By the same argument it can also be seen that $A$ is weakly kernel flat, translation kernel flat if and only if $A_i$ has these properties for every $i \in I$. □

## References


[1] S. Bulman-Fleming, Pullback flat acts are strongly flat, *Canad. Math. Bull.* **34** (1991), no. 4, 456–461.

[2] S. Bullman-Fleming, M. Kilp, and V. Laan, Pullbacks and flatness properties of acts II, *Comm. Algebra* **29** (2001), no. 2, 851–878.

[3] S. Bulman-Fleming and K. McDowell, Absolutely flat semigroups, *Pacific J. Math.* **107** (1983), 319–333.

[4] J. M. Howie, *Fundamentals of Semigroup Theory*, London Mathematical Society Monographs, OUP, 1995.

[5] V. Laan, Pullbacks and flatness properties of acts I, *Comm. Algebra* **29** (2001), no. 2, 829–850.

[6] P. Normak, On equalizer-flat and pullback-flat acts, *Semigroup Forum* **36** (1987), 293–313.

# ON PROOFS IN FINITELY PRESENTED GROUPS

GEORGE HAVAS and COLIN RAMSAY[1]

ARC Centre for Complex Systems
School of Information Technology and Electrical Engineering
The University of Queensland, Queensland 4072, Australia
Email: havas@itee.uq.edu.au, cram@itee.uq.edu.au

## Abstract

Given a finite presentation of a group $G$, proving properties of $G$ can be difficult. Indeed, many questions about finitely presented groups are unsolvable in general. Algorithms exist for answering some questions while for other questions algorithms exist for verifying the truth of positive answers. An important tool in this regard is the Todd–Coxeter coset enumeration procedure. It is possible to extract formal proofs from the internal working of coset enumerations. We give examples of how this works, and show how the proofs produced can be mechanically verified and how they can be converted to alternative forms. We discuss these automatically produced proofs in terms of their size and the insights they offer. We compare them to hand proofs and to the simplest possible proofs. We point out that this technique has been used to help solve a longstanding conjecture about an infinite class of finitely presented groups.

*Keywords*: finitely presented group, proof, Todd–Coxeter coset enumeration, Hilbert's 24th problem, Fibonacci group, van Kampen diagram, trivial group, Andrews–Curtis conjecture.

## 1 Introduction

Many theorems in group theory are based on the results of computations. Indeed, often, the computations are now done on machines. Comprehensive details about computing with finitely presented groups appear in [32, 17]. Of particular relevance to our considerations are the chapters on coset enumeration and also a brief review of computability issues.

Proofs of theorems which include machine computations are opaque. This is especially true for proofs which rely on the collapse of a coset enumeration to one coset. Indeed, one of the most likely consequences of an error in a hand computation or of a bug in a computer application of coset enumeration is an incorrect total collapse. Leech [22] discusses this difficulty.

We address this lack of transparency of proofs based on direct coset enumeration by studying them carefully and deriving more usual proofs. We compare and contrast various machine and human proofs. As regards coset enumeration, we focus on the least transparent situation, where total collapse occurs.

[1]Both authors were partially supported by the Australian Research Council.

Questions regarding the length and elegance of proofs are difficult in general. Thiele [33] tells us that Hilbert considered including another problem in his famous list.

> The twenty-fourth problem belongs to the realm of foundations of mathematics. In a nutshell, it asks for the simplest proof of any theorem.

Grattan-Guinness [9] suspects it was ultimately omitted since "simplicity is an extremely *complicated* notion to capture in a *general* way".

In the context of a limited range of proofs in finitely presented groups we can address this problem with reasonable success. We introduce the notion of a proof certificate for some results based on coset enumeration. This enables us to classify proofs with respect to sensible criteria.

Finally, the Andrews–Curtis [1, 2] conjecture addresses the nature of proofs of triviality for balanced presentations and has been studied computationally [13, 14]. Work on this conjecture provides alternative proof methods. As a result we are able to guarantee some proofs are simplest in specific terms.

We give examples of how our ideas work by investigating various proofs of results about some interesting finitely presented groups. Throughout this paper, we use the convention that lower case letters denote group generators and upper case letters their inverses; thus, $A = a^{-1}$, etc. Coset number 1 will always represent the subgroup and all cosets are right cosets.

## 2 Coset enumeration

Coset enumeration is the basis of important techniques for investigating finitely presented groups with manifold applications. Its use dates back at least to Moore [26]. Given defining relations for a group $G$ and words generating a subgroup $H$ of finite index, coset enumeration programs implement systematic procedures for enumerating the cosets of $H$ in $G$. Computer implementations are based on methods initially described by Todd and Coxeter [35]. Many details are given in [32, 17]. Implementations are available in the computer algebra systems GAP [8] and MAGMA [3] and as a stand-alone program, the ACE coset enumerator, [12]. A particularly useful tool for experimenting with coset enumerations is the Interactive Todd Coxeter package, ITC [7].

The end result of a successful coset enumeration is a coset table which gives a permutation representation for $G$, corresponding to the action by multiplication of $G$ on the cosets of $H$. When the index is not one we can readily check that the nontrivial representation obtained is indeed a representation for the group, which gives significant comfort that the process worked correctly. However, when the index is one, nothing reassuring is left except for our confidence in the process.

Implicit in the underlying working of an enumeration are formal proofs that particular words in the generators are in the subgroup, as shown by Leech [23]. The utility PEACE [29] (proof extraction after coset enumeration) has been developed to automate the production of such proofs. PEACE produces *proof-words* which can be regarded as certificates attesting to subgroup membership.

The natural measure of simplicity for a coset enumeration based proof is the total number of cosets defined in the process. Once we have a proof-word, then natural measures are based on its length.

## 3 Proof certificates

A successful coset enumeration, in its workings, embodies proofs of many embedded results. However its very mechanical nature makes it both unpalatable to read and also error-prone. In spite of this, very many computer implementations are regularly used to correctly build coset tables for all sorts of purposes.

Our focus here is on proving theorems based upon coset enumeration. Given a successful coset enumeration of the cosets of the subgroup $H$ in the group $G$, we may claim that this proves that some word $\omega$ is in $H$. (The fact that $\omega \in H$ is checked by applying coset 1 to $\omega$ and tracing it through the coset table back to coset 1.) This claim rests on the validity of the coset enumeration strategy employed and the correctness of its implementation.

Instead of making such a claim on this basis we can provide a stand-alone proof, as follows. We extract from the workings of a coset enumeration a *proof-word* which can be verified mechanically and which explicity gives $\omega$ in terms of the generators of $H$. This proof-word, together with the presentation for $G$ and the generators of $H$, forms a *(proof) certificate* that $\omega \in H$. The validity of this certificate is independent of how it was generated, and does not depend on anything other than the group axioms and the definitions of $G$ and $H$. Such proof-words are implicit in the workings of a coset enumeration, and their generation by considering circuits in the Schreier diagram is explained by Leech [23]. Many further details are given in [29], where the PEACE package is documented.

A fully expanded proof-word consists of a product of subgroup generators and of conjugates of group relators (by group generators). The subgroup generators appear as given in the presentation for $H$, or as the formal inverses thereof. There is no such requirement on relators for $G$, and they or their formal inverses may be cycled in proofs. By construction, $\omega$ and the proof-word are equivalent and, since conjugates of relators are trivial in $G$, the proof-word is also equivalent to a product of subgroup generators. Thus, free reduction of the proof word produces $\omega$, while reduction after cancelling the conjugates of relators produces a product of subgroup generators. Such reductions are easy to verify by hand for short proof-words and by computer for longer words.

Although the content of a proof-word depends on the details of how the coset enumeration is performed, it stands alone as a proof. Its validity depends only on the presentation used for the group, the generators of the subgroup, the word to be proved, and the general axioms of group theory. It is easy to understand and, in this sense, provides a simple proof that a word is in a subgroup (and gives an explicit construction). The PEACE package includes a stand-alone verification utility, which both checks that a proof-word is properly formed and performs the two required reductions.

## 4 Pruned enumeration

Starting with a successful coset enumeration where the total number of cosets used, $T$, exceeds the subgroup index, $I$, it is often possible to *prune* the sequence of $T-1$ coset definitions by eliminating cosets which do not contribute to the final table. (It would be nice to reduce $T$ to $I$ but, in general, this is not possible.) Early workers pruned enumerations by hand, but ITC automates the process by repeatedly reordering the sequence of definitions and rerunning the enumeration, while omitting definitions which become redundant. Such pruning techniques have been incorporated into PEACE and, combined with the ability to consider equivalent presentations (that is, presentations where the relators have been cycled, inverted and/or reordered), have proved very effective at generating short enumerations.

The pruned enumerations that ITC and PEACE yield are frequently the shortest known. To prove that no shorter enumeration exists an exhaustive search technique can be used, although this is only feasible for small values of $T$. PEACE has the ability to generate and test all possible *valid* sequences of definitions of a given length, in a *standardised* form; e.g., ordered by coset number and coset-table column number. See §6.2 for more details, and an example.

We are now ready to consider some proofs. We focus on four examples which use coset enumerations that collapse to one coset. This is the case where the proof based on straight coset enumeration is least perspicuous.

## 5 Some Fibonacci groups

The Fibonacci group (see [20]) $F(r,n)$ is generated by $n$ generators $x_1,\ldots,x_n$, which satisfy $n$ relations $x_j x_{j+1} \cdots x_{j+r-1} = x_{j+r}$, $1 \leqslant j \leqslant n$, where the subscripts are reduced modulo $n$ to lie in the range $1,\ldots,n$. Not all of the Fibonacci groups have been identified, see [34]. They have been a fertile area for investigation. Some are known to be cyclic, and it is from these that we draw our examples since many proofs by coset enumeration for these groups are based on total collapses.

### 5.1 $F(2,5)$

Conway [4] raised interest in the Fibonacci groups by asking for a proof that the Fibonacci group $F(2,5) = \langle a,b,c,d,e \mid abC, bcD, cdE, deA, eaB\rangle$ is cyclic of order 11. Various solutions are mentioned in [5], each utilising a few lines of algebraic manipulation and some group theory.

A coset enumeration in $F(2,5)$ over a cyclic subgroup yields a single coset, and forms the basis of a short proof of the result. Such an enumeration is easily carried out by hand; for example, using the subgroup $\langle a\rangle$, with cosets 2 and 3 defined as $1b$ and $1B$, the required collapse to a single coset occurs after processing some nine deductions. It is straightforward to confirm that at least two definitions (not counting coset 1) are required for a successful enumeration. Thus $F(2,5)$ is cyclic, and hence abelian, and its order is easily found. For example, the presentation's exponent sum matrix, along with its Hermite normal form, is shown in the first

Figure 1. Exponent sums and Hermite normal forms

| Group | Exponent sums | Hermite normal form |
|---|---|---|
| $F(2,5)$ | $\begin{bmatrix} 1 & 1 & -1 & 0 & 0 \\ 0 & 1 & 1 & -1 & 0 \\ 0 & 0 & 1 & 1 & -1 \\ -1 & 0 & 0 & 1 & 1 \\ 1 & -1 & 0 & 0 & 1 \end{bmatrix}$ | $\begin{bmatrix} 1 & 0 & 0 & 0 & 7 \\ 0 & 1 & 0 & 0 & 6 \\ 0 & 0 & 1 & 0 & 2 \\ 0 & 0 & 0 & 1 & 8 \\ 0 & 0 & 0 & 0 & 11 \end{bmatrix}$ |
| $F(3,5)$ | $\begin{bmatrix} 1 & 1 & 1 & -1 & 0 \\ 0 & 1 & 1 & 1 & -1 \\ -1 & 0 & 1 & 1 & 1 \\ 1 & -1 & 0 & 1 & 1 \\ 1 & 1 & -1 & 0 & 1 \end{bmatrix}$ | $\begin{bmatrix} 1 & 0 & 0 & 0 & 13 \\ 0 & 1 & 0 & 0 & 7 \\ 0 & 0 & 1 & 0 & 19 \\ 0 & 0 & 0 & 1 & 17 \\ 0 & 0 & 0 & 0 & 22 \end{bmatrix}$ |
| $F(2,7)$ | $\begin{bmatrix} 1 & 1 & -1 & 0 & 0 & 0 & 0 \\ 0 & 1 & 1 & -1 & 0 & 0 & 0 \\ 0 & 0 & 1 & 1 & -1 & 0 & 0 \\ 0 & 0 & 0 & 1 & 1 & -1 & 0 \\ 0 & 0 & 0 & 0 & 1 & 1 & -1 \\ -1 & 0 & 0 & 0 & 0 & 1 & 1 \\ 1 & -1 & 0 & 0 & 0 & 0 & 1 \end{bmatrix}$ | $\begin{bmatrix} 1 & 0 & 0 & 0 & 0 & 0 & 5 \\ 0 & 1 & 0 & 0 & 0 & 0 & 4 \\ 0 & 0 & 1 & 0 & 0 & 0 & 9 \\ 0 & 0 & 0 & 1 & 0 & 0 & 13 \\ 0 & 0 & 0 & 0 & 1 & 0 & 22 \\ 0 & 0 & 0 & 0 & 0 & 1 & 6 \\ 0 & 0 & 0 & 0 & 0 & 0 & 29 \end{bmatrix}$ |

line of Figure 1. This not only tells us that $F(2,5) \cong C_{11}$, but also that $a = E^7$, $b = E^6$, etc.

In the context of the above 5-generator, 5-relator presentation, the 'best' proof-word we have been able to extract from any coset enumeration is

$$\begin{aligned} \Psi \;=\; & [E](eaB)[E][E](eDC)c(deA)(abC) \quad (cDb)C(cBA) \\ & [E](eaB)[E][E](eDC)c(deA)(abC) \quad (cdE)(eAd)(Dbc)C. \end{aligned}$$

($\Psi$ is a single string, which has been written as shown to illustrate its structure; see below). Items in square brackets, $[\cdot]$, are subgroup generators (in this case inverted); items in round brackets, $(\cdot)$, are relators (perhaps inverted or cycled); and the remaining items are conjugating group generators. The word $\Psi$ was extracted from an enumeration over $\langle e \rangle$, and proves that $b = E^6$ (i.e., $b \in \langle e \rangle$). The entire word, freely reduced, equals $b$, while if the relators are first deleted (being trivial in $F(2,5)$), reduction yields $E^6$.

Although our coset enumeration proves that $F(2,5)$ is cyclic, our proof-word by itself does not. However, substituting $b = E^6$ into the other relations rapidly leads to that conclusion and also enables us to find the group order.

The overall length of the proof-word compares favorably with that of the other published proofs (see §7). It is not a conventional proof, but we can convert it in

Figure 2. van Kampen diagram for $c\beta$

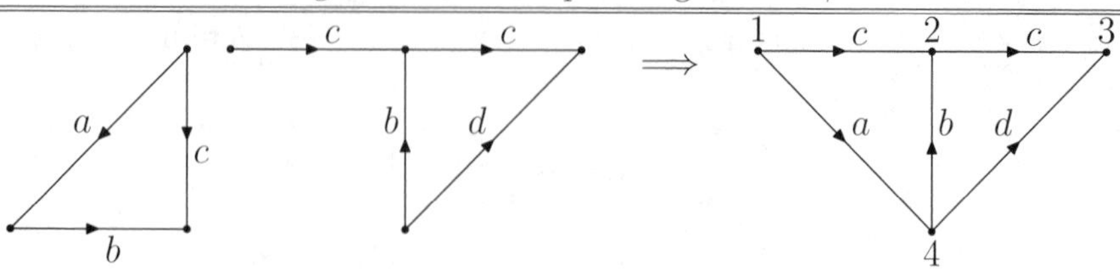


various ways. For this example, first note that $\Psi$ can be written as $\alpha\beta\alpha\gamma$, where

$$\begin{aligned}\alpha &= [E](eaB)[E][E](eDC)c(deA)(abC),\\ \beta &= (cDb)C(cBA),\\ \gamma &= (cdE)(eAd)(Dbc)C.\end{aligned}$$

Now $\alpha$, $\beta$ and $\gamma$ prove, respectively, that $aC = E^3c$, $cDA = C$ and $cdAb = C$. So $\alpha\beta$ proves $aDA = E^3$, while $\alpha\gamma$ proves $adAb = E^3$. Thus $\Psi$ can be split into a series of lemmas, which can be combined to yield the result. To rewrite, say, $\beta$ as a more usual type of proof, start with $cDA$ and insert the trivial relator $bB$ to give $cDbBA$. Now $cDb$ is trivial in $F(2,5)$, being a cyclic conjugate of the relator $bcD$, and so we can cancel to $BA$. Now insert the trivial relator $Cc$ to give $CcBA$, and cancel the relator $cBA$ (inverse of the relator $abC$) to yield $C$.

Note that $\alpha$, $\beta$ and $\gamma$ are *not* proof-words (since the conjugating symbols are not properly paired within them). However, $\alpha C$, $c\beta$ and $c\gamma$ are; they prove, respectively, $aC^2 = E^3$, $c^2DA = 1$ and $c^2dAb = 1$. This is equivalent to rewriting $\Psi$ as $\Psi_1 = \alpha Cc\beta\alpha Cc\gamma$. Now $c\beta$ and $c\gamma$ make no use of $[e]$ or $[E]$, and so can be regarded as proof-words in $F(2,5)$ over the trivial subgroup, while $\alpha C$ makes use of the subgroup $\langle e\rangle$. However, if the square brackets are deleted, the word can be rewritten as the proof-word $e^3\alpha C$, which proves that $e^3aC^2$ is trivial in $F(2,5)$. We can now rewrite $\Psi$ as $\Psi_2 = e^3\alpha Cc\beta E^3e^3\alpha Cc\gamma e^3$, proving that $e^3be^3$ is trivial in $F(2,5)$.

Various graphical representations of a group have been described, and these have numerous applications [31, 25]. One useful pictorial representation of proofs in a group $G = \langle X \mid R\rangle$ is provided by *van Kampen diagrams*. These can be thought of as portions of the Cayley diagram of $G$ drawn in the plane, where the words corresponding to the boundaries of internal faces are members of $R$. Such a diagram demonstrates that the boundary of the graph is trivial in $G$ (see [18, 19]).

Each relator or conjugated relator yields a component of such a graph, as shown in the left part of Figure 2 for $c\beta$. Then the (conjugated) relator can be recovered by tracing the boundary of the component and recording the arc labels, where a traversal against the arrow yields the inverse. These components are combined into a single *reduced* diagram (the right part of the figure), where there is at most one in-arc and one out-arc at each node for each generator. The proof-word $c\beta$ corresponds to the traversal 123421241 of the diagram; note that internal edges are

Figure 3. Composite diagram for $\Psi_2$

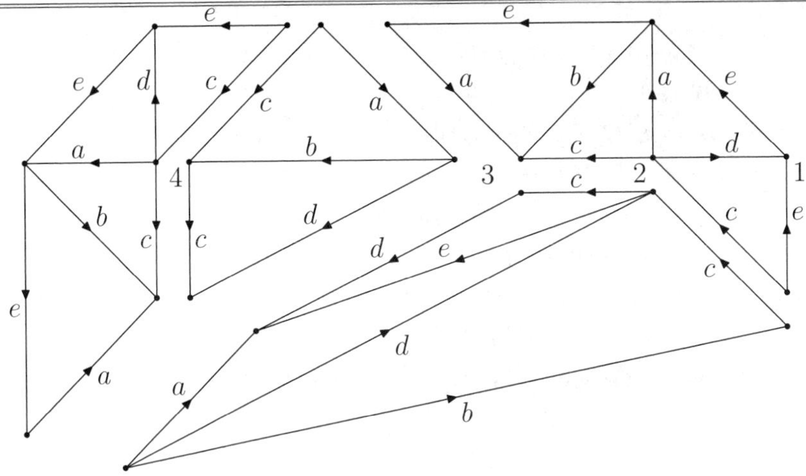


traversed an equal number of times in opposite directions, while boundary arcs are traversed once more in the direction of the external traversal than against it.

Van Kampen diagrams for $e^3\alpha C$ and $c\gamma$ can be constructed similarly. Copies of all these diagrams can be combined to yield a composite diagram for $\Psi_2$, which provides an alternative proof that $be^6$ is trivial in $F(2,5)$. Figure 3 is an 'exploded' version of such a diagram, illustrating how the various sub-diagrams (i.e., lemmas) combine to yield the result.

Since $F(2,5) \cong C_{11}$, an enumeration over the trivial subgroup must make at least ten definitions (not counting coset 1). We do not know how few suffice, and our best result is fifteen definitions. One example is the sequence: $1c$, $2c$, $1A$, $4B$, $1C$, $6e$, $7b$, $7C$, $9e$, $6d$, $11A$, $2B$, $4E$, $2e$, $2b$. That is: coset 2 is defined as coset 1 times $c$, coset 3 is defined $\ldots$, coset 16 is defined as coset 2 times $b$. Note that there is a coset table involving eleven cosets implicit in the diagram of Figure 3. However, this table is incomplete, and only contains the table entries corresponding to the diagram's edges (after making all possible deductions). Furthermore, two of the nodes represent the same coset. (Since the group is abelian, the relator $bcD$ implies that $Dcb$ — the path 1234 — should trace out a cycle, which it does not.)

We were unable to find any proof-word showing that $F(2,5)$ is cyclic which used less than thirteen relators of our initial presentation. We found many proof-words with thirteen relators, and $\Psi$ has the smallest number of subgroup generators plus group generators (six and four respectively) and the 'nicest' structure. There are many possible ways to extract a proof which readily implies the group is cyclic. For example, we can enumerate over the trivial subgroup and prove that, say, $ae^7$ or $be^6$ is trivial, or that the commutator $[a,b]$ or $[a,c]$ is trivial. Or we could enumerate over a subgroup $\langle e^m \rangle$, for some $m$, and extract a proof that $a$ or $b$ is in the subgroup. We tried a variety of these, and none yielded a better proof-word than $\Psi$. We did find various other proofs using only thirteen relators, but these

Figure 4. Proof-word components for $F(3,5)$

$$\delta = (deaB)(bCea)A[E]c(bAED)(deAc)Caa(BAEc)(CaED)(deaB)A(abcD)$$
$$(deaB)(bCea)A[E]c(bAED)(deAc)Caa(BAEc)(CBAd)A(aEDC)A[E]$$
$$D(deaB)(bcdE)(Dabc)d(DCBe)[E][E](eAcd)(DCBe)[E]b(deAc)Cd(Dabc)$$
$$(CBeD)(abCe)D(dEbc)cB(bCea)(AEDb)B(dCBA)(AEcB)bC(eDCB)$$

$$\varepsilon = ad(eabC)(dEbc)(CBAd)DAc(CaED)(deaB)be(Dabc)EB[e](Ebcd)DB$$
$$(bCea)(AdCB)bc(DCBe)E(Ebcd)e(EDCa)C(Dabc)d(DCBe)[E][E](eAcd)$$
$$(DCBe)[E]b(deaB)(bCea)B(bAED)dc(AEcB)(bAED)(deAc)C(Ebcd)D$$
$$(CBAd)Da(eaBd)(DCBe)E(bcDa)(Ebcd)eA(aEDC)B(eAcd)(DCBe)Eb$$
$$(deAc)B(AEDb)(BAdC)(cBAE)e(abcD)b(Bdea)(bAED)d[e]a(BAEc)$$

were all the same as $\Psi$, in the sense that they proved that $b = E^6$, or an equivalent of this under the action of the cyclic automorphism of the generators.

## 5.2 $F(3,5)$

The Fibonacci group $F(3,5) = \langle a,b,c,d,e \mid abcD, bcdE, cdeA, deaB, eabC\rangle$ is cyclic of order 22. The proof of this relies on coset enumeration, and Johnson [18] noted that "some 200 cosets are required and no short manual proof has yet been found".

A coset enumeration in $F(3,5)$ over one of the generators gives index one, and establishes that $F(3,5)$ is cyclic. The exponent sum and normal form matrices can now be used to complete the proof (see Figure 1). PEACE can complete such an enumeration using 30 definitions. One such sequence of definitions, over $\langle a\rangle$, is: $1d$, $1c$, $3d$, $4A$, $3D$, $6A$, $6c$, $1b$, $9d$, $6b$, $11C$, $9b$, $13b$, $1C$, $15d$, $6D$, $17c$, $13c$, $19d$, $20a$, $19A$, $9A$, $23b$, $24A$, $11c$, $26d$, $27b$, $3E$, $29E$, $30a$. This is a substantial improvement on the number of cosets used by Johnson.

The shortest proof-words which we have been able to extract via PEACE all contained 103 occurrences of the group's relators. They all have a similar structure, and the example chosen came from a (pruned) enumeration in 46 total cosets; it proves that $a \in \langle e\rangle$. Consider the words $\delta$ and $\varepsilon$ shown in Figure 4. These prove, respectively, that $CB = AE^6C$ and $dcc = cEa$. Now consider the proof-word $a\delta\varepsilon\delta(bCea)(AEDb)c$. This proves $a = E^{13}$, as required, and the proof can now be completed by substitution in the presentation. Note that there is considerable further structure in the proof-word which we have not exploited; e.g., the first two-thirds of the first two lines of $\delta$ are the same.

We also attempted to extract proofs that some commutators are trivial from enumerations over the trivial subgroup. The best we found had 126 group relators, and did not have any obvious substructure. However we did note that the coset enumeration over the trivial subgroup could be done using a total of only 98 cosets.

### 5.3 $F(2,7)$

The Fibonacci group

$$F(2,7) = \langle a, b, c, d, e, f, g \mid abC, bcD, cdE, deF, efG, fgA, gaB \rangle$$

is cyclic of order 29. The first hand proof of this was given by Havas [10]. His proof that $F(2,7)$ is cyclic was extracted from the workings of a series of index 1 coset enumerations over a one-generator subgroup. The first of these took 327 total cosets, and subsequent enumerations incorporated the results of previous ones as additional (redundant) subgroup generators. Taking into account the multiplicity with which the various results are used, Havas's proof (implicitly) uses the group's relators a total of some 24742 times. Subsequently, using a single enumeration in 55 cosets, Edeson [6] was able to extract a similar proof which used the relators only 2278 times.

At the time of writing, the best proof-word we have found using PEACE and a cyclic subgroup uses the group's relators a total of 595 times, and was derived from an enumeration using a total of 74 cosets. The proof-word produced by PEACE contains many repeated substrings and could be rewritten as a series of lemmas, in a similar fashion to the proofs given by Havas and Edeson. Although much shorter than the previous proofs, our PEACE proof is no less opaque, so we refrain from reproducing it here.

Edeson's proof is some four times longer than ours, despite the fact that the enumeration she used is approximately three-quarters the length of ours. Edeson had available other, shorter enumerations, but she rejected these since "the path to the primary coincidence was longer and more convoluted in the smaller sets". It is interesting to note that Leech [24] also extracted, but did not publish, a proof from a 55 coset enumeration, but its length only bettered that of Havas's proof by a factor of about two.

The developers of ITC used $F(2,7)$ as one of their test cases, and they were able to complete a coset enumeration in $F(2,7)$ over one of the generators with 50 cosets (i.e., 49 definitions). We were able to duplicate this value using PEACE, and one sequence of 49 definitions, for an enumeration over $\langle b \rangle$, is: $1c$, $2a$, $3F$, $3B$, $5E$, $6g$, $7B$, $8F$, $1a$, $10B$, $11C$, $12d$, $1A$, $14F$, $15C$, $16g$, $17E$, $11D$, $19a$, $20B$, $12C$, $22E$, $1F$, $14g$, $10c$, $14E$, $1D$, $28F$, $1E$, $30C$, $30B$, $11B$, $11f$, $11a$, $28E$, $36A$, $36d$, $10g$, $39F$, $40a$, $41F$, $39D$, $10C$, $44A$, $45E$, $40C$, $36C$, $45D$, $15c$.

It is straightforward to verify this definition sequence works. For example, this can be done independently using ITC. The result now follows from the exponent sum and normal form matrices (see Figure 1). Given the relative sizes of the coset table involved in this enumeration and of the proof-word, it is not clear which of these is the better proof. The sequence of definitions needed to complete the coset enumeration is certainly much shorter than the proof-word. However, verifying that the enumeration completes correctly is more involved than checking the reductions of the proof-word. Neither proof seems to offer any insight, so their relative merit in that sense is a moot point.

## 6 The trivial group

The presentation of the group $E_1 = \langle a, b, c \mid CacAA, AbaBB, BcbCC\rangle$ is a well-known example for the trivial group. That $E_1$ is trivial is not obvious; see Higman [16] and Neumann [27]. Neumann [28] notes that this presentation was very difficult for early computer implementations of the Todd–Coxeter process, and that it can be used to build an infinite series of presentations for the trivial group of ever-increasing difficulty (see [11] for more details and solutions for the next presentation in the series, $E_2$).

### 6.1 $E_1$ and 2-generator subgroups

Higman [16] proved that $E_1$ is trivial by expressing one of the generators in terms of the other two, and then using an argument based on derived groups. Replacing his $a_1$, $a_2$, $a_3$ by our $a$, $b$, $c$, the first part of his proof runs as follows. $AbaBB = 1$ implies that $baB = ab$, and conjugating this by $c$ gives $CbcCacCBc = CacCbc$. Now $CacAA = 1$ and $BcbCC = 1$ imply that $Cac = aa$ and $Cbc = bC$, and substituting these yields $bCaacB = aabC$. Finally, $CacAA = 1$ also implies that $Ca = aaC$ and $ac = caa$, and substitution now gives $baaCcaaB = baaaaB = aabC$. Thus, $c = bAAAABaab$. (Note that Higman's paper says $c = ba^4b^{-1}a^2b$, but this is presumably just a typographical error.)

Coset enumerations in $E_1$ over two-generator subgroups yield a single coset and can be completed with two definitions. (For the subgroup $\langle a, b\rangle$, there are precisely two such definitions sequences: $1C$, $2b$ and $1C$, $2C$.) The best proof-words we have been able to extract from such enumerations have eight group relators, and a typical example is the word

$$\begin{aligned}\Upsilon \;=\; & (CbccB)[b]\ [A][A](aaCAc)\ [A][A](aaCAc)\ [B](bCCBc)\\ & Cb(caaCA)(aBBAb)Bc\ (CacAA)[a][a]\ (CbccB)[b].\end{aligned}$$

This word freely reduces to $c$ and, after deleting the relators, reduces to the product $bAAAABaab$ of subgroup generators $a$ and $b$. Higman's proof also contains eight uses of the relators and, in fact, each of the relators presenting $E_1$ is used the same number of times in Higman's proof as in $\Upsilon$. We obtained various other proof-words of a similar length, but these were all analogues of $\Upsilon$, over different subgroups and with different arrangements of the relators and varying amounts of conjugation.

Given a balanced presentation of the trivial group (i.e., with the same number of relators as generators), one way of proving it to be trivial is to reduce it to the *standard* presentation $\langle X \mid X\rangle$ by a sequence of relator inversions, conjugations and multiplications. Andrews and Curtis [1, 2] have conjectured that this is always possible. This conjecture is still open, and $E_1$ is a long-standing potential counter-example for the 3-generator case; in fact, the shortest 3-generator possibility.

In some circumstances we can extract from a proof-word a sequence of Andrews–Curtis moves on the relators which prove the same result (see [14] for more details). Since $\Upsilon$ contains the relator $AbaBB$ once only, the simple technique described in [14] can be used. We first rewrite $\Upsilon$ as $\Upsilon BAAbaaaaB$ so that it is a proof,

Figure 5. Andrews–Curtis moves for $E_1$

| move | relator $r$ | relator $s$ | relator $t$ |
|---|---|---|---|
| - | $CacAA$ | $AbaBB$ | $BcbCC$ |
| $r \to r^C$ | $acAAC$ | $AbaBB$ | $BcbCC$ |
| $s \to s^A$ | $acAAC$ | $baBBA$ | $BcbCC$ |
| $s \to sr$ | $acAAC$ | $baBBcAAC$ | $BcbCC$ |
| $s \to s^b$ | $acAAC$ | $aBBcAACb$ | $BcbCC$ |
| $s \to st$ | $acAAC$ | $aBBcAAbCC$ | $BcbCC$ |
| $s \to s^{-1}$ | $acAAC$ | $ccBaaCbbA$ | $BcbCC$ |
| $s \to s^a$ | $acAAC$ | $AccBaaCbb$ | $BcbCC$ |
| $s \to s^B$ | $acAAC$ | $bAccBaaCb$ | $BcbCC$ |
| $s \to st$ | $acAAC$ | $bAccBaabCC$ | $BcbCC$ |
| $s \to s^b$ | $acAAC$ | $AccBaabCCb$ | $BcbCC$ |
| $s \to st$ | $acAAC$ | $AccBaabCbCC$ | $BcbCC$ |
| $s \to s^A$ | $acAAC$ | $ccBaabCbCCA$ | $BcbCC$ |
| $s \to sr$ | $acAAC$ | $ccBaabCbCAAC$ | $BcbCC$ |
| $s \to s^c$ | $acAAC$ | $cBaabCbCAA$ | $BcbCC$ |
| $s \to s^a$ | $acAAC$ | $AcBaabCbCA$ | $BcbCC$ |
| $s \to sr$ | $acAAC$ | $AcBaabCbAAC$ | $BcbCC$ |
| $s \to s^A$ | $acAAC$ | $cBaabCbAACA$ | $BcbCC$ |
| $s \to sr$ | $acAAC$ | $cBaabCbAAAAC$ | $BcbCC$ |
| $s \to s^c$ | $acAAC$ | $BaabCbAAAA$ | $BcbCC$ |

over the trivial subgroup, that $cBAAbaaaaB$ is trivial (cf. Section 5.1). It is now straightforward to build up the proof-word, one relator at a time, by first using $AbaBB$ and thereafter multiplying by appropriately conjugated versions of the other two relators. Counting conjugation by a single group generator as a single move, one sequence of moves produced was 34 moves long, and transformed $(CacAA, AbaBB, BcbCC)$ into $(CacAA, bAAAABaabC, bCCBc)$.

A much shorter sequence can be found by searching directly for a sequence of Andrews–Curtis moves which produce a relator of the desired form. As part of the work described in [13] we produced a utility ACME (Andrews–Curtis move enumerator) which performs a breadth-first search through a tree of Andrews–Curtis equivalent presentations. It is a simple matter to modify this to check each new relator as it is generated to see whether it contains some group generator once only. If the intermediate presentations generated are allowed to grow to a total length of 21, then ACME can produce a 20-move sequence. For a length bound of 22, ACME can produce the 19-move sequence illustrated in Figure 5. (These searches took about 92 and 517 minutes, respectively, of CPU time on a 733 MHz Itanium 1 machine, and used about 3 and 12 gigabytes of memory.)

ACME was able to generate all cyclic permutations and inversions of the word $BaabCbAAAA$, as well as the equivalent proofs for $a$ and $b$ (60 proofs in all), but was unable to produce any shorter word or any presentation of shorter total length. (In [13] we were able to show that some move sequences had minimal length. Here, the searches were non-exhaustive, so we cannot say whether or not any shorter word can be generated, or whether there is a shorter sequence of moves.) The sequence of moves in Figure 5 uses each of the relators the same number of times as Higman's proof and as $\Upsilon$, and is easily converted to a proof-word similar to $\Upsilon BAAbaaaaB$ (in fact, slightly shorter).

Note that Higman's proof that $c = bAAAABaab$ can be extended by using the relator $AbaBB$ twice to yield $c = bAABBB$. (However, Andrews–Curtis and ACME proofs cannot be simply extended to reproduce this, since the relator $AbaBB$ has been 'forgotten' by the time it is needed.) Continuing on, if we conjugate $c = bAABBB$ by $b$ and then use the relator $BcbCC$, we obtain $c^2 = A^2B^2$ (cf. Neumann's proof in Section 6.2).

We note that Rapaport [30] also reduced the problem from a presentation on three generators to one on two generators; however her proof used automorphisms as well as Andrews–Curtis moves. Automorphisms can always be eliminated from a successful reduction to the standard presentation, but not necessarily from a partial reduction. As far as we are aware, our reduction is the first direct proof that the Andrews–Curtis conjecture for $E_1$ can be reduced to a two generator problem. (Of course, this result is already implicit in Higman's work, which predated the conjecture.)

The various methods we have illustrated all yield essentially the same result; i.e., a particular relation which holds in $E_1$ and which proves that one of the generators can be written in terms of the other two. This raises the question as to whether or not this relation is the shortest of its kind in $E_1$. Of course, since $E_1$ is trivial *any* relation is true in it, so this not a well-formed question. However, it is meaningful to ask whether $\Upsilon$ is a shortest proof-word for this relation, or for any relation of this kind. $\Upsilon$ was found as part of a large, non-deterministic search and, in principle, we can establish whether or not it is a minimal proof-word via an exhaustive search of all proof-words containing fewer than 53 group generator symbols. However, the search-space is very large, and we have not investigated this further.

## 6.2 $E_1$ and cyclic subgroups

Neumann [27] proved $E_1$ trivial by first showing that $c^2 = b^3$, and so $b$ and $c^2$ commute. With this, $BcbCC = 1$ implies that $c$ is trivial, and the result follows. We elaborate the details here to enable easy counting of relator use. To establish $c^2 = b^3$, Neumann started by noting that $B^iab^i = aB^i$ (1). The relator $AbaBB$ yields that $aBB = Ba$ (2) and $aB = Bab$ (3). Conjugating (3) by $b$ yields $Ba = BBabb$, and substitution using (2) gives $aBB = BBabb$. Repeated conjugation now gives (1) for general $i$. Similarly, $C^ibc^i = bC^i$ (4).

Now consider $CacAA = 1$, in the form $Cac = aa$, and conjugate by $b$ to give $BCacb = Baab$ (5). The left-hand side $BC.a.cb = CCBabcc$ using the relator $BcbCC$, and (1) now gives $CCaBcc$. Now $CacAA = 1$ implies $Ca = aaC$, and substituting repeatedly for $Ca$ implies $CCaBcc = aaaaCCBcc$. Now use (4), in the form $C^2Bc^2 = c^2B$, to give $aaaaccB$. For the right-hand side of (5), $Baab = BabBab = aBaB = aaBBB$, using (1) and (2). Thus $aaaaccB = aaBBB$, and so $cc = AABB$ (6).

Now take (4), in the form $bc^2 = c^2bC^2$, and substitute using (6). This gives $bcc = AABBbbbaa = A.Aba.a = Abba$, using the relator $AbaBB$. But $Abba = Aba.Aba = bbbb$, again using the relator $AbaBB$. Thus, $bcc = bbbb$ and so $cc = bbb$.

Neumann's proof suggests that we should attempt to extract proofs of the general form $W \in \langle x^i \rangle$, where $x$ is a generator of $E_1$, and $W$ is a word such as $y^j$ or $y^j z^k$. His proofs that $c^2 = A^2B^2$ and $c^2 = b^3$ use the group relators a total of 12 and 29 times, respectively, and provide convenient metrics against which to compare proof-words produced by PEACE. (Given $c^2 = b^3$, direct substitution in $BcbCC$ yields $c = b^3$, and so a single relator use completes the proof that $c$ is trivial.)

The shortest proof-words that PEACE was able to extract from enumerations over cyclic subgroups had 24 or 26 occurrences of the relators and proved that $c = b^3$ (or an equivalent). Although not the shortest proofs, the ones with 26 relators have an interesting structure. A typical example is $\Phi = [B]c(CbccB)\zeta\eta\zeta^{-1}C$, where

$$\begin{aligned}\zeta &= bbb(aaCAc)B(bABab)BAb(aCAca)B(bABab)aB(bABab)(BCbcc)\\ &\quad CC(BAbaB)bAB(acAAC)(cbCCB)b(caaCA)aBc(CbccB),\\ \eta &= [b]A(AbaBB)a[b](BAbaB)[b][b](BAbaB).\end{aligned}$$

The words $\zeta$ and $\eta$ prove, respectively, that $bbbaaB = C$ and $bAAbaaB = bbbb$, so $\zeta\eta\zeta^{-1}$ proves $b = Cb^4c$. The eleven relators in $\zeta$ match those in the extended version of Higman's proof given in Section 6.1, while the conjugation of $\eta$ by $\zeta^{-1}$ mimics a key step in Neumann's proof.

Our shortest proof-words, with 24 relators, are 20% shorter than Neumann's 30, and seem to have a variety of different structures. We have not been able to find any 'nice' characterisation of these proofs, and we simply report an example with a minimal number of group generator symbols. This has 163 symbols, and proves that $a = c^3$.

$$\begin{aligned}&aCBCBA(abbAB)B(bbABa)(ACacA)ba(ABabb)(CBcbC)bc(CBcbC)bc\\ &(CBcbC)A(acAAC)[c]B(baBBA)ab(baBBA)B(baBBA)Ab[C](cBCbc)[C]\\ &B(caaCA)a(cbCCB)A(abbAB)bac(CBcbC)(ccBCb)C(cBCbc)CB(ccBCb)\\ &bcAAC(caaCA)(aBBAb)c(CBcbC)(caaCA)a[c][c](CBcbC)[c][c].\end{aligned}$$

### 6.3 Proof variability

To illustrate the extent of variation in the proofs produced by PEACE, we collected statistics from one set of runs for $E_1$ over cyclic subgroups. This involved 1000 coset enumerations over each of $\langle a \rangle$, $\langle b \rangle$ and $\langle c \rangle$, where the relators were randomly permuted, cycled and inverted before each enumeration. For each enumeration the number of cosets was pruned, and then proofs that each of the other two group generators are trivial (i.e., in the subgroup) were extracted. In the 3000 enumerations, the initial number of cosets varied from 17 to 112, and the pruned number varied from 10 to 56; the ratio of initial to pruned varied from 1.125 to 7.385. In the 6000 proofs, the number of occurrences of group relators varied from 26 to 39721; the median value was 961, and the lower and upper quartiles were 393 and 1669 respectively. There were two proofs with 26 relators, and these both had a pruned coset count of 19. Thus enumerations and proofs are very variable, while succinct proofs are uncommon and are not necessarily associated with short enumerations.

The shortest enumerations we were able to find used ten cosets; i.e., made nine definitions. To check that this is the best possible, we need to enumerate and test all different eight-definition sequences. Now each of these definitions must fall within the first eight rows in the coset-table and, since the table has six columns (there are no involutions), must fall in one of 48 table positions. Each definition uses two positions (the definition and its inverse), and processing the subgroup generator fills an initial pair; thus we have 23 pairs to choose from. Since each pair can be filled in two ways, the total number of possible *sets* of eight definitions is $2^8\binom{23}{8} = 125520384$. Each such set corresponds to a unique standardised definition sequence, if we order the definitions. (For example, first by coset number and then by column number: $1a$, $1A$, $1b$, $1B$, $1c$, $1C$, $2a$, ..., $8c$, $8C$. We have to take care that each definition uses an already defined coset, but, other than that, reordering the sequence simply relabels the cosets.) A similar calculation yields a total of 1599769600 possible sets for nine-definition sequences.

These counts are overestimates, since many of the sequences are not valid. For example, each definition in a valid sequence must use an already defined coset, so the first definition must be drawn from $\{1a, 1A, 1b, 1B, 1c, 1C\}$. It is easy to enumerate valid sequences in order using a backtrack search, and then to test whether or not they yield a complete coset table.

There are 16112057 valid standardised definition sequences of eight definitions (for each choice of subgroup generator), and to generate and test them took about 6 minutes of CPU time on a 400 MHz SPARC machine. None of these sequences generated a complete coset-table. For nine definitions, the corresponding figures are 167710664 sequences and about 68 minutes. 564 of these sequences generate a complete coset-table, and the first and last definition sequences in order are: $1B$, $1C$, $2a$, $2C$, $3C$, $4C$, $5C$, $6C$, $8b$ and $1C$, $2C$, $3C$, $4C$, $5B$, $6c$, $7c$, $8B$, $9C$ (for enumerations over $\langle a \rangle$ with columns ordered $a$, $A$, $b$, $B$, $c$, $C$).

Leech [21, §12] gave an example of a coset enumeration wherein, to achieve an enumeration with the smallest number of definitions, a particular definition had to be avoided. He also noted that it would be difficult to "devise a computer procedure" to find such a sequence. It is interesting to note that none of the 564 sequences (for each of $\langle a \rangle$, $\langle b \rangle$ and $\langle c \rangle$) our searches generated used the definitions $1a$, $1b$ or $1c$, implying that these definitions must be avoided if a shortest definition sequence is required.

### 6.4 $E_1$ over the trivial subgroup

For comparison with the previous results, we also performed coset enumerations over the trivial subgroup. The best enumerations, after pruning, completed with only 56 definitions; the best previous result seems to be 59 [7, §6.1]. We also built a variety of proof-words. The shortest that we found contained 49 relators, and proved that one of the group's generator is trivial. Not surprisingly, we obtained no new insights into why the group is trivial.

## 7 Conclusions

We have described tools which enable us to investigate proofs in finitely presented groups. We have used these to throw light on various kinds of proofs, in particular proofs based upon coset enumeration. Our aim was to clarify the nature of some proofs and to start addressing Hilbert's 24th problem about simplest proofs.

We introduced the notion of a proof-certificate for proofs which show that a specific word is constructively in a subgroup of a finitely presented group. We used the length of the certificate as a measure of simplicity. This is readily compared to the total number of cosets defined in an enumeration as an alternative measure. We also showed how proof certificates can be converted into other forms of proof.

We studied four specific examples for which coset enumeration based proofs rely on total collapse. We comment on each of these in turn.

For the Fibonacci group $F(2,5)$ we illustrated a range of proofs that could be derived from coset enumerations using the initial presentation and we gave some attractive proof-words. We also showed how to translate these into alternative proofs. The best coset enumerations use only 2 definitions, so are indeed quite simple. We have not determined the length of the shortest useful proof-word for this case, but give a proof word which is both concise and has some meaningful structure. This proof-word uses the relators 13 times and contains only 49 generator symbols. We conjecture that it is a shortest useful proof-word for this situation.

To compare it with human proofs we see that Conway [5] explicitly gave two solutions to the problem, by J.A. Wenzel and by R.C. Lyndon, and acknowledged 50 more (one of which explicitly used coset enumeration). Presumably he chose the published solutions for their elegance. We note that both published solutions start by reducing the 5-generator, 5-relator presentation to a 2-generator, 2-relator presentation. If we count relator use in the explicit way it is done for proof-words, this first step uses the original relators 10 times and precedes some more manipulations which take a few lines. All up, they (implicitly) use the relators more than our proof-word. By that measure, our proof is simpler.

A 2-generator, 2-relator presentation for $F(2,5)$ is $\langle a,b \mid abaBabb, abbabAbab\rangle$. With this as a basis for coset enumeration, PEACE produces the following proof-word for $a = B^8$ from a coset enumeration over $\langle b\rangle$:

$$
\begin{array}{lllll}
\;\;\,[B]aabb & (abaBabb) & BBAA & & \\
[B][B]A & (abbabAbab) & \qquad\;\; a & [B]abb(BBAbABA)BBA & \\
[B][B]AAbab & (abbabAbab) & BABaa & [B]abb(BBAbABA)BBA & [B]
\end{array}
$$

It uses its relators five times and fits easily on two lines, but is shown spaced out over three to highlight its structure.

For the Fibonacci group $F(3,5)$ we illustrated how a more difficult problem could be addressed in the same kind of way. Indeed we obtain concise coset definition sequences and moderate length proof-words. For the Fibonacci group $F(2,7)$, a harder problem again, we demonstrated how much better coset enumeration based proofs could now be done than before. However we obtained no new insights into why these groups are cyclic.

Consideration of the trivial group $E_1$ introduces additional proof techniques related to the Andrews–Curtis conjecture. We observe that our best proof-word based on 2-generator subgroups corresponds closely to a long established proof by Higman. We conjecture that these indeed provide simplest proofs. We showed that the Andrews–Curtis conjecture for $E_1$ is not a totally genuine 3-generator problem in the sense that resolution of the conjecture for a 2-generator example would resolve it for $E_1$. Moving on to cyclic subgroups, we showed that our best proof-word is substantially simpler than Neumann's corresponding proof, in terms of its use of relators.

We used coset enumerations over cyclic subgroups in $E_1$ to investigate shortest definition sequences. We were able to determine the length of these by a brute force calculation. On the other hand we have not determined the length of the shortest possible proof-word. Rather we showed that the lengths of proof-words produced by PEACE are very variable and that short ones are hard to find. In spite of this we were successful in finding some quite short proof-words.

Our experiments show that shortest proof-words are not necessarily obtained from shortest coset enumerations. The sizes of the enumerations and proof-words produced by PEACE are very variable. To extract short enumerations and proof-words we had to run large numbers of enumerations, while varying the enumeration parameters, the presentation's form, the enumeration pruning, and the word to be proved. As yet we have no good methods for directly producing short enumerations or for predicting which enumerations will yield good proofs. However we have been successful in producing short proofs which perhaps meet Hilbert's simplicity criteria, even when they do not provide natural insights as to why they work.

Finally, we point out that proof certificates can do more than just give us access to another method of proof and to some simple proofs. They can sometimes provide important insights. Automatic generation of proof certificates was a key step in resolving a long-standing conjecture [15] about an infinite family of groups. By studying proof certificates for particular cases we were able to develop completely general proofs.

## References


[1] J. J. Andrews and M. L. Curtis, Free groups and handlebodies, *Proc. Amer. Math. Soc.* **16** (1965) 192–195.

[2] J. J. Andrews and M. L. Curtis, Extended Nielsen operations in free groups, *Amer. Math. Monthly* **73** (1966) 21–28.

[3] W. Bosma, J. Cannon and C. Playoust, The Magma algebra system I: the user language, *J. Symbolic Comput.* **24** (1997) 235–265.
See also `http://magma.maths.usyd.edu.au/magma/`

[4] J. H. Conway, Advanced problem 5327, *Amer. Math. Monthly* **72** (1965) 915.

[5] J. H. Conway et al., Generators and relations for cyclic groups (solution to advanced problem 5327), *Amer. Math. Monthly* **74** (1967) 91–93.

[6] Margaret Edeson, *Investigations in Coset Enumeration*, Master's thesis, Canberra College of Advanced Education, 1989.

[7] Volkmar Felsch, Ludger Hippe and Joachim Neubüser, *GAP package ITC Interactive Todd–Coxeter*, 2004. `http://www.gap-system.org/Packages/itc.html`

[8] The GAP Group, *GAP – Groups, Algorithms, and Programming, Version* 4.4, 2004. `http://www.gap-system.org/`

[9] I. Grattan-Guinness, Response to Moore's letter, *Notices Amer. Math. Soc.* **48** (2001) 167.

[10] George Havas, Computer aided determination of a Fibonacci group, *Bull. Austral. Math. Soc.* **15** (1976) 297–305.

[11] George Havas and Colin Ramsay, Proving a group trivial made easy: a case study in coset enumeration, *Bull. Austral. Math. Soc.* **62** (2000) 105–118.

[12] George Havas and Colin Ramsay, *Coset enumeration: ACE version 3.001*, 2001; `http://www.itee.uq.edu.au/~havas/ace3001.tar.gz`

[13] George Havas and Colin Ramsay, Breadth-first search and the Andrews–Curtis conjecture, *Internat. J. Algebra Comput.* **13** (2003) 61–68.

[14] George Havas and Colin Ramsay, Andrews–Curtis and Todd–Coxeter proof words, in *Groups St Andrews 2001 in Oxford*, London Math. Soc. Lecture Note Ser. **304**, 232–237, Cambridge University Press, 2003.

[15] George Havas, Edmund F. Robertson and Dale C. Sutherland, The $F^{a,b,c}$ conjecture is true, II, *J. Algebra* **300** (2006), 57–72.

[16] Graham Higman, A finitely generated infinite simple group, *J. London Math. Soc.* **26** (1951) 61–64.

[17] Derek F. Holt, Bettina Eick and Eamonn A. O'Brien. *Handbook of Computational Group Theory*, CRC Press, 2005.

[18] D. L. Johnson, *Presentations of Groups*, London Math. Soc. Lecture Note Ser. **22**, 1976.

[19] D.L. Johnson, *Topics in the Theory of Group Presentations*, London Math. Soc. Lecture Note Ser. **42**, 1980.

[20] D. L. Johnson, J. W. Wamsley and D. Wright, The Fibonacci groups, *Proc. London Math. Soc. (3)* **29** (1974) 577–592.

[21] John Leech, Coset enumeration, In John Leech (ed.), *Computational Problems in Abstract Algebra*, pp. 21–35. Pergamon Press, 1970.

[22] John Leech, What is a proof? *Computer Bulletin* Sep. 1976, pp. 17, 29.

[23] John Leech, Computer proof of relations in groups, In Michael P.J. Curran (ed.), *Topics in Group Theory and Computation*, pp. 38–61. Academic Press, 1977.

[24] John Leech, Coset enumeration, In Michael D. Atkinson (ed.), *Computational Group Theory*, pp. 3–18. Academic Press, 1984.

[25] Roger C. Lyndon and Paul E. Schupp, *Combinatorial Group Theory*, Springer-Verlag, 1977.

[26] E. H. Moore, Concerning the abstract groups of order $k!$ and $\frac{1}{2}k!$ holohedrically isomorphic with the symmetric and the alternating substitution-groups on $k$ letters, *Proc. London Math. Soc. (1)* **28** (1897) 357–366.

[27] B. H. Neumann, An essay on free products of groups with amalgamations, *Philos. Trans. Roy. Soc. London (Ser. A)* **246** (1954), 503–554.

[28] B. H. Neumann, Proofs, *Math. Intelligencer* **2** (1979) 18–19.

[29] Colin Ramsay, PEACE 1.100: Proof Extraction After Coset Enumeration, Technical Report **22**, Centre for Discrete Mathematics and Computing, The University of Queensland, 2003.

[30] Elvira Strasser Rapaport, Groups of order 1: some properties of presentations, *Acta Math.* **121** (1968) 127–150.

[31] Paul E. Schupp, Groups and graphs, *Math. Intelligencer* **1** (1979) 205–208, 212–214.

[32] C. C. Sims, *Computation with finitely presented groups*, Cambridge University Press, 1994.

[33] Rüdiger Thiele, Hilbert's twenty-fourth problem, *Amer. Math. Monthly* **110** (2003) 1–24.

[34] Richard M. Thomas, The Fibonacci groups revisited, in C. M. Campbell and E. F. Robertson (eds.), *Groups St Andrews 1989, Volume 2*, pp. 445–454. London Mathematical Society Lecture Note Series **160**, 1991.

[35] J. A. Todd and H. S. M. Coxeter, A practical method for enumerating cosets of finite abstract groups, *Proc. Edinb. Math. Soc. (2)* **5** (1936) 26–34.

# COMPUTING WITH 4-ENGEL GROUPS

GEORGE HAVAS*[1] and M. R. VAUGHAN-LEE†

*ARC Centre for Complex Systems,
School of Information Technology and Electrical Engineering,
The University of Queensland, Queensland 4072, Australia
Email: havas@itee.uq.edu.au

†Christ Church, Oxford OX1 1DP, England
Email: michael.vaughan-lee@christ-church.oxford.ac.uk

## Abstract

We have proved that 4-Engel groups are locally nilpotent. The proof is based upon detailed computations by both hand and machine. Here we elaborate on explicit computer calculations which provided some of the motivation behind the proof. In particular we give details on the hardest coset enumerations now required to show in a direct proof that 4-Engel $p$-groups are locally finite for $5 \leq p \leq 31$. We provide a theoretical result which enables us to do requisite coset enumerations much better and we also give a new, tight bound on the class of 4-Engel 5-groups. In addition we give further information on use of the Knuth–Bendix procedure for verifying nilpotency of a finitely presented group.

## 1 Introduction

A group $G$ is said to be an $n$-Engel group if

$$[x, \underbrace{y, y, \ldots, y}_{n}] = 1$$

for all $x, y \in G$. Here $[x, y]$ denotes $x^{-1}y^{-1}xy$, and $[x, y, \ldots, y]$ denotes the left-normed commutator $\Big[[\ldots[x, y], \ldots], y\Big]$.

Questions about nilpotency of groups satisfying Engel conditions have been long considered and in 1936 Zorn proved that finite Engel groups are nilpotent. In [6] (where there is more historical background) we proved

**Theorem 1.1** *4-Engel groups are locally nilpotent.*

This result came after a number of results pointing in this direction. Traustason [9] showed that if 4-Engel groups of exponent $p$ are locally finite, then 4-Engel $p$-groups are locally finite. Vaughan-Lee [13] proved that 4-Engel groups of exponent 5 are locally finite, and it follows from this and Traustason's work that 4-Engel 5-groups are locally finite. Then Traustason [11] proved that 2-generator, 4-Engel groups are nilpotent.

[1]The first author was partially supported by the Australian Research Council.

We started by aiming to extend the results of Vaughan-Lee [13], which showed that 4-Engel 5-groups are locally finite, to $p$-groups for other primes. Vaughan-Lee's proof uses the $p$-quotient algorithm and coset enumeration. Here we give details on a key step in proving local finiteness for primes $p$ up to 31.

These results convinced us that the general result is true and provided enough motivation for us to carry through our general proof. The proofs for general 4-Engel groups and for 4-Engel $p$-groups follow the same line of argument as is used in [13] for 5-groups. Given our general proof and other results, here we focus on one particular part of a direct proof for small $p$.

## 2 4-Engel 5-groups

The proof for 4-Engel 5-groups is based on the following theorem which is proved in [13].

**Theorem 2.1** *4-Engel groups of exponent* 5 *satisfy the identical relation*

$$[x, [y, z, z, z], [y, z, z, z], [y, z, z, z]] = 1.$$

This means that it suffices to consider 3-generator groups. The proof in [13] relies on a number of coset enumerations, and on a number of calculations with the $p$-quotient algorithm in 2- and 3-generator groups. Enough information is given to allow verification of the computations. In view of Traustason's result that 2-generator, 4-Engel groups are nilpotent and the straightforward nature of the $p$-quotient calculations, we focus on the most challenging 3-generator coset enumeration and its analogue for general $p$.

The computer tools that we use include the computer algebra systems GAP [2] and MAGMA [1]. We use the ACE coset enumerator (Havas and Ramsay [4]) either as available in GAP or MAGMA, or as a stand-alone program for some more difficult cases. For an up to date description of coset enumeration see Sims [8]. We need to complete various coset enumerations. To do so we investigate presentations in a similar way that other problems are addressed in [3]. Enumeration strategies are discussed in detail in [5] but not considered in this paper. Here we simply use an enumeration strategy which worked well enough in practice.

In Lemma 9 of [13] it is shown that the subgroup $\langle uv, vw\rangle$ has finite index, $5^4$, in the group $G = \langle u, v, w \mid R_1\rangle$ where

$$R_1 = \begin{cases} u^5 = v^5 = w^5 = 1 \\ (u^r v^s w^t)^5 = 1 \text{ for } r = 1, 2 \text{ and } s, t = \pm 1, \pm 2 \\ [u, v] = 1 \\ [w, u, u] = [u, w, w, w] = 1 \\ [v, w, w, w] = [v, w, w, v] = [v, w, v, w] = [v, w, v, v] = 1 \\ [w, u, w, v, v] = 1 \end{cases}$$

The coset enumeration is moderately difficult. Advances in coset enumeration mean that more recent enumerators define a total of 136926 cosets to complete the enumeration, in comparison to 276037 (with the same MAGMA code shown in [13] but using MAGMA V2.11, which incorporates an updated coset enumerator, instead of MAGMA V1.01). Notice that here we impose a collection of 5-th powers and some commutator relations, but no explicit 4-Engel relations. (We give actual coset enumeration performance figures throughout this paper, but be aware that the figures depend critically on exact details of presentations and enumeration strategies used. It should not surprise if related computations differ in statistics such as total cosets.)

## 3 4-Engel $p$-groups; theory

The proof for 5-groups can be extended to other $p$-groups by following a similar line of argument. In view of our result that 4-Engel groups are locally nilpotent we omit most of the details. Briefly, for an analogous sequence of groups we need to show that specific subgroups have finite index. The corresponding problem which we need to solve to obtain the analogue of Lemma 9 for general $p$ requires us to replace 5-th powers in $R_1$ by $p$-th powers.

Our initial attempts at this were unsuccessful and it is instructive to understand why we failed. We started with the same kinds of relations, adding extra $p$-th powers, but the coset enumerations did not complete.

We were trying to enumerate the cosets of the subgroup $\langle uv, vw\rangle$ in the group $G = \langle u, v, w \mid R,\ \text{exponent } p\rangle$ where

$$R = \begin{cases} [u, v] = 1 \\ [w, u, u] = [u, w, w, w] = 1 \\ [v, w, w, w] = [v, w, w, v] = [v, w, v, w] = [v, w, v, v] = 1 \\ [w, u, w, v, v] = 1 \end{cases}$$

For $p = 7$ we tried various sets of 7-th powers in an effort to obtain the hypothetical index, $7^4$, always without success. We were in fact studying the group

$$H_p = \langle u, v, w \mid R,\ \text{exponent } p\rangle$$

instead of the group

$$G_p = \langle u, v, w \mid R,\ \text{exponent } p,\ \text{4-Engel}\rangle.$$

The largest nilpotent quotient of $G_p$ has class 4 and order $p^{12}$ for $p \geq 5$. The same is true for $H_5$, but for $H_7$ it has class 6 and order $7^{17}$ and for $H_p$, $p \geq 11$, it has class 7 and order $p^{19}$. Our results [6] and those of Traustason [12] already imply that $G_p$ is indeed nilpotent, and a tough coset enumeration can show that the same is true for $H_5$. The subgroup $\langle v, w\rangle$, which is easily shown to have order $5^5$, can be shown to have index $5^7$ using the presentation $\langle u, v, w \mid R_1\rangle$ of [13]. That enumeration uses a total of 122623883 cosets, but no doubt could be done much better using other presentations for preimages of $H_5$. However we do not know whether $H_p$ is nilpotent or not for $p \geq 7$.

This realization led us to use extra relations in addition to exponent relations for $p = 7$. Whereas exponent 5 in the context of $R$ was adequate for our purposes, this was not so for $p = 7$, the next case. The initial solution, once found, is easy. Impose only three $p$-th powers, but also impose extra 4-Engel relations. That we need only three $p$-th powers is a consequence of the following theorem.

**Theorem 3.1** *Let $G$ be a 4-Engel group generated by elements of prime order $p$, where $p \geq 5$. Then $G$ has exponent $p$.*

**Proof** We know from Traustason's result [11] that 2-generator, 4-Engel groups are nilpotent. The nilpotent quotient algorithm then shows that 2-generator, 4-Engel groups have class at most 6. Now let $G$ be a 4-Engel group generated by elements of order $p$ for $p \geq 5$. To show that $G$ has exponent $p$ we need to show that if $a, b \in G$ have order $p$ then $(ab)^p = 1$. The remarks above show that the subgroup $\langle a, b\rangle$ has class at most 6, and so if $p \geq 7$ then $\langle a, b\rangle$ is a regular $p$-group, and has exponent $p$. In particular $(ab)^p = 1$. When $p = 5$ the following simple MAGMA program shows that $(ab)^p = 1$.

```
G := Group<a, b | a^5, b^5>;
Q := NilpotentQuotient(G, 7: Engel := 4);
(Q.1*Q.2)^5;
```

□

**Remark 3.2** The argument used in the proof above shows that 4-Engel $p$-groups are regular for $p \geq 7$. It is of interest to note that 4-Engel 5-groups are also regular. It is easy to verify using the nilpotent quotient algorithm that in a 4-Engel group

$$\begin{aligned}(ab)^5 &= a^5b^5[b,a]^{10}[b,a,a]^{10}[b,a,b]^{30}[b,a,a,a]^5[b,a,b,a]^{35}[b,a,b,b]^{35}\\ &\quad [b,a,b,a,a]^{10}[b,a,b,a,b]^5[b,a,b,b,a]^{170}[b,a,b,b,a,a]^{330}.\end{aligned}$$

These and related considerations enable us to provide tight bounds on the class of 4-Engel 5-groups. Corollary 2 of [6] states that if $G$ is an $m$-generator, 4-Engel group then $G$ is nilpotent of class at most $4m$, and if $G$ has no elements of order 2, 3 or 5 then $G$ is nilpotent of class at most 7. Given our result that 4-Engel groups are locally nilpotent, this corollary follows from Traustason's work [9, 10] on the class of locally nilpotent 4-Engel groups. But if we restrict our attention to $p$-groups then we can sharpen this corollary. Traustason's work shows that if $p > 5$ then a 4-Engel $p$-group has class at most 7. He also shows that in a 4-Engel 3-group the normal closure of an element is nilpotent of class at most 3, and this implies that an $m$-generator, 4-Engel 3-group has class at most $3m$. The bound of $4m$ in Corollary 2 of [6] comes from Traustason's result that in locally nilpotent 2-groups and 5-groups the normal closure of an element is nilpotent of class at most 4. This bound is sharp, since in the free 4-Engel group of rank 3 the element $[x, y, x, x, z, x]$ has order 10. Despite this, it turns out that we can do better than $4m$ in the case of 5-groups.

**Theorem 3.3** *Let $G$ be an $m$-generator, 4-Engel 5-group. If $m = 2$ then the class of $G$ is at most 6, if $m = 3$ then the class of $G$ is at most 8, and if $m > 3$ then the class of $G$ is at most $2m$. Furthermore, these class bounds are attained if $G$ is the free 4-Engel group of exponent 5 and rank $m$.*

**Proof** Newman and Vaughan-Lee [7] proved that the free 4-Engel group of exponent 5 and rank $m$ has class 6 if $m = 2$, class 8 if $m = 3$, and class $2m$ if $m > 3$. They first established that the associated Lie rings of free 4-Engel groups of exponent 5 are free Lie rings in the variety of Lie rings determined by the following multilinear Lie identities

$$\begin{aligned} 5x &= 0, \\ \sum_{\sigma \in \mathrm{Sym}(4)} [x, y_{1\sigma}, y_{2\sigma}, y_{3\sigma}, y_{4\sigma}] &= 0, \\ \sum_{\sigma \in \mathrm{Sym}(4)} [y_{1\sigma}, x_1, x_2, y_{2\sigma}, y_{3\sigma}, y_{4\sigma}] &= 0, \\ \sum_{\sigma \in \mathrm{Sym}(4)} [y_{1\sigma}, x_1, x_2, x_3, y_{2\sigma}, y_{3\sigma}, y_{4\sigma}] &= 0. \end{aligned}$$

They then computed the class and the order of the free Lie ring of rank $m$ in this variety. Now let $G$ be an $m$-generator, 4-Engel 5-group, and let $L$ be the associated Lie ring of $G$. We will show that $L/5L$ satisfies these 4 multilinear identities. So Newman and Vaughan-Lee's result shows that $L/5L$ has class at most 6 if $m = 2$, class at most 8 if $m = 3$, and class at most $2m$ if $m > 3$. But $L/5L$, $L$ and $G$ all have the same nilpotency class, and so this proves the theorem.

It remains to show that $L/5L$ satisfies these 4 identities. Clearly $L/5L$ satisfies the identity $5x = 0$. The other 3 Lie identities are direct consequences of the group identity $[x, y, y, y, y] = 1$ in $G$. This follows from Wall's theory of multilinear Lie relators in varieties of groups [14], but we can also see it directly. The identity $\sum_{\sigma \in \mathrm{Sym}(4)} [x, y_{1\sigma}, y_{2\sigma}, y_{3\sigma}, y_{4\sigma}] = 0$ comes from expanding the group commutator

$$[x, y_1y_2y_3y_4, y_1y_2y_3y_4, y_1y_2y_3y_4, y_1y_2y_3y_4]$$

and picking out the terms which involve all the elements $y_1, y_2, y_3, y_4$. Since $G$ is a 4-Engel group, if $x, y_1, y_2, y_3, y_4 \in G$ we obtain the group relation

$$\prod_{\sigma \in \mathrm{Sym}(4)} [x, y_{1\sigma}, y_{2\sigma}, y_{3\sigma}, y_{4\sigma}] \in \gamma_6(G),$$

and hence the Lie identity

$$\sum_{\sigma \in \mathrm{Sym}(4)} [x, y_{1\sigma}, y_{2\sigma}, y_{3\sigma}, y_{4\sigma}] = 0$$

in the associated Lie ring of $G$.

Next, consider the identity $[x_1x_2, y, y, y, y] = 1$ which holds in 4-Engel groups. Expanding, and using the fact that the normal closure of $y$ in a 4-Engel group is nilpotent of class at most 4, we obtain

$$[x_1, y, y, y, y][x_1, y, x_2, y, y, y][x_2, y, y, y, y] = 1,$$

which gives the identity

$$[x_1, y, x_2, y, y, y] = 1.$$

We obtain the Lie identity

$$\sum_{\sigma\in\mathrm{Sym}(4)} [y_{1\sigma}, x_1, x_2, y_{2\sigma}, y_{3\sigma}, y_{4\sigma}] = 0$$

by substituting $y_1y_2y_3y_4$ for $y$, and expanding as above.

We similarly obtain the Lie identity

$$\sum_{\sigma\in\mathrm{Sym}(4)} [y_{1\sigma}, x_1, x_2, x_3, y_{2\sigma}, y_{3\sigma}, y_{4\sigma}] = 0$$

from the group identity

$$[x_1x_2x_3, y, y, y, y] = 1.$$

□

## 4 4-Engel $p$-groups; coset enumerations

Initially we started with the 24 4-Engel relations

$$[u, x, x, x, x],\ [v, x, x, x, x],\ [w, x, x, x, x]$$

where $x = u^{\pm1}v^{\pm1}w^{\pm1}$. Using these relations together with $u^p$, $v^p$, $w^p$ and $R$ to give a presentation ${}_{24}G_p$ for a preimage of $G_p$ we can readily enough find that $\langle uv, vw\rangle$ has index $p^4$ in $G_7$ by coset enumeration, in a total of 1631060 cosets. A MAGMA program to do this kind of computation is in Appendix A.

With a limit of 6 million cosets, the equivalent enumerations work for $p = 11$ and $p = 13$, but overflow for greater primes. For more information see Table 1. The successful enumerations in this Table for ${}_{24}G_p$ are already enough to suggest that the result holds for all $p$.

However, having succeeded with these enumerations, we can both simplify and generalize them. What we really are trying to do here is to find preimages of

$$\langle u, v, w \mid R,\ \text{exponent } p,\ \text{4-Engel}\rangle$$

in which $\langle uv, vw\rangle$ can be shown to have index $p^4$ by coset enumeration and with correct maximal $p$-quotient.

The $p$-quotient part of this is easy. Determining how to solve coset enumeration problems well is more an art than a science. Two major issues arise: finding appropriate presentations and finding appropriate enumeration strategies.

In the present case we found better presentations for our purposes which also revealed a surprise. Start with the 35 relations comprising $u^p$, $v^p$ and $w^p$ together with $R$ plus the 24 4-Engel relations above and simplify them by using Tietze transformations (readily done using either GAP or MAGMA; we give a MAGMA program which does this and investigates the presentations in Appendix B). Then take the first 19 relations obtained this way (comprising 3 power relations, 8 consequences of $R$ and 8 consequences of 4-Engel relations) to give a presentation for a preimage, ${}_8P_p$, of ${}_{24}G_p$. The group defined by ${}_8P_p$ still has the correct maximal

Table 1. Coset enumeration performance: ${}_{24}G_p$ and ${}_8P_p$

| $p$ | Index | Total cosets ${}_{24}G_p$ | Total cosets ${}_8P_p$ | cpu seconds ${}_{24}G_p$ | cpu seconds ${}_8P_p$ |
|---|---|---|---|---|---|
| 7 | 2401 | 1631060 | 127028 | 86 | 3 |
| 11 | 14641 | 2058258 | 632081 | 123 | 13 |
| 13 | 28561 | 5410257 | 2461035 | 339 | 51 |
| 17 | 83521 | 48875422 | 12201625 | 3720 | 325 |
| 19 | 130321 | 115063303 | 23231584 | 8781 | 702 |
| 23 | 279841 | 356330189 | 83165369 | 18574 | 2450 |
| 29 | 707281 | $>$ 500000000 | 370372846 | | 7383 |
| 31 | 923521 | $>$ 500000000 | 484110102 | | 10876 |

$p$-quotient but the presentatation is much better for coset enumeration as shown in Table 1. The choice of 8 relators on top of the power relators and $R$ was made somewhat arbitrarily: it worked relatively well. Not only were the total numbers of cosets significantly better for ${}_8P_p$ than for ${}_{24}G_p$ but in addition the cpu time improvements were even greater. Because the presentations are shorter, processing time per coset is reduced. Thus, for $p = 23$ the enumeration took 2450 versus 18574 cpu seconds on a SparcV9 1200MHz processor.

Experiments were undertaken with fewer extra relators revealing interesting results for $5 \le p \le 31$. We denote the simplified presentation obtained from the 3 $p$-th powers and $R$ plus the first (in the order given by our Magma code) $i$ 4-Engel consequences by ${}_iP_p$. Naturally enough we started with $p = 7$, the first unknown case at the time. Total cosets for successful enumerations were as follows — ${}_8P_7$: 127028; ${}_7P_7$: 123554; ${}_6P_7$: 117447; ${}_5P_7$: 173953; ${}_4P_7$: 175140; ${}_3P_7$: 1763201; ${}_2P_7$: 1664225; and the enumerations failed to complete for ${}_1P_7$ and ${}_0P_7$ with 400 million total cosets.

Further investigation showed that with $i = 3$ or more extra relations the maximal $p$-quotient of ${}_iP_7$ is the same as for $G_7$. However for ${}_2P_7$ the class of the maximal $p$-quotient went up by one to 5 and the order up by a factor of $p$ to $p^{13}$. For ${}_1P_7$ the class is also 5 but this time the order is up by a factor of $p^2$. For ${}_0P_7$ (just the powers and $R$), the class is 7 and the order is $p^{19}$. However, even though the maximal $p$-quotients vary in class and size, in all cases the image of the subgroup $\langle uv, vw\rangle$ has index $p^4$ in the maximal $p$-quotient.

Having observed this we repeated the experiments for the other primes in our range. For all of them we discovered the same story as far as $p$-quotients and subgroup index in the $p$-quotients is concerned. But some surprises were revealed in the coset enumerations. For all $11 \le p \le 31$ and all $0 \le i \le 8$ the subgroup $\langle uv, vw\rangle$ has index $p^4$ in ${}_iP_p$ and can be found by coset enumeration, with performance figures given in Table 2. Notice that the total number of cosets generally decreases with decreasing number of extra relations. Extra 4-Engel relations hinder rather than help coset enumeration prove that the relevant subgroup has finite index. The enumeration for ${}_0P_{23}$ took 525 versus 2450 cpu seconds for ${}_8P_{23}$ on a SparcV9

Table 2. Coset enumeration performance: ${}_0P_p$ to ${}_8P_p$

| Group | Total cosets | | | | | |
|---|---|---|---|---|---|---|
| | ${}_iP_{11}$ | ${}_iP_{13}$ | ${}_iP_{17}$ | ${}_iP_{19}$ | ${}_iP_{23}$ | ${}_iP_{29}$ |
| ${}_8P_p$ | 632081 | 2461035 | 12201625 | 23231584 | 83165369 | 370372846 |
| ${}_7P_p$ | 601107 | 2275056 | 11613947 | 25005869 | 64094145 | 294718450 |
| ${}_6P_p$ | 563866 | 2093938 | 10921180 | 20777684 | 70388182 | 268403697 |
| ${}_5P_p$ | 526937 | 2043525 | 10121438 | 19799079 | 62701875 | 227600531 |
| ${}_4P_p$ | 489386 | 1885669 | 9602477 | 18768587 | 58598226 | 259035552 |
| ${}_3P_p$ | 473244 | 1773915 | 8936948 | 17420731 | 59346654 | 215577073 |
| ${}_2P_p$ | 440993 | 1611026 | 8252655 | 17767232 | 48378123 | 189715507 |
| ${}_1P_p$ | 398821 | 1492580 | 7883161 | 15219869 | 52330359 | 186096576 |
| ${}_0P_p$ | 369260 | 1384228 | 7299078 | 13783117 | 50234568 | 170431569 |

1200MHz processor. Finally, for $p = 31$ we have ${}_0P_{31}$ uses 289274269 cosets (3438 seconds); ${}_1P_{31}$, 332261302; ${}_2P_{31}$, 329376595; ${}_3P_{31}$, 358424103; ${}_4P_{31}$, 397744988; ${}_5P_{31}$, 401683049; ${}_6P_{31}$, 425874010; ${}_7P_{31}$, 457785384; and ${}_8P_{31}$, 484110102 (10876 seconds).

In §3, we imposed extra 4-Engel relations to move from consideration of $H_p$ to $G_p$. Now we see that we need to do this for the coset enumeration for $p = 7$, but not for larger primes. This leads us back to $p = 5$ where we find that ${}_{24}G_5$ uses 1332997 cosets; ${}_8P_5$, 261641; ${}_7P_5$, 259354; ${}_6P_5$, 256839; ${}_5P_5$, 253161; ${}_4P_5$, 250230; ${}_3P_5$, 247629; and the enumerations fail to complete for ${}_2P_5$, ${}_1P_5$ and ${}_0P_5$ in 400 million cosets.

We can draw the following conclusions from these computations. First and most important, we can prove a key step in a direct proof that 4-Engel $p$-groups are locally finite for $5 \leq p \leq 31$ by practical coset enumeration, but some care is needed to do so. (Note that the performance figures for $p = 5$ indicate that 5-th powers as used in [13] enable coset enumeration to work more easily than 4-Engel relations.)

There are a number of unresolved questions which arise. Our calculations end up showing that finitely generated 4-Engel $p$-groups are finite. However the status of some groups we investigated in the process is unclear. Is the group $H_p$ finite for $p \geq 7$? Which groups constructed along the lines of the ${}_iP_p$ are finite? (Finiteness of 3-generator, 4-Engel $p$-groups implies that finite presentations like this do exist.) Why do the coset enumerations for ${}_1P_7$, ${}_0P_7$, ${}_2P_5$, ${}_1P_5$ and ${}_0P_5$ fail to complete?

## 5 Proving $T$ nilpotent

The chronologically last step that we completed for our proof that 4-Engel groups are locally nilpotent is explained in [6, §4]. There we show that the group $T$ presented by

$$\langle u, v, w \mid [u, v], [w, u, u, u], [w, u, u, w], [w, u, w, w], [w, v, v], [w, v, w], \text{4-Engel}\rangle$$

is nilpotent. Our proof used difficult computations with implementations of the Knuth–Bendix procedure. Subsequently, Traustason [12] found a very clever proof of the nilpotence of $T$ which does not use the Knuth–Bendix procedure.

In our proof, we prefaced use of the Knuth–Bendix procedure by determining separately some additional relations which hold in $T$. Further study of the relevant computations, aided by suggestions from Charles Sims, reveals that we can use the Knuth–Bendix procedure to prove nilpotence without explicitly adding these extra relations.

Briefly, by adding more redundant generators and by altering the sequence of Knuth–Bendix iterations we can first obtain all of the required additional relations and subsequently deduce that $T$ is nilpotent. This makes this part of the proof both shorter and faster.

## References

[1] W. Bosma, J. Cannon and C. Playoust, The Magma algebra system I: the user language, *J. Symbolic Comput.* **24** (1997) 235–265.
See also `http://magma.maths.usyd.edu.au/magma/`

[2] The GAP Group, *GAP – Groups, Algorithms, and Programming, Version* 4.4, 2004; `http://www.gap-system.org/`

[3] George Havas, M. F. Newman, Alice C. Niemeyer and Charles C. Sims, Computing in groups with exponent six, in *Computational and Geometric Aspects of Modern Algebra*, London Mathematical Society Lecture Note Series **275**, 87–100, Cambridge University Press, 2000.

[4] George Havas and Colin Ramsay, *Coset enumeration: ACE version* 3.001, 2001; `http://www.itee.uq.edu.au/~havas/ace3001.tar.gz`

[5] George Havas and Colin Ramsay, Experiments in coset enumeration, in *Groups and Computation III*, Ohio State University Mathematical Research Institute Publications **8**, 183–192, de Gruyter, 2001.

[6] George Havas and M. R. Vaughan-Lee, 4-Engel groups are locally nilpotent, *Internat. J. Algebra Comput.* **15** (2005) 649–682.

[7] M. F. Newman and Michael Vaughan-Lee, Engel-4 groups of exponent 5 II. Orders, *Proc. London Math. Soc.* (3) **79** (1999) 283–317.

[8] C. C. Sims, *Computation with finitely presented groups*, Cambridge University Press, 1994.

[9] G. Traustason, On 4-Engel groups, *J. Algebra* **178** (1995) 414–429.

[10] Gunnar Traustason, Locally nilpotent 4-Engel groups are Fitting groups, *J. Algebra* **270** (2003) 7–27.

[11] Gunnar Traustason, Two generator 4-Engel groups, *Internat. J. Algebra Comput.* **15** (2005) 309–316.

[12] Gunnar Traustason, A note on the local nilpotence of 4-Engel groups, *Internat. J. Algebra Comput.* **15** (2005) 757–764.

[13] M.R. Vaughan-Lee, Engel-4 groups of exponent 5, *Proc. London Math. Soc.* **74** (1997) 306–334.

[14] G.E. Wall, Multilinear Lie relators for varieties of groups, *J. Algebra* **157** (1993) 341–393.

## A MAGMA program for coset enumeration

The following program computes the index of $\langle uv, vw\rangle$ in ${}_{24}G_p$ and provides extra information including details of its maximal $p$-quotient.

```
// Edit the following line for different primes
p := 7;
G := Group<u, v, w | u^p, v^p, w^p,
// Commutator relations R
(u,v), (w,u,u), (u,w,w,w), (w,u,w,v,v),
(v,w,w,w), (v,w,w,v), (v,w,v,w), (v,w,v,v),
// 24 4-Engel relations
[ (u,x,x,x,x): x in
  [ u^e1 * v^e2 *w^e3 :e1 in [1,-1], e2 in [1,-1], e3 in [1,-1] ] ],
[ (v,x,x,x,x): x in
  [ u^e1 * v^e2 *w^e3 :e1 in [1,-1], e2 in [1,-1], e3 in [1,-1] ] ],
[ (w,x,x,x,x): x in
  [ u^e1 * v^e2 *w^e3 :e1 in [1,-1], e2 in [1,-1], e3 in [1,-1] ] ] >;

Q := NilpotentQuotient(G,8);
"Max NQ of G has class", NilpotencyClass(Q),
            "and order", FactoredOrder(Q);
H := sub<Q | Q.1*Q.2, Q.2*Q.3>;
"In the max NQ the subgroup image has index", FactoredIndex(Q,H);
H := sub<G | G.1*G.2, G.2*G.3>;
I,_,M,T := ToddCoxeter(G, H : CosetLimit:=6000000,
                      Strategy:=<1000,1>, SubgroupRelations:=1);
"Index", I,"/",M,"/",T ;
```

## B MAGMA program for presentation simplification

The following program performs coset enumerations for the groups defined by the simplified presentations ${}_iP_p$ and provides extra information including details of maximal $p$-quotients and information on presentation lengths.

```
// Edit the following line for different primes
p := 7;
G := Group<u, v, w | u^p, v^p, w^p,
// Commutator relations R
(u,v), (w,u,u), (u,w,w,w), (w,u,w,v,v),
(v,w,w,w), (v,w,w,v), (v,w,v,w), (v,w,v,v),
// 24 4-Engel relations
[ (u,x,x,x,x): x in
  [ u^e1 * v^e2 *w^e3 :e1 in [1,-1], e2 in [1,-1], e3 in [1,-1] ] ],
[ (v,x,x,x,x): x in
  [ u^e1 * v^e2 *w^e3 :e1 in [1,-1], e2 in [1,-1], e3 in [1,-1] ] ],
[ (w,x,x,x,x): x in
  [ u^e1 * v^e2 *w^e3 :e1 in [1,-1], e2 in [1,-1], e3 in [1,-1] ] ]
>;
```

```
Q := NilpotentQuotient (G,9);
"Max NQ of G has class", NilpotencyClass(Q),
            "and order", FactoredOrder(Q);
H := sub<Q | Q.1*Q.2, Q.2*Q.3>;
"In the max NQ the subgroup image has index", FactoredIndex(Q,H);
"For G, plen and #rel", PresentationLength(G), #Relations(G);

for l in [11..19] do
   S := Simplify(G);
   "For", l, "rels: S, plen and #rel",
             PresentationLength(S), #Relations(S);
   d := Ngens (S);
   F := FreeGroup(d);
   r := Relations (S);
   srels := [LHS (x) * RHS (x)^-1: x in r];
   prels := [ F!Eltseq (x): x in srels[1 .. l] ];
   P := quo <F | prels>;
   "For P, plen and #rel", PresentationLength(P), #Relations(P);
   Q:=NilpotentQuotient(P,9);
   "Max NQ of P has class", NilpotencyClass(Q),
               " and order", FactoredOrder(Q);
   H := sub<Q | Q.1*Q.2, Q.2*Q.3>;
   "In the max NQ the subgroup image has index", FactoredIndex(Q,H);

   H := sub<P | P.1*P.2, P.2*P.3>;
   I,_,M,T := ToddCoxeter(P, H : CosetLimit:=6000000,
                           Strategy:=<1000,1>, SubgroupRelations:=1);
   "Index", I,"/",M,"/",T ;
end for;
```

# ON THE SIZE OF THE COMMUTATOR SUBGROUP IN FINITE GROUPS

MARCEL HERZOG*, GIL KAPLAN† and ARIEH LEV†[1]

*School of Mathematical Sciences, Tel-Aviv University, Tel-Aviv, Israel

†School of Computer Sciences, The Academic College of Tel-Aviv-Yaffo, 4 Antokolski St., Tel-Aviv 64044, Israel

## Abstract

We study conditions under which the commutator subgroup of a finite group $G$ must be "large" in comparison with $G$.

## 1 Introduction

All groups in this paper are finite. For a group $G$, we denote the center of $G$ and the Frattini subgroup of $G$ by $Z(G)$ and $\Phi(G)$, respectively.

In this paper we survey recent results of the authors treating the following general problem:

**Problem 1.1** *Study conditions under which the commutator subgroup $G'$ of a group $G$ must be "large" in comparison with $G$.*

The authors proved the following result in [4]:

**Theorem 1.1** *Let $G \neq 1$ be a group such that $\Phi(G) = Z(G) = 1$. Then $|G'| > |G|^{1/2}$.*

As will be shown in Section 2, both conditions $\Phi(G) = 1$ and $Z(G) = 1$ are necessary for the result of Theorem 1.1. Furthermore, as the examples of Frobenius groups of order $q(q-1)$ ($q$ a power of a prime) show, for any $\epsilon > 0$ there exists a group $G \neq 1$ with $\Phi(G) = Z(G) = 1$, such that $|G'| \leq |G|^{1/2+\epsilon}$. Thus the constant $1/2$ can not be replaced by a larger constant in Theorem 1.1. We remark further that the main result in [4] is more general. Indeed, we prove there that if $G$ is a non-abelian group satisfying $\Phi(G) = 1$, then $|G'| > [G : Z(G)]^{1/2}$.

In Section 2 we survey some further results of the authors dealing with groups satisfying $\Phi(G) = Z(G) = 1$. In these results we consider not only $G'$, but also the nilpotent residual of $G$. We recall that the *nilpotent residual* of a group $G$, denoted by $U(G)$, is the smallest normal subgroup of $G$ such that $G/U(G)$ is nilpotent. Clearly, $U(G) \leq G'$. Further information on $U(G)$ can be found in [1] (it is called there the *hypercommutator* subgroup of $G$).

In Section 3 we treat some special cases in which the condition $\Phi(G) = 1$ may be dropped. For example, if $G \neq 1$ is a group with all Sylow subgroups abelian and $Z(G) = 1$, then $|G'| > |G|^{1/2}$. Section 4 deals with the question: What can

[1]Corresponding author.

we say on any group when both conditions $\Phi(G) = 1$ and $Z(G) = 1$ are dropped? We survey results of [5] stating that any non-nilpotent group must posses certain factors $K/M$ with large commutator subgroups, where $M$ is nilpotent.

## 2 Groups with $\Phi(G) = Z(G) = 1$

We note first that $G'$ *cannot* be replaced by the nilpotent residual $U(G)$ in our Theorem 1.1, as the following example shows.

**Example 2.1** Let $H$ be a Sylow 2-subgroup of $GL_2(3)$. Then $H$ is a nonabelian 2-group of order 16. Let $V$ be the space of all row vectors of length 2 over the field of 3 elements, and note that $H$ acts on $V$ by right multiplication. Let $G$ be the semidirect product of $V$ by $H$. Then $V \triangleleft G$, $|G| = 16 \times 9 = 144$, $\Phi(G) = Z(G) = 1$. As Theorem 1.1 ensures, we have $|G'| > |G|^{1/2}$. However, $|U(G)| = |V| = 9 < |G|^{1/2}$.

However, we have the following result on $U(G)$.

**Theorem 2.1** *Let $G \neq 1$ be a group such that $\Phi(G) = Z(G) = 1$. Let $p = \min(\pi(G))$. Then $|U(G)| > |G|^{(p-1)/(2p-1)}$.*

The result of Theorem 2.1 for *solvable* groups follows from Theorem A in [3] and the remark on [5], p. 199. The proof for arbitrary groups, which is given below, applies Proposition 2.4 in [4] (this proposition was proved by using the classification of the finite simple groups).

**Proof of Theorem 2.1** We prove by induction on $|G|$. Assume first that $\mathrm{Fit}(G) > 1$. Then the claim follows by the same arguments like in the proof of Theorem A in [3] and the remark in [5], p. 199. We note that the solvability assumption is not needed here: we need only the property $\mathrm{Fit}(G) > 1$ and the inductive hypothesis. Thus, it remains to consider the case $\mathrm{Fit}(G) = 1$. For that, we apply Proposition 2.4 in [4]. This proposition says that for a group $G \neq 1$ satisfying $\mathrm{Fit}(G) = 1$, we have $|\mathrm{Res}(G)| > |G|^{1/2}$, where $\mathrm{Res}(G)$ denotes the solvable residual of $G$. Since $\mathrm{Res}(G) \leq U(G)$, and since $(p-1)/(2p-1) < 1/2$, the case $\mathrm{Fit}(G) = 1$ is proved by this result. □

The following theorem is an immediate corollary of Theorem 2.1 (compare to Corollary A1 and Corollary A2 in [3], where the solvability assumption was still included).

**Theorem 2.2** *Let $G \neq 1$ be a group such that $\Phi(G) = Z(G) = 1$. Then $|U(G)| > |G|^{1/3}$. Moreover, if, in addition, $G$ has an odd order, then $|U(G)| > |G|^{2/5}$.*

We notice that both conditions $\Phi(G) = 1$ and $Z(G) = 1$ are necessary in Theorem 1.1 and in the results above. Indeed, by taking a direct product of a "large" elementary abelian group with a non-nilpotent group $S$, with $\Phi(S) = 1$, we obtain, for every $\epsilon > 0$, examples of a non-nilpotent group $G$, such that $\Phi(G) = 1$, $Z(G) > 1$ and $|G'| \leq |G|^{\epsilon}$. Furthermore, the following example (see [3], [2]), shows that for each $\epsilon > 0$ there exists a group $G$ with $\Phi(G) > 1$, $Z(G) = 1$, satisfying $|G'| \leq |G|^{\epsilon}$.

**Example 2.2** Let $F$ be a field of order $p$, where $p$ is an odd prime, and let $m$ and $n$ be positive integers. Let $K$ be the subset of $GL(m+n, F)$ consisting of all the matrices of the form

$$\pm \begin{pmatrix} I_m & A \\ 0 & I_n \end{pmatrix}$$

where $I_m$ and $I_m$ are respectively the $m \times m$ and $n \times n$ identity matrices and $A$ is an arbitrary $m \times n$ matrix. It is easy to see that $K$ is an abelian group of order $2p^{mn}$ and the Sylow $p$-subgroup of $K$ is elementary abelian. Now, let $V$ be the space of row vectors of length $m+n$ over $F$, and note that $K$ acts on $V$ by right multiplication. Let $G = VK$ be the semidirect product with respect to this action. Clearly $|G| = 2p^{mn+m+n}$.

Since the action of $K$ on $V$ is faithful, we see that $C_G(V) = V$, and thus $Z(G) \leq V$. But the $-1$ scalar matrix $c$ lies in $K$, and it fixes no non-identity element of $V$, so we conclude that $Z(G) = 1$. Finally, since $K$ is abelian, we have $G' \leq V$, and since $[V, c] = V$, $G' = V$ is of order $p^{m+n}$. now $|G/G'| = 2p^{mn+m+n}/p^{m+n} = 2p^{mn}$, and it follows that for every $\epsilon > 0$ there exists a group $G$, with $Z(G) = 1$, such that $|G'| \leq |G|^{\epsilon}$.

## 3 Omitting the condition $\Phi(G) = 1$

It was shown in Example 2.2 that the condition $\Phi(G) = 1$ cannot be omitted in Theorems 1.1, 2.1 and 2.2. However, the example given there is a group $G$ of order $2p^r$, where $p$ is a prime, and the Sylow $p$-subgroup of $G$ is nonabelian (it is actually metabelian). So, it is natural to ask whether we may drop the condition $\Phi(G) = 1$ when all the Sylow subgroups of $G$ are abelian (one easily checks that such a group satisfies $G' = U(G)$). The answer to this question is affirmative, as given in the following theorem (see [6]).

**Theorem 3.1** *Let $G \neq 1$ be a group with all Sylow subgroups abelian, where $Z(G) = 1$. Then $|G'| > |G|^{1/2}$.*

In the following theorem we provide another type of condition (see [3], Corollary B1).

**Theorem 3.2** *Let $G \neq 1$ be a group such that $Z(G) = 1$, and assume that $G/G'$ is cyclic. Then $|G'| > |G|^{1/2}$.*

We remark that, like in Theorem 3.1, the assumptions of Theorem 3.2 imply $G' = U(G)$. Indeed, since $G/G'$ is cyclic, we obtain that $(G/U(G))/\Phi(G/U(G))$ is cyclic, implying that $G/U(G)$ is cyclic and $G' = U(G)$.

## 4 Factors and subgroups of non-nilpotent groups

As was shown in Section 2, both assumptions $\Phi(G) = 1$ and $Z(G) = 1$ are necessary for the results there. In the current section we obtain results on "large" commutator subgroups without assuming $\Phi(G) = 1$ or $Z(G) = 1$. Our only assumption is that $G$ is not nilpotent. In this general setting we can not claim that $G'$ must be large in comparison with $G$. However, we may obtain the existence of certain factors $K/M$ with large commutator subgroups, where $M$ is nilpotent. In the sequel, for any group $K$ with Frattini subgroup $\Phi(K)$ we shall denote the quotient $K/\Phi(K)$ by $K^*$. Our first result is the following ([5], Theorem A):

**Theorem 4.1** *Let $G$ be a non-nilpotent group. Then there exists a subnormal subgroup $K \leq G$ such that*

$$|U(K^*)| > |K^*|^{1/2}.$$

*equivalently, for a non-nilpotent group $G$ there exists a subnormal subgroup $K \leq G$, such that*

$$[U(K) : U(K) \cap \Phi(K)] > [K : \Phi(K)]^{1/2}.$$

Let $G = D_{24}$, the dihedral group of size 24. Notice that $U(G)$ is the unique Sylow 3-subgroup of $G$, and that $|\Phi(G)| = 2$. As one can check, $G$ does not have *normal* subgroups $K$ satisfying $|U(K^*)| > |K^*|^{1/2}$. This shows that the word "subnormal" cannot be replaced by "normal" in Theorem 4.1. The following is obtained when we require $\Phi(G) = 1$ ([5], Corollary A2):

**Corollary 4.1** *Let $G$ be a non-abelian group such that $\Phi(G) = 1$. Then there exists a subnormal subgroup $K \leq G$ such that $|K'| > |K|^{1/2}$.*

Our next result is taken from [5], Theorem B:

**Theorem 4.2** *Let $G$ be a non-nilpotent group. Then there exist characteristic subgroups $M, K$ of $G$ such that $M < K$, $M$ nilpotent, satisfying $|U(K/M)| > |K/M|^{1/2}$ and $Z(K/M) = 1$.*

When we limit the discussion to the commutator subgroup, we obtain Theorem 4.3 below ([5], Theorem C). We need first the following notation and definition. For a group $G$ let $W(G)$ denote the preimage of $Z(G/\Phi(G))$ in $G$, i.e., $W(G)/\Phi(G) = Z(G/\Phi(G))$. We summarize some basic properties of $W(G)$ for any group $G$ (see [5] for further details):

1. $W(G)$ is a characteristic nilpotent subgroup of $G$.
2. $Z(G/W(G)) = \Phi(G/W(G)) = 1$.
3. If $G$ is non-nilpotent then also $G/W(G)$ is non-nilpotent.

**Theorem 4.3** *Let $G$ be a non-nilpotent group. Then there exists a characteristic subgroup $K$ of $G$ such that $|(K/W(K))'| > |K/W(K)|^{1/2}$.*

When we require $\Phi(G) = 1$ we have the following ([5], Corollary C1):

**Corollary 4.2** *Let $G$ be a non-abelian group such that $\Phi(G) = 1$. Then there exists a characteristic subgroup $K$ of $G$ such that $|(K/Z(K))'| > |K/Z(K)|^{1/2}$.*

Notice that for a Frobenius group $G$ of order $p(p+1)$, where $p$ is a Mersenne prime, we have $|U(G)| = |G'| = p+1$, and each proper subnormal subgroup of $G$ is nilpotent. Consequently (notice $\Phi(G) = Z(G) = 1$), $G$ itself is the only subnormal (resp., characteristic) subgroup $K$ satisfying the results of Theorem 4.1 and Theorem 4.3. Furthermore, $M = 1$, $K = G$ is the unique choice for the subgroups $M, K$ of Theorem 4.2. Thus, if there exist infinitely many Mersenne primes (this is still an open problem) then we obtain an infinite family of examples showing that we are not able to replace 1/2 by a larger constant in Theorems 4.1, 4.2 and 4.3.

## References

[1] R. Baer, Group elements of prime power index, *Trans. Amer. Math. Soc.* **75** (1953), 20–47.

[2] M. Bianchi, A. Gillio, H. Heineken and L. Verardi, Groups with big centralizers, *Instituto Lombardo (Rend. Sc.) A* **130** (1996), 25–42.

[3] M. Herzog, G. Kaplan and A. Lev, On the commutator and the center of finite groups, *J. Algebra* **278** (2004), 494–501.

[4] M. Herzog, G. Kaplan and A. Lev, The size of the commutator subgroup of finite groups, submitted.

[5] G. Kaplan and A. Lev, The existence of large commutator subgroups in factors and subgroups of non-nilpotent groups, *Arch. Math.* **85** (2005), 197–202.

[6] G. Kaplan and A. Lev, On groups satisfying $|G'| > [G : Z(G)]^{1/2}$, *Beiträge zur Algebra und Geometrie*, to appear.

# GROUPS OF INFINITE MATRICES

WALDEMAR HOŁUBOWSKI

Institute of Mathematics, Silesian University of Technology,
Kaszubska 23, 44-100 Gliwice, Poland
Email: w.holubowski@polsl.pl

## Abstract

We show that in the group of infinite invertible column-finite matrices over an associative ring $R$, every element is a product of a row- and column-finite matrix and a unitriangular matrix. Moreover we prove that its subgroup of banded matrices is generated by strings (block-diagonal matrices, with finite blocks along the main diagonal).

## 1 Introduction

Let $R$ be an associative ring with 1. Let $GL_c(\infty, R)$ denote the group of infinite $\mathbb{N} \times \mathbb{N}$ column-finite matrices over $R$, and $GL_{rc}(\infty, R)$ its subgroup of row- and column-finite matrices. A systematic study of normal subgroups of $GL_c(\infty, R)$ in the case of division rings was initiated by A. Rosenberg [8]. Research continued in works of Maxwell [6], Robertson [7], Arrell [1], Arrell and Robertson [2], Hausen [3], Thomas [9] and others. We refer to [4] for a comprehensive survey.

In this paper we are interested in results concerning generators of subgroups and a special form of an element in the case of an arbitrary associative ring of coefficients. We prove:

**Theorem 1.1** *Every element of $GL_c(\infty, R)$ is a product of an invertible row- and column-finite matrix and an upper unitriangular matrix.*

The matrix $a \in GL_c(\infty, R)$ is called $n$-banded if $a_{ij} = 0$ for all $i, j$ such that $|j - i| > n$ and either $a_{i+n,i} \neq 0$ or $a_{i,i+n} \neq 0$ for at least one index $i$. We say that $a$ is *banded* if $a$ is $n$-banded for some $n$. All matrices $a$ from $GL_c(\infty, R)$ such that $a$ and $a^{-1}$ are banded form the subgroup of banded matrices, denoted by $GL_b(\infty, R)$. We find a specific set of generators for this subgroup.

A block-diagonal matrix $a = \operatorname{diag}(a_1, a_2, a_3, \ldots) \in GL_c(\infty, R)$ with finite blocks along the main diagonal is called *a string.* Of course, the inverse of the string $a$ is the string $a^{-1} = \operatorname{diag}(a_1^{-1}, a_2^{-1}, a_3^{-1}, \ldots)$. However, the product of two strings need not be a string. The set of all finite products of strings form a subgroup of $GL_{rc}(\infty, R)$ (we refer the reader to [4] for a detailed study of its properties). A string $b = \operatorname{diag}(1, \ldots, 1, b_n, 1, \ldots)$ with only one nontrivial block is called *a bead.*

We note here that if $R = K$ is a field, then the group $GL_{rc}(\infty, K)$ is generated by strings (see for example [9]). The problem of when the same is true in the case of any associative ring $R$ is open, however we can give an affirmative answer for the subgroup $GL_b(\infty, R)$.

**Theorem 1.2** *Every element a of $GL_b(\infty, R)$ is a product of strings.*

The proof describes an algorithm for writing any $a \in GL_b(\infty, R)$ as a product of strings. The question arises:

> Is it true that the number of factors in such a product is uniformly bounded by some natural number $k$?

We note here, that our proof works without change in the more general setting of all matrices having no more than $k$ nonzero elements in each block of the block-tridiagonal form considered in the proof.

## 2 Proofs of main results

**Proof of Theorem 1:** Let $g \in GL_c(\mathbb{N}, R)$. Let $g_{m1}$ be the last nonzero entry in the first column. Since the first row is unimodular there exist $r_1, \ldots, r_m \in R$ such that $r_1 g_{11} + \ldots + r_m g_{m1} = 1$. Multiplying $g$ from the left by the corresponding matrix we can add to the $(m+1)$-th row $r_1$ times the first row, $r_2$ times the second row, $\ldots$, $r_m$ times the $m$-th row. In the first column and $(m+1)$-th row we obtain 1. Now we subtract the $(m+1)$-th row $g_{11}$ times from the first row, $g_{21}$ times from the second row, and so on. Next, we permute the first and $(m+1)$-th rows to get 1 at the top of the first column. It is clear that all operations above can be carried out by multiplying $g$ from the left by the corresponding $(m+1) \times (m+1)$ bead. Let $g_{k,2}$ be the last nonzero entry in the second column. After excluding the first row, the second column is unimodular. So, we can make the $(k+1, 2)$ entry equal to 1, by adding corresponding multiples of rows $2, 3, \ldots, k$. Using this 1 we make zeros in the rows above excluding the first row. We interchange next, the second and $(k+1)$-th rows.

So, there exists a bead $a_1 \in GL_f$ such that $a_1 \cdot g$ has its first column the same as the unit matrix $e$. There exists a matrix $a_2$ such that the second column of $a_2(a_1 g)$ has 1 on the main diagonal and zeros below. We continue this process. At each step we have $a_n(a_{n-1}(\ldots(a_1 g)\ldots) = (a_n \cdot \ldots \cdot a_1)g$. So we obtain by this process

$$\prod_{i=1}^{\infty} a_i \cdot g = u,$$

where $u$ is upper unitriangular. The matrix $\prod_{i=1}^{\infty} a_i$ is row-finite. Moreover, it is also column-finite, because $\prod_{i=1}^{\infty} a_i = u \cdot g^{-1} \in GL_c(\infty, R)$. □

**Remark** We have proved that the inverse of $\prod_{i=1}^{\infty} a_i$ is column-finite, however we have not been able to prove that $\prod_{i=1}^{\infty} a_i$ is an element of $GL_{rc}(\infty, R)$.

**Proof of Theorem 2:** As was shown in the proof of Theorem 1, we can assume that $a = \prod_{i=1}^{\infty} a_i$. We find a finite sequence of strings such that their product is $(\prod_{i=1}^{\infty} a_i)^{-1}$.

Now we describe the process of reducing $\prod_{i=1}^{\infty} a_i$ to the identity matrix by using multiplication by invertible strings. The steps are:

**I.** Finding a block-tridiagonal form of $\prod_{i=1}^{\infty} a_i$.

**II.** Reducing one column in the above block-form to the identity column by multiplying by beads.

**III.** Main process of simultaneous transformations on $\prod_{i=1}^{\infty} a_i$ reducing it to the identity matrix.

**I.** Choose any $n_1 \geq 1$. Let $m_1$ be a minimal number such that all the nonzero entries in the first $n_1$ columns are in the first $m_1$ rows, and all nonzero entries in the first $n_1$ rows are in the first $m_1$ columns. We put $n_2 = \max\{n_1 + 1, m_1\}$. Let $m_2$ be the minimal number such that all nonzero entries in the first $n_2$ columns are in the first $m_2$ rows and all nonzero entries in the first $n_2$ rows are in the first $m_2$ columns. We put $n_3 = \max\{n_2 + 1, m_2\}$. We repeat this procedure and obtain an infinite sequence $n_1 < n_2 < n_3 < \ldots$.

Now, we can represent $\prod_{i=1}^{\infty} a_i$ in a block-tridiagonal form

$$\prod_{i=1}^{\infty} a_i = \begin{pmatrix} b_1 & c_2 & & & & \\ d_1 & b_2 & c_3 & & & \\ & d_2 & b_3 & c_4 & & \\ & & d_3 & b_4 & c_5 & \\ & & & \ddots & \ddots & \ddots \end{pmatrix}$$

On the main diagonal we have square matrices $b_1, b_2, b_3, \ldots$ of sizes $n_1 \times n_1$, $(n_2 - n_1) \times (n_2 - n_1)$, $(n_3 - n_2) \times (n_3 - n_2)$, $\ldots$. Matrices $d_1, d_2, \ldots$ have sizes $(n_2 - n_1) \times n_1$, $(n_3 - n_2) \times (n_2 - n_1)$, $\ldots$ and matrices $c_2, c_3, \ldots$ have corresponding sizes $n_1 \times (n_2 - n_1)$, $(n_2 - n_1) \times (n_3 - n_2)$, $\ldots$. Moreover, all nonzero entries of $\prod_{i=1}^{\infty} a_i$ occur in blocks $b_1$, $d_1$, $c_i$, $b_i$, $d_i$, $i \geq 2$, as in the diagram above.

**II.** 1) We look now at the $k$-th block-column $c_k, b_k, d_k$. This block-column covers columns $n_{k-1} + 1$, $n_{k-1} + 2$, $\ldots$, $n_k$. Since every column is unimodular we can add all the nonzero rows from the $n_{k-1}+1$ column to the $n_{k+1}+1$ row to obtain 1 as the $(n_{k-1}+1) \times (n_{k+1}+1)$ entry. Next, using this 1, we make zeros in the $n_{k+1}+1$ row in columns $n_{k-1}+2, n_{k-1}+3, \ldots, n_k$. Similarly, we add all rows to the $n_{k+1}+2$ row to obtain 1 in the $n_{k-1}+2$ column. Next, using this 1, we make zeros in the $n_{k+1}+2$ row in columns $n_{k-1} + 1$, $n_{k-1} + 3$, $n_{k-1} + 4$, $\ldots$, $n_k$. Continuing this process we obtain below $d_k$ a square identity matrix $e$ of size $(n_k - n_{k-1}) \times (n_k - n_{k-1})$ as in the diagram below:

$$\begin{array}{|c|}\hline c_k \\ \hline b_k \\ \hline d_k \\ \hline 0 \\ \hline 0 \\ \hline \end{array} \to 1) \to \begin{array}{|c|}\hline c_k \\ \hline b_k \\ \hline d_k \\ \hline e \\ \hline 0 \\ \hline \end{array} \to 2) \to \begin{array}{|c|}\hline 0 \\ \hline 0 \\ \hline 0 \\ \hline e \\ \hline 0 \\ \hline \end{array} \to 3) \to \begin{array}{|c|}\hline 0 \\ \hline e \\ \hline 0 \\ \hline 0 \\ \hline 0 \\ \hline \end{array}$$

2) We use the block $e$ to create zeros above it. We subtract 1 in the $(n_{k+1} + 1)$-th row from all rows above. We subtract 1 in the $(n_{k+1} + 2)$-th row from all rows above. We complete this process after creating zeros in all the blocks above $e$.

3) We interchange the $n_{k+1}+1$ row with the $n_{k-1}+1$ row, $n_{k+1}+2$ row with $n_{k-1}+2$ row, .... This puts the block $e$ into the position formerly occupied by block $b_k$.
4) Now using $e$ we create zeros in rows $n_{k-1}+1, \ldots, n_k$ to the left of $e$ and next to the right.

All steps in our process can be achieved by multiplication from the left (or right) by beads, which are elementary matrices.

**III.** We describe the process of reducing $\prod_{i=1}^{\infty} a_i$ to the identity matrix by simultaneous transformations which can be performed on $\prod_{i=1}^{\infty} a_i$ with multiplication by strings from the left and right. This will show that $\prod_{i=1}^{\infty} a_i$ is a product of strings.
1) We look now at block-column $c_5, b_5, d_5$.
a) Using procedure II we make a block $e_5$ (identity matrix of size $n_5 - n_4$) below $d_5$.
b) We make zeros to the left and right side of $e_5$.
c) We make zeros above block $e_5$.
d) Interchanging block rows we put $e_5$ onto the main diagonal.

It is clear that all such operations can be performed using multiplication by the corresponding beads. Moreover, we can simultaneously carry this out for blocks with numbers 1, 5, $9 = 5+4$, $13 = 5+8$, $17 = 5+12$ and so on. Each corresponds to multiplication by strings.

$$\begin{pmatrix}
e_1 & 0 & & & & & & & & & & & \\
0 & \star & \star & & & & & & & & & & \\
 & \star & \star & \star & & & & & & & & & \\
 & & \star & \star & 0 & & & & & & & & \\
 & & & 0 & e_5 & 0 & & & & & & & \\
 & & & & 0 & \star & \star & & & & & & \\
 & & & \star & 0 & \star & \star & \star & & & & & \\
 & & & & & & \star & \star & 0 & & & & \\
 & & & & & & & 0 & e_9 & 0 & & & \\
 & & & & & & & & 0 & \star & \star & & \\
 & & & & & & & \star & 0 & \star & \star & \star & \\
 & & & & & & & & & & & & \ddots
\end{pmatrix}$$

2) We repeat the procedure from 1) applying it to blocks with numbers 2, 6, 10, .... We obtain:

$$
\begin{pmatrix}
e_1 &&&&&&&&&&&& \\
& e_2 &&&&&&&&&&& \\
&& \star & \star &&&&&&&&& \\
&& \star & \star &&&&&&&&& \\
&&& 0 & e_5 &&&&&&&& \\
&&& 0 && e_6 &&&&&&& \\
&&& \star &&& \star & \star &&&&& \\
&&&&&& \star & \star &&&&& \\
&&&&&&& 0 & e_9 &&&& \\
&&&&&&& 0 && e_{10} &&& \\
&&&&&&& \star &&& \star & \star & \\
&&&&&&&&&&&& \ddots
\end{pmatrix}
$$

3) a) First we make in the seventh block column a block $e_7$ (identity matrix of size $n_7 - n_6$) below the last nonzero block in a similar way to I.
b) We make zeros to the left and right side of $e_7$ and then above $e_7$.
c) We interchange the seventh block row with rows corresponding to $e_7$ and we obtain $e_7$ in correct place.
d) Using two nonzero blocks in the ninth block row we reproduce the left upper corner of $e_9$ (destroyed by previous operations).
e) Using $e_9$ we make zeros in the ninth block row.

It is clear that all operations can be achieved using multiplication by corresponding beads. Moreover, we can simultaneously carry it out for blocks with numbers 3, 7, 11, 15, 19 and so on. Each corresponds to multiplication by strings.

We obtain as the result the following invertible string:

$$\operatorname{diag}(e_1, e_2, e_3, x_4, e_5, e_6, e_7, x_8, e_9, \ldots).$$

Now, multiplying by the inverse string we get the identity matrix as required. □

## References


[1] D. G. Arrell, The subnormal subgroup structure of the infinite general linear group, *Proc. Edinburgh Math. Soc. (2)* **25** (1982), no. 1, 81–86.
[2] D. G. Arrell and E. F. Robertson, Infinite dimensional linear groups, *Proc. Roy. Soc. Edinburgh Sect. A* **78** (1977/78), no. 3–4, 237–240.
[3] J. Hausen, Infinite general linear groups over rings, *Arch. Math. (Basel)* **39** (1982), no. 6, 510–524.
[4] W. Hołubowski and N. A. Vavilov, Infinite general linear group, in preparation.
[5] B. Laschinger, Automorphismengruppen freier Moduln von unendlichem Rang, *J. Algebra* **122** (1989), no. 1, 15–63.
[6] G. Maxwell, Infinite general linear groups over rings, *Trans. Amer. Math. Soc.* **151** (1970), 371–375.
[7] E. F. Robertson, On certain subgroups of $GL(R)$, *J. Algebra* **15** (1970), 293–300.
[8] A. Rosenberg, The structure of the infinite general linear group, *Ann. of Math. (2)* **68** (1958), 278–294.
[9] S. Thomas, The cofinalities of the infinite-dimensional classical groups, *J. Algebra* **179** (1996), no. 3, 704–719.

# TRIPLY FACTORISED GROUPS AND NEARRINGS

PETER HUBERT

Johannes-Gutenberg-Universität Mainz, Fachbereich 08, Physik, Mathematik, Informatik, Staudingerweg 9, D-55099 Mainz, Germany
Email: hubert@mathematik.uni-mainz.de

## Abstract

A group $G$ is called *triply factorised* if it can be written as $G = A \ltimes M = B \ltimes M = AB$ for two subgroups $A$ and $B$ and a normal subgroup $M$ of $G$. It is shown how such triply factorised groups can be constructed using nearrings. Moreover, if $G = A \ltimes M = B \ltimes M = AB$ is any triply factorised group with $A \cap B = 1$, there exists a nearring by which the group $G$ can be constructed. Finally, some structural properties of nearrings are described.

## 1 Introduction

A group $G$ is called *factorised* if $G = AB$ is the product of two subgroups $A$ and $B$ of $G$. If, in addition, there exists a normal subgroup $M$ of $G$ such that $G = A \ltimes M = B \ltimes M = AB$ is a semidirect product of $A$ resp. $B$ and $M$, then $G$ is called *triply factorised* by $A$, $B$, and $M$. In this case, $A$ and $B$ are complements of $M$ and hence isomorphic.

**Examples 1.1**

1. Let $G$ be an arbitrary group, and let $A = B = G$ and $M = 1$. Then $G$ is triply factorised by $A$, $B$, and $M$. In the following these trivial triply factorisations will not be considered.
2. Let $H$ be an arbitrary group and $G = H \times H = \{(g, h) \mid g, h \in H\}$. Consider the subgroups $A = \{(a, 1) \mid a \in H\}$, $B = \{(b, b) \mid b \in H\}$, and $M = \{(1, m) \mid m \in H\}$ of $G$. Then $G$ is triply factorised by $A$, $B$, and $M$.
3. Let $D_{12} = \langle a, c \mid a^2 = c^6 = 1, c^a = c^{-1} \rangle$ be the dihedral group of order 12. If $A = \langle a, c^2 \rangle$, $B = \langle ac, c^2 \rangle$, and $M = \langle c^3 \rangle$, then $G$ is triply factorised by $A$, $B$, and $M$.

In the theory of factorised groups, triply factorised groups play an important rôle, since many problems in this area can be reduced to questions about triply factorised groups (cf. Amberg, Franciosi, de Giovanni [1]). Thus it is desirable to have examples of triply factorised groups. Sysak [11] describes a method to construct triply factorised group using radical rings.

### 1.1 Radical rings

Let $R$ be an associative ring, not necessarily with an identity element. Then $R$ forms a semi-group under the "*circle-operation*" $a \circ b = ab + a + b$ for all $a, b \in R$.

The group of all invertible elements of this semi-group is called the *adjoint group* $R^\circ$ of $R$. Following Jacobson [8], a ring $R$ is called *radical* if $R = R^\circ$, i.e., $R$ coincides with its Jacobson radical $\mathcal{J}(R)$. In particular, the Jacobson radical of every ring is a radical ring. Obviously, non-trivial radical rings do not have an identity element.

**Example 1.2** Let $\mathbb{Q}_p$ be the set of all rational numbers whose denominators are not divisible by the prime $p$. Then $p\mathbb{Q}_p$ is a radical ring, since $p\mathbb{Q}_p = \mathcal{J}(\mathbb{Q}_p)$.

## 1.2 A connection between certain triply factorised groups and radical rings

The following construction of triply factorised groups was first described in [11]; this preprint was never published in a mathematical journal, but its main results can be found in [1, Section 6].

**Construction 1.3** Let $R$ be a radical ring, embedded in an arbitrary way into any ring $R_1$ with identity element. Then the adjoint group $R^\circ$ of $R$ is isomorphic to the subgroup $R+1$ of the group of units of $R_1$.

Let $U$ be a left ideal of $R$ and $M = R/U$ as a left $R$-module. Then the group $A = R+1$ operates on $M$ via

$$(l+U)^{(m+1)} = (m+1)^{-1}l + U$$

for all $l$, $m \in R$. In the semidirect product $G = G(R) = A \ltimes M$,

$$B = \left\{\left((l+1)^{-1}, l+U\right) \,\middle|\, l \in R\right\}$$

is a complement of $M$ with the property

$$G = A \ltimes M = B \ltimes M = AB,$$

i.e., $G$ is triply factorised by $A$, $B$, and $M$.

This construction raises the question if for every triply factorised group $G = A \ltimes M = B \ltimes M = AB$ there exists always a radical ring $R$ such that $G$ can be constructed using $R$ in the above way. In the case when $A$, $B$, and $M$ are abelian and the intersection of $A$ and $B$ is trivial, this was established in [11] (cf. also [1, Proposition 6.1.4]).

**Theorem 1.4** *If $G = A \ltimes M = B \ltimes M = AB$ is a triply factorised group with abelian subgroups $A$, $B$, and $M$, and with $A \cap B = \{1\}$, then there exists a commutative radical ring $R$ such that $G \cong G(R)$.*

Moreover, simple examples show that Theorem 1.4 is not true if $A$ and $B$ are nilpotent of class 2 (cf. [3, Example 2.4]).

## 2 Nearrings

In the triply factorised groups $G = A \ltimes M = B \ltimes M = AB$ obtained from Construction 1.3, the normal subgroup $M$ is always abelian, since it is the additive group of a ring module. For further investigations of factorised groups it is desirable to have also examples of triply factorised groups with non-abelian $M$. To obtain such examples, in the following nearrings will be used in the place of of radical rings.

First recall some definitions. Further information about nearrings can be found, e.g., in Meldrum [10].

**Definition 2.1** A set $(R, +, \cdot)$ with two binary operations, addition and multiplication, is called a *(left) nearring* if
(N1) $(R, +)$ is a (not necessarily abelian) group;
(N2) $(R, \cdot)$ is a semi-group;
(N3) for all $x$, $y$, $z \in R$: $x \cdot (y + z) = x \cdot y + x \cdot z$.
If $R$ contains a multiplicative identity 1, $R$ is called a nearring *with identity.* In this case the set of multiplicatively invertible elements of $R$ forms a group under multiplication which is denoted by $R^\times$. The additive group of $R$ will be written as $R^+$. The nearring $R$ is called *zero-symmetric*, if $0r = 0$ for all $r \in R$.

**Example 2.2** Let $G$ be a (not necessarily abelian) additively written group, and let
$$M(G) = \{\alpha : G \to G\}$$
be the set of all mappings from $G$ in $G$. Then $M(G)$ forms a nearring under pointwise addition of mappings and multiplication by composition.

### 2.1 A connection between triply factorised groups and nearrings

To generalise Construction 1.3 using nearrings, the following concept is needed.

**Definition 2.3** Let $R$ be a nearring with identity 1. Let $U \leq R^+$ such that $(U + 1) \leq R^\times$. Then $U$ is called a *construction subgroup* of $R$.

**Proposition 2.4** *Let $R$ be a nearring with identity and $U$ a construction subgroup of $R$. Then $(U + 1)U \subseteq U$.*

**Proof** Let $A = U + 1 \leq R^\times$. Since $U$ is an additive group, for every $a$, $b \in A$ one has $a - b = a - 1 + 1 - b = (a - 1) - (b - 1) \in U$, since $a - 1$, $b - 1 \in U$. Now let $u$, $v \in U$ with $u = a - 1$ and $v = b - 1$ for suitable elements $a$, $b \in A$. Then $(u + 1)v = a(b - 1) = ab - a \in U$, since $ab$, $a \in A$. □

**Examples 2.5**

1. Let $R$ be a nearring with identity. Then the trivial subgroup $\{0\}$ is a construction subgroup.
2. Let $R$ be a ring with identity element. Then the Jacobson radical $\mathcal{J}(R)$ is a construction subgroup of $R$.

3. Let $R$ be a nearring with identity element, in which the set $L_R$ of elements which are not right-invertible forms an additive subgroup. Then $R$ is called a *local nearring* and $L_R$ is a construction subgroup of $R$ (cf. Maxson [9]).

Using construction subgroups it is possible to construct triply factorised groups in a similar way as in Construction 1.3. In fact, it is not difficult to see that Construction 1.3 is a special case of the following construction (cf. [7]).

**Construction 2.6** Let $R$ be a nearring with identity element, $U$ a construction subgroup of $R$, and $N^+ \trianglelefteq U^+$ a normal subgroup of $U^+$ with $(U+1)\,N \subseteq N$. Let $M = U^+/N$ and $A = (U+1)^\times$. Then $A$ operates on $M$ via the rule

$$(u+N)^{(v+1)} = (v+1)^{-1}\,u + N$$

for all $u + N \in M$ and all $v + 1 \in A$. Next, form the semidirect product $G = G(R, U, N) = A \ltimes M = \{(u+1, v+N) \mid u, v \in U\}$ and let

$$B = \left\{\left((u+1)^{-1}, u+N\right) \,\middle|\, u \in U\right\}.$$

Then $G = A \ltimes M = B \ltimes M = AB$ is a triply factorised group. In the following, $G(R,U)$ will be written instead of $G(R, U, \{0\})$.

**Example 2.7** Let $R = \mathbb{Z}/8\mathbb{Z}$ be the ring of integers modulo 8 and let $U = 2R$. It is not difficult to see that $U$ is a construction subgroup of $R$ and that $U + 1 = R^\times$. Then the group $G(R,U)$ is the non-abelian semidirect product of Klein's Four Group with the cyclic group of order 4. In particular, this group is triply factorized.

In Theorem 1.4 it was shown that a triply factorised group $G = A \ltimes M = B \ltimes M = AB$ with $A \cap B = 1$ can always be constructed using a radical ring if $A$, $B$, and $M$ are abelian groups. Within nearring theory, one can even show that every triply factorised group with $A \cap B = 1$ and an arbitrary subgroup $M$ can be obtained from a construction subgroup of the nearring $M(M)$.

Let $G = A \ltimes M = B \ltimes M = AB$ a triply factorised group with $A \cap B = 1$. In the following, the elements of $G$ will be written as tuples $(a, m)$ with $a \in A$ and $m \in M$. For $A$, $B$, and $G$ the multiplicative notation will be used, whereas $M$ will be written additively. Furthermore, let $\tilde{\pi} : G \to A$, $(a, m) \mapsto a$, be the canonical epimorphism from $G$ onto $A$, and let $\pi = \tilde{\pi}|_B$ be the restriction of $\tilde{\pi}$ on $B$. Finally, let $\tilde{\mu} : G \to M$, $(a, m) \mapsto m$ and $\mu = \tilde{\mu}|_B$. (Clearly, $\mu$ need not be a homomorphism in general.) It can be shown (cf. [7, Chapter 3]) that $\mu$ is a bijection between $B$ and $M$ and that $\pi$ is an isomorphism from $B$ onto $A$. Next let $\delta = \pi^{-1}\mu$, which is a bijection from $A$ to $M$. Define the mapping $\gamma : A \to M(M)$, $a \mapsto \gamma_a$ with $\gamma_a : M \to M$, $x \mapsto \left(a^{-1}\left(x\delta^{-1}\right)\right)\delta$. Then $\gamma$ is a group monomorphism from $A$ in $M(M)^\times$. In particular, $A \cong \mathrm{Im}(\gamma)$. Finally, let $V = \mathrm{Im}(\gamma)$ and $U = V - \mathrm{id}_M$. The following lemma can be proved.

**Lemma 2.8 (cf. [7, Lemma 3.2.4])**

1. *$U$ is a group with respect to addition.*

2. *The mapping* $\xi : M \to U$, $m \mapsto \gamma_{(m\delta^{-1})^{-1}} - \mathrm{id}_M$, *is a group isomorphism.*
3. $U$ *is a construction subgroup of* $M(M)$.
4. ${\gamma_a}^{-1}(m\xi) = (m^a)\xi$ *for all* $a \in A$ *and all* $m \in M$.

By Construction 2.6, $U$ leads to a triply factorised group $V \ltimes U$, which is isomorphic to $G$ via the group isomorphism $\alpha : G \to V \ltimes U$, $(a, m) \mapsto (\gamma_a, m\xi)$. In summary it follows that the group $G$ can be constructed from the construction subgroup $U \leq M(M)$. Thus the following theorem holds.

**Theorem 2.9** *If* $G = A \ltimes M = B \ltimes M = AB$ *is a triply factorised group with* $A \cap B = 1$, *then the nearring* $M(M)$ *contains a construction subgroup* $U$ *with* $G \cong G(R, U)$.

**Remark 2.10** Example 1.1.1 shows that the condition $A \cap B = 1$ is necessary here, since for $M = 1$ also $|M(M)| = 1$. Hence in this case $M(M)^\times$ cannot contain a subgroup isomorphic to $A$, if $|A| \geq 2$. In fact, if $A \cap B \neq 1$, the mapping $\mu$ defined above will not be a bijection.

The question remains whether Theorem 2.9 still holds if $A \cap B \neq 1$.

## 3 Nearrings with non-abelian construction subgroups

As mentioned above, a nearring $R$ is called *local* if the set $L_R$ of elements of $R$ which are not right invertible forms a group with respect to addition. In a local nearring $R$ the group $L_R$ forms a construction subgroup.

The following method shows how to find a local nearring $R$ whose subgroup $L_R$ contains a given $p$-group $N$ of finite exponent as a subgroup (cf. [6] and also [5]).

Let $N$ be a $p$-group of finite exponent $p^\ell$ for the prime $p$, not necessarily abelian, but written additively. The zero nearring (i.e., the nearring over $N$ with $ab = 0$ for all $a, b \in N$) embeds into a nearring with identity as follows (cf. Clay [4, Theorem 1.3.27]).

Let $G = N \oplus \mathbb{F}_p$, where $\mathbb{F}_p$ is the field of order $p$, and consider the following mapping: $\phi : N \to M(G)$, $n \mapsto \theta_n$, where $M(G) = \{\alpha : G \to G \mid 0\alpha = 0\}$ is the nearring of all mappings from $G$ in $G$, and for all $g \in G$ the mapping $\theta_n$ is defined by

$$g\theta_n = \begin{cases} n, & g \notin N \\ 0, & g \in N. \end{cases}$$

Then $N\phi$ is a subnearring of $M(G)$ which is isomorphic to the zero nearring over the group $N$. Now consider the set

$$R = \left\{ \sum_{j=1}^{r} (\varepsilon a_j + \theta_{n_j}) \;\middle|\; a_j \in \mathbb{Z}_{p^\ell},\, n_j \in N,\, r \in \mathbb{N}_0 \right\},$$

where $\varepsilon$ is the identity map on $G$, $\mathbb{Z}_{p^\ell}$ is the ring of integers modulo $p^\ell$, and $\mathbb{N}_0$ is the set of non-negative integers. Then it can be shown that $R$ is a local nearring with

$$L_R = \left\{ \sum_{k=1}^{s} (\varepsilon a_k + \theta_{n_k}) \in R \;\middle|\; \sum_{k=1}^{s} a_k \equiv 0 \pmod{p} \right\},$$

where $N\phi$ is contained in $L_R$ (cf. [6, Theorem 3.5]).

## 4 More on nearrings

### 4.1 Prime rings

The prime field of a field is the subfield generated by the identity element. For nearrings with identity, this concept may be generalised. One cannot expect that the subnearring generated by the identity element in general is a field, but it turns out that it is a commutative ring. The proofs of the statements in this section can be found in [7].

**Definition 4.1** Let $R$ be a nearring with identity element 1. Then define $E_R = \langle 1\rangle^+$, the subgroup of $R^+$ generated by 1.

Furthermore, let $P_R = \{nm^{-1} \mid n \in E_R,\, m \in E_R \cap R^\times\}$, called the *prime ring* of $R$.

**Lemma 4.2** *Let $R$ be a nearring with identity element* 1.

(a) *$E_R$ and $P_R$ are commutative rings.*

(b) *If $o^+(1) = n < \infty$, then $E_R = P_R \cong \mathbb{Z}/n\mathbb{Z}$.*

(c) *If $o^+(1) = \infty$, then $E_R \cong \mathbb{Z}$. Furthermore, there exists a set $\pi_R$ of primes such that an element $n \in E_R$ is invertible in $R$ if and only if no prime $p \in \pi_R$ is a divisor of $n$. $P_R \cong \mathbb{Z}D^{-1}$, where $D = \mathbb{Z} \setminus (\bigcup_{p\in\pi_R} p\mathbb{Z})$, i.e.,*

$$P_R \cong \left\{ \frac{n}{m} \in \mathbb{Q} \;\middle|\; \forall p \in \pi_R : p \nmid m \right\}.$$

*Note that if $\pi_R$ contains all prime numbers, then $P_R$ is isomorphic to $\mathbb{Z}$.*

For the description of the construction subgroups of prime rings the following is needed. Let $n$ be a positive integer with $n = \prod_{i=1}^k p_i^{\alpha_i}$, where the primes $p_i$ are pairwise distinct, $\alpha_i > 0$, and $k \geq 0$. Then $\kappa(n) = \prod_{i=1}^k p_i$ is the greatest square-free divisor of $n$. For negative integers let $\kappa(n) = \kappa(-n)$.

**Theorem 4.3** *Let $R$ be a nearring with identity.*

(a) *If $o^+(1) = n < \infty$ then $U \leq {P_R}^+$ is a construction subgroup of $P_R$ if and only if $U^+ = \langle k\rangle^+$ with $k \mid n$ and $\kappa(k) = \kappa(n)$.*

(b) *If $o^+(1) = \infty$ and $\pi_R$ is the set of all prime numbers in $E_R$ which are not invertible in $R$, then $U$ is a construction subgroup of $P_R$ if and only if there is an integer $x$ with $p \mid x$ for all $p \in \pi_R$ and*

$$U \cong \mathbb{Q}_x = \left\{ \frac{n}{m} \in \mathbb{Q} \;\middle|\; x \mid n,\ (x, m) = 1 \right\}.$$

*In particular, if $\pi_R$ is finite and $q = \prod_{p\in\pi_R} p$, then $\mathbb{Q}_q = qP_R$ is the unique maximal construction subgroup of $P_R$. If $\pi_R$ is infinite, there is no non-trivial construction subgroup of $P_R$.*

**Remark 4.4** If $R$ is a local nearring, it is easy to see that the prime ring $P_R$ is also local.

## 4.2 Local nearrings whose groups of units are dihedral

In [2] local nearrings whose groups of units are dihedral are investigated. Since nearfields, i.e., nearrings $R$ with $R^\times = R \setminus \{0\}$, are special cases of local nearrings, the following lemma considers nearfields with dihedral groups of units.

**Lemma 4.5** *If $R$ is a nearfield with dihedral multiplicative group, then $|R| = 3$ and hence $R \cong \mathbb{F}_3$.*

It is proved in [2] that a local nearring $R$ with dihedral group of units is always finite. If $R$ is zero-symmetric, it follows from [9] that $R^+$ is a $p$-group for a prime $p$. This statement is generalised to arbitrary finite local nearrings in [7].

**Theorem 4.6** (a) *Let $R$ be a finite local nearring of odd order. If the multiplicative group $R^\times$ is dihedral, then either $R \cong \mathbb{F}_3$ or $R^+$ is an elementary abelian group of order 9.*

(b) *There are exactly two local nearrings over the elementary abelian group of order 9 whose multiplicative groups are dihedral, one of which is zero-symmetric.*

The investigation of local nearrings of odd order whose groups of units are dihedral is more difficult. In [2] it is shown that such a nearring $R$ is of order at most 32. Considering all groups of order 32, it is proved in [7] that no group of order 32 can occur as the additive group of a local nearring with a dihedral group of units. This leads to the following result.

**Theorem 4.7** *If $R$ is a local nearring with dihedral group of units, then $|R| \leq 16$.*

## References

[1] B. Amberg, S. Franciosi and F. de Giovanni, *Products of groups*, Oxford Mathematical Monographs, Clarendon Press, Oxford, 1992.

[2] B. Amberg, P. Hubert, and Ya. Sysak, Local nearrings with dihedral multiplicative group, *J. Algebra* **273** (2004), 700–717.

[3] B. Amberg and Ya. Sysak, Radical rings and products of groups, in C. M. Campbell et al. (eds.), *Groups St Andrews 1997 in Bath. Selected papers of the international conference, Bath, UK, July 26–August 9, 1997, Vol. 1*, Cambridge, Cambridge University Press, London Math. Soc. Lecture Note Ser. 260, 1–19 , 1999.

[4] J. R. Clay, *Nearrings: geneses and applications*, Oxford Science Publications, Oxford University Press, Oxford, 1992.

[5] P. Hubert, *Lokale Fastringe und eine Konstruktion dreifach faktorisierter Gruppen*, Diploma Thesis, Johannes-Gutenberg-Universität, Mainz, 2000.

[6] P. Hubert, Local near-rings and triply factorized groups, *Comm. Algebra* **32** (2004), 1229–1235.
[7] P. Hubert, *Nearrings and a Construction of Triply Factorized Groups*, Dissertation, Johannes-Gutenberg-Universität, Mainz, 2005.
[8] N. Jacobson, *Structure of rings.* Colloquium Publications, American Mathematical Society, 1956, Vol. 37.
[9] C. J. Maxson, On local near-rings. *Math. Z.* **106** (1968), 197–205.
[10] J. D. P. Meldrum, *Near-rings and their links with groups*, Pitman, London, 1985.
[11] Ya. Sysak, Products of infinite groups, Preprint 82.53, Akad. Nauk. Ukrainy, Inst. Mat. Kiev, 1982.

# ON THE SPACE OF CYCLIC TRIGONAL RIEMANN SURFACES OF GENUS 4

MILAGROS IZQUIERDO[1] and DANIEL YING

Matematiska institutionen, Linköpings universitet, 581 83 Linköping, Sweden
Email: miizq@mai.liu.se, dayin@mai.liu.se

## Abstract

A closed Riemann surface which can be realized as a 3-sheeted covering of the Riemann sphere is called trigonal, and such a covering is called a trigonal morphism. If the trigonal morphism is a cyclic regular covering, the Riemann surface is called a cyclic trigonal Riemann surface. Using the characterization of cyclic trigonality by Fuchsian groups, we describe the structure of the space of cyclic trigonal Riemann surfaces of genus 4.

## 1 Introduction

A closed Riemann surface $X$ which can be realized as a 3-sheeted covering of the Riemann sphere is said to be *trigonal,* and such a covering will be called a *trigonal morphism.* This is equivalent to the fact that $X$ is represented by a curve given by a polynomial equation of the form:

$$y^3 + yb(x) + c(x) = 0.$$

If $b(x) \equiv 0$ then the trigonal morphism is a cyclic regular covering and the Riemann surface is called *cyclic trigonal.* Trigonal Riemann surfaces have been recently studied (see [2] and [8]).

By Lemma 2.1 in [1], if the surface $X$ has genus $g \geq 5$, then the trigonal morphism is unique. The Severi-Castelnouvo inequality is used in order to prove such uniqueness, but this technique is not valid for small genera.

We develop some ad hoc methods to deal with small genera. Using the characterization of trigonality by means of Fuchsian groups [3], we obtain all possible cyclic trigonal Riemann surfaces of genus four. We also find that there is a uniparametric family of Riemann surfaces of genus four admitting several cyclic trigonal morphisms. The last result was shown with different calculations in [4], but we state it here for the sake of completeness.

## 2 Trigonal Riemann surfaces and Fuchsian groups

Let $X_g$ be a compact Riemann surface of genus $g \geq 2$. The surface $X_g$ can be represented as a quotient $X_g = \mathcal{D}/\Gamma$ of the complex unit disc $\mathcal{D}$ under the action of a (cocompact) Fuchsian group $\Gamma$, that is, a discrete subgroup of the group

[1]Partially supported by the Swedish Research Council (VR)

$G = \mathrm{Aut}(\mathcal{D})$ of conformal automorphisms of $\mathcal{D}$. The algebraic structure of a Fuchsian group and the geometric structure of its quotient orbifold are given by the signature of $\Gamma$:

$$s(\Gamma) = (g;\ m_1, \ldots, m_r). \tag{1}$$

The orbit space $\mathcal{D}/\Gamma$ is an orbifold with underlying surface of genus $g$, having $r$ cone points. The integers $m_i$ are called the periods of $\Gamma$ and they are the orders of the cone points of $\mathcal{D}/\Gamma$. The group $\Gamma$ is called the *fundamental group* of the orbifold $\mathcal{D}/\Gamma$.

A group $\Gamma$ with signature (1) has a *canonical presentation:*

$$\langle x_1, \ldots, x_r, a_1, b_1, \ldots, a_g, b_g \mid x_i^{m_i}, i = 1, \ldots, r,\ x_1 \ldots x_r a_1 b_1 a_1^{-1} b_1^{-1} \ldots a_g b_g a_g^{-1} b_g^{-1} \rangle \tag{2}$$

The last relation in the above presentation is called the long relation. The generators $x_1, \ldots, x_r$, are called the *elliptic generators.* Any elliptic element in $\Gamma$ is conjugated to a power of some of the elliptic generators.

The hyperbolic area of the orbifold $\mathcal{D}/\Gamma$ coincides with the hyperbolic area of an arbitrary fundamental region of $\Gamma$ and equals:

$$\mu(\Gamma) = 2\pi\left(2g - 2 + \sum_{i=1}^{r}\left(1 - \frac{1}{m_i}\right)\right), \tag{3}$$

Given a subgroup $\Gamma'$ of index $N$ in a Fuchsian group $\Gamma$, one can calculate the structure of $\Gamma'$ in terms of the structure of $\Gamma$ and the action of $\Gamma$ on the $\Gamma'$-cosets:

**Theorem 1 ([12])** *Let $\Gamma$ be a Fuchsian group with signature* (1) *and canonical presentation* (2). *Then $\Gamma$ contains a subgroup $\Gamma'$ of index $N$ with signature*

$$s(\Gamma') = (h; m'_{11}, m'_{12}, \ldots, m'_{1s_1}, \ldots, m'_{r1}, \ldots, m'_{rs_r}). \tag{4}$$

*if and only if there exists a transitive permutation representation $\theta : \Gamma \to \Sigma_N$ satisfying the following conditions:*

1. *The permutation $\theta(x_i)$ has precisely $s_i$ cycles of lengths less than $m_i$, the lengths of these cycles being $m_i/m'_{i1}, \ldots, m_i/m'_{is_i}$.*
2. *The Riemann–Hurwitz formula*

$$\mu(\Gamma')/\mu(\Gamma) = N. \tag{5}$$

The map $\theta : \Gamma \to \Sigma_N$ is the *monodromy* of the covering $f : \mathcal{D}/\Gamma' \to \mathcal{D}/\Gamma$. Moreover $\Gamma' = \theta^{-1}(\mathrm{Stb}(1))$, where $1, 2, \ldots, N$ are the labels of the sheets of the covering $f$. See Chapter 8 in [11] .

A Fuchsian group $\Gamma$ without elliptic elements is called a *surface group* and it has signature $(h; -)$. Given a Riemann surface represented as the orbit space $X = \mathcal{D}/\Gamma$, with $\Gamma$ a surface Fuchsian group, a finite group $G$ is a group of automorphisms of $X$ if and only if there exists a Fuchsian group $\Delta$ and an epimorphism $\theta : \Delta \to G$ with $\ker(\theta) = \Gamma$. The Fuchsian group $\Delta$ is the lifting of $G$ to the universal covering $\pi : \mathcal{D} \to \mathcal{D}/\Gamma$ and is called the *universal covering transformations group* of $(X, G)$.

Let $\Gamma$ be a Fuchsian group with signature (1). Then the Teichmüller space $T(\Gamma)$ of $\Gamma$ is homeomorphic to a complex ball of dimension $d(\Gamma) = 3g - 3 + r$ (see [10]). Let $\Gamma' \leq \Gamma$ be Fuchsian groups, the inclusion mapping $\alpha : \Gamma \to \Gamma'$ induces an embedding $T(\alpha) : T(\Gamma) \to T(\Gamma')$ defined by $[r] \mapsto [r\alpha]$. See [7], [10] and [13].

The modular group $\mathrm{Mod}(\Gamma)$ of $\Gamma$ is the quotient $\mathrm{Mod}(\Gamma) = \mathrm{Aut}(\Gamma)/\mathrm{Inn}(\Gamma)$ where $\mathrm{Inn}(\Gamma)$ is the normal subgroup of $\mathrm{Aut}(\Gamma)$ consisting of all inner automorphisms of $\Gamma$. The *moduli space* of $\Gamma$ is the quotient $M(\Gamma) = T(\Gamma)/\mathrm{Mod}(\Gamma)$ endowed with the quotient topology.

A Fuchsian group $\Gamma$ such that there does not exist any other Fuchsian group containing it with finite index is called a *finite maximal* Fuchsian group.

To decide whether a given finite group can be the full group of automorphism of some compact Riemann surface we will need all pairs of signatures $s(\Gamma)$ and $s(\Gamma')$ for some Fuchsian groups $\Gamma$ and $\Gamma'$ such that $\Gamma' \leq \Gamma$ and $d(\Gamma) = d(\Gamma')$. The full list of such pairs of groups was obtained by Singerman in [13].

**Definition 2** A Riemann surface $X$ is said to be *trigonal* if it admits a three sheeted covering $f : X \to \widehat{\mathbb{C}}$ onto the Riemann sphere. If $f$ is a cyclic regular covering then $X$ is called cyclic trigonal. The covering $f$ will be called the (cyclic) trigonal morphism.

The following result gives us a characterization of cyclic trigonal Riemann surfaces using Fuchsian groups:

**Theorem 3 ([3])** *Let $X_g$ be a Riemann surface, $X_g$ admits a cyclic trigonal morphism $f$ if and only if there is a Fuchsian group $\Delta$ with signature $(0; \overbrace{3, \ldots, 3}^{g+2})$ and an index three normal surface subgroup $\Gamma$ of $\Delta$, such that $\Gamma$ uniformizes $X_g$.*

By Lemma 2.1 in [1], if the surface $X_g$ has genus $g \geq 5$, then the trigonal morphism is unique. In this case, the cyclic trigonal morphism $f$ is induced by a normal subgroup $C_3$ in $\mathrm{Aut}(X_g)$. From now on we shall consider surfaces with genera smaller than 5.

## 3 Non-unique cyclic trigonal morphisms on Riemann surfaces

For genus four we use Theorems 1 and 3 to obtain the following method to find cyclic trigonal Riemann surfaces. Let $G$ be the full automorphisms group of the surface $X_4$.

Let $X_4 = \mathcal{D}/\Gamma$ be a Riemann surface of genus 4 uniformized by the surface Fuchsian group $\Gamma$, $X_4$ admits a cyclic trigonal morphism $f$ if and only if there is a maximal Fuchsian group $\Delta$ with signature $(0; m_1, \ldots, m_r)$, a trigonal automorphism $\varphi : X_4 \to X_4$, such that $\langle\varphi\rangle \leq G$ and an epimorphism $\theta : \Delta \to G$ with $\ker(\theta) = \Gamma$ in such a way that $\theta^{-1}(\langle\varphi\rangle)$ is a Fuchsian group with signature $(0; 3, 3, 3, 3, 3, 3)$. Notice that the condition $\Gamma$ to be a surface Fuchsian group imposes that the order of the image under $\theta$ of an elliptic generator $x_i$ of $\Delta$ is the same as the order of $x_i$. Furthermore the trigonal morphism $f$ is unique if and only if $\langle\varphi\rangle$ is normal in $G$.

Now, let $\theta : \Delta \to G$ be such an epimorphism and let $|G| = N$. Then $s(\Delta) = (0; m_{1(N)}, \ldots, m_{r(N)})$, where $m_{i(N)}$ runs over the divisors of $N$. Applying the Riemann–Hurwitz formula we have that

$$2(g + N - 1) = \sum_{1(N)}^{r(N)} N \frac{(m_{i(N)} - 1)}{m_{i(N)}}. \tag{6}$$

Equation (6) and the list of maximal signatures ([13]) yield the following list of allowed signatures for genus 4.

**Lemma 4** *Let $X_4 = \mathcal{D}/\Gamma$ be a Riemann surface of genus 4 uniformized by the surface group $\Gamma$, $X_4$ admitting a cyclic trigonal morphism $f$. Then the maximal Fuchsian group $\Delta$ uniformizing the orbifold $X/G$ must have one of the following signatures:*

| $\|G\|$ | $s(\Delta)$ | $\|G\|$ | $s(\Delta)$ | $\|G\|$ | $s(\Delta)$ |
|---|---|---|---|---|---|
| 3 | $(0; 3, 3, 3, 3, 3, 3)$, | 6 | $(0; 2, 6, 6, 6)$, | 6 | $(0; 2, 2, 3, 3, 3)$, |
| 6 | $(0; 2, 2, 2, 3, 6)$, | 6 | $(0; 2, 2, 2, 2, 2, 2)$, | 12 | $(0; 2, 2, 2, 2, 2)$, |
| 12 | $(0; 2, 2, 3, 6)$, | 12 | $(0; 2, 3, 3, 3)$, | 12 | $(0; 4, 6, 12)$, |
| 18 | $(0; 2, 2, 2, 6)$, | 24 | $(0; 3, 4, 6)$, | 24 | $(0; 2, 2, 2, 4)$, |
| 36 | $(0; 2, 4, 12)$, | 36 | $(0; 2, 2, 2, 3)$, | 42 | $(0; 2, 3, 42)$, |
| 48 | $(0; 2, 3, 24)$, | 54 | $(0; 2, 3, 18)$, | 60 | $(0; 2, 3, 15)$, |
| 72 | $(0; 2, 4, 6)$, | 72 | $(0; 2, 3, 12)$, | 90 | $(0; 2, 3, 10)$, |
| 108 | $(0; 2, 3, 9)$, | 120 | $(0; 2, 4, 5)$, | 144 | $(0; 2, 3, 8)$, |
| 252 | $(0; 2, 3, 7)$, | | | | |

We study the existence of epimorphisms satisfying the diagram below

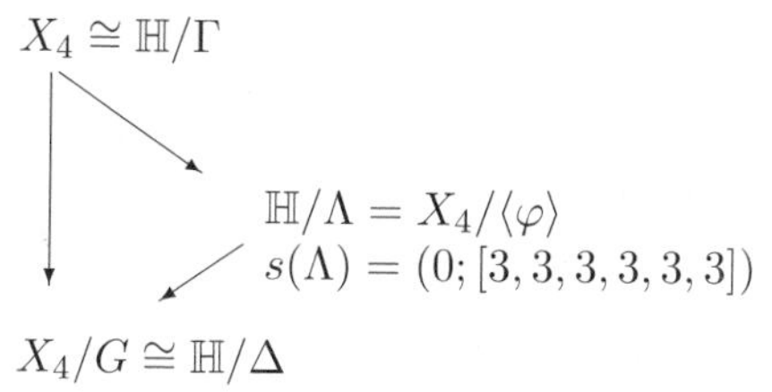


## Existence of cyclic trigonal Riemann surfaces of genus 4

In the following we calculate all possible epimorphisms $\theta : \Delta \to G$. We separate the cases according to the order of the group $G$.

We use Theorem 1 to calculate $\theta^{-1}(\langle\varphi\rangle)$. The Fuchsian groups $\Delta$ have signatures as in Lemma 4 and presentation (2).

1. $|G| = 3$. There are epimorphisms $\theta : \Delta \to C_3 = \langle a \mid a^3 = 1\rangle$ where $s(\Delta) = (0; 3, 3, 3, 3, 3, 3)$. For instance, $\theta_1(x_{2i}) = a$ and $\theta_1(x_{2i-1}) = a^{-1}$, $1 \le i \le 3$, and $\theta_2(x_i) = a$, $1 \le i \le 6$.

**Remark** The cyclic trigonal Riemann surfaces of genus 4 form a space $\mathcal{M}_4^3$ of (complex) dimension $d(\Delta) = 0-3+6 = 3$. By Theorem 2 in [6] this space consists of two disconnected components. One connected component is given by epimorphisms $\theta_1$, that is, half the stabilizers of the cone points rotate in opposite directions. The other component is given by epimorphisms $\theta_2$, that is, the stabilizers of all the cone points rotate in the same direction.

2. $|G| = 6$. First, consider the signature $s(\Delta_1) = (0; 2, 6, 6, 6)$. There are epimorphisms $\theta : \Delta \to C_6 = \langle a \mid a^6 = 1\rangle$. The cone points in $\mathcal{D}/\ker(\theta)$ are given by the action of $\theta(x_2)$, $\theta(x_3)$ and $\theta(x_4)$ on the $\langle a^2\rangle$-cosets. Each such element gives one cone point of order 3. Thus, the signature of $\theta^{-1}(\langle a^2\rangle) = (1; 3, 3, 3)$ and the surfaces $\mathcal{D}/\ker(\theta)$ are not trigonal.

   Consider now the signature $s(\Delta_2) = (0; 2, 2, 3, 3, 3)$. There are possible epimorphisms from $\Delta_2$ onto both $D_3$ and $C_6$. In each case, each of $\theta(x_3)$, $\theta(x_4)$ and $\theta(x_5)$ induces two cone points in $\mathcal{D}/\ker(\theta)$. Thus $s(\theta^{-1}(C_3)) = (1; 3, 3, 3)$, therefore these surfaces are trigonal with unique trigonal morphisms.

   Consider the signature $s(\Delta_3) = (0; 2, 2, 2, 3, 6)$. There are epimorphisms $\theta : \Delta_3 \to C_6$. $\theta(x_4)$ induces two cone points, while $\theta(x_5)$ induces one conical point in $\mathcal{D}/\ker(\theta)$. These surfaces are not trigonal since $s(\theta^{-1}(\langle a^2\rangle)) = (1; 3, 3, 3)$.

   The signature $s(\Delta_4) = (0; 2, 2, 2, 2, 2, 2)$ does not produce any trigonal surface since the orders of the elliptic elements of $\Delta_4$ are relative prime to 3.

**Remark** The cyclic trigonal surfaces $X_4$ of genus 4 with automorphism group of order 6 are determined by the Fuchsian groups with $s(\Delta_2) = (0; 2, 2, 3, 3, 3)$, the universal covering transformation groups of $(X_4, \mathrm{Aut}(X_4))$. Therefore these surfaces form a disconnected subspace of $\mathcal{M}_4^3$ with (complex) dimension $0-3+5 = 2$. Moreover, the surfaces with automorphism group $\mathrm{Aut}(X_4) = D_3$ lie in one of the connected components of $\mathcal{M}_4^3$ while the surfaces with $\mathrm{Aut}(X_4) = C_6$ lie in the other connected component.

3. $|G| = 12$. First, consider the signature $s(\Delta_1) = (0; 4, 6, 12)$. There are epimorphisms $\theta : \Delta \to C_{12} = \langle a \mid a^{12} = 1\rangle$. The cone points in $\mathcal{D}/\ker(\theta)$ are given by the action of $\theta(x_2)$ and $\theta(x_3)$ on the $\langle a^4\rangle$-cosets. $\theta(x_2)$ induces two cone points, while $\theta(x_3)$ induces one cone point in $\mathcal{D}/\ker(\theta)$. Thus, the signature is $s(\theta^{-1}(\langle a^4\rangle)) = (1; 3, 3, 3)$ and the surface $\mathcal{D}/\ker(\theta)$ is not trigonal.

   Consider the signature $s(\Delta_2) = (0; 2, 3, 3, 3)$. The only group of order 12 generated by three elements of order 3 is $A_4 = \langle a, s \mid a^3 = s^2 = (as)^3 = 1\rangle$ (see [5]). There are epimorphisms $\theta : \Delta_2 \to A_4$. For instance $\theta(x_1) = s$, $\theta(x_2) = a$, $\theta(x_3) = as$ and $\theta(x_4) = sas$. Any element of order 3 in $A_4$ leaves just one coset fixed when acting on the $\langle a\rangle$, $\langle sa\rangle$, $\langle as\rangle$ or the $\langle sas\rangle$-cosets, since all of them are conjugated. Then $\theta^{-1}(C_3)$ has signature $(1; 3, 3, 3)$ and the corresponding surfaces are not trigonal.

   Now, consider the signature $s(\Delta_3) = (0; 2, 2, 3, 6)$. There are possible epimorphisms from $\Delta_3$ onto both $D_6$ and $C_6 \times C_2$. In each case, $\theta(x_3)$ induces four cone points and $\theta(x_4)$ induces two cone points in $\mathcal{D}/\ker(\theta)$. Thus, these surfaces are trigonal with unique trigonal morphisms.

The signature $s(\Delta_4) = (0; 2, 2, 2, 2, 2)$ does not produce any trigonal surface since the orders of the elliptic elements of $\Delta_4$ are relative prime to 3.

**Remark** The cyclic trigonal surfaces $X_4$ of genus 4 with automorphism group of order 12 are determined by the Fuchsian groups with $s(\Delta_2) = (0; 2, 2, 3, 6)$, the universal covering transformation groups of $(X_4, \mathrm{Aut}(X_4))$. Therefore these surfaces form a disconnected subspace of $\mathcal{M}_4^3$ with (complex) dimension $0 - 3 + 4 = 1$. Moreover, the surfaces with automorphism group $\mathrm{Aut}(X_4) = D_6$ lie in one of the connected components of $\mathcal{M}_4^3$ while the surfaces with $\mathrm{Aut}(X_4) = C_6 \times C_2$ lie in the other connected component.

4. $|G| = 18$. Consider the signature $s(\Delta) = (0; 2, 2, 2, 6)$. There is no group of order 18 generated by three involutions and containing elements of order 6 ([5]). Hence there are no epimorphsims $\theta$.

5. $|G| = 24$. First of all, the signature $s(\Delta_1) = (0; 2, 2, 2, 4)$ cannot give trigonal surfaces since the orders of the elliptic generators of $\Delta_1$ are relative prime to 3.

   For the signature $s(\Delta_2) = (0; 3, 4, 6)$: The only groups of order 24 generated by one element of order 3 and one element of order 6 are $A_4 \times C_2$ and $\langle 2, 3, 3\rangle = Q \rtimes C_3 = \langle a, s, t \mid a^3 = t^4 = s^4 = (st)^4 = 1, s^2 = t^2, a^2sa = t, a^2ta = st\rangle$, the binary tetrahedral group, see [5]. There are no possible epimorphisms from $\Delta_2$ onto $A_4 \times C_2$ since this group has no elements of order 4. There are epimorphisms $\theta : \Delta_2 \to \langle 2, 3, 3\rangle$, for instance $\theta(x_1) = sta$, $\theta(x_2) = s$ and $\theta(x_3) = s^2a^2$. Now, the group $\langle 2, 3, 3\rangle$ contains just one conjugacy class of subgroups of order 3, and one conjugacy class of elements of order 6, with representatives $sta$ and $s^2a^2$ respectively. Therefore it is sufficient to study the action of $sta$ and $s^2a^2$ on the $\langle a\rangle$-cosets. The first action has orbits $\{[a], [s], [t^3]\}$, $\{[t], [s^2], [s^3]\}$, $\{[ts]\}$ and $\{[st]\}$, the second action has orbits $\{[a], [s^2]\}$ and $\{[s], [t^3], [st], [s^3], [t], [ts]\}$. Thus the signature of $\theta^{-1}(\langle a\rangle)$ is $(1; 3, 3, 3)$ and the corresponding surface is not trigonal.

6. $|G| = 36$. First, consider Fuchsian groups $\Delta_1$ with signature $(0; 2, 4, 12)$. The only groups of order 36 containing elements of order 12 are $C_{36}$, $C_{12} \times C_3$, $(C_3 \times C_3) \rtimes_1 C_4 = \langle a, b, t \mid a^3 = b^3 = t^4 = [a, b] = 1, t^3at = a^{-1}, t^3bt = b^{-1}\rangle$ and $C_3 \rtimes C_{12} = \langle a, b / a^3 = b^{12} = b^{-1}aba = 1\rangle$. None of these groups are generated by elements of order 2 and 4. Hence there is no epimorphism from $\Delta_1$ onto a group of order 36 and there are no cyclic trigonal Riemann surfaces of genus 4 such that the quotient $X_4/\mathrm{Aut}(X_4)$ is uniformized by $\Delta_1$.

   Secondly, consider Fuchsian groups $\Delta_2$ with signature $(0; 2, 2, 2, 3)$. The only group of order 36 generated by 3 involutions is $D_3 \times D_3 = \langle a, b, s, t \mid a^3 = b^3 = s^2 = t^2 = [a, b] = [s, b] = [t, a] = (sa)^2 = (tb)^2 = 1\rangle$. The non-normal order three subgroups in $D_3 \times D_3$ are $\langle ab\rangle$ and $\langle a^2b\rangle$. Consider the epimorphism $\theta : \Delta_2 \to D_3 \times D_3$ defined by $\theta(x_1) = s$, $\theta(x_2) = tb$,

$\theta(x_3) = sta$ and $\theta(x_4) = a^2b$. The action of $\theta(x_4) = a^2b$ on the $\langle ab\rangle$-cosets has the following orbits: $\{[1], [a], [b]\}$, $\{[s]\}$, $\{[sb]\}$, $\{[sa]\}$, $\{[t]\}$, $\{[tb]\}$, $\{[ta]\}$, $\{[st], [stb], [sta]\}$. Then $s(\theta^{-1}(\langle ab\rangle))$ contains six periods equal to 3 and by the Riemann–Hurwitz formula $s(\theta^{-1}(\langle ab\rangle)) = (0; 3, 3, 3, 3, 3, 3)$. In the same way the action of $\theta(x_4) = a^2b$ on the $\langle a^2b\rangle$-cosets has the following orbits: $\{[1]\}$, $\{[b]\}$, $\{[ab]\}$, $\{[s], [sa^2], [sb]\}$, $\{[t], [tb], [ta^2]\}$, $\{[st]\}$, $\{[stb]\}$, $\{[sta^2]\}$. Again, $s(\theta^{-1}(\langle a^2b\rangle))$ contains six periods equal to 3 and then $s(\theta^{-1}(\langle a^2b\rangle)) = (0; 3, 3, 3, 3, 3, 3)$. Thus the Riemann surfaces uniformized by $\ker(\theta)$ are cyclic trigonal Riemann surfaces that admit two different trigonal morphisms $f_1 : \mathcal{D}/\ker(\theta) \to \hat{\mathbb{C}}$ and $f_2 : \mathcal{D}/\ker(\theta) \to \hat{\mathbb{C}}$ induced by the subgroups $\langle ab\rangle$ and $\langle a^2b\rangle$ of $D_3 \times D_3$. The dimension of the family of surfaces $\mathcal{D}/\ker(\theta)$ is given by the dimension of the space of groups $\Delta_2$ with $s(\Delta_2) = (0; 2, 2, 2, 3)$. This (complex)-dimension is $3(0)-3+4 = 1$. By theorem 1 in [6] this is the unique conjugacy class of order 3 subgroups of $D_3 \times D_3$ inducing trigonal morphisms on the surfaces.

7. $|G| = 42$. There is no such epimorphism, since the only possible epimorphism from a Fuchsian group with signature $(0; 2, 3, 42)$ should be onto $C_{42}$, which is not generated by elements of order 2 and 3.

8. $|G| = 48$. There is no epimorphism from a Fuchsian group $\Delta$ with signature $(0; 2, 3, 24)$ onto a group of order 48. Otherwise there would be an epimorphism from a Fuchsian group with signature $(0; 3, 3, 12)$ onto a group of order 24 generated by two elements of order 3. This is impossible because the only such group is the binary tetrahedral group, which does not have elements of order 12. See [5] and [13].

9. $|G| = 54$. There is no epimorphism from a Fuchsian group $\Delta$ with signature $(0; 2, 3, 18)$ onto a group of order 54. Otherwise it would be an epimorphism from a Fuchsian group with signature $(0; 3, 3, 9)$ onto $C_9 \rtimes C_3$. This group is not generated by order three elements. See [5] and [13].

10. $|G| = 60$. Clearly there is no epimorphism from a Fuchsian group $\Delta$ with signature $(0; 2, 3, 15)$ onto a group of order 60, since the only group of order 60 generated by elements of order 2 and 3 is $A_5$.

11. $|G| = 72$.
i) Consider Fuchsian groups $\Delta_1$ with signature $(0; 2, 3, 12)$. The group $\Delta_1$ contains the group $\Lambda_1$ with signature $s(\Lambda_1) = (0; 3, 3, 6)$ as a subgroup of index 2 (see [13]). Any epimorphism $\theta_1 : \Delta_1 \to G_{72}$ is an extension of an epimorphism $\phi_1 : \Lambda_1 \to G_{36}$. The existence of epimorphisms $\phi_1$ obliges the group $G_{36}$ to be generated by two elements of order 3. The only group of order 36 generated by elements of order 3 is $A_4 \times C_3 = \langle a, b, s \mid a^3 = b^3 = s^2 = [a, b] = [s, b] = (as)^3 = 1\rangle$. It contains four conjugacy classes of subgroups of order 3 with representatives $\langle a\rangle$, $\langle ab\rangle$, $\langle a^2b\rangle$ and $\langle b\rangle$, where $\langle b\rangle$ is central. Consider the epimorphism $\phi_1 : \Lambda_1 \to A_4 \times C_3$ defined as $\phi_1(z_1) = ba$, $\phi_1(z_2) = bsa$, $\phi_1(z_3) = bsas$. As $\langle b\rangle$ is central, the action of any element of

order 3 on the $\langle b\rangle$-cosets gives four orbits, each with three cosets. The action of any element of order 6 has six orbits, with 2 cosets each. Then $\phi_1(z_3) = bsas$ induces six periods of order 3 in $\phi_1^{-1}(\langle b\rangle)$. By the Riemann–Hurwitz formula $s(\phi_1^{-1}(\langle b\rangle)) = (0; 3, 3, 3, 3, 3, 3)$. Therefore, a cyclic trigonal Riemann surface $X_4$ will be uniformized by $\ker\theta_1$, where $\theta_1 : \Delta_1 \to G_{72}$ as above and $G_{72}$ an extension of degree 2 of $A_4 \times C_3$ containing elements of order 12. This extension is $\Sigma_4 \times C_3 = \langle a, b, \overline{s} \mid a^3 = b^3 = (\overline{s})^2 = [a, b] = (a\overline{s})^4 = [t, b] = 1\rangle$, with $\langle b\rangle$ central in $\Sigma_4 \times C_3$. The epimorphism $\theta_1 : \Delta_1 \to \Sigma_4 \times C_3$ given by $\theta_1(x_1) = \overline{s},\ \theta_1(x_2) = ab,\ \theta_1(x_3) = a^2\overline{s}b$ yields a trigonal Riemann surface $X_4 = \mathcal{D}/\ker(\theta_1)$ with trigonal morphism induced by $\langle b\rangle$. But by Theorem 1 in [6] this trigonal morphism is unique.

ii) Consider Fuchsian groups $\Delta_2$ with signature $(0; 2, 4, 6)$. The group $\Delta_2$ contains the group $\Lambda_2$ with signature $s(\Lambda_2) = (0; 3, 4, 4)$ as a subgroup of index 2 (see [13]). Any epimorphism $\theta_2 : \Delta_2 \to G_{72}$ is an extension of an epimorphism $\phi_2 : \Lambda_2 \to G_{36}$. The existence of epimorphisms $\phi_2$ obliges the group $G_{36}$ to be generated by two elements of order 4. The only group of order 36 generated by elements of order 4 is $(C_3 \times C_3) \rtimes_2 C_4 = \langle a, b, t \mid a^3 = b^3 = t^4 = [a, b] = 1, t^3at = b, t^3bt = a^{-1}\rangle$. It contains two conjugacy classes of subgroups of order 3 with representatives $\langle a\rangle$ and $\langle ab\rangle$. Consider the epimorphism $\phi_2 : \Lambda_2 \to (C_3 \times C_3) \rtimes C_4$ defined as $\phi_2(y_1) = a,\ \phi_2(y_2) = at,\ \phi_2(y_3) = bt^3$. The action of $\phi_2(y_1) = a$ on the $\langle a\rangle$-cosets has the following orbits: $\{[1]\}$, $\{[b]\}$, $\{[t^2]\}$, $\{[t], [at], [a^2t]\}$, $\{[t^3], [at^3], [a^2t^3]\}$, $\{[bt^2]\}$, $\{[b^2]\}$, $\{[b^2t^2]\}$. The action of $a$ on the $\langle b\rangle$-cosets has the same cycle-structure. Then $\phi_2^{-1}(\langle a\rangle)$ and $\phi_2^{-1}(\langle b\rangle)$ have six periods of order 3, induced by $\phi_2(y_1) = a$. By the Riemann–Hurwitz formula $s(\phi_2^{-1}(\langle a\rangle)) = s(\phi_2^{-1}(\langle b\rangle)) = (0; 3, 3, 3, 3, 3, 3)$. Therefore a cyclic trigonal Riemann surface $Y_4$ with non-unique trigonal morphism will be uniformized by $\ker(\theta_2)$, where $\theta_2 : \Delta_2 \to G_{72}$ as above and $G_{72}$ is an extension of degree 2 of $(C_3 \times C_3) \rtimes C_4$ containing just two conjugacy classes of subgroups of order 3. This extension is $(C_3 \times C_3) \rtimes D_4 = \langle a, b, t, s \mid a^3 = b^3 = t^4 = s^2 = (st)^2 = [a, b] = (sa)^2 = (sb)^2 = 1, t^3at = b, t^3bt = a^2\rangle$. The epimorphism $\theta_2 : \Delta_2 \to (C_3 \times C_3) \rtimes D_4$ given by $\theta_2(x_1) = s,\ \theta_2(x_2) = ta,\ \theta_2(x_3) = stb$ yields the required trigonal Riemann surface $Y_4 = \mathcal{D}/\ker(\theta_2)$ with non-unique trigonal morphisms. The trigonal morphisms $f_1 : \mathcal{D}/\ker(\theta_2) \to \hat{\mathbb{C}}$ and $f_2 : \mathcal{D}/\ker(\theta_2) \to \hat{\mathbb{C}}$ are induced by the conjugated subgroups $\langle a\rangle$ and $\langle b\rangle$ of $(C_3 \times C_3) \rtimes D_4$.

12. $|G| = 90$. There is no epimorphism from a Fuchsian group $\Delta$ with signature $(0; 2, 3, 10)$ onto a group of order 90. Otherwise it would be an epimorphism from a Fuchsian group $\hat{\Delta}$ with signature $(0; 5, 5, 5)$ onto $C_{15}$, which is impossible. Notice that groups with signature $(0; 5, 5, 5)$ are normal subgroups of index 6 in groups with signature $(0; 2, 3, 10)$ (see [13]).

13. $|G| = 108$. As in Case 9, there is no epimorphism $\theta$ from a Fuchsian group $\Delta$ with signature $(0; 2, 3, 9)$ onto a group of order 108. Notice that groups with signature $(0; 3, 3, 9)$ are non-normal subgroups of index 4 in groups with sig-

nature $(0; 2, 3, 9)$ (see [13]).

14. $|G| = 120$. The signature $s(\Delta) = (0; 2, 4, 5)$ cannot give trigonal surfaces since the orders of the elliptic generators of $\Delta$ are relative prime to 3.

15. $|G| = 144$. There is no epimorphism $\Delta$ with signature $(0; 2, 3, 8)$ onto a group of order 144 since there is no epimorphism $\theta : \hat{\Delta} \to C_{24}$ where $s(\hat{\Delta}) = (0; 2, 8, 8)$. $C_{24}$ is the only group of order 24 containing elements of order 8 (see [5]). Notice that the groups with signature $s(\hat{\Delta}) = (0; 2, 8, 8)$ are non-normal subgroups of index 6 in groups $\Delta$ with $s(\Delta) = (0; 2, 3, 8)$ (see [13]).

16. $|G| = 252$. The are no Riemann surfaces of genus 4 with exactly 252 automorphisms, otherwise there would be an epimorphism $\theta : \Delta \to G_{252}$, with $s(\Delta) = (0; 2, 3, 7)$. A possible epimorphism $\theta : \Delta \to G_{252}$, is an extension of a epimorphism $\phi : \hat{\Delta} \to C_7 \rtimes C_4$ where $s(\hat{\Delta}) = (0; 2, 7, 7)$ (see [13]). Such an epimorphism $\phi$ does not exist since no group of order 28 is generated by two elements of order 7 (see [5]).

We summarize the above results in the following theorems:

**Theorem 5 ([4])** *There is a uniparametric family of Riemann surfaces $X_4(\lambda)$ of genus 4 admitting several cyclic trigonal morphisms. The surfaces $X_4(\lambda)$ have $G = \mathrm{Aut}(X_4(\lambda)) = D_3 \times D_3$ and the quotient Riemann surfaces $X_4(\lambda)/G$ are uniformized by the Fuchsian groups $\Delta$ with signature $s(\Delta) = (0; 2, 2, 2, 3)$. There is one Riemann surface $Y_4$ in the family with automorphism group $\mathrm{Aut}(Y_4) = (C_3 \times C_3) \rtimes D_4$ and the quotient Riemann surface $Y_4/\mathrm{Aut}(Y_4)$ is uniformized by the Fuchsian group $\overline{\Delta}$ with signature $s(\overline{\Delta}) = (0; 2, 4, 6)$.*

**Theorem 6** *The space $\mathcal{M}_4^3$ of cyclic trigonal Riemann surfaces of genus 4 form a disconnected subspace of dimension 3 of the moduli space $\mathcal{M}_4$.*

1. *The subspace ${}^6\mathcal{M}_4^3$ of $\mathcal{M}_4^3$ formed by Riemann surfaces of genus 4 with automorphism group of order 6 is a disconnected space of dimension 2 determined by the Fuchsian groups $\Delta''$ with $s(\Delta'') = (0; 2, 2, 3, 3, 3)$. The automorphism group of the Riemann surfaces is either $C_6$ or $D_3$.*
2. *The subspace ${}^{12}\mathcal{M}_4^3$ of $\mathcal{M}_4^3$ formed by Riemann surfaces of genus 4 with automorphism group of order 12 is a disconnected space of dimension 1 determined by the Fuchsian groups $\Delta'$ with $s(\Delta') = (0; 2, 2, 3, 6)$. The automorphism group of the Riemann surfaces is either $C_2 \times C_6$ or $D_6$.*
3. *The subspace ${}^{36}\mathcal{M}_4^3$ of $\mathcal{M}_4^3$ formed by Riemann surfaces $X_4(\Delta)$ of genus 4 with automorphism group of order 36 is a space of dimension 1 detemined by the Fuchsian groups $\Delta$ with $s(\Delta) = (0; 2, 2, 2, 3)$. The automorphism group of the Riemann surfaces is $D_3 \times D_3$ and the surfaces admit non-normal trigonal morphisms.*
4. *There are exactly 2 cyclic trigonal Riemann surfaces $X_4$ and $Y_4$ of genus 4 with automorphism groups of order 72.*
   - (i) *$X_4$ has a normal trigonal morphism, $\mathrm{Aut}(X_4) = S_4 \times C_3$ and $X_4/S_4 \times C_3$ uniformized by the Fuchsian group $\Delta_1$ with $s(\Delta_1) = (0; 2, 3, 12)$.*

(ii) $Y_4$ *has non-normal trigonal morphisms,* $\mathrm{Aut}(Y_4) = (C_3 \times C_3) \rtimes D_4$ *and* $Y_4/(C_3 \times C_3) \rtimes D_4$ *uniformized by the Fuchsian group* $\Delta_2$ *with* $s(\Delta_2) = (0; 2, 4, 6)$.

**Theorem 7** *The different subspaces in Theorem 6 satisfy the following inclusion relations:*

1. *The space* ${}^{12}\mathcal{M}_4^3$ *is a subspace of* ${}^{6}\mathcal{M}_4^3$.
2. *The space* ${}^{36}\mathcal{M}_4^3$ *is a subspace of* ${}^{6}\mathcal{M}_4^3$.
3. *The surface* $X_4$ *given in Theorem 6.4 belongs to the space* ${}^{12}\mathcal{M}_4^3$.
4. *The surface* $Y_4$ *given in Theorem 6.4 belongs to* ${}^{36}\mathcal{M}_4^3 \cap {}^{12}\mathcal{M}_4^3$.

**Proof**

1. In fact the Riemann surfaces $\mathcal{D}/\Delta''$ with $s(\Delta'') = (0; 2, 2, 3, 3, 3)$ are double coverings of the the Riemann surfaces $\mathcal{D}/\Delta'$ with $s(\Delta') = (0; 2, 2, 3, 6)$. Consider the map $\phi'' : \Delta' \to \Sigma_2$ defined by $\phi''(x_1') = (1, 2)$, $\phi''(x_2') = \phi''(x_3') = 1_d$, $\phi''(x_4') = (1, 2)$. By Theorem 1, $\phi''(x_1')$ induces no cone points, $\phi''(x_2')$ induces two cone points of order 2, $\phi''(x_3')$ two cone points of order 3 and $\phi''(x_4')$ one of order 3. Thus $\phi''$ is the required monodromy of the covering $\mathcal{D}/\Delta'' \to \mathcal{D}/\Delta'$, with $\Delta'' = \phi''^{-1}(\mathrm{Stb}(1))$.
2. In fact the Riemann surfaces $\mathcal{D}/\Delta''$ with $s(\Delta'') = (0; 2, 2, 3, 3, 3)$ are 6-sheeted coverings of the the Riemann surfaces $\mathcal{D}/\Delta$ with $s(\Delta) = (0; 2, 2, 2, 3)$. Consider the maps $\phi_1 : \Delta \to \Sigma_2$ defined by $\phi_1(x_1) = \phi_1(x_2) = (1, 2)$, $\phi_1(x_3) = \phi_1(x_4) = 1_d$ and $\phi_2 : \Lambda \to \Sigma_3$ defined by $\phi_2(y_1) = (1, 2)$, $\phi_2(y_2) = (1, 3)$, $\phi_2(y_3) = (1, 2, 3)$, $\phi_2(y_4) = 1_d$, where $\Lambda = \phi_1^{-1}(\mathrm{Stb}(1))$. By Theorem 1, $s(\Lambda) = (0; 2, 2, 3, 3)$. Now $\phi_2(y_1)$ and $\phi_2(y_2)$ induce one cone point of order 2 each, and $\phi_2(y_4)$ induces three cone points of order 3. Therefore the composition map $\phi_2 \cdot \phi_1$ is the required monodromy of the covering $\mathcal{D}/\Delta'' \to \mathcal{D}/\Delta$. Again $\Delta'' = \phi_2^{-1}(\mathrm{Stb}(1))$. The monodromy $\phi_2 \cdot \phi_1$ yields the action of $D_3 \times D_3$ on the $\langle ab, st\rangle$-cosets, where $ab, st$ are defined in Case 6.2

   The space ${}^{36}\mathcal{M}_4^3$ belongs to the connected component of the space $\mathcal{M}_4^3$ consisting of the Riemann surfaces with half the stabilizers of the cone points rotating in opposite directions since $\langle ab, st\rangle = D_3$. We conjecture here that ${}^{36}\mathcal{M}_4^3$ is a connected subspace.
3. The Riemann surface $\mathcal{D}/\Delta_1$ with $s(\Delta_1) = (0; 2, 3, 12)$ is a 6-sheeted covering of the Riemann surfaces $\mathcal{D}/\Delta'$ with $s(\Delta') = (0; 2, 2, 3, 6)$. Consider the representation $\phi : \Delta_1 \to \Sigma_6$ defined by $\phi(\overline{x}_1) = (2, 4)(3, 5)$, $\phi(\overline{x}_2) = (1, 2, 3)(4, 5, 6)$, $\phi(\overline{x}_3) = (1, 5, 6, 2)(3, 4)$. By Theorem 1, $\phi(\overline{x}_1)$ induces two cone points of order 2, $\phi(\overline{x}_2)$ induces no cone points and $\phi(\overline{x}_3)$ induces one cone point of order 3 and one of order 6, then $s(\Delta') = s(\phi^{-1}(\mathrm{Stb}(1))) = (0; 2, 2, 3, 6)$. Thus, the map $\phi$ is the required monodromy of the covering $\mathcal{D}/\Delta' \to \mathcal{D}/\Delta_1$. The monodromy $\phi$ yields the action of $\Sigma_4 \times C_3$ on the $C_6 \times C_2$-cosets, where $C_6 \times C_2 = \langle b, \overline{s}, (a\overline{s})^2\rangle$ as in Case 11.i).

   As $C_6 \times C_2$ is a subgroup of $\Sigma_4 \times C_3$ this surface belongs to the connected component of the space $\mathcal{M}_4^3$ consisting of the Riemann surfaces with the stabilizers of all the cone points rotating in the same direction.

4. First of all the Riemann surface $Y_4 = \mathcal{D}/\Delta_2$ with $s(\Delta_2) = (0;2,4,6)$ is a 6-sheeted covering of the Riemann surfaces $\mathcal{D}/\Delta'$ with $s(\Delta') = (0;2,2,3,6)$. Consider the representation $\phi : \Delta_2 \to \Sigma_6$ defined by $\phi(\overline{y}_1) = (1,4)(2.3)(5,6)$, $\phi(\overline{y}_2) = (2,4,5,6)(1,3)$, $\phi(\overline{y}_3) = (1,2,5)(3,4)$. By Theorem 1, $\phi(\overline{y}_1)$ induces no cone points, $\phi(\overline{y}_2)$ induces one cone point of order 2 and $\phi(\overline{y}_3)$ induces one cone point of order 3, one cone point of order 6 and one of order 2, then $s(\Delta') = s(\phi^{-1}(\mathrm{Stb}(1))) = (0;2,2,3,6)$. Thus, the map $\phi$ is the required monodromy of the covering $\mathcal{D}/\Delta' \to \mathcal{D}/\Delta_2$. The monodromy $\phi$ yields the action of $(C_3 \times C_3) \rtimes D_4$ on the $D_6$-cosets, where $D_6 = \langle a, s, t^2 \rangle$ as in Case 11.ii).

   Secondly, the Riemann surface $Y_4 = \mathcal{D}/\Delta_2$ with $s(\Delta_2) = (0;2,4,6)$ is a double covering of the the Riemann surfaces $\mathcal{D}/\Delta$ with $s(\Delta) = (0;2,2,2,3)$. By Theorem 1, the map $\tau : \Delta \to \Sigma_2$ defined by $\tau(\overline{x}_1) = 1_d$, $\tau(\overline{x}_2) = \tau(\overline{x}_3) = (1,2)$ is the required monodromy of the covering $\mathcal{D}/\Delta \to \mathcal{D}/\Delta_2$ with $\Delta = \tau^{-1}(\mathrm{Stb}(1))$. Notice that $\tau(\overline{x}_1)$ induces two cone points of order 2, $\tau(\overline{x}_2)$ the third cone point of order 2 and $\tau(\overline{x}_3)$ one cone point of order 3. The monodromy $\tau$ yields the action of $(C_3 \times C_3) \rtimes D_4$ on the $D_3 \times D_3$-cosets, where $D_3 \times D_3 = \langle a, b, s, t^2 \rangle$ as in Case 11.ii).

□

**Remark** The Riemann surfaces $\mathcal{D}/\Delta'$ with $s(\Delta') = (0;2,2,3,3,3)$ cannot be coverings of the the Riemann surfaces $\mathcal{D}/\Delta$ with $s(\Delta) = (0;2,2,2,3)$. Hence the space ${}^{36}\mathcal{M}_4^3$ is not a subspace of ${}^{12}\mathcal{M}_4^3$.

Some results in this paper form part of Ying's Licentiate Thesis [14].
We thank A. F. Costa for useful comments.

## 4 Appendix: Groups of order 36 and 72

### Groups of order 36

1. $C_{36} = \langle u \mid u^{36} = 1 \rangle$
2. $C_{18} \times C_2 = \langle s, u \mid s^2 = u^{18} = [s,u] = 1 \rangle$
3. $C_{12} \times C_3 = \langle s, u \mid s^3 = u^{12} = [s,u] = 1 \rangle$
4. $C_6 \times C_6 = \langle s, u \mid s^6 = u^6 = [s,u] = 1 \rangle$
5. $(C_2 \times C_2) \rtimes C_9 = \langle a, s, t \mid a^9 = s^2 = t^2 = [s,t] = 1, a^8sa = t, a^8ta = st \rangle$
6. $A_4 \times C_3 = \langle a, b, s \mid a^3 = b^3 = s^2 = [a,b] = [s,b] = (as)^3 = 1 \rangle$
7. $C_9 \rtimes C_4 = \langle a, t \mid a^9 = t^4 = 1, t^3at = a^8 \rangle$
8. $D_{18} = \langle a, s \mid a^{18} = s^2 = (sa)^2 \rangle$
9. $(C_3 \times C_3) \rtimes_1 C_4 = \langle a, b, t \mid a^3 = b^3 = t^4 = [a,b] = 1, t^3at = a^2, t^3bt = b^2 \rangle$
10. $T \times C_3 = C_{12} \rtimes C_3 = \langle a, b, t \mid a^3 = b^4 = t^3 = [a,t] = [b,t] = 1, b^3ab = a^2 \rangle = \langle a, b | a^3 = b^{12} = b^{-1}aba = 1 \rangle$
11. $(C_3 \times C_3) \rtimes_2 C_4 = \langle a, b, t \mid a^3 = b^3 = t^4 = [a,b] = 1, t^3at = b, t^3bt = a^2 \rangle$
12. $\langle 3,3,3,2 \rangle \times C_2 = \langle a, b, s, t \mid a^3 = b^3 = s^2 = t^2 = [a,b] = [a,t] = [b,t] = [s,t] = 1,\ sas = a^2, sbs = b^2 \rangle$

13. $D_3 \times C_6 = \langle a, s, t \mid a^6 = s^2 = t^3 = (sa)^2 = [a, t] = [s, t] = 1\rangle$
14. $D_3 \times D_3 = \langle a, b, s, t \mid a^3 = b^3 = s^2 = t^2 = (st)^2 = [a, b] = [a, t] = [b, s] = (sa)^2 = (tb)^2\rangle$

## Groups of order 72

There are 50 groups of order 72. However, we only list the groups that are interesting for this work:

1. $\Sigma_4 \times C_3 = \langle t, a, b \mid t^2 = a^3 = b^3 = (ta)^4 = [t, b] = [a, b] = 1\rangle$
2. $(C_3 \times C_3) \rtimes_8 C_8 = \langle a, b, t \mid a^3 = b^3 = [a, b] = t^8 = 1, t^7at = b, t^7bt = ab\rangle$
3. $((C_3 \times C_3) \rtimes_1 C_4) \times C_2 = \langle a, b, t, s \mid a^3 = b^3 = [a, b] = t^4 = s^2 = [a, s] = [b, s] = [t, s] = 1, t^3at = a^2, t^3bt = b^2\rangle$
4. $((C_3 \times C_3) \rtimes_2 C_4) \times C_2 = \langle a, b, t, s \mid a^3 = b^3 = [a, b] = t^4 = s^2 = [a, s] = [b, s] = [t, s] = 1, t^3at = b, t^3bt = a^2\rangle$
5. $(C_3 \times C_3) \rtimes_1 D_4 = \langle a, b, s, t \mid a^3 = b^3 = [a, b] = t^4 = s^2 = (st)^2 = [s, a] = [s, b] = 1, t^3at = b, t^3bt = a\rangle = T \rtimes C_6$
6. $(C_3 \times C_3) \rtimes D_4 = \langle a, b, s, t \mid a^3 = b^3 = [a, b] = t^4 = s^2 = (st)^2 = (sa)^2 = [s, b] = 1, t^3at = b, t^3bt = a^2\rangle$
7. $(C_3 \times C_3) \rtimes_1 Q = \langle a, b, s, t \mid a^3 = b^3 = s^4 = t^4 = [a, b] = [b, t] = [a, s] = (ts)^2 = 1, t^2 = s^2, t^{-1}at = a^{-1}, s^{-1}bs = b^{-1}\rangle$
8. $(C_3 \times C_3) \rtimes_2 Q = \langle a, b, s, t \mid a^3 = b^3 = s^4 = t^4 = [a, b] = [a, s] = [b, t] = 1, t^2 = s^2, s^{-1}ts = t^{-1}, s^{-1}bs = b^{-1}, t^{-1}at = a^{-1}\rangle$
9. $(C_3 \times C_3) \rtimes_3 Q = \langle a, b, s, t \mid a^3 = b^3 = s^4 = t^4 = [a, b] = [a, b] = [b, t] = [a, s] = 1, t^2 = s^2, s^{-1}ts = t^{-1}, s^{-1}bs = b^{-1}\rangle$
10. $(C_3 \times C_3) \rtimes_4 Q = \langle a, b, s, t \mid a^3 = b^3 = s^4 = t^4 = [a, b] = [a, t] = [b, t] = 1, t^2 = s^2, s^{-1}bs = b^{-1}, s^{-1}as = a^{-1}\rangle$

## References

[1] R. D. M. Accola, On cyclic trigonal Riemann surfaces, I, *Trans. Amer. Math. Soc.* **283** (1984), 423–449.
[2] R. D. M. Accola, A classification of trigonal Riemann surfaces, *Kodai Math. J.* **23** (2000), 81–87.
[3] A. F. Costa and M. Izquierdo, On real trigonal Riemann surfaces, *Math. Scand.*, to appear.
[4] A. F. Costa, M. Izquierdo and D. Ying, On trigonal Riemann surfaces with non-unique morphisms, submitted.
[5] H. S. M. Coxeter and W. O. J. Moser, *Generators and relations for discrete groups*, Springer-Verlag, Berlin, 1957.
[6] G. González-Díez, On prime Galois covering of the Riemann sphere, *Ann. Mat. Pure Appl.* **168** (1995), 1–15.
[7] L. Greenberg, Maximal groups and signatures, *Ann. of Math. Studies* **79** (1974), 207–226.
[8] T. Kato and R. Horiuchi, Weierstrass gap sequences at the ramification points of trigonal Riemann surfaces, *J. Pure Appl. Alg.* **50** (1988), 271–285.

[9] W. Magnus, *Noneuclidean Tessellations and Their Groups*, Academic Press, New York, 1974

[10] S. Nag, *The complex analytic theory of Teichmüller spaces*, Wiley-Interscience Publication (1988)

[11] H. Seifert and W. Threfall, *A Textbook of Topology*, Academic Press, New York, 1980.

[12] D. Singerman, Subgroups of Fuchsian groups and finite permutation groups, *Bull. London Math. Soc.* **2** (1970), 319–323.

[13] D. Singerman, Finitely maximal Fuchsian groups. *J. London Math. Soc.* **6** (1972), 29–38.

[14] D. Ying, *Cyclic trigonal Riemann Surfaces of genus* 4, Linköping University Electronic Press, Licentiate Thesis Number 1125, 2004.

# ON SIMPLE $K_n$-GROUPS FOR $n = 5, 6$

A. JAFARZADEH and A. IRANMANESH*

Department of Mathematics, Tarbiat Modarres University, P.O.Box: 14115–137, Tehran, Iran

*Email: iranmana@modares.ac.ir

## Abstract

A finite nonabelian simple group is called a simple $K_n$-group if the order of $G$ has exactly $n$ distinct prime factors. M. Herzog and W. J. Shi gave a characterization of simple $K_n$-group for $n = 3, 4$, respectively. In this paper, we characterize all simple $K_n$-groups for $n = 5, 6$.

## 1 Introduction

First we need some notation. Given a natural number $n$ and a finite simple group $G$, we denote by $\pi(n)$ and $\Pi(G)$ the number of distinct prime factors of $n$ and the set of distinct prime factors of $|G|$, respectively. We say that $G$ is a *simple $K_n$-group* if $|\Pi(G)| = n$. Also when $a, b$ are two natural numbers, by $(a, b)$ we mean $\gcd(a, b)$. The rest of notation is standard and you can find them for example in [2].

Huppert in [5] studied the following conjecture:

**Conjecture 1** *Let $H$ be a finite nonabelian simple group and denote by $cd(H)$ the set of the degrees of the irreducible complex characters of $H$. If $cd(H) = cd(G)$ for some finite group $G$, then $G \cong H \times A$ with $A$ abelian.*

He proved this conjecture for some $H$ by the following procedure:

In the first step he showed that $G' = G''$. In a second step he proved that whenever $G'/M$ is a chief factor of $G$, then $G'/M \cong H$. To prove this critical step assume that $G'/M \cong S_1 \times \cdots \times S_k$ with $S_i \cong S$ being simple and let $\Pi$ be the set of primes which divides some degree of $G$. Then the degrees of $S$ are only divisible by primes in $\Pi$. Hence by the theorem of Ito and Michler, $S$ is a $\Pi$-group. For this step (and finally the proof of the conjecture) we need to know all the simple $\Pi$-groups. If we could characterize all of simple $K_n$-groups for arbitrary $n$, then we know all the simple $\Pi$-groups, when $|\Pi| = n$.

Herzog in [4] proved that there are eight simple $K_3$-groups

$$A_5, A_6, L_2(7), L_2(8), L_2(17), L_3(3), U_3(3) \text{ and } U_4(2).$$

Also Shi in [7] and Bugeaud, Cao and Mignotte in [1] gave a characterization of all simple $K_4$-groups and showed a simple $K_4$-group is isomorphic to one of $L_2(q)$, where $q$ is a prime power satisfying $q(q^2-1) = (2, q-1)2^a3^bp^cr^d$ with $p, r > 3$ distinct primes and $a, b, c, d$ natural numbers, or one of the 31 other simple groups

$$A_7, A_8, A_9, A_{10}, M_{11}, M_{12}, J_2, L_3(4), L_3(5), L_3(7), L_3(8), L_3(17), L_4(3), O_5(4),$$

$$O_5(5), O_5(7), O_5(9), O_7(2), O_8^+(2), G_2(3), U_3(4), U_3(5), U_3(7), U_3(8),$$
$$U_3(9), U_4(3), U_5(2), Sz(8), Sz(32), {}^3D_4(2), {}^2F_4(2)'.$$

In this paper we characterize all simple $K_n$-groups for $n = 5, 6$. In fact we prove the following Theorems:

**Theorem A** *Each simple $K_5$-group is isomorphic to one of $L_2(q)$ where $q$ satisfies $\pi(q^2-1) = 4$, $L_3(q)$ where $\pi((q^2-1)(q^3-1)) = 4$, $U_3(q)$ where $\pi((q^2-1)(q^3+1)) = 4$, $O_5(q)$ where $\pi(q^4-1) = 4$, $Sz(2^{2m+1})$ where $\pi((2^{2m+1}-1)(2^{4m+2}+1)) = 4$, $R(q)$ where $q$ is an odd power of 3 and $\pi(q^2-1) = 3$ and $\pi(q^2-q+1) = 1$ or one of the 30 other simple groups*

$$A_{11}, A_{12}, M_{22}, J_3, HS, He, McL, L_4(4), L_4(5), L_4(7), L_5(2), L_5(3), L_6(2), O_7(3),$$
$$O_9(2), PSp_6(3), PSp_8(2), U_4(4), U_4(5), U_4(7), U_4(9), U_5(3), U_6(2), O_8^+(3),$$
$$O_8^-(2), {}^3D_4(3), G_2(4), G_2(5), G_2(7), G_2(9).$$

**Theorem B** *Each simple $K_6$-group is isomorphic to one of $L_2(q)$ where $\pi(q^2-1) = 5$, $L_3(q)$ where $\pi((q^2-1)(q^3-1)) = 5$, $L_4(q)$ where $\pi((q^2-1)(q^3-1)(q^4-1)) = 5$, $U_3(q)$ where $\pi((q^2-1)(q^3+1)) = 5$, $U_4(q)$ where $\pi((q^2-1)(q^3+1)(q^4-1)) = 5$, $O_5(q)$ where $\pi(q^4-1) = 5$, $G_2(q)$ where $\pi(q^6-1) = 5$, $Sz(2^{2m+1})$ where $\pi((2^{2m+1}-1)(2^{4m+2}+1)) = 5$, $R(3^{2m+1})$ where $\pi((3^{2m+1}-1)(3^{6m+3}+1)) = 5$ or one of the 38 other simple groups*

$$A_{13}, A_{14}, A_{15}, A_{16}, M_{23}, M_{24}, J_1, Suz, Ru, Co_2, Co_3, Fi_{22}, HN, L_5(7), L_6(3),$$
$$L_7(2), O_7(4), O_7(5), O_7(7), O_9(3), PSp_6(4), PSp_6(5), PSp_6(7), PSp_8(3),$$
$$U_5(4), U_5(5), U_5(9), U_6(3), U_7(2), F_4(2), O_8^+(4), O_8^+(5), O_8^+(7),$$
$$O_{10}^+(2), O_8^-(3), O_{10}^-(2), {}^3D_4(4), {}^3D_4(5).$$

After collecting some preliminary results from elementary number theory which play the most important roles to prove Theorems A and B in this section, the proof of Theorem A is carried out in section 2 in a series of lemmas and the proof of Theorem B is carried out in section 3. Finally we list some open problems of interest.

**Theorem 1.1** **([8])** *If $p$ is a prime and $n \geq 2$, then there exists a prime $z$ such that $z \mid p^n - 1$ and $z$ does not divide $p^m - 1$ for $1 \leq m < n$ unless either*

1. *$p = 2$ and $n = 6$ or*
2. *$p = 2^q - 1$ is a Mersenne prime and $n = 2$.*

**Lemma 1.2** **([6])** *Let $p$ and $q$ be two primes and $m, n$ two natural number such that*

$$p^m = q^n + 1$$

*then one of the following holds:*

1. $q = 2$, $p = 3$, $n = 3$ *and* $m = 2$.
2. $q = 2$, $m = 1$, $n$ *is a power of* 2 *and* $p = q^n + 1$ *is a Fermat prime.*
3. $p = 2$, $n = 1$ *and* $q = p^m - 1$ *is a Mersenne prime; in particular,* $m$ *is prime.*

**Lemma 1.3** *Let* $q = p^n$ *be a prime power. Then the following hold:*

1. $q^2 - 1$ *has at most two different prime divisors if and only if* $q \in \{2, 3, 4, 5, 7, 8, 9, 17\}$.
2. $q^4 - 1$ *has at most three different prime divisors if and only if* $q \in \{2, 3, 4, 5, 7, 9\}$.
3. $q^6 - 1$ *has at most three different prime divisors if and only if* $q \in \{2, 3\}$.
4. $q^8 - 1$ *has at most four different prime divisors if and only if* $q \in \{2, 3, 4, 5, 7\}$.
5. $q^6 - 1$ *has exactly four different prime divisors if and only if* $q \in \{4, 5, 7, 8\}$.

**Proof** Assume first that $p = 2$ and $n \geq 4$. Hence $2^n - 1 = r^\alpha$ and $2^n + 1 = s^\beta$ with prime numbers $r$ and $s$ and $\alpha, \beta \geq 1$. By Lemma 1.2, $n$ is a prime and also $n$ is a power of 2; i.e., $n = 2$, a contradiction.

Hence we assume that $p$ is odd. Since $(p^n - 1, p^n + 1) = 1$, we get $p^n - 1 = 2^m$ or $p^n + 1 = 2^m$ for some $m \geq 1$.

If $p^n - 1 = 2^m$ and $p^n \notin \{3, 9\}$, then by Lemma 1.2 we get $n = 1$ and $p = 2^m + 1$ is a Fermat prime with $m = 2^k \geq 2$. Moreover, we have $p + 1 = 2r^\alpha$ for some prime $r$ and some $\alpha \geq 1$; hence $r^\alpha = 2^{m-1} + 1$. By Lemma 1.2 again we obtain $r^\alpha = 3^2$ or $m - 1 = 2^l$ for some $l \geq 0$. The first case yields $p = 17$ and the other case yields $m = 2$ and thus $p = 5$.

Next if $p^n + 1 = 2^m \geq 8$, then by Lemma 1.2, $n = 1$ and $p = 2^m - 1$ is a Mersenne prime; in particular $m$ is prime. Moreover, $p - 1 = 2(2^{m-1} - 1) = 2r^\alpha$ for some prime $r$ and some $\alpha \geq 1$. Hence we have $r^\alpha = 2^{m-1} - 1$. Applying Lemma 1.2 again we get $\alpha = 1$ and $m - 1$ is prime. Since $m$ is prime, we get $m = 3$ and $p^n = 7$. This proves the claims in part (1).

In order to prove parts (2) and (3), let $r \in \{4, 6\}$ and note that $q^r - 1$ has a prime divisor which does not divide $q^2 - 1$ by Theorem 1.1. Therefore $q \in \{2, 3, 4, 5, 7, 8, 9, 17\}$ by part (1). Now a straightforward check implies $q \in \{2, 3\}$ if $r = 6$ and $q \in \{2, 3, 4, 5, 7, 9\}$ if $r = 4$.

To prove part (4), with the same proof as above, $q^8 - 1$ has a prime divisor which does not divide $q^4 - 1$. Therefore $q \in \{2, 3, 4, 5, 7, 9\}$ by part (2). Now simple calculation proves the claim in this part.

Finally to prove part (5), we have $q \neq 2$. There are distinct prime factors $z_1$ and $z_2$ of $q^6 - 1$ and $q^3 - 1$, respectively, such that $z_1 \nmid q^k - 1$ for $1 \leq k \leq 5$ and $z_2 \nmid q^k - 1$ for $1 \leq k \leq 2$. Hence $q^2 - 1$ has at most two different prime divisors. Therefore $q \in \{2, 3, 4, 5, 7, 8, 9, 17\}$ by part (1), and a simple calculation completes the proof. □

## 2 Simple $K_5$-groups

A straightforward inspection reveals the following Lemma:

**Lemma 2.1** *Suppose that $G$ is a finite simple group of alternating or sporadic type with $|\Pi(G)| = 5$. Then $G$ is isomorphic to $A_n$ for some $n \in \{11, 12\}$ or to $M_{22}$, $J_3$, $HS$, $He$ or $McL$.*

In view of the last result our main concern in this section and the next section will be the groups of Lie type. The strategy will be to deal with the various types by induction on the Lie rank. In doing so we shall make frequent use of well known formulae for the various groups of Lie type. For these and for coincidence of isomorphism types of groups of Lie type we refer the reader to [3].

**Lemma 2.2** *Suppose that $G$ is a simple group of type $A_n(q)$ and $|\Pi(G)| = 5$. Then $n \le 5$ and one of the following holds:*

1. *$n = 1$ and $q^2 - 1$ has exactly four different prime divisors.*
2. *$n = 2$ and $(q^2 - 1)(q^3 - 1)$ has exactly four different prime divisors.*
3. *$n = 3$ and $q \in \{4, 5, 7\}$.*
4. *$n = 4$ and $q \in \{2, 3\}$.*
5. *$n = 5$ and $q = 2$.*

**Proof** For $n = 1$ and $n = 2$, the statements (1) and (2) are obviously true.

In order to prove the Lemma for $n = 3$, we note that $|A_3(q)| = \frac{1}{(4,q-1)}q^6(q^2-1)(q^3-1)(q^4-1)$. By Theorem 1.1, there is a prime number $z$ such that $z \mid q^3 - 1$ and $z \nmid q^k - 1$ for $k = 1, 2$ and therefore $z \mid q^2 + q + 1$. On the other hand, for every natural number $m$, we have $(m^2+m+1, m^3+m^2+m+1) = 1$. Therefore $z \nmid q^3 + q^2 + q + 1$ and so $z \nmid q^4 - 1$. Hence we have $\pi(q^4 - 1) \le 3$. By part (2) of Lemma 1.3, $q$ belongs to $\{2, 3, 4, 5, 7, 9\}$. Now simple calculation proves the statement.

For $n = 4$, we have $|A_4(q)| = \frac{(4,q-1)}{(5,q-1)}q^4(q^5 - 1)|A_3(q)|$. By Theorem 1.1, $q^5 - 1$ has a prime divisors which does not divide $q^k - 1$ for $k = 1, 2, 3, 4$. Therefore $|\Pi(A_3(q))| \le 4$ and so $q$ belongs to $\{2, 3\}$. A simple calculation now finishes the statement.

For $n = 5$, we have $|A_5(q)| = \frac{1}{(6,q-1)}q^{15}(q^2 - 1)(q^3 - 1)(q^4 - 1)(q^5 - 1)(q^6 - 1)$ So if $q \ne 2$, then by Theorem 1.1, $|A_5(q)|$ has at least 6 different prime divisors. Now with the same proof we get $|\Pi(A_6(q))| \ge 6$ for all $q$ and because $|A_6(q)| \mid |A_n(q)|$ for all $n \ge 6$, we have $|\Pi(A_n(q))| \ge 6$ and this completes the proof. □

**Lemma 2.3** *Suppose that $G$ is a simple group of type $B_n(q)$ or $C_n(q)$ with $n \ge 2$ and $|\Pi(G)| = 5$. Then $n \le 4$ and one of the following holds:*

1. *$n = 2$ and $q^4 - 1$ has exactly four different prime divisors.*
2. *$n = 3$ and $q = 3$.*
3. *$n = 4$ and $q = 2$.*

**Proof** Note that for all $q$, we have $B_2(q) \cong C_2(q)$ and if $n \ge 3$, then $|B_2(q)| = |C_2(q)|$. Now the statement for $n = 2$ is obviously true.

For $n = 3$, we have $|B_3(q)| = q^5(q^6 - 1)|B_2(q)|$. Suppose $q \neq 2$, so $q^6 - 1$ has a prime divisor which does not divide $|B_2(q)|$ and therefore $|\Pi(B_2(q))| \leq 4$. Hence $q$ belongs to $\{3, 4, 5, 7, 9\}$. Now straightforward check implies the statement.

For the case $n = 4$, we have $|B_3(q)| \mid |B_4(q)|$ and so we have $|\Pi(B_3(q))| \leq 5$, therefore $q$ belongs to $\{2, 3\}$. A straightforward calculation shows that only $q = 2$ is acceptable. We have $|B_4(q)| \mid |B_5(q)|$ and $|\Pi(B_5(2))| = 7$, therefore $n \neq 5$. Now for $n \geq 5$ we have $|B_5(q)| \mid |B_n(q)|$; hence $n \ngeq 5$ and the proof is completed. □

**Corollary 2.4** *Suppose that $G$ is a simple group of type $D_n(q)$ with $n \geq 4$ and $|\Pi(G)| = 5$. Then $n = 4$ and $q = 3$.*

**Proof** For $n = 4$, we have $|A_3(q)| \mid |D_4(q)|$, thus $q$ belongs to $\{2, 3, 4, 5, 7\}$. A straightforward check gets $q$ must be 3. With the same method for $n = 5$ and $n = 6$, we have $|A_4(q)| \mid |D_5(q)|$ and $|A_5(q)| \mid |D_6(q)|$ and therefore $q$ belongs to $\{2, 3\}$ and $\{2\}$, respectively which are impossible. For $n \geq 7$ also we have $|A_6(q)| \mid |D_n(q)|$ and so $|\Pi(D_n(q))| \geq 6$ and the proof is complete. □

**Lemma 2.5** *Suppose that $G$ is a simple group of type $E_n(q)$ with $n \in \{6, 7, 8\}$, then $|\Pi(G)| \geq 8$.*

**Proof** We have $|E_6(q)| = \frac{3}{q-1} q^{36}(q^{12} - 1)(q^9 - 1)(q^8 - 1)(q^6 - 1)(q^5 - 1)(q^2 - 1)$ and therefore $q^k - 1 \mid |E_6(q)|$ for all $k \in \{1, 3, 4, 5, 8, 9, 12\}$. By Theorem 1.1, we have $|\Pi(E_6(q))| \geq 8$. The proof for the cases $n = 7$ and 8 is the same. □

**Lemma 2.6** *Suppose that $G$ is a simple group of type $F_4(q)$, then $|\Pi(G)| > 5$.*

**Proof** Similar to the proof of Lemma 2.5, we have $q^k - 1 \mid |F_4(q)|$ for all $k \in \{1, 3, 4, 8, 12\}$. Thus $|\Pi(F_4(q))| \geq 6$. □

**Corollary 2.7** *Suppose that $G$ is a simple group of type $G_2(q)$ and $|\Pi(G)| = 5$. Then $q \in \{4, 5, 7, 9\}$.*

**Proof** We have $|G_2(q)| = q^6(q^2 - 1)(q^6 - 1)$. Now the result immediately follows by part (5) of Lemma 1.3. □

**Lemma 2.8** *Suppose that $G$ is a simple group of type ${}^2A_n(q)$ with $n \geq 2$ and $|\Pi(G)| = 5$. Then $n \leq 5$ and one of the following cases holds:*

1. *$n = 2$ and $(q^2 - 1)(q^3 + 1)$ has exactly four different prime divisors.*
2. *$n = 3$ and $q \in \{4, 5, 7, 9\}$.*
3. *$n = 4$ and $q = 3$.*
4. *$n = 5$ and $q = 2$.*

**Proof** The result is obvious for $n = 2$. Assume that $n = 3$ and $q \neq 2$. We have $|{}^2A_3(q)| = \frac{4}{q+1} q^6(q^2 - 1)(q^3 + 1)(q^4 - 1)$ and so by Theorem 1.1, there is a prime number $z$ that $z \mid q^6 - 1$ and $z \nmid q^k - 1$ for $1 \leq k \leq 5$; particularly $z \nmid q^3 - 1$ and therefore $z \mid q^3 + 1$. Hence $q^4 - 1$ has at most three prime factors, thus $q$ belongs to $\{3, 4, 5, 7, 9\}$. With a simple calculation we get the result.

Assume $n = 4$. we have $|^2A_3(q)| \mid |^2A_4(q)|$ so we should have $|\Pi(^2A_3(q))| \leq 5$ and thus $q$ belongs to $\{2, 3, 4, 5, 7, 9\}$ and a straightforward check implies the result.

Assume $n = 5$. Similar to the last part, we have $|^2A_4(q)| \mid |^2A_5(q)|$ so we should have $|\Pi(^2A_4(q))| \leq 5$ and thus $q$ belongs to $\{2, 3\}$ and a simple calculation implies the result.

Also $|^2A_5(q)| \mid |^2A_6(q)|$ so we have $|\Pi(^2A_5(q))| \leq 5$ and thus $q = 2$. But $|\Pi(^2A_6(2))| > 5$ and because $|^2A_6(q)| \mid |^2A_n(q)|$ for all $n \geq 6$, we have $|\Pi(^2A_n(q))| > 5$ and therefore $n \leq 5$ and the proof is completed. □

**Lemma 2.9** *Suppose that $G$ is a simple group of type $^2B_2(q)$ with $q = 2^{2m+1}$ and $|\Pi(G)| = 5$. Then $2^{2m+1} - 1$ has at most two different prime divisors.*

**Proof** Assume $r = 2^m$. We have $|^2B_2(q)| = q^2(q-1)(q+2r+1)(q-2r+1)$. Numbers $q$, $q-1$, $q+2r+1$ and $q-2r+1$ are relatively prime numbers and therefore $q-1$ must have at most two different prime divisors. □

**Lemma 2.10** *Suppose that $G$ is a simple group of type $^2D_n(q)$ with $n \geq 4$ and $|\Pi(G)| = 5$. Then $n = 4$ and $q = 2$.*

**Proof** If $n \geq 6$, then $q^k - 1 \mid |^2D_n(q)|$ for $k \in \{1, 3, 4, 8, 10, 12\}$ and so $|\Pi(^2D_n(q))| \geq 7$. For $n = 5$, we have $q^k - 1 \mid |^2D_5(q)|$ for $k \in \{1, 3, 4, 8\}$; in addition, $q^5 + 1 \mid |^2D_5(q)|$ and by Theorem 1.1, there is a prime $z$ such that $z \mid q^{10} - 1$ and $z \nmid q^k - 1$ for $1 \leq k \leq 9$ and therefore $z \mid q^5 + 1$. Hence $|\Pi(^2D_5(q))| \geq 6$. Thus $n = 4$ and since $q^8 - 1 \mid |^2D_4(q)|$, by part (4) of the Lemma 1.3, $q$ belongs to $\{2, 3, 4, 5, 7\}$. A straightforward check now completes the proof. □

**Lemma 2.11** *Suppose that $G$ is a simple group of type $^3D_4(q)$ and $|\Pi(G)| = 5$. Then $q = 3$.*

**Proof** We have $|^3D_4(q)| = Q^{12}(q^2-1)(q^6-1)(q^8+q^4+1)$. By Theorem 1.1, there is a prime $z$ such that $z \mid q^{12} - 1$ and $z \nmid q^k - 1$ for $1 \leq k \leq 11$; particularly $z \nmid q^4 - 1$. Since $q^{12} - 1 = (q^4-1)(q^8+q^4+1)$, $z \mid q^8 + q^4 + 1$. But also $z \nmid q^6 - 1$ and therefore $q^6 - 1$ has at most three different prime divisors and so by part (3) of Lemma 1.3, $q$ belongs to $\{2, 3\}$. Now a straightforward check completes the proof. □

**Lemma 2.12** *Suppose that $G$ is a simple group of type $^2E_6(q)$, then $|\Pi(G)| > 6$.*

**Proof** With a simple calculation we get $|\Pi(^2E_6(2))| \geq 7$. If $q \neq 2$, then $q^k - 1 \mid |^2E_6(q)|$ for all $k \in \{1, 3, 4, 6, 8, 12\}$ and by Theorem 1.1, we get $|\Pi(^2E_6(q))| \geq 7$. □

**Lemma 2.13** *If $G$ is a simple group of type $^2F_4(q)$ with $q = 2^{2m+1}$ and $|\Pi(G)| \leq 6$, then $q = 2$ and $G \cong {}^2F_4(2)'$ and $|\Pi(G)| = 4$.*

**Proof** Similar to the proof of Lemma 2.2, there are primes $z_1, z_2$ such that $z_1 \mid q^6 - 1$ and $z_2 \mid q^3 - 1$, but $z_1, z_2 \nmid q^4 - 1$. Thus $\pi(q^4 - 1) \leq 3$ and so $q$ belongs to $\{2, 3, 4, 5, 7, 9\}$. Since $q$ is a power of 2, by a simple calculation we get $q = 2$. □

**Lemma 2.14** *Suppose that $G$ is a simple group of type ${}^2G_2(q)$ with $q = 3^{2m+1}$ and $|\Pi(G)| = 5$. Then $q^2 - 1$ has exactly three different prime divisors.*

**Proof** We have $|{}^2G_2(q)| = q^3(q^2 - 1)(q^2 - q + 1)$. Since $q$ is a multiple of 3, we have the following relations:

$$(q^2 - q + 1, q - 1) = 1, \quad (q^2 - q + 1, q + 1) = (3, q + 1) = 1$$

Therefore $(q^2 - q + 1, q^2 - 1) = 1$ and so $q^2 - 1$ has at most three different prime divisors. Now if $\pi(q^2 - 1) \leq 2$, then $q$ belongs to $\{2, 3, 4, 5, 7, 8, 9, 17\}$. Since $q$ is an odd power of 3, only $q = 3$ can be acceptable that is impossible. So we have $\pi(q^2 - 1) = 2$ and therefore $q^2 - q + 1$ is a prime power. □

Now Lemmas 2.1–2.14 complete the proof of Theorem A.

## 3 Simple $K_6$-groups

Another straightforward inspection reveals the following Lemma:

**Lemma 3.1** *Suppose that $G$ is a finite simple group of alternating or sporadic type with $|\Pi(G)| = 6$. Then $G$ is isomorphic to $A_n$ for some $n \in \{13, 14, 15, 16\}$ or to $M_{23}$, $M_{24}$, $J_1$, $Suz$, $Ru$, $Co_2$, $Co_3$, $Fi_{22}$ or $HN$.*

**Lemma 3.2** *Suppose that $G$ is a simple group of type $A_n(q)$ and $|\Pi(G)| = 6$. Then $n \leq 6$ and one of the following holds:*

1. *$n = 1$ and $q^2 - 1$ has exactly five different prime divisors.*
2. *$n = 2$ and $(q^2 - 1)(q^3 - 1)$ has exactly five different prime divisors.*
3. *$n = 3$ and $(q^2 - 1)(q^3 - 1)(q^4 - 1)$ has exactly five different prime divisors.*
4. *$n = 4$ and $q = 7$.*
5. *$n = 5$ and $q = 3$.*
6. *$n = 6$ and $q = 2$.*

**Proof** For $n = 1, 2, 3$, the statements (1), (2) and (3) are obviously true.

In order to prove the Lemma for $n = 4$, we note that $(q^5 - 1)|A_3(q)| \mid |A_4(q)|$. By Theorem 1.1, there is a prime number $z$ such that $z \mid q^5 - 1$ and $z \nmid |A_3(q)|$. Therefore we must have $|\Pi(A_3(q))| \leq 5$ and so $q$ belongs $\{2, 3, 4, 5, 7\}$ by Lemma 2.2. Now a simple calculation completes the proof.

Assume $n = 5$ and $q \neq 2$. Similar to the last case, we must have $|\Pi(A_4(q))| \leq 5$ and so $q$ belongs $\{2, 3\}$ by Lemma 2.2. Now a simple calculation completes the proof.

For $n = 6$, proof is similar to the last two cases and if $n \geq 7$ and $|\Pi(A_n(q))| = 6$ then we must have $|\Pi(A_{n-1}(q))| \leq 5$ which has no answer. □

**Lemma 3.3** *Suppose that $G$ is a simple group of type $B_n(q)$ or $C_n(q)$ with $n \geq 2$ and $|\Pi(G)| = 6$. Then $n \leq 4$ and one of the following holds:*

1. *$n = 2$ and $q^4 - 1$ has exactly five different prime divisors.*
2. *$n = 3$ and $q \in \{4, 5, 7\}$.*
3. *$n = 4$ and $q = 3$.*

**Proof** The statement for $n = 2$ is obviously true because $|B_2(q)| = \frac{1}{(2,q-1)} q^4(q^2 - 1)(q^4 - 1)$.

For $n = 3$, we have $|B_3(q)| = \frac{1}{(2,q-1)} q^9(q^2 - 1)(q^4 - 1)(q^6 - 1)$. There is a prime number $z$ such that $z \mid q^4 - 1$ and $z \nmid q^k - 1$ for $k = 1, 2, 3$. We claim that $z \nmid q^6 - 1$. For the proof, we have $z \nmid q^2 - 1$ and so $z \mid q^2 + 1$ and therefore $z \mid q^3 + q$. If $z \mid q^6 - 1$, then we must have $z \mid q^3 + 1$ and therefore $z \mid (q^3 + q) - (q^3 + 1) = q - 1$, a contradiction. Hence $\pi(q^6 - 1) \leq 4$ and $q$ belongs to $\{2, 3, 4, 5, 7, 8\}$. Now a straightforward check completes the proof.

For the case $n = 4$, we have $(q^8 - 1)|B_3(q)| \mid |B_4(q)|$. Therefore by Theorem 1.1, there is a prime $z$ such that $z \mid |B_4(q)|$ and $z \nmid |B_3(q)|$ and so we must have $|\Pi(B_3(q))| \leq 5$, therefore $q$ belongs to $\{2, 3\}$. A straightforward calculation now shows that the only solution is $q = 3$.

If $n \geq 5$, then there is a prime $z$ such that $z \mid |B_n(q)|$ and $z \nmid |B_{n-1}(q)|$ and so we must have $|\Pi(B_{n-1}(q))| \leq 5$, therefore $n - 1 = 4$ and $q = 2$. But $|\Pi(B_5(2))| = 7$. Hence $n < 5$ and the proof is completed. □

**Lemma 3.4** *Suppose that $G$ is a simple group of type $D_n(q)$ with $n \geq 4$ and $|\Pi(G)| = 6$. Then $n = 4$ and $q \in \{4, 5, 7\}$ or $n = 5$ and $q = 2$.*

**Proof** For $n = 4$, we have $|D_4(q)| = \frac{1}{(4,q^4-1)} q^{12}(q^2 - 1)(q^4 - 1)^2(q^6 - 1)$. Similar to the proof of Lemma 2.2 and by Theorem 1.1, there are prime numbers $z_1, z_2$ such that $z_1 \mid q^6 - 1$ and $z_2 \mid q^3 - 1$ but $z_1, z_2 \nmid q^4 - 1$. Thus $\pi(q^4 - 1) \leq 3$ and so $q$ belongs to $\{2, 3, 4, 5, 7, 9\}$. A simple calculation completes the proof.

For $n = 5$, we have $|A_4(q)| \mid |D_5(q)|$, therefore $q$ belongs to $\{2, 3, 7\}$. A simple calculation now completes the proof.

For $n = 6$, we have $|D_5(q)| \mid |D_6(q)|$, therefore $|\Pi(D_5(q))| \leq 6$ which its only solution is $q = 2$ and it is impossible. Also $|D_6(q)| \mid |D_7(q)|$ and therefore $|\Pi(D_7(q))| \geq 7$. Now define $f(q) = q^{56}\Pi_{i=1}^{7}(q^{2i} - 1)$. Hence for all $n \geq 8$, we have $|A_7(q)| \mid f(q) \mid |D_n(q)|$ and since $|\Pi(A_7(q))| \geq 7$ we have $|\Pi(D_n(q))| \geq 7$ and the proof is completed. □

**Lemma 3.5** *Suppose that $G$ is a simple group of type $F_4(q)$ and $|\Pi(G)| = 6$, then $q = 2$.*

**Proof** We have $|\Pi(F_4(2))| = 6$ and if $q \neq 2$, then similar to the proof of Lemma 2.5 we have $q^k - 1 \mid |F_4(q)|$ for all $k \in \{1, 3, 4, 6, 8, 12\}$ and by Theorem 1.1 we have $|\Pi(F_4(q))| \geq 7$. □

The next Lemma is obvious by the order of the group:

**Lemma 3.6** *Suppose that $G$ is a simple group of type $G_2(q)$ and $|\Pi(G)| = 6$. Then $\pi(q^6 - 1) = 5$.*

**Lemma 3.7** *Suppose that $G$ is a simple group of type ${}^2A_n(q)$ with $n \geq 2$ and $|\Pi(G)| = 6$. Then $n \leq 6$ and one of the following cases holds:*

1. *$n = 2$ and $(q^2 - 1)(q^3 + 1)$ has exactly five different prime divisors.*
2. *$n = 3$ and $(q^2 - 1)(q^3 + 1)(q^4 - 1)$ has exactly five different prime divisors.*
3. *$n = 4$ and $q \in \{4, 5, 9\}$.*
4. *$n = 5$ and $q = 3$.*
5. *$n = 6$ and $q = 2$.*

**Proof** The result is obvious for the cases $n = 2, 3$. Assume $n = 4$ and $q \neq 2$. We have $|{}^2A_4(q)| = \frac{5}{q+1} q^{10}(q^2 - 1)(q^3 + 1)(q^4 - 1)(q^5 + 1)$. By Theorem 1.1, there are prime numbers $z_1, z_2$ such that $z_1 \mid q^{10} - 1$, $z_2 \mid q^6 - 1$, $z_1 \nmid q^k - 1$ for $1 \leq k \leq 9$ and $z_2 \nmid q^l - 1$ for $1 \leq l \leq 5$; particularly $z_1 \nmid q^5 - 1$ and $z_2 \nmid q^3 - 1$ and therefore $z_1 \mid q^5 + 1$ and $z_2 \mid q^3 + 1$. Hence $q^4 - 1$ has at most three prime factors, thus $q$ belongs to $\{3, 4, 5, 7, 9\}$. With a simple calculation we get the result.

Assume $n = 5$ and $q \neq 2$. We have $(q^6 - 1)|{}^2A_4(q)| \mid |{}^2A_5(q)|$, therefore there is a prime number $z$ such that $z \mid |{}^2A_5(q)|$ but $z \nmid |{}^2A_4(q)|$. Thus we must have $|\Pi({}^2A_4(q))| \leq 5$ and thus $q$ belongs to $\{2, 3\}$ and a straightforward check implies the result.

Assume $n = 6$. Similar to the last case we have $(q^7 + 1)|{}^2A_5(q)| \mid |{}^2A_6(q)|$ so we should have $|\Pi({}^2A_5(q))| \leq 5$ and thus $q$ could be 2 and a simple calculation implies the result.

Also we have $(q^8 - 1)|{}^2A_6(q)| \mid |{}^2A_7(q)|$ so we must have $|\Pi({}^2A_6(q))| \leq 5$ which has no solution, thus $n \neq 7$, and because $|{}^2A_7(q)| \mid |{}^2A_n(q)|$ for all $n \geq 7$ we have $|\Pi({}^2A_n(q))| > 6$ and therefore $n \leq 6$ and the proof is completed. □

**Lemma 3.8** *Suppose that $G$ is a simple group of type ${}^2B_2(q)$ with $q = 2^{2m+1}$ and $|\Pi(G)| = 6$. Then $2^{2m+1} - 1$ has at most three different prime divisors.*

**Proof** The proof is similar to the case $|\Pi(G)| = 5$. □

**Lemma 3.9** *Suppose that $G$ is a simple group of type ${}^2D_n(q)$ with $n \geq 4$ and $|\Pi(G)| = 6$. Then $n = 4$ and $q = 3$ or $n = 5$ and $q = 2$.*

**Proof** For $n \geq 6$, we refer to the proof of Lemma 2.10.

For $n = 5$, $q = 2$ is a solution and if $q \neq 2$, then we have $q^k - 1 \mid |{}^2D_5(q)|$ for $k \in \{1, 3, 4, 6, 8\}$; in addition, $q^5 + 1 \mid |{}^2D_5(q)|$ and by Theorem 1.1, there is a prime $z$ such that $z \mid q^{10} - 1$ and $z \nmid q^k - 1$ for $1 \leq k \leq 9$ and necessarily $z \mid q^5 + 1$. Hence $|\Pi({}^2D_5(q))| \geq 7$.

If $n = 4$ and $q \neq 2$, then since $q^k - 1 \mid |{}^2D_4(q)|$ for $k = 4, 6, 8$, we must have $\pi(q^4 - 1) \leq 3$. Now by part (2) of the Lemma 1.3, $q$ belongs to $\{2, 3, 4, 5, 7, 9\}$. A straightforward check now gives us $q = 3$ and the proof is completed. □

**Lemma 3.10** *Suppose that $G$ is a simple group of type ${}^3D_4(q)$ and $|\Pi(G)| = 6$. Then $q \in \{4, 5\}$.*

**Proof** Similar to the case $|\Pi(G)| = 5$, $q^6 - 1$ has at most four different prime divisors and so by parts (3) and (5) of Lemma 1.3, $q$ belongs to $\{2, 3, 4, 5, 7, 8\}$. Now a straightforward check completes the proof. □

**Lemma 3.11** *Suppose that $G$ is a simple group of type ${}^2G_2(q)$ with $q = 3^{2m+1}$ and $|\Pi(G)| = 6$. Then $q^2 - 1$ has at most four different prime divisors.*

**Proof** The proof is similar to the case $|\Pi(G)| = 5$. □

Lemmas 3.1–3.11 together with Lemmas 2.5, 2.12 and 2.13 proves Theorem B.

At the end, we list two interesting open problems related to $K_n$-simple groups:

**Problem 3.12** For which power primes $q$ does $q^2 - 1$ have at most five different prime divisors?

**Problem 3.13** For which natural numbers $m$ does $2^{2m+1} - 1$ have at most three different prime divisors?

## References

[1] Y. Bugeaud, Z. Cao and M. Mignotte, On simple $K_4$-groups, *J. Algebra* **241** (2001), 658–668.
[2] J. H. Conway, R. T. Curtis, S. P. Norton, R. A. Parker and R. A. Wilson, *Atlas of finite groups*, Oxford Univ. Press (Clarendon), Oxford 1985.
[3] D. Gorenstein, R. Lyons, R. Solomon, *The classification of finite simple groups, Vol. 1*, Amer. Math. Soc., Math. Surveys and Monographs **40**, 1994.
[4] M. Herzog, On finite simple groups of order divisible by three primes only, *J. Algebra* **10** (1968), 383–388.
[5] B. Huppert, Some simple groups which are determined by the set of their character degrees I, *Illinois J. Math.* **44** (2000), no. 4, 828–842.
[6] B. Huppert and W. Lempken, Simple groups of order divisible by at most four primes, preprint.
[7] W. J. Shi, On simple $K_4$-groups, *Chinese Sci. Bull.* **36** (1991), no. 17, 1281–1283.
[8] K. Zsigmondy, Zur theorie der potenzreste, *Monatshefte Math. Physik* **3** (1892), 265–284.

# PRODUCTS OF SYLOW SUBGROUPS AND THE SOLVABLE RADICAL

GIL KAPLAN and DAN LEVY

The School of Computer Sciences, The Academic College of Tel-Aviv-Yaffo,
4 Antokolsky st., Tel-Aviv 64044, Israel
Email: danlevy@trendline.co.il

## Abstract

We study products of the form $P_1 \cdots P_m$, where $P_i$ is any Sylow $p_i$-subgroup of a given finite group $G$, and $p_1, \ldots, p_m$ are all of the distinct prime divisors of $|G|$. We give a solvability criterion for $G$ in terms of such products, and discuss the relation between the intersection of all of these products and the solvable radical of $G$.

## 1 Introduction

Let $G$ be a finite group[1] and let $\pi(G) = \{p_1, \ldots, p_m\}$ be the set of all distinct prime divisors of $|G|$. It is well-known that $G$ is nilpotent if and only if it is a direct product of its Sylow subgroups, one for each prime divisor of $|G|$. Moreover, the nilpotent radical of an arbitrary $G$, i.e., the Fitting subgroup of $G$, has a direct description in terms of Sylow subgroups of $G$:

$$\mathrm{Fit}(G) = \prod_{i=1}^{m} O_{p_i}(G) = \prod_{i=1}^{m} \Big( \bigcap_{P_i \in \mathrm{Syl}_{p_i}(G)} P_i \Big).$$

Here we will present a solvability criterion (Theorem A below) which is also formulated in terms of products of Sylow subgroups, one for each prime divisor of $|G|$. In analogy to the nilpotent radical, the solvability criterion leads to a description of the solvable radical of an arbitrary $G$. We will show that in general the solvable radical of $G$ is closely related to

$$H(G) \stackrel{\mathrm{def}}{=} \bigcap_{\substack{\text{all orderings of } \pi(G) \\ \text{all choices of } P_i \in \mathrm{Syl}_{p_i}(G)}} \left( \prod_{i=1}^{m} P_i \right), \tag{1}$$

and review the classes of groups for which we are able to prove that the solvable radical is equal to $H(G)$.

## 2 Complete Sylow Products and Solvability

**Definition 1** A *complete Sylow sequence* of a group $G$ is a sequence of the form $\mathcal{P} = (P_1, \ldots, P_m)$, where $P_i$ is a Sylow $p_i$-subgroup of $G$, and $p_1, \ldots, p_m$ are all

[1] Unless stated otherwise, all groups under discussion are assumed finite.

the distinct members of $\pi(G)$ given in any order. The corresponding product $\Pi(\mathcal{P}) = P_1 \cdots P_m$, which is a subset of $G$, is called a *complete Sylow product* of $G$.

We have the following characterization of solvability ([2] Theorem A):

**Theorem A** *The following three conditions on a group $G$ are equivalent:*

a. *$G$ is solvable.*

b. *Each complete Sylow product of $G$ equals $G$. Equivalently, $G = H(G)$.*

c. *For a fixed ordering of $\pi(G)$, $G$ is equal to each of its complete Sylow products in that order.*

Our proof that (b) implies (a) relies on Corollary 3 to Thompson's classification theorem of $N$-groups [5]. This corollary states that a group $G$ is solvable if and only if it does not contain three nonidentity elements $a, b, c$ of pairwise coprime orders such that $abc = 1_G$.

If $G$ is not solvable then, by Theorem A, for any ordering of $\pi(G)$, there exists at least one complete Sylow product which is not equal to $G$. For example, the smallest non-solvable group is $A_5$, having $\pi(A_5) = \{2, 3, 5\}$. Some of the complete Sylow products $P_2P_3P_5$ of $A_5$ are equal to $A_5$, while some are not. On the other hand, none of the products $P_2P_5P_3$ is equal to $A_5$. A more extreme example is provided by the special unitary group $SU(3, 3)$ which is not equal to any of its complete Sylow products (see [1] for a discussion of these and other examples). We shall say that a group $G$ is *Sylow factorizable* if it is equal to at least one of its complete Sylow products. We believe that the class of Sylow factorizable groups, which includes the solvable groups, is an interesting object and deserves further study.

## 3 Intersection of Sylow Products and the Solvable Radical

In this section we describe the basic properties of $H(G)$ which, by Definition 1, is the intersection of all complete Sylow products of $G$. In particular, we describe the connection between $H(G)$ and the solvable radical of $G$, which from now on we denote by $R(G)$. Since the Sylow products $P_1 \cdots P_m$ are not necessarily subgroups, it is not apparent from the definition of $H(G)$ that it is a subgroup. This fact follows from the following, more general result ([2] Section 3).

**Proposition 2** *Let $A = A_1, \ldots, A_m$ be an arbitrary sequence of subgroups of a group $G$. Denote*

$$M = \bigcap_{x_1,\ldots,x_m \in G} A_1^{x_1} \cdots A_m^{x_m}.$$

*Then $M$ is a normal subgroup of $G$.*

The following properties of $H(G)$ are proved in [2] Theorem **B**.

**Theorem B** *Let $G$ be a group. Then:*

a. *$H(G)$ is a characteristic subgroup of $G$.*

b. *Every complete Sylow product is a union of cosets of* $H(G)$.

c. $R(G) = R(H(G))$, *and in particular,* $R(G)$ *is a characteristic subgroup of* $H(G)$.

d. $H(G/H(G)) = 1$. *In particular,* $R(G/H(G)) = 1$.

e. *Define* $H^{(0)}(G) = G$ *and* $H^{(j)}(G) = H(H^{(j-1)}(G))$ *for* $j \geq 1$. *Then there exists a non-negative integer* $n$ *such that* $H^{(n)}(G) = R(G)$.

Note that part (e) of the last theorem provides a description of the solvable radical in terms of intersections of Sylow products. An immediate question which arises in view of this result is how fast does the sequence described by part (e) of Theorem B reach $R(G)$? This is an open question. At present we do not know of any $G$ which does not satisfy $H(G) = R(G)$. Note, for instance, that an immediate consequence of Theorem A and Theorem B part (a) is that $H(G) = R(G)$ whenever $G$ is solvable or simple. Further results on this question are presented in the next section.

## 4 Sufficient Conditions for $H(G) = R(G)$

In this section we present some conditions on a group $G$ which guarantee that $H(G) = R(G)$. We begin with an arithmetical condition on the order of $G$ ([2] Theorem C).

**Theorem C** *Let* $G$ *be a group such that none of the divisors of* $|G|$ *belongs to the following set:*

$$\{2^{24}, 7 \cdot 2^{23}, 3 \cdot 2^{22}, 7^5 \cdot 2^{19}, 7^6 \cdot 2^{18}, 7^5 \cdot 3 \cdot 2^{17}, 3^5 \cdot 2^{14}, 7 \cdot 3^5 \cdot 2^{13}, 3^6 \cdot 2^{12}\}.$$

*Then* $H(G) = R(G)$.

Our next condition ([3] Corollary A1) involves the concept of Sylow factorizable groups defined earlier.

**Theorem D** *Let* $G$ *be a Sylow factorizable group. Then* $H(G) = R(G)$.

Notice that Sylow factorizability of $G$ is not necessary for $H(G) = R(G)$, as is shown by the example of $SU(3,3)$.

Finally, we also have a "local" condition ([3] Theorem B):

**Theorem E** *Let* $G$ *be a group and let* $P$ *be a Sylow 2-subgroup of* $H(G)$. *Then:*

a. $G$ *is Sylow factorizable if and only if* $N_G(P)$ *is Sylow factorizable. In particular, if* $N_G(P)$ *is Sylow factorizable then* $H(G) = R(G)$.

b. $H(N_G(P)) = N_{H(G)}(P)$, *and, in particular,* $H(N_G(P))$ *is solvable.*

## 5 Further Results and Open Problems

In this section we mention other results which are related to the topics discussed above, and list some open problems.

Let $G$ be a group and let $\sigma$ be any non-empty subset of $\pi(G)$. In analogy to $R(G)$ we can define $R_\sigma(G)$ to be the the unique maximal solvable normal $\sigma$-subgroup of $G$. It turns out that one can also define $H_\sigma(G)$ in analogy to $H(G)$ and much of the discussion on the relations between $R(G)$ and $H(G)$ generalizes to $R_\sigma(G)$ and $H_\sigma(G)$. For further details we refer the reader to [3].

Another possible look at $H(G)$ is offered by the following construction. Given a group $G$, define $K(G)$ to be the subset of all $x \in G$ such that for any complete Sylow sequence $P_1, \ldots, P_m$ and any $1 \leq j \leq m$:

$$|P_1 \cdots P_m| = |P_1 \cdots P_{j-1} P_j^x P_{j+1} \cdots P_m|.$$

Then one can prove ([4]) that $K(G)$ is a characteristic subgroup of $G$ which contains $H(G)$, and moreover, if $G$ is Sylow factorizable then $K(G) = R(G)$.

We believe that products of Sylow subgroups provides a fruitful subject for investigation. We regard the results and ideas which we described here as initial steps in this direction and hope that they will be carried much further ahead. We end this paper with a list of open problems. We do not attempt to estimate the difficulty of the following questions nor their importance.

1. Which groups $G$ satisfy $R(G) = H(G)$? What can be said about the "$\sigma$ variation", $R_\sigma(G) = H_\sigma(G)$, of this problem?
2. Which groups $G$ satisfy $H(G) = K(G)$?
3. Is there a general lower bound on the size of a complete Sylow product? Is there a general lower bound on the maximal size of a complete Sylow product? In relation to the second question, we mention that if $G$ has a complete Sylow sequence $\mathcal{P}$ such that $|\Pi(\mathcal{P})| > \frac{1}{2}|G|$ then $R(G) = H(G)$ [4].
4. For which groups $G$ does the following property hold: For every ordering of the primes in $\pi(G)$ there exists a complete Sylow product, taken in this order, which is equal to $G$. One can view this last property as a strengthening of Sylow factorizability. We recall that $A_5$ is Sylow factorizable but does not have the property above.

### References

[1] D. F. Holt and P. Rowley, On products of Sylow subgroups in finite groups, *Arch. Math.* **60** (1993), no. 2, 105–107.

[2] G. Kaplan and D. Levy, Sylow Products and the Solvable Radical, *Arch. Math.* (to appear).

[3] G. Kaplan and D. Levy, The Solvable Radical of Sylow Factorizable Groups, *Arch. Math.* (to appear).

[4] G. Kaplan and D. Levy, work in preparation.

[5] J. G. Thompson, Nonsolvable finite groups all of whose local subgroups are solvable, *Bull. Amer. Math. Soc.* **74** (1968), no. 3, 383–437.

# ON COMMUTATORS IN GROUPS

LUISE-CHARLOTTE KAPPE* and ROBERT FITZGERALD MORSE†

*Department of Mathematical Sciences, State University of New York at Binghamton, Binghamton, NY 13902-6000 USA

†Department of Electrical Engineering and Computer Science, University of Evansville, Evansville, IN 47722 USA

Email: `rm43@evansville.edu`

*Dedicated to Hermann Heineken on the occasion of his 70th birthday*

## Abstract

Commutators originated over 100 years ago as a by-product of computing group characters of nonabelian groups. They are now an established and immensely useful tool in all of group theory. Commutators became objects of interest in their own right soon after their introduction. In particular, the phenomenon that the set of commutators does not necessarily form a subgroup has been well documented with various kinds of examples. Many of the early results have been forgotten and were rediscovered over the years. In this paper we give a historical overview of the origins of commutators and a survey of different kinds of groups where the set of commutators does not equal the commutator subgroup. We conclude with a status report on what is now called the Ore Conjecture stating that every element in a finite nonabelian simple group is a commutator.

## 1 Origins of commutators

> "In a group the product of two commutators need not be a commutator, consequently the commutator group of a given group cannot be defined as the set of all commutators, but only as the group generated by these. There seems to exist very little in the way of criteria or investigations on the question when all elements of the commutator group are commutators."

This is what Oystein Ore says in 1951 in the introduction to his paper "Some remarks on commutators" [57]. Since Ore made his comments, numerous contributions have been made to this topic and they are widely scattered over the literature. Many results have been rediscovered and republished. A case in point is Ore himself. The main result of [57] is that the alternating group on $n$ letters, $n \geq 5$, consists entirely of commutators. This was already proved by G. A. Miller [54] over half a century earlier. The two authors of this paper almost got into a similar situation after rediscovering one of the major results in this area. Fortunately we realized this before publication and then concluded that a survey of the major questions and results in this area was needed, together with a historical overview of the origins of commutators.

Commutators came into the world 125 years ago as a by-product of Dedekind's first foray into determining group characters of nonabelian groups. In an 1896 letter to Frobenius, Dedekind revealed his ideas and results for the first time. Here is what Frobenius says in [16]:

> "Das Element $F$, das sich mittelst der Gleichung $BA = ABF$ aus $A$ und $B$ ergiebt, nenne ich nach DEDEKIND den Commutator von $A$ und $B$." [1]

According to Frobenius, Dedekind proved in 1880 that the conjugate of a commutator is again a commutator, and therefore that the commutator subgroup generated by the commutators of a group is a normal subgroup of the group. Furthermore, Dedekind proved that any normal subgroup with abelian quotient contains the commutator subgroup, and that the commutator subgroup is trivial if and only if the group is abelian. However, these results were first published by G. A. Miller in [52].

The motivating force behind Dedekind's introduction of commutators was his goal of extending group characters from abelian to nonabelian groups. The central object of investigation for Frobenius and Dedekind was the group determinant and its factorization, out of which arose the theory of group characters. For the definition of the group determinant and further details we refer to [9], since we are only interested in a by-product of this concept, namely commutators.

Dedekind had spent the early years of his career at the ETH Zürich (1858–62). In 1880 he revisited Zürich and became personally acquainted with Frobenius, 18 years his junior, who was at the time a professor at the ETH. This was the starting point of an on-again-off-again, sometimes intense, correspondence between the two over many years as detailed by Hawkins in [32] and [33]. Around 1880 Dedekind was motivated by his studies of the discriminant in a normal field to consider the group determinant. One of the earliest results he obtained was that for a finite abelian group of order $n$ the group determinant factored into $n$ linear factors with the characters as coefficients of the linear factors. In his correspondence with Frobenius, Dedekind conjectured that for a nonabelian group $G$ the number of linear factors of the group determinant was equal to the index of the commutator subgroup $G'$ in $G$, with coefficients corresponding to those of the abelian group $G/G'$, and in this context commutators and the commutator subgroup made their appearance.

A good deal of the correspondence between Dedekind and Frobenius deals with the group determinant, its factorization, and Dedekind's conjecture stated above. Dedekind determined the group determinant and its factorization for the symmetric group $S_3$ and the quaternions of order 8, and in turn, Frobenius did the same for the dihedral group of order 8. Finally, in his 1896 paper [16] Frobenius proves Dedekind's conjecture as part of the general theorem on the factorization of the group determinant for finite nonabelian groups. For the details of this result we refer the interested reader to Theorem 3.4 in [9].

[1] "Following Dedekind, I am calling the element $F$, which is obtained from $A$ and $B$ with the help of the equation $BA = ABF$, the commutator of $A$ and $B$."

Dedekind himself never published anything concerning the group determinant nor its connection with commutators. However, according to Hawkins [33], Dedekind decided to pursue some group theoretic research of his own that allowed him to use his commutators. Earlier, Dedekind had studied normal extensions of the rational field with all subfields normal. Some years later these investigations suggested to him the related problem: Characterize those groups with the property that all subgroups are normal — he called such groups Hamiltonian. Dedekind found, by making use of commutators, that determining the answer was relatively simple, and he communicated this to his friend Heinrich Weber, an editor of the Mathematische Annalen, who urged him to publish the result there. Dedekind eventually published his results in [11], but only after checking with Frobenius, who assured him that this result was significant and not a consequence of known results.

As was already mentioned, G. A. Miller was the first to publish the essential results on the commutator subgroup in [52]. However, he does not attach the label "commutator" to Dedekind's correction factor $F$. The headline of the section in which he deals with commutators is simply "On the operation $sts^{-1}t^{-1}$". Miller's motivation in [52] for using the commutator concept was the classification of groups of order less than 48 up to isomorphism. In his two later publications addressing commutators, namely [53] and [54], he uses the label commutator and attributes it to Dedekind. In [53] Miller further expands the basic properties of the commutator subgroup, and he introduces the derived series of a group. He also shows that the derived series is finite and ends with the identity if and only if $G$ is solvable.

In his 1899 paper [54] Miller deals with commutators as objects that are of interest in their own right. He first develops a formula that shows that under certain conditions the product of two commutators is again a commutator. In modern notation, he is showing that $[tb, a][a, b] = [t^b, a^b]$ for $a, b, t$ in a group $G$. With the help of this identity he shows in Theorem I that every element of the alternating group on $n$ letters, $n \geq 5$, is a commutator, a result rediscovered over 50 years later by Ito [37] and Ore [57]. In Theorem II, Miller shows that in the holomorph of a cyclic group $C_n$ the commutator subgroup consists entirely of commutators and is equal to $C_n$, if $n$ is odd, and equal to the subgroup of index 2 in $C_n$ in case $n$ is even. This foreshadows later results by Macdonald [44] and others who investigate groups with cyclic commutator subgroup in which not all elements of the commutator subgroup are commutators.

In his publications Miller never addresses the central issue in our context of whether the commutator subgroup always consists entirely of commutators. In [53] he states that for generating the commutator subgroup not all commutators are needed, and he says that a rather small portion will suffice for this purpose. On the other hand, he shows in Theorem I and II of [54] that for certain groups the set of commutators is equal to the commutator subgroup. However, as we will see when discussing [15] below, this question can not have been far from his mind. The first explicit statement of this question is found in Weber's 1899 textbook [74], which is the first textbook to introduce commutators and the commutator subgroup. After referring to Dedekind's definition of commutators, Weber states that the set of commutators is not necessarily a subgroup, but does not provide an example to prove his claim. He does prove that the commutator subgroup is generated by the set of commutators and this subgroup forms a normal subgroup. It is Fite [15] who provides the first such example in his paper "On metabelian groups". It should be mentioned that the metabelian groups in the title are what we now call groups of nilpotency class two, or, as Fite states it, a group with an abelian group of inner automorphisms. Fite constructs a group $G$ of order 1024 and nilpotency class 2 in which not all elements of the commutator subgroup are commutators. He attributes this example to G. A. Miller. In addition he provides a homomorphic image of $G$ that has order 256, in which the set of commutators is not equal to the commutator subgroup. We discuss this in detail in Section 5.

To conclude our early history of commutators we mention a 1903 paper by Burnside [5]. As detailed earlier, commutators arose out of the development of group characters. Burnside uses characters to obtain a criterion for when an element of the commutator subgroup is the product of two or more commutators. So we have come full circle! This criterion was later extended by Gallagher [17]. We discuss this in detail in Section 6.

There seemed to be little interest in the topic of commutators for the 30 years following 1903. It should be kept in mind that the familiar notation for commutators had not yet been developed and its absence apparently stifled further development. The first occurrence of the commutator notation we could find is in Levi and van der Waerden's seminal paper on the Burnside groups of exponent 3 [41]. They denote the commutator of two group elements $i, j$ as $(i, j) = iji^{-1}j^{-1}$ and make creative use of this notation in their proofs. The first textbook using the new notation is by Zassenhaus [76]. There he gives familiar commutator identities, for example, the expansion formulas for products, but not the Jacobi identity. However, Zassenhaus states that in a group with abelian commutator subgroup the following "strange" (merkwürdig) rule holds: $(a, b, c)(b, c, a)(c, a, b) = e$. As the source for the definitions, notation and formulas in his section on commutators, Zassenhaus refers to Philip Hall's paper [31], which appeared after [41].

The new notation made it possible to develop a commutator calculus to solve a variety of group theoretic problems that had not been previously accessible. In turn, the extended use of commutators as a tool brought about renewed interest in questions about commutators themselves, in particular the question on when the set of commutators is a subgroup. The remainder of the paper focuses on this question.

There are significant topics about commutators that we do not cover in this paper. These topics include: viewing the commutator operation as a binary operation; Levi's characterization of groups in which the commutator operation is associative [40]; conditions for when a product of commutators is guaranteed to be a commutator [34]; and investigations into an axiomatic treatment of the commutator laws by Macdonald and Neumann ([48], [49], and [50]) and by Ellis [14].

In the following two sections we give a survey of conditions which imply that either the set of commutators is equal to the commutator subgroup or unequal to it. Sufficient conditions for equality are rather scarce and not very powerful. As

Macdonald acknowledges in [46], a forerunner of [47], there are fundamental logical difficulties in this area, for example, the main theorem of [2] implies that there is no effective algorithm for deciding whether an element is a commutator when $G$ is a finitely presented group. However, there are necessary and sufficient conditions for an element of a finite group to be a commutator using the irreducible characters of the group. Hence, from the character table of a finite group we can read off if every element of the commutator subgroup is a commutator. Details of this can be found in Section 6.

We introduce the following notation to facilitate our discussion. For a group $G$ let $K(G) = \{[g,h] \mid g,h \in G\}$ be the set of commutators of $G$ and set $G' = \langle K(G) \rangle$, the commutator subgroup of $G$. We say that the group element $g$ is a commutator if it is an element of $K(G)$ and a noncommutator otherwise.

In Section 4 we construct various minimal examples of groups such that the commutator subgroup contains a noncommutator. These examples are minimal with respect to the order of the group $G$ and the order of $G'$, respectively. With the help of GAP [19] we construct minimal examples $G$ and $H$ where $G$ is a perfect group $G$ such that $K(G) \neq G'$ and $H$ is a group in which $H' \cap Z(H)$ is generated by noncommutators.

As mentioned earlier, the first examples of groups with the set of commutators not equal to the commutator subgroup are finite nilpotent 2-groups of class 2. In Section 5 we develop a general construction for nilpotent $p$-groups of class 2 such that the commutator subgroup contains a noncommutator. This construction is obtained by finding various *covering groups* $\tilde{A}$ of an elementary abelian $p$-group $A$ of rank $n \geq 4$. By a counting argument it is always the case that $K(\tilde{A}) \neq \tilde{A}'$. We look at homomorphic images of two covering groups resulting in groups of order $p^8$ with exponent $p$ and $p^2$, respectively, such that the set of commutators is unequal to the commutator subgroup. These groups appear in the literature ([47] and [67]) and various ad-hoc methods are used to show that the commutator subgroup contains a noncommutator. The question arises: What is the smallest integer $n$ such that for a given prime $p$ there exists a group $G$ of order $p^n$ with $G' \neq K(G)$? We conclude Section 5 with an answer to this question.

For a group $G$ the function $\lambda(G)$ denotes the smallest integer $n$ such that every element of $G'$ is a product of $n$ commutators. This function was introduced by Guralnick in his dissertation [23]. The statement $K(G) \neq G'$ is then equivalent to $\lambda(G) > 1$. In Section 6 we consider conditions for upper and lower bounds for $\lambda(G)$, as well as provide conditions and examples when $\lambda(G)$ can be specified exactly. Some of these results involve character theory, in particular, to provide a necessary and sufficient condition on a finite group $G$ such that $\lambda(G) = n$. In this section we include a well known example by Cassidy [8]. This is a group of nilpotency class 2 and it is claimed there is no bound on the number of commutators in the product representing an element of the commutator subgroup. However, a typographical error impacts the verification of this claim made in [8]. We include a slightly more general proof of the claim.

For most problems one encounters in group theory the solution in the cyclic case is trivial. Not so here, where the situation for cyclic commutator subgroups is a microcosm for the complexity of the general case. In Section 7 we survey groups with cyclic commutator subgroup in which the commutator subgroup contains a noncommutator.

Many results in Sections 2 through 7 have been extended to higher terms of the lower central series. A survey of these results is the topic of Section 8.

The topic of the final section is a report on the current status on what has been called in the literature the Ore Conjecture (see [1], [4], [13], [22], [72] and [73]), which states that every element in a nonabelian finite simple group is a commutator. The Ore Conjecture is still open for some of the finite simple groups of Lie type over small fields. The details are given in a table at the end of the paper. There are many contributions on the Ore Conjecture in the literature concerning various types of semisimple and infinite simple groups (see for example [58] and [59]). These contributions go beyond the scope of this survey.

## 2 Conditions for equality

In this section we discuss mostly conditions implying that the set of commutators is equal to the commutator subgroup. There are two types of such conditions. Those of the first type are conditions on the structure of the group or the commutator subgroup that allow us to conclude the commutator subgroup contains only commutators. Those of the second type are restrictions on the order of a group or its commutator subgroup. These restrictions are mainly derived from the structural conditions of the first type. Showing that the restrictions on the orders are best possible leads to the minimal examples discussed in Sections 4 and 5.

We start with conditions on the structure of the group. One of the most versatile results is due to Spiegel.

**Theorem 2.1** **([68])** *Suppose the group $G$ contains a normal abelian subgroup $A$ with cyclic factor group $G/A$. Then $K(G) = G'$.*

Motivated by results in [43], Liebeck in [42] gives a necessary and sufficient condition that an element of the commutator subgroup is a commutator provided the group has nilpotency class 2. Using this condition, he shows that for a group $G$ it follows $K(G) = G'$ whenever $G' \subseteq Z(G)$ and $\mathrm{d}(G') \leq 2$, where $\mathrm{d}(G')$ denotes the minimal number of generators of $G'$, and he gives an example that this cannot be extended to rank 4 or greater. Rodney in [62] extends these results. In particular, he shows the following.

**Theorem 2.2** **([62])** *The following two conditions on a group $G$ imply $G' = K(G)$:*

(i) *$G$ is nilpotent of class two and the minimal number of generators of $G'$ does not exceed three;*

(ii) *$G'$ is elementary abelian of order $p^3$.*

The following result by Guralnick generalizes one of Rodney in [62].

**Theorem 2.3 ([28])** *Let $P$ be a Sylow $p$-subgroup of $G$ with $P^* = P \cap G'$ abelian and* $\mathrm{d}(P^*) \leq 2$. *Then $P^* \subseteq K(G)$.*

Similarly, Guralnick obtains the following result if $p > 3$.

**Theorem 2.4 ([28])** *If $G'$ is an abelian $p$-subgroup of $G$ with $p > 3$ and* $\mathrm{d}(G') \leq 3$, *then $G' = K(G)$.*

For nilpotent groups, in particular for finite $p$-groups, the following conditions of the first type are useful results for arriving at sufficient conditions of the second type.

**Theorem 2.5 ([61])** *If $G$ is nilpotent and $G'$ is cyclic, then $G' = K(G)$.*

**Theorem 2.6 ([39])** *Let $G$ be a finite $p$-group with $G'$ elementary abelian of rank less than or equal to three. Then $K(G) = G'$.*

With the exception of some additional conditions of type one (in the case of cyclic commutator subgroups) that we will consider in a later section, Theorems 2.1 – 2.6 are the tools currently available for arriving at sufficient conditions on the orders of $G$ and $G'$ that imply $K(G) = G'$. We start with sufficient conditions on the orders of $G$ and $G'$, which are shown to be best possible in Section 4.

**Theorem 2.7 ([26])** *Let $G$ be a group. If* (i) *$G'$ is abelian and $|G| < 128$ or $|G'| < 16$ or* (ii) *$G'$ is nonabelian and $|G| < 96$ or $|G'| < 24$, then $K(G) = G'$.*

Many examples of groups whose commutator subgroup contains a noncommutator are groups of prime power order. The question arises: For a $p$-group $G$ of order $p^n$, what is the largest $n$ such that we can guarantee that $K(G) = G'$? As we show in Section 5, the following result is best possible.

**Theorem 2.8 ([39])** *Let $p$ be a prime and $G$ a group of order $p^n$. Then $G' = K(G)$ if $n \leq 5$ for odd $p$ and $n \leq 6$ for $p = 2$.*

## 3 Conditions for inequality

In this section we discuss conditions that lead to the conclusion that the set of commutators is not equal to the commutator subgroup. However, in some cases restrictions are imposed on the structure of the group or the commutator subgroup, and then the conditions for inequality turn out to be necessary and sufficient under these restrictions. These sufficient conditions often lead to the construction of families of groups in which the commutator subgroup always contains a noncommutator. Often the objective is to find minimal examples in a certain class of groups. The conditions discussed in this section come mainly from [47], [36], [29], and [39]. The selection is based on their relevance in the next two sections, which includes the discussion of minimal examples.

We start with an almost obvious criterion that one obtains by comparing the number of possible distinct commutators with the number of elements in the commutator subgroup. The condition is stated formally for the first time in [47], but earlier applications can be found in [43] and [18].

**Theorem 3.1** **([47])** *If $G$ is any group and if $|G : Z(G)|^2 < |G'|$, then there are elements in $G'$ that are not commutators.*

As Macdonald observes, the criterion is very well suited for groups with central commutator subgroups. With the help of Theorem 3.1, Macdonald constructs a large family of groups of nilpotency class 2 with the property that the set of commutators is not equal to the commutator subgroup. Isaacs' motivation in [36] for stating his criterion is similar to Macdonald's. He says that it is well known for a group $G$ that not every element of $G'$ need be a commutator, but what is less well known is a convenient source of finite groups that are examples of this phenomenon. His examples are wreath products satisfying the following criterion.

**Theorem 3.2** **([36])** *Let $U$ and $H$ be finite groups with $U$ abelian and $H$ nonabelian. Let $G = U \wr H$ be the wreath product of $U$ and $H$. Then $G'$ contains a noncommutator if*

$$\sum_{A\in\mathcal{A}} \left(\frac{1}{|U|}\right)^{[H:A]} \leq \frac{1}{|U|},$$

*where $\mathcal{A}$ is the set of maximal abelian subgroups of $H$. In particular, this inequality holds whenever $|U| \geq |\mathcal{A}|$.*

The above construction yields both solvable and nonsolvable groups with the set of commutators not equal to the commutator subgroup. Choosing $H$ simple and $U$ large enough leads to perfect groups, that is, groups such that $G' = G$, with the desired property.

Guralnick's goal in [29] is to determine bounds on a group $G$ and its commutator subgroup $G'$ such that $G' = K(G)$ always holds whenever the respective orders are below these bounds. The following criterion rules out groups with a "large" abelian commutator subgroup.

**Theorem 3.3** **([29])** *Suppose that $A$ is an abelian group of even order. Then there exists a group $G$ with $G' \cong A$ and $G' \neq K(G)$ if and only if $A \cong C_2 \times A_1 \times A_2 \times A_3$ or $A \cong C_{2^\alpha} \times A_1 \times A_2$, where the $A_i$ are nontrivial abelian groups and $\alpha \geq 2$.*

In [39] sufficient conditions on the nilpotency class and certain elements belonging to the center of a group $G$ are established that guarantee that $K(G) \neq G'$. This leads to three classes of groups with the property that the commutator subgroup contains a noncommutator. The groups of smallest order in these classes are finite $p$-groups and appear as minimal examples in Section 5. As it turns out, we need to establish different criteria depending on whether $p > 3$, $p = 3$, or $p = 2$, respectively, which form the three classes of groups.

**Proposition 3.4 ([39])** *Let $p \geq 5$ be a prime and $H = \langle a, b \rangle$ be a nilpotent group of class exactly 4 with $[b, a, b] \in Z(H)$ and $\exp(H') = p$. Then $K(H) \neq H'$.*

**Proposition 3.5 ([39])** *Let $H = \langle a, b \rangle$ be a nilpotent group of class exactly 4 with $a^3$, $b^9$, $[b, a, b] \in Z(H)$. Then $K(H) \neq H'$.*

**Proposition 3.6 ([39])** *Let $H = \langle a, b, c \rangle$ be a group of class 3 precisely. If $a^4$, $b^2$, $c^2$, $[a, c]$, $[b, c]$ and $(ab)^2 \in Z(H)$, then $K(H) \neq H'$.*

## 4 Some minimal examples

MacHale in [51] lists 47 conjectures about groups that are known to be false and asks for a minimal counterexample for each. Conjecture 7 in his paper states "In any group $G$, the set of all commutators forms a subgroup". MacHale indicates that a minimal counterexample is known. In fact this is the topic of Guralnick's Ph.D. dissertation [23] and several subsequent papers, in particular [26] and [28].

In [26] examples are constructed or cited to show that the conditions of Theorem 2.7 are tight. In this section and subsequent sections we consider the following classes of groups and find groups of minimal order in each:

Groups such that the commutator subgroup is not equal to the set of commutators and

(i) the commutator subgroup is abelian;

(ii) the commutator subgroup is abelian of order 16;

(iii) the commutator subgroup is nonabelian;

(iv) the commutator subgroup is nonabelian of order 24;

(v) the intersection of the commutator subgroup and the center is generated by noncommutators;

(vi) the group is perfect.

In Section 5 we construct a metabelian group $G$ of order $2^7$ such that $G'$ has order 16 and $K(G) \neq G'$. This group is of minimal order in classes (i) and (ii). It follows from Theorem 2.6 that $G'$ is not cyclic and in fact in any minimal example $H$ in classes (i) and (ii) the commutator subgroup $H'$ can not be cyclic. Hence we consider the following variant of classes (i) and (ii):

Groups such that the commutator subgroup is not equal to the set of commutators and

(i′) the commutator subgroup is cyclic;

(ii′) the commutator subgroup is cyclic of order 60.

The group $G$ constructed in Example 7.10 has order 240 such that its commutator subgroup is cyclic of order 60 and $K(G) \neq G'$. This group is of minimal order in classes (i′) and (ii′).

Of course MacHale was interested in the smallest group order such that the commutator subgroup contains a noncommutator. As we see below, such groups are in class (iii). Minimal examples for each class (i) – (vi), (i′) and (ii′) can be found using GAP by searching its small groups library. For example, Rotman [65] states that via computer search the smallest examples $G$ such that $K(G) \neq G'$ have order 96. Rotman was apparently unaware of Guralnick's earlier work. The following example gives explicit constructions of the two nonisomorphic groups of order 96 whose commutator subgroups contain noncommutators.

**Example 4.1** **([23])** There are exactly two nonisomorphic groups $G$ of order 96 such that $K(G) \neq G'$. In both cases $G'$ is nonabelian of order 32 and $|K(G)| = 29$.

(a) Let $G = H \rtimes \langle y \rangle$, where $H = \langle a \rangle \times \langle b \rangle \times \langle i, j \rangle \cong C_2 \times C_2 \times Q_8$ and $\langle y \rangle \cong C_3$. Let $y$ act on $H$ as follows: $a^y = b$, $b^y = ab$, $i^y = j$ and $j^y = ij$.

(b) Let $H = N \rtimes \langle c \rangle$, where $N = \langle a \rangle \times \langle b \rangle \cong C_2 \times C_4$ and $\langle c \rangle \cong C_4$. Let $c$ act on $N$ by $a^c = a$ and $b^c = ab$. Let $G = H \rtimes \langle \gamma \rangle$ with $\langle \gamma \rangle \cong C_3$, where $a^\gamma = c^2b^2$, $b^\gamma = cba$, $c^\gamma = ba$.

The group of Example 4.1(a) appears in Dummit and Foote's textbook [12] as an example of a group $G$ with $K(G) \neq G$. No claim of minimality is made.

Guralnick in [26] describes the following class of groups that yields many examples of groups $G$ in which $K(G) \neq G'$. Let

$$G_1 = \langle a, b, x \mid a^4 = b^4 = x^3 = 1,\ xax^{-1} = b,\ xbx^{-1} = ab, \\ a^2 = b^2,\ aba^{-1} = b^{-1} \rangle. \tag{4.1.1}$$

Then $G_1 = \langle H_1, x \rangle$, where $H_1 = G_1' = \langle a, b \rangle \cong Q_8$. Choose $G_2$ to be any nonabelian group with normal abelian subgroup $H_2$ of index 3. Then there exists $y \in G_2$ such that $G_2 = \langle H_2, y \rangle$. Let $G$ be the subgroup of $G_1 \times G_2$ generated by $H_1 \times H_2$ and the element $(x, y)$. Then $G$ has order $24|H_2|$ and $K(G) \neq G'$. Note that $G' = G_1' \times G_2'$. In particular, for any $1 \neq g \in G_2'$ the element $(a^2, g)$ is not a commutator. Our next three examples arise from this construction for particular choices of $G_2$.

**Example 4.2** Let $G_1$ be defined as in (4.1.1) and take $G_2 = A_4$ and $H_2 = G_2' \cong C_2 \times C_2$. Then $G$ has order 96 and $G' \cong Q_8 \times C_2 \times C_2$.

The group constructed in Example 4.2 is isomorphic to the group in Example 4.1(a). However, Example 4.1(b) cannot be constructed in this manner, since there is not another nonabelian group of order 12 with a normal subgroup of order 4.

The following example gives a group of minimal order in class (iv).

**Example 4.3** Let $G_1$ be the group defined in (4.1.1). Take $G_2$ to be any nonabelian group of order 27 so that $|H_2| = 9$ and $G_2' \cong C_3$. Then $G' \cong Q_8 \times C_3$, which is nonabelian of order 24. The order of $G$ is $24 \cdot 9 = 216$.

We finish this section by giving minimal examples of groups $G$ such that $K(G) \neq G'$ with some additional property. The first example is a group $G$ in which $G' \cap Z(G)$ is generated by noncommutators. The question of whether such a group exists was asked by R. Oliver and an example was given by Caranti and Scopolla [6]. Their example has order $p^{14}$, where $p$ is an odd prime, but it is noted in the paper that a smaller example exists. Our example with this property has order 216 which by a search of the small groups library in GAP is the smallest such example.

**Example 4.4** Set $G_1 = \langle H_1, x\rangle$ as in (4.1.1) and set

$$G_2 = \langle c, y \mid c^9 = y^3 = 1,\ y^{-1}cy = c^4\rangle \cong C_9 \rtimes C_3$$

where $H_2 = C_9 = \langle c\rangle$ and $C_3 = \langle y\rangle$. Let $G = \langle H_1 \times H_2, \{(x, y)\}\rangle$. Then $Z(G) = \langle (a^2, c^3)\rangle \cong C_6$. However, since $c^3 \in G_2'$, we have that $(a^2, c^3)$ is a noncommutator, as needed.

To find a perfect group $G$ in which $K(G) \neq G'$ we can use the group construction and criterion from Theorem 3.2 due to Isaacs [36]. The smallest perfect group obtainable from this construction is the following.

**Example 4.5** Let $G = C_2 \wr A_5$. Then $G$ is a perfect group with $|G| = 2^{60} \cdot 60$. To see that $G' \neq K(G)$ we note that the maximal abelian subgroups of $A_5$ are its Sylow subgroups. There are ten maximal abelian subgroups of order 3, five maximal abelian subgroups of order 4, and six maximal abelian subgroups of order 5. This gives

$$10\left(\frac{1}{3}\right)^{20} + 5\left(\frac{1}{4}\right)^{15} + 6\left(\frac{1}{5}\right)^{12} \leq \frac{1}{60},$$

which by Theorem 3.2 shows that $K(G) \neq G'$.

A search of the perfect groups in GAP shows that the smallest perfect group $G$ that contains an element that is not a commutator has order 960. This group can be visualized in the following way. Let $H = C_2^5 \rtimes A_5$, where we think of $A_5$ having a "wreath action" on $C_2^5$. Set $G = H/Z(H)$. Then $G$ is a perfect group of order $2^5 \cdot 60 \cdot \frac{1}{2}$ and has the property that $K(G) \neq G'$.

## 5 p-Groups

The earliest examples of groups $G$ in the literature for which $K(G) \neq G'$ are 2-groups. Fite [15] attributes the following group of nilpotency class 2 and order 1024 to G. A. Miller, represented here as a subgroup of $S_{24}$:

$$\begin{aligned} G = \langle &(1,3)(5,7)(9,11), (1,2)(3,4)(13,15)(17,19),\\ &(5,6)(7,8)(13,14)(15,16)(21,23),\\ &(9,10)(11,12)(17,18)(19,20)(21,22)(23,24)\rangle. \end{aligned}$$

He states that $G'$ contains 36 commutators and 28 noncommutators. Fite then considers a homomorphic image of $G$, represented here as a subgroup of $S_{16}$:

$$\begin{aligned} H = \langle &(1,3)(5,7)(9,11), (1,2)(3,4)(13,15),\\ &(5,6)(7,8)(13,14)(15,16), (9,10)(11,12)\rangle. \end{aligned}$$

The group $H$ has order 256 with $|K(H)| = 15$ and $|H'| = 16$. The group $H$ of Fite is used in several textbooks, for example, Carmichael [7] and an early edition of Rotman [64], as an example of a group in which the commutator subgroup is not equal to the set of commutators.

Miller's group $G$ above is the basis for our study of $p$-groups of nilpotency class 2 for which $K(G) \neq G'$. For some normal subgroups $S$ of $G'$ of order 4 the property $K(G/S) \neq (G/S)'$ holds; the group $H$ given by Fite is an example. In fact there are several such subgroups of order 4 in $G$ for which this is true, and explicit constructions of such quotients can be found in the literature, for example, in [45].

We now adapt the construction of the 2-group $G$ above to a $p$-group $\tilde{A}$ that is the covering group of an elementary abelian $p$-group that has the property that $\tilde{A}' \neq K(\tilde{A})$. [2]

**Definition 5.1** Let $Q$ be a finite group and let $M(Q)$ be the Schur multiplier of $Q$. A group $\tilde{Q}$ is called the *covering group* of $Q$ if $\tilde{Q}$ contains a normal subgroup $N$ such that $N \subseteq Z(\tilde{Q}) \cap \tilde{Q}'$, $N \cong M(Q)$, and $\tilde{Q}/N \cong Q$.

Schur in [66] showed that every finite group has a covering group. Now let $A$ be an elementary abelian $p$-group of rank $n \geq 2$. Then it can be shown that $M(A)$ is elementary abelian of order $p^{n(n-1)/2}$ and hence that the covering group $\tilde{A}$ has order $p^{n(n+1)/2}$. The center of $\tilde{A}$ equals $\tilde{A}'$; therefore $\tilde{A}$ is nilpotent of class 2 and $M(A) \cong \tilde{A}'$.

For $n \geq 6$, where $n$ is the rank of $A$, we have $|\tilde{A}/Z(\tilde{A})| = p^n$. Hence in this case $|\tilde{A}/Z(\tilde{A})|^2 < |\tilde{A}'|$, and we can use Theorem 3.1 to see that $K(\tilde{A}) \neq \tilde{A}'$. We can actually take $n \geq 4$ and note that the covering group contains elements in $\tilde{A}'$ that are not in $K(\tilde{A})$. This is because the number of nonidentity commutators in $K(\tilde{A})$ is exactly

$$|K(\tilde{A})| - 1 = \frac{(p^n - 1)(p^{n-1} - 1)}{p^2 - 1}. \tag{5.1.1}$$

For $n \geq 4$ we have that $|K(\tilde{A})| < |\tilde{A}'|$. Moreover, there are homomorphic images $\mathcal{H}$ of $\tilde{A}$ such that $K(\mathcal{H}) \neq \mathcal{H}'$. This is exactly the case for Miller's group $G$ and Fite's group $H$ above for $p = 2$ with $G = \tilde{A}$ and $H = \mathcal{H}$.

An explicit construction of various covering groups of the elementary abelian $p$-group of rank $n \geq 6$ can be found in Macdonald [47]. Macdonald constructs these groups explicitly for the purpose of finding groups $G$ for which $G' \neq K(G)$. As noted above, the covering groups for the elementary abelian $p$-group $A$ of rank 4 have the property that $K(\tilde{A}) \neq \tilde{A}'$. In the following example we construct a covering group of exponent $p$ for $n = 4$ and $p > 2$. This group can be found in [67] and is attributed to W. P. Kappe. Because the group has exponent $p$ it is straightforward to show explicitly that a particular element of the commutator subgroup is not a commutator and one does not need the counting argument of (5.1.1). The details can be found in [67].

[2] The authors would like to thank R. Gow and R. Quinlan for pointing out the connection between covering groups and this analysis of nilpotent groups of class 2.

**Example 5.2** Let $G = \langle g_1, g_2, g_3, g_4 \rangle$ be the free nilpotent group of class 2 and exponent $p$, where $p > 2$ is a prime. The group $G$ has order $p^{10}$, $G' = Z(G) \cong C_p^6$ and $G/G' \cong C_p^4$. Hence $G$ is the covering group for $C_p^4$ as needed, and we have $K(G) \neq G'$ by (5.1.1).

Our next example is also a covering group of $A$ for $n = 4$ and $p$ a prime. This group has exponent $p^2$.

**Example 5.3** Let $H = \langle g_1, g_2, g_3, g_4 \rangle$ be the free nilpotent group of class 2 and let $p$ be an odd prime. Consider the following relations:

$$R = \{g_1^{p^2} = g_2^{p^2} = g_3^{p^2} = g_4^{p^2} = 1,$$
$$g_1^p = [g_1, g_3],\ g_2^p = [g_2, g_4],\ g_3^p = [g_1, g_2],\ g_4^p = [g_2, g_3]\}.$$

Let $G = H/N$, where $N$ is the normal closure of $R$. Then $G$ is a covering group for $C_p^4$, and $K(G) \neq G'$ by (5.1.1).

For Examples 5.2 and 5.3 we construct a homomorphic image of each group of order $p^8$. The first quotient, given in Example 5.4, is the smallest nilpotent group of class 2 and exponent $p$ such that the set of commutators is unequal to the commutator subgroup. This group can also be found in [67].

**Example 5.4** Let $G$ be the group in Example 5.2 and let $S$ be the normal subgroup of $G$ generated by $[g_3, g_4]$ and $[g_2, g_4]$. The group $Q = G/S$ is a group of order $p^8$ and exponent $p$ in which $K(Q) \neq Q'$.

The following group is a quotient of the group of Example 5.3. It is isomorphic to the group constructed by Macdonald in [45] as an example of a $p$-group in which the set of commutators is unequal to the commutator subgroup.

**Example 5.5** Let $G$ be the group in Example 5.3. Let $S$ be the normal subgroup generated by $[g_1, g_4]$ and $[g_3, g_4]$. Then $Q = G/S$ has order $p^8$, $\exp(Q) = p^2$, and $Q$ has the property $K(Q) \neq Q'$.

It is an open question whether every nilpotent group $G$ of class 2 such that $K(G) \neq G'$ is a homomorphic image of a cover group of some abelian group.

The nilpotent groups of class 2 do not provide examples of $p$-groups of order $p^n$ with $n$ minimal such that $K(G) \neq G'$. As we can see from (i) of Theorem 2.2, any proper homomorphic image $H/L$ of $H$ of order less than $p^8$ has the property $K(H/L) = (H/L)'$. We finish this section with three examples to show that the bounds on $n$ given in Theorem 2.8 are sharp.

**Example 5.6 ([39])** Let $p$ be a prime with $p > 3$ and let $V = \langle u \rangle \times \langle v \rangle \times \langle w \rangle \times \langle z \rangle$ be an elementary abelian $p$-group of rank 4. Let $B = V \rtimes \langle b \rangle$, the semidirect product of $V$ with a cyclic group $\langle b \rangle$ of order $p$. The defining relations of $B$ are those of $V$ along with

$$b^p = 1, \quad [u, b] = w, \quad \text{and} \quad [v, b] = [w, b] = [z, b] = 1.$$

Similarly, let $G = B \rtimes \langle a \rangle$ be the semidirect product of $B$ with a cyclic group $\langle a \rangle$ of order $p$. The defining relations of $G$ are those of $B$ along with

$$[b, a] = u, \quad [u, a] = v, \quad [v, a] = z, \quad a^p = [w, a] = [z, a] = 1.$$

It can be verified that $G$ has order $p^6$, nilpotence class 4, and $\exp(G) = p$. Furthermore, $u = [b, a]$, $v = [b, a, a]$, $w = [b, a, b]$ and $z = [b, a, a, a]$. Thus $G$ satisfies the conditions of Proposition 3.4. We conclude that $K(G) \neq G'$.

The next two examples are minimal examples from the classes of groups found in Propositions 3.5 and 3.6, respectively. These examples are different than those found in [39] and show that minimal examples in these classes are not unique.

**Example 5.7** Let $A = \langle a \rangle \times \langle b \rangle \cong C_3 \times C_9$. Let $B = A \rtimes \langle x \rangle$ be the semidirect product of $A$ with a cyclic group $\langle x \rangle$ of order 9. The defining relations of $B$ are those of $A$ along with $x^9 = [a, x] = 1$, $[b, x] = a^2$. We form $G = B \rtimes \langle y \rangle$ as a semidirect product of $B$ with a cyclic group $\langle y \rangle$ of order 3. The relations of $G$ are those of $B$ along with $y^3 = [a, y] = 1$, $[b, y] = x$ and $[x, y] = b^6$. The order of $G$ is $3^6$ and $G = \langle b, y \rangle$, since $x = [b, y]$ and $a = [x, b]$. The group is nilpotent of class 4, since $[b, y, y, y] = [x, y, y] = [b^6, y] = x^6 \neq 1$. Now $b^9$ and $y^3$ are in the center of $G$ because $b^9 = y^3 = 1$. It can be verified that $[b, y, b]$ is also in $Z(G)$. It follows by Proposition 3.5 that $K(G) \neq G'$.

**Example 5.8** Let $A = \langle a \rangle \times \langle b \rangle \cong C_2 \times C_8$. Let $G = A \rtimes \langle x, y \rangle$ be the semidirect product of $A$ with $Q_8 = \langle x, y \rangle$, the group of quaternions. The defining relations of $G$ are those of $A$ and $Q_8$ along with $[a, x] = [a, y] = 1$, $[b, x] = b^6$, $[b, y] = ab^2$. The group $G$ has order 128, is nilpotent of class 3 and $G = \langle x, bx^3, y \rangle$, since $[y, b] = a$. It is readily verified that $G$ satisfies the conditions of Proposition 3.6, and hence $K(G) \neq G$.

The group $G$ constructed in Example 5.8 is metabelian with $|G'| = 16$. The only groups $G$ of order less than 128 for which $G' \neq K(G)$ are the two groups of order 96 found in Example 4.1. These two groups have nonabelian commutator subgroups. Hence the group constructed in Example 5.8 is a minimal group in the classes (i) and (ii) listed in Section 4.

## 6 The function $\lambda(G)$

We define the value $\lambda(G)$ for a group $G$ to be the the smallest positive integer $n$ such that every element of $G'$ is a product of $n$ commutators and if $n$ is unbounded then we define $\lambda(G) = \infty$ [23]. The condition $K(G) \neq G'$ is equivalent to $\lambda(G) > 1$. In this section we consider conditions for upper and lower bounds for $\lambda(G)$, as well as providing conditions and examples when $\lambda(G)$ can be specified exactly.

Obtaining an arbitrary lower bound for $\lambda(G)$ does not require a complicated structure for $G'$. The following result is due to Macdonald [44] and is a corollary to Theorem 7.1.

**Theorem 6.1** (**[44]**) *For any positive integer $n$ there is a group $G$ such that $G'$ is cyclic and $\lambda(G) > n$.*

Guralnick in [27] also investigates groups with cyclic commutator subgroup of order $m$ and gives necessary and sufficient conditions such that $\lambda(G) > f(m)$. The details can be found in the next section (Theorem 7.8).

In [63] and [25] finite upper bounds on $\lambda(G)$ are applied to prove a classical result of Schur that if $[G : Z(G)]$ is finite then $G'$ is finite. For $a, b$ in $G$ and $u, v$ in $Z(G)$ we have $[au, bv] = [a, b]$. Hence $G$ has at most $[G : Z(G)]^2$ commutators. When $[G : Z(G)]$ is finite it follows that $G'$ is finite exactly when $\lambda(G)$ is finite. Setting $[G : Z(G)] = n$, Rosenlicht [63] shows $\lambda(G) \leq n^3$. Guralnick in [25] proves $\lambda(G) < \tau(n)/2$, where $\tau(n)$ is the number of divisors of $n$ and improves this bound to $\lambda(G) < 3\rho(n)/2$, where $\rho(n)$ is the number of prime divisors of $n$ (counting multiplicity).

Under certain conditions on the group $G$ the bounds on $\lambda(G)$ can be improved as follows.

**Theorem 6.2** (**[25]**) *Let $G$ be a group.*

(i) *If $G$ is nilpotent and $G/Z(G)$ is generated by $n$ elements, then $\lambda(G) \leq n$. If, in addition, $G' \subseteq Z(G)$, then $\lambda(G) \leq \lfloor \frac{n}{2} \rfloor$.*

(ii) *If $G$ is finitely generated and is nilpotent-by-nilpotent, then $\lambda(G)$ is finite.*

Nikolov and Segal in [56] prove that every subgroup of finite index in a (topologically) finitely generated profinite group is open. The next theorem is a special case of this result.[3]

**Theorem 6.3** (**[56]**) *There is a function $g$ defined on the positive integers such that if $G$ is a finite group generated by $d$ elements, then $\lambda(G) \leq g(d)$.*

In the following example Guralnick constructs groups $G$ with $\lambda(G) = n$ for every positive integer $n$.

**Example 6.4** (**[25]**) Let $p$ be a prime and let $n$ be a positive integer. Let

$$H = H(n, p) = \langle x_1, \ldots, x_{2n} \mid x_i^p = [x_i, [x_j, x_k]] = 1,\ 1 \leq i, j, k \leq 2n \rangle$$

and

$$N = N(n, p) = \langle [x_i, x_j] \mid i + j > 2n + 1 \rangle.$$

Set $G = G(n, p) = H/N$. Then $\lambda(G) = n$.

**Proof** We observe that $H$ is nilpotent of class 2 and $|H| = p^{2n + \binom{2n}{2}}$ with $\exp H = p$ for $p$ odd. We have $N \lhd H$, since $N \subseteq Z(H)$ and $|N| = p^{n^2 - n}$. Thus $|G| = |H/N| = p^{2n + n^2}$. We observe $H' = Z(H)$ and $G' = Z(G)$, as well as $[H : Z(H)] = [G : Z(G)] = p^{2n}$. Thus by (i) of Theorem 6.2 it follows that $\lambda(G) \leq n$. Lemma 5.1 in [25] yields that there exists an element in $G'$ which is the product of exactly $n$ nontrivial commutators. □

[3] The authors would like to thank R. Guralnick for pointing out this result to us.

We conclude this discussion on bounds for $\lambda(G)$ by considering a well-known example of Cassidy [8] that replaced Fite's example of order 256 [15] as an example of a group in which the set of commutators is not a subgroup (see, e.g., [65]). Our context here is different. It is well known that in free groups, even of finite rank, $\lambda(G)$ is not bounded (see [60]). Cassidy's example, which is not finitely generated, but is nilpotent of class 2, shows that the assumption of being finitely generated cannot be omitted from Theorem 6.2(ii). As already mentioned in the introduction, the proof given in [8] contains a major typographical error. Our example below addresses a slightly more general situation.[4]

**Example 6.5** Let $f$ and $g$ be polynomials over a field $K$ in $x$ and $y$, respectively, and let $h$ be a polynomial in $x$ and $y$ with coefficients in $K$. Let $m(f,g,h)$ be the matrix

$$\begin{pmatrix} 1 & f(x) & h(x,y) \\ 0 & 1 & g(y) \\ 0 & 0 & 1 \end{pmatrix}.$$

Then the set of matrices $m(f,g,h)$ forms a group $G$ under matrix multiplication. The group $G$ is nilpotent of class 2 with $G' = Z(G)$ and $\lambda(G) = \infty$.

**Proof** It is easy to see that the set of matrices forms a group as claimed and that the center of $G$ consists of the matrices of the form $m(0,0,h)$. An arbitrary commutator in $G$ has the form

$$[m(f_1,g_1,h_1), m(f_2,g_2,h_2)] = m(0,0,f_1g_2 - f_2g_1). \tag{6.5.1}$$

It follows that $G' \subseteq Z(G)$. Conversely, we have

$$m(0,0,\sum a_{ij}x^iy^j) = \prod [m(a_{ij}x^i,0,0), m(0,y^j,0)],$$

hence $Z(G) \subseteq G'$, and our claim follows.

To show $\lambda(G) = \infty$, let $n$ be a positive integer and $h(x,y) = \sum_{i=0}^{2n} x^iy^{2n-i}$. We will show that $m(0,0,h(x,y))$ is not the product of $n$ commutators. Assume otherwise. Without loss of generality we can write

$$m(0,0,h(x,y)) = \prod_{j=1}^{n} [m(s_j(x),t_j(x),0), m(u_j(x),v_j(y),0)],$$

where $s_j(x) = \sum_i a_{ij}x^i$, $u_j(x) = \sum_i b_{ij}x^i$, and the $a_{ij}$ and $b_{ij}$ are elements of $K$. By (6.5.1) it follows that

$$h(x,y) = \sum_{j=1}^{n} (s_j(x)v_j(y) - u_j(x)t_j(y)). \tag{6.5.2}$$

[4] Special thanks go to W. P. Kappe for providing the authors with this proof.

In (6.5.2) we compare the coefficients of the powers of $x$ which are polynomials in $y$ and obtain the following $2n+1$ linear relations in the vector space of polynomials in $y$ over $K$

$$y^{2n-i} = \sum_{j=1}^{n}(a_{ij}v_j(y) - b_{ij}t_j(y)) \quad \text{for } i = 0, 1, \ldots, 2n. \tag{6.5.3}$$

Set $V = \operatorname{span}\{1, y, \ldots, y^{2n}\}$ and $W = \operatorname{span}\{t_1(y), \ldots, t_n(y), v_1(y), \ldots, v_n(y)\}$. By (6.5.3) it follows $V \subseteq W$. We observe $\dim W \leq 2n$ and $\dim V = 2n+1$, since $\{1, y, \ldots, y^{2n}\}$ is a linearly independent set, so $2n+1 \leq 2n$, a contradiction. □

A few remarks on the nature of the typographical error in [8] are in order. The critical element in the commutator subgroup is defined as $m(0, 0, h)$, where $h(x, y) = \sum_{i=0}^{2n+1} x^i y^j$. No specification for $j$ is given. According to P. J. Cassidy[5], the $j$ should be replaced by $i$. After some minor adjustments of indices, the proof can be carried out as indicated in [8].

Character theory for finite groups provides necessary and sufficient conditions as to when an element is a commutator, and allows one to compute $\lambda(G)$. Oft quoted is Honda [35], but these ideas were known to Burnside [5] and, even earlier, to Frobenius.

**Theorem 6.6** *Let $G$ be a finite group and $g$ an element of $G$. Consider the following function $\sigma : G \to \mathbb{C}$ defined by*

$$\sigma(g) = \sum_{\chi \in \mathrm{Irr}(G)} \frac{\chi(g)}{\chi(1)}. \tag{6.6.1}$$

*Then $g$ is a commutator if and only if $\sigma(g) \neq 0$.*

A brute force method for testing whether or not $\lambda(G) > 1$ is to compute the set $S = \{g \in G \mid \sigma(g) \neq 0\}$ and check whether $|S| = |G'|$. Since the irreducible characters are functions on the conjugacy classes, we know that all elements of a conjugacy class are either commutators or are not commutators. Hence in general we can refine our test by computing the following sum:

$$\sum_{c \in C} |c| \cdot \sigma'(c(g)),$$

where $C$ is the set of conjugacy classes, $c(g)$ is a representative of the conjugacy class $c$, and

$$\sigma'(g) = \begin{cases} 0 & \text{if } \sigma(g) = 0 \\ 1 & \text{otherwise.} \end{cases}$$

Our observations now lead to the following necessary and sufficient criterion such that $\lambda(G) > 1$ in a finite group $G$. This result is a consequence of Theorem 6.6 and until now has not been explicitly stated in the literature.

---

[5]Personal communication with P. J. Cassidy by the authors.

**Corollary 6.7** *Let $G$ be a finite group. Then $\lambda(G) > 1$ if and only if*

$$\sum_{c \in C} |c|\sigma'(c(g)) \neq |G'|. \tag{6.7.1}$$

For perfect groups the test is even simpler. If $G$ is a perfect group, then $\lambda(G) > 1$ if and only if any $\sigma(c(g))$ is equal to zero.

Let $f_\chi$ be the degree of the irreducible character $\chi$. Define $m(G)$ to be the cardinality of the set $\{f_\chi \mid \chi \in \mathrm{Irr}(G)\}$. Guralnick gives an upper bound for $\lambda(G)$ using $m(G)$.

**Theorem 6.8 ([25])** *If $G$ is finite, then $\lambda(G) < m(G)$.*

Guralnick also states a lemma based on a result in [17], from which $\lambda(G)$ can be exactly computed for a finite group $G$.

**Lemma 6.1 ([25])** *Let $G$ be a finite group with irreducible characters $\chi_1, \ldots, \chi_h$. Consider the expression*

$$S(k, g) = \sum_{i=1}^{h} f_i^{1-2k} \chi_i(g). \tag{6.8.1}$$

*Then $\lambda(G) = n$ if and only if $S(n, g) = 0$ for all $g \in G'$ and $S(n-1, g) \neq 0$ for some $g \in G'$.*

The criterion of Lemma 6.1 can be implemented in GAP. It turns out that every group $G$ of order less than 1000 has $\lambda(G) \leq 2$. The question arises: What is the smallest group $G$ such that $\lambda(G) = 3$, and more generally, for which $\lambda(G) = n$?

## 7 Cyclic commutator subgroups

In [44], I. D. Macdonald begins the discussion on groups with cyclic commutator subgroups by showing that the commutator subgroup of such groups need not have a generating commutator. There are three follow-up papers, [61], [20] and [27], with more or less the same title, which expand on Macdonald's earlier result. These four papers are the topic of discussion in this section.

Macdonald's result is summarized in the following theorem.

**Theorem 7.1 ([44])** *If $G'$ is cyclic and either $G$ nilpotent or $G'$ is infinite, then $G'$ is generated by a suitable commutator. However, for any given positive integer $n$ there is a group $G$ in which $G'$ is cyclic and generated by no set of fewer than $n$ commutators.*

It should be mentioned here that by a result of Honda [35], for any $g \in G'$ every generator of $\langle g \rangle$ is the product of the same number of commutators.

The main result of Rodney's paper is a sufficient condition for every element of $G'$ to be a commutator, where $G$ is a group with finite cyclic commutator subgroup generated by a commutator.

**Theorem 7.2 ([61])** *Let $G'$ be cyclic of finite order with $4 \nmid |G'|$. Suppose $G' = \langle c \rangle$ with $c = [a, b]$. Let $\mu$ and $\nu$ be integers such that $c^a = c^\mu$ and $c^b = c^\nu$. If one of the following four conditions fails to hold for every prime divisor $p$ of $|G'|$, then $G'$ consists entirely of commutators:*

I. $\mu - 1 \equiv 0 \mod p, \quad \nu - 1 \equiv 0 \mod p;$

II. $\mu - 1 \equiv 0 \mod p, \quad \nu - 1 \not\equiv 0 \mod p;$

III. $\mu - 1 \not\equiv 0 \mod p, \quad \nu - 1 \equiv 0 \mod p;$

IV. $\mu - 1 \not\equiv 0 \mod p, \quad \nu - 1 \not\equiv 0 \mod p.$

As one can observe, the four conditions are mutually exclusive and generate a partition on the set of primes dividing $|G'|$. The statement of the theorem then says that $K(G) = G'$ if at least one of the equivalence classes is empty. This leads to the following corollary not stated in [61].

**Corollary 7.3** *If $G' = \langle [a, b] \rangle$, $|G'|$ is finite and at most three primes divide $|G'|$, but $4 \nmid |G'|$, then $K(G) = G'$.*

We observe that the assumption of being generated by a commutator can be dropped in the case $G'$ is cyclic of $p$-power order. Finally, Rodney obtains the following corollary to Theorems 7.1 and 7.2.

**Corollary 7.4 ([61])** *If $G'$ is cyclic and either $G$ is nilpotent or $G'$ is infinite, then $G' = K(G)$.*

In [20], Gordon, Guralnick and Miller determine all integers $n$ such that there is a group $G$ in which the set of commutators has fewer than $n$ elements and that set generates a cyclic subgroup of order $n$. At the same time, they obtain a sufficient condition on $n$ such that for cyclic $G'$ of order $n$ we have the equality $K(G) = G'$. Both results are under the assumption that $G'$ is generated by a commutator.

**Theorem 7.5 ([20])** *Let $p$ and $q_i$ be primes. Suppose that either*

(i) $n = p^\alpha \cdot q_0^{\alpha_0} \cdot q_1^{\alpha_1} \cdots q_h^{\alpha_h}$, *$h \geq p$ and $q_i \equiv 1 \mod p$ for $i = 0, 1, \ldots, p$, or*

(ii) $n = 2^\alpha \cdot q_0^{\alpha_0} \cdot q_1^{\alpha_1} \cdots q_h^{\alpha_h}$, $\alpha \geq 2$, $h \geq 1$.

*Then there exists a group $G$ such that $G' = \langle a \rangle$ is cyclic of order $n$, $a$ is a commutator but not every element of $G'$ is a commutator.*

**Theorem 7.6 ([20])** *Let $p$ and $p_i$ be primes. Suppose that $G' = \langle a \rangle$, where $a$ is a commutator. Assume that $a$ has order $n$, where either*

(i) $n = p_1^{\alpha_1} p_2^{\alpha_2} \cdots p_r^{\alpha_r}$ *with $4 \nmid n$ and $|\mathcal{J}_i| \leq p_i$, where $\mathcal{J}_i = \{j \mid p_j \equiv 1 \mod p_i\}$, or*

(ii) $n = 2^\alpha \cdot p^\beta$, $\alpha \geq 2$, $\beta \geq 0$.

*Then every element of $G'$ is a commutator.*

Unaware of [61], Gordon et al. show in [20] that for cyclic $G'$ either of infinite or $p$-power order, or for $G$ nilpotent, it follows that $K(G) = G'$. That paper concludes with the following interesting result.

**Corollary 7.7 ([20])** *Suppose that $G$ is a commutator subgroup of a group $H$ and that $G'$ is cyclic. Then every element of $G'$ is a commutator.*

In [27] all pairs of integers $(m, n)$ are determined for which there exists a group $G$ with $G'$ cyclic of order $n$ and $\lambda(G) > m$. The results are summarized in the following theorem.

**Theorem 7.8 ([27])** *Let $p$ and $p_i$ be primes.*

(a) *Given the ordered pair $(n, m)$ with $m \geq 2$, there exists a group $G$ with $G'$ cyclic of order $n$ and $\lambda(G) > m$ if and only if*
  (i) $n = p_1^{\alpha_1} \cdots p_\nu^{\alpha_\nu}$, $\nu \geq 2^{2m+1} - 1$, *or*
  (ii) $n = 2^\alpha p_2^{\alpha_2} \cdots p_\nu^{\alpha_\nu}$, $\nu \geq 2^{2m+1} - 3$.

(b) *There exists a group $G$ with $G'$ cyclic of order $n$ and $\lambda(G) > 1$ if and only if*
  (i) $n = p_1^{\alpha_1} \cdots p_\nu^{\alpha_\nu}$, $\nu \geq 7$,
  (ii) $n = 3^\alpha p_2^{\alpha_2} \cdots p_\nu^{\alpha_\nu}$, $p_i \equiv 1 \pmod 3$ *for* $i = 1, 2, 3, 4$,
  (iii) $n = 2p_2^{\alpha_2} \cdots p_\nu^{\alpha_\nu}$, $\nu \geq 3$, *or*
  (iv) $n = 2^\alpha p_2^{\alpha_2} \cdots p_\nu^{\alpha_\nu}$, $\nu \geq 2$, $\alpha \geq 2$.

In this context Guralnick takes up the original theme of Macdonald [44], that is the study of groups with cyclic commutator subgroup in which the generators are noncommutators.

**Theorem 7.9 ([27])** *Let $p$ and $p_i$ be primes. Consider a natural number $n$ satisfying one of the following conditions:*
  (i) $n = p_1^{\alpha_1} \cdots p_r^{\alpha_r}$, $r \geq 2^{2m+1} - 1$;
  (ii) $n = 2^\alpha p_2^{\alpha_2} \cdots p_r^{\alpha_r}$, $r \geq 2^{2m+1} - 3$.
*Then there exists a group $G$ with $G'$ cyclic of order $n$ such that no generator of $G'$ is a product of $m$ commutators. Conversely, let $H = \langle a \rangle$ be cyclic of order $n$. If there exists a group $G$ with $G' = H$ and no generator of $H$ is a product of $m$ commutators, then $n$ satisfies condition* (i) *or* (ii).

According to Theorem 7.9 the minimal order for a cyclic commutator subgroup with none of the generators being a commutator is 2810. The smallest $n$ and the smallest group order for which there exists a group with cyclic commutator subgroup of order $n$, where the set of commutators is not equal to the commutator subgroup, is given in [20], and follows from Theorem 7.8. Similar examples are given in [61], but no claim of minimality is made.

**Example 7.10 ([20])** Let $G = \langle x, y, a \rangle$ with relations $a^{60} = x^8 = y^6 = 1$, $x^{-1}ax = a^{29}$, $y^{-1}ay = a^{19}$, $[x, y] = a$, $x^2 = a^{15}$, $y^2 = a^{40}$. Then $|G| = 240$, $G' = \langle a \rangle$ and $|G'| = 60$. This group has minimal order in both classes (i′) and (ii′) defined in Section 4.

## 8 Higher commutators

The topic of this section is the relationship between the set of $r$-fold simple commutators and the $r$-th term of the lower central series, the subgroup generated by them. This relationship is studied in [18], [38], [10], [29], and [30]. Since each paper extends and generalizes some of the results of the preceding ones, we discuss them in chronological order. The notation in these papers is not uniform. Thus we adopt the following notation. For $x_1, x_2, \dots, x_r \in G$, let $[x_1, x_2] = x_1^{-1}x_2^{-1}x_1x_2$ and recursively define $[x_1, \dots, x_{r-1}, x_r] = [[x_1, \dots, x_{r-1}], x_r]$ as the $r$-fold simple commutator of $x_1, \dots, x_r$. Let $K_r(G) = \{[x_1, \dots, x_r] \mid x_1, \dots, x_r \in G\}$, the set of $r$-fold simple commutators of $G$, and let $\gamma_r(G) = \langle K_r(G) \rangle$, which is the $r$-th term of the lower central series. Observe that $K_2(G) = K(G)$ and $G' = \gamma_2(G)$. In this section we deal with the case when $r > 2$. We also extend the function $\lambda(G)$, introduced in Section 6, to higher terms of the lower central series. We say that $\lambda_r(G)$ is the smallest integer $n$ such that every element in $\gamma_r(G)$ is a product of $n$ elements in $K_r(G)$ and if $n$ is unbounded we define $\lambda_r(G) = \infty$. Note that $\lambda(G) = \lambda_2(G)$.

In [18], Gallagher extends results obtained in [17] for the commutator subgroup to higher terms of the lower central series. As in [17], the proofs involve intricate but elementary character calculations. His main result is the following.

**Theorem 8.1 ([18])** *Let $G$ be a group. If*

$$n > \left(\frac{2^{r-1}}{3}\right)^{1/2} \log(2|\gamma_r(G)| - 2),$$

*then each element of $\gamma_r(G)$ is a product of $n$ commutators.*

In [38], the results of Macdonald [44] on cyclic commutator subgroups are extended to the terms of the lower central series. Specifically, the following analogue of Theorem 7.1 is proved.

**Theorem 8.2 ([38])** *If $\gamma_r(G)$ is cyclic and either $G$ is nilpotent or $\gamma_r(G)$ is infinite, then $\gamma_r(G)$ is generated by a suitable commutator of weight $r$. For any given integer $n$, however, there is a group $G$ in which $\gamma_r(G)$ is cyclic and generated by no set of fewer than $n$ commutators of weight $r$.*

The examples mentioned in the above theorem are the same as given by Macdonald in [44]. The next theorem is a partial extension of Spiegel's result [68] (see Theorem 2.1) to higher terms of the lower central series.

**Theorem 8.3 ([38])** *Let $G$ be a metabelian group. If $\gamma_r(G)$ and the automorphism group induced on $\gamma_r(G)$ are both cyclic, then $K_r(G) = \gamma_r(G)$.*

In [10] Dark and Newell extend some of the results in [44], [61], [42], [24], and [38] on commutator subgroups with a small number of generators to the higher terms of the lower central series. They give examples to show that some results for $\gamma_2(G)$ do not necessarily hold for higher terms of the lower central series. The next theorem extends results of [61] and [38] (see Corollary 7.4 and Theorem 8.2).

**Theorem 8.4 ([10])** *If $\gamma_r(G)$ is cyclic and either $G$ is nilpotent or $\gamma_r(G)$ is infinite, then $\gamma_r(G) = K_r(G)$.*

It should be mentioned here that, as shown in [29], the assumption in the above theorem that $G$ is nilpotent can be replaced by the weaker condition that $\gamma_\infty(G)$ is finite and cyclic of $p$-power order, where

$$\gamma_\infty(G) = \bigcap_{i=2}^{\infty} \gamma_i(G).$$

The next theorem establishes that Rodney's results [62] (see Theorem 2.2) can only be extended in a limited way to higher terms of the lower central series. It is obvious from Theorem 8.4 that for cyclic and central $\gamma_r(G)$ we have $\gamma_r(G) = K_r(G)$ for all $r$. However, as the next theorem shows, this cannot be extended to the case that $\mathrm{d}(\gamma_r(G)) \geq 2$, if $r \geq 3$. (Recall that for a group $G$ we denote the minimal number of generators of $G$ by $\mathrm{d}(G)$.)

**Theorem 8.5 ([10])** *For every integer $r \geq 3$ there is a metabelian group $G$ such that $\gamma_{r+1}(G) = 1$, $\mathrm{d}(\gamma_r(G)) = 2$, and $\gamma_r(G) \neq K_r(G)$.*

If $|\gamma_r(G)|$ is finite and central, then (i) of Theorem 2.2 can be extended to $\mathrm{d}(\gamma_r(G)) = 2$, if $r \geq 3$. But the theorem is not true if $\mathrm{d}(\gamma_r(G)) = 3$, in particular (ii) of Theorem 2.2 does not hold, if $r > 2$.

**Theorem 8.6 ([10])** *If $\gamma_{r+1}(G) = 1$ and $\gamma_r(G)$ is finite with $\mathrm{d}(\gamma_r(G)) = 2$, then $\gamma_r(G) = K_r(G)$. However, for every integer $r \geq 3$ and every prime $p$, there is a metabelian group $G$ such that $\gamma_{r+1}(G) = 1$, $\gamma_r(G)$ is an elementary abelian $p$-group of rank 3, and $\gamma_r(G) \neq K_r(G)$.*

Guralnick in [29] quantifies the results for $r \geq 3$ of [10] in the same way as this is done for $r = 2$ in [20] and [27] for Macdonald's results in [44]. The main result of [29] is the following.

**Theorem 8.7 ([29])** *Suppose $r \geq 3$. There exists a group $G$ with $\gamma_r(G)$ cyclic of order $n$ and $\lambda_r(G) > k$ if and only if $n = p_1^{\alpha_1} \ldots p_m^{\alpha_m}$, where the $p_i$ are distinct primes and $m \geq 2^{k+1} - 1$.*

The conditions for the case $r = 2$, discussed in [27], are much more complicated than those of the above theorem for $r > 2$ (see Theorem 7.8). As shown in [20], a generating commutator for $\gamma_2(G)$ does not imply that $\lambda_2(G) = 1$ (see Theorem 7.5). Similar examples can be constructed for $r > 2$. The following result is of interest in this context.

**Theorem 8.8 ([29])** *For any $r \geq 2$, if $\gamma_r(G) = \langle a \rangle$ and $a \in (K_r(G))^e$, then $\gamma_r(G) = (K_r(G))^{e+1}$.*

Guralnick in [30] extends various results on Sylow subgroups of the commutator subgroup to the Sylow subgroups $S$ of higher terms of the lower central series. The main result is summarized in the following theorem.

**Theorem 8.9 ([30])** *Suppose $\gamma_r(G)$ is finite and $P \in \mathrm{Syl}_p(\gamma_r(G))$ with $P$ abelian of rank at most 2. If any of the following conditions hold then $P \subseteq K_r(G)$:*

(i) $p \geq 5$;

(ii) $P$ *is cyclic;*

(iii) $\exp(P) = p$;

(iv) $P \cap \gamma_\infty(G) \neq 1$;

(v) $P \cap \gamma_{r+1}(G) = 1$;

(vi) $r \leq 2$.

The result for $r = 2$ can be found in [62] and [28] (see Theorems 2.3 and 2.4). The main idea of the proof of Theorem 8.9 is to reduce to the case where $P = \gamma_r(G)$. With this additional hypothesis, the proofs of (iii) and (iv) are given in [29], while the proofs for (ii) and (v) can be found in [10]. The condition (i) is new here. An example is given in [30] showing that the condition $p \geq 5$ cannot be replaced by $p = 2$, and possibly not by $p = 3$. Examples in [10] and [29] show that rank 2 cannot be replaced by rank 3. Guralnick's paper concludes with some results on the more general problem of when $P \subseteq (K_r(G))^k$.

## 9 Ore's conjecture

After proving that every element in the alternating group $A_n$, $n \geq 5$, is a commutator, Ore [57] states the following: "It is possible that a similar theorem holds for any simple group of finite order, but it seems that at present we do not have the necessary methods to investigate the question." Now over fifty years later, the question is still not answered, no counterexample has been found, but for most families of finite simple groups what is now called Ore's Conjecture has been verified. The open cases are finite simple groups of Lie type over small fields. In this section we give a full account of which cases are settled and which are still open.

Although the classification of finite simple groups did not result in a general method for proving his conjecture, as Ore had hoped, it led to a better understanding of finite simple groups, allowing for a piecemeal approach and the potential to know that at some point all cases have been covered. Investigations of finite simple groups led to the following stronger conjecture attributed to John Thompson (see, e.g., [1] and [13]), which states that every finite simple group $G$ contains a conjugacy class $C$ such that $C^2 = G$. Thompson's Conjecture implies Ore's Conjecture but the converse does not hold. To see that Ore's Conjecture is weaker, note that the infinite restricted alternating group $A$ has no class $C$ such that $C^2 = A$. But every element of $A$ has finite support and thus is clearly a commutator [4]. The status of Thompson's Conjecture is the same as that of Ore's Conjecture.

The fact that every element in the alternating group on five or more letters is a commutator seems to be one of the most rediscovered and republished results in this area. As already mentioned in the introduction, G. A. Miller [54] proved this result in 1899. Ito [37] published it simultaneously with Ore in 1951. Yet another proof can be found in [73]. Hsü (Xu) in [75] proved Thompson's Conjecture for the alternating groups.

There are some early results for certain sporadic simple groups, for example, the Ore Conjecture is verified for the Mathieu groups in [73]. As announced in [55], Neubüser et al. verified Thompson's (and consequently Ore's) Conjecture for all sporadic simple groups using computer aided calculations. A year after [55], the Ore Conjecture was verified in [1] for the sporadic groups using classical methods.

This leaves the finite simple groups of Lie type to be discussed. In our account we use the notation found in [21]. This differs slightly from the one used in [13] with regards to norming of the parameter $q$. This difference is only an issue when we have to identify cases for which Ore's Conjecture is open. Fortunately, for those families where the notations differ, there are no open cases and so we do not have to address this issue.

R. C. Thompson in [69], [70], and [71] proves Ore's Conjecture for the entire family $A_n(q)$, $n \geq 1$. Tseng and Hsü in [72] do the same for the Suzuki groups, that is the entire class ${}^2B_2(q)$, $q = 2^{2m+1}$.

There are various partial solutions of the Thompson Conjecture for some families of finite simple groups that have since been overridden by more general results in [13]. In a recent paper Gow derives sufficient conditions in terms of character theory that a simple group of Lie type must satisfy so that the Ore conjecture holds [22]. We should mention here that with the help of computer calculations Karni verified Thompson's Conjecture for all finite simple groups of order less than $10^6$ [1].

In his dissertation Bonten [3] proves the following interesting result, which gives an asymptotic solution to Ore's Conjecture: Let $G(q) = X_n(q)$, ${}^lX_n(q)$ be a series of groups of Lie type. Then there exists a constant $q_0$, depending on $n$ and $l$, such that every element in $G(q)$ is a commutator if $q > q_0$. In [3] only the existence of such numbers $q_0$ is proved, but the methods allowed Bonten to calculate an estimate for $q_0$ in some cases. These were good enough for groups of small Lie rank so that, together with computer calculations for small $q$, Bonten was able to prove Ore's Conjecture for the following families of finite simple groups of Lie type: $G_2(q)$, ${}^2G_2(q)$, ${}^3D_4(q)$, $F_4(q)$ and ${}^2F_4(q)$.

The most far reaching results on Ore's Conjecture to date were obtained by Ellers and Gordeev in [13]. They prove Thompson's Conjecture for all finite simple groups of Lie type over fields with more than 8 elements. In fact, the result is somewhat stronger than this, since for most of the families of these groups, field sizes smaller than 8 suffice as the lower bound for establishing the validity of the conjecture. The details can be found in Table 1, which gives the current status of Ore's Conjecture for finite simple groups of Lie type.

**Acknowledgments** The authors would like to thank E. Ellers, R. Guralnick, K. Johnson, W. P. Kappe, M. Mazur and M. Ward for their helpful insights and for reading earlier versions of this paper and the anonymous referee for his thoughtful comments. The authors are especially grateful to R. Blyth for proof reading the final manuscript.

| Name | Verified cases | References | Open cases |
|---|---|---|---|
| $A_n(q)$, $n \geq 1$ | all cases | [69], [70], [71] | none |
| $B_2(q)$ | all cases | [13], [1] | none |
| $B_n(q)$, $n \geq 3$ | $q \geq 7$ | [13] | $q = 2, 3, 4, 5$ |
| $C_n(q)$, $n > 2$ | $q \geq 4$ | [13] | $q = 2, 3$ |
| $D_{2n}(q)$, $n \geq 2$ | $q \geq 5$ | [13] | $q = 2, 3, 4$ |
| $D_{2n+1}(q)$, $n \geq 2$ | $q \geq 4$ | [13] | $q = 2, 3$ |
| $G_2(q)$ | all cases | [3] | none |
| $F_4(q)$ | all cases | [3] | none |
| $E_6(q)$ | $q \geq 7$ | [13] | $q = 2, 3, 4, 5$ |
| $E_7(q)$ | $q \geq 5$ | [13] | $q = 2, 3, 4$ |
| $E_8(q)$ | $q \geq 7$ | [13] | $q = 2, 3, 4, 5$ |
| ${}^2A_{2l-1}(q)$, $l > 1$ | $q \geq 8$ | [13] | $q = 2, 3, 4, 5, 7$ |
| ${}^2A_{2l}(q)$, $l \geq 1$ | $q \geq 4$ | [13] | $q = 2, 3$ |
| ${}^2B_2(q)$, $q = 2^{2m+1}$ | all cases | [73] | none |
| ${}^2D_n(q)$, $n > 3$ | $q \geq 7$ | [13] | $q = 2, 3, 4, 5$ |
| ${}^3D_4(q)$ | all cases | [3] | none |
| ${}^2G_2(q)$ | all cases | [3] | none |
| ${}^2F_4(q)$ | all cases | [3] | none |
| ${}^2E_6(q)$ | $q \geq 8$ | [13] | $q = 2, 3, 4, 5, 7$ |

Table 1. Status of Ore's Conjecture for finite simple groups of Lie Type

## References


[1] Z. Arad and M. Herzog, editors, *Products of conjugacy classes in groups*, Lecture Notes in Math. **1112**, Springer-Verlag, Berlin, 1985.

[2] Gilbert Baumslag, W. W. Boone and B. H. Neumann, Some unsolvable problems about elements and subgroups of groups, *Math. Scand.* **7** (1959), 191–201.

[3] O. Bonten, *Über Kommutatoren in endlichen einfachen Gruppen*, Aachener Beiträge zur Mathematik **7**, Augustinus Buchhandlung, Aachen, 1993.

[4] J. L. Brenner, Covering theorems for finasigs. X. The group $G = \mathrm{PSL}(n, q)$ had a class $C$ such that $CC = G$, *Ars Combin.* **16** (1983), 57–67.

[5] W. Burnside, On the arithmetical theorem connected with roots of unity and its application to group characteristics, *Proc. London Math. Soc.* **1** (1903), 112–116.

[6] A. Caranti and C. M. Scoppola, Central commutators, *Bull. Austral. Math. Soc.* **30** (1984), no. 1, 67–71.

[7] Robert D. Carmichael, *Introduction to the theory of groups of finite order*, Dover Publications Inc., New York, 1956.

[8] Phyllis Joan Cassidy, Products of commutators are not always commutators: an example, *Amer. Math. Monthly* **86** (1979), no. 9, 772.

[9] Charles W. Curtis, *Pioneers of representation theory: Frobenius, Burnside, Schur, and Brauer*, History of Mathematics **15**, American Mathematical Society, Providence, RI, 1999.

[10] R. S. Dark and M. L. Newell, On conditions for commutators to form a subgroup, *J. London Math. Soc. (2)* **17** (1978), no. 2, 251–262.

[11] Richard Dedekind, Über Gruppen, deren sämtliche Teiler Normalteiler sind, *Math. Ann.* **48** (1897), 548–561.

[12] David S. Dummit and Richard M. Foote, *Abstract Algebra*, Third edition, John Wiley and Sons, Inc., Hoboken, NJ, 2004.

[13] Erich W. Ellers and Nikolai Gordeev, On the conjectures of J. Thompson and O. Ore, *Trans. Amer. Math. Soc.* **350** (1998), no. 9, 3657–3671.

[14] Graham J. Ellis, On five well-known commutator identities, *J. Austral. Math. Soc. Ser. A* **54** (1993), no. 1, 1–19.

[15] William Benjamin Fite, On metabelian groups, *Trans. Amer. Math. Soc.* **3** (1902), no. 3, 331–353.

[16] Ferdinand Georg Frobenius, Über die Primfactoren der Gruppendeterminante, *Sitzungberichte der Königlich Preußischen Akademie der Wissenschaften zu Berlin*, 1896, 1343–1382.

[17] P. X. Gallagher, Group characters and commutators, *Math. Z.* **79** (1962), 122–126.

[18] P. X. Gallagher, The generation of the lower central series, *Canad. J. Math.* **17** (1965), 405–410.

[19] The GAP Group, *GAP – Groups, Algorithms, and Programming, Version 4.4*, 2004, (`http://www.gap-system.org`).

[20] Basil Gordon, Robert M. Guralnick and Michael D. Miller, On cyclic commutator subgroups, *Aequationes Math.* **17** (1978), no. 2–3, 241–248.

[21] Daniel Gorenstein, *Finite groups*, Harper & Row Publishers, New York, 1968.

[22] Rod Gow, Commutators in finite simple groups of Lie type, *Bull. London Math. Soc.* **32** (2000), no. 3, 311–315.

[23] R. Guralnick, *Expressing group elements as products of commutators*, Ph.D. thesis, UCLA, 1977.

[24] Robert M. Guralnick, On groups with decomposable commutator subgroups, *Glasgow Math. J.* **19** (1978), no. 2, 159–162.

[25] Robert M. Guralnick, On a result of Schur, *J. Algebra* **59** (1979), no. 2, 302–310.

[26] Robert M. Guralnick, Expressing group elements as commutators, *Rocky Mountain J. Math.* **10** (1980), no. 3, 651–654.

[27] Robert M. Guralnick, On cyclic commutator subgroups, *Aequationes Math.* **21** (1980), no. 1, 33–38.

[28] Robert M. Guralnick, Commutators and commutator subgroups, *Adv. Math.* **45** (1982), no. 3, 319–330.

[29] Robert M. Guralnick, Generation of the lower central series, *Glasgow Math. J.* **23** (1982), no. 1, 15–20.

[30] Robert M. Guralnick, Generation of the lower central series. II, *Glasgow Math. J.* **25** (1984), no. 2, 193–201.

[31] P. Hall, A contribution to the theory of groups of prime-power order, *Proc. London Math. Soc. (2)* **36** (1934), 29–95.

[32] T. Hawkins, The origins of the theory of group characters, *Arch. Hist. Exact Sci.* **7** (1971), 142–170.

[33] T. Hawkins, New light on the theory of group characters, *Arch. Hist. Exact Sci.* **12** (1974), 217–243.

[34] P. Hegarty and D. MacHale, Products of group commutators, *Irish Math. Soc. Bull.* **34** (1995), 14–21.

[35] Kin'ya Honda, On commutators in finite groups, *Comment. Math. Univ. St. Paul* **2** (1953), 9–12.

[36] I. M. Isaacs, Commutators and the commutator subgroup, *Amer. Math. Monthly* **84** (1977), no. 9, 720–722.

[37] Noboru Ito, A theorem on the alternating group $\mathfrak{A}_n (n \geqq 5)$, *Math. Japonicae* **2** (1951),

59–60.

[38] Luise-Charlotte Kappe, Groups with a cyclic term in the lower central series, *Arch. Math. (Basel)* **30** (1978), no. 6, 561–569.

[39] Luise-Charlotte Kappe and Robert Fitzgerald Morse, On commutators in $p$-groups, *J. Group Theory* **8** (2005), no. 4, 415–429.

[40] F. W. Levi, Groups in which the commutator operation satisfies certain algebraic conditions, *J. Indian Math. Soc. (N.S.)* **6** (1942), 87–97.

[41] F. W. Levi and B. L. van der Waerden, Über eine besondere Klasse von Gruppen, *Abh. Math. Seminar der Universität Hamburg* **9** (1933), 154–158.

[42] Hans Liebeck, A test for commutators, *Glasgow Math. J.* **17** (1976), no. 1, 31–36.

[43] I. D. Macdonald, On a set of normal subgroups, *Proc. Glasgow Math. Assoc.* **5** (1962), 137–146.

[44] I. D. Macdonald, On cyclic commutator subgroups, *J. London Math. Soc.* **38** (1963), 419–422.

[45] I. D. Macdonald, *The theory of groups*, Clarendon Press, Oxford, 1968.

[46] I. D. Macdonald, Commutators, Unpublished manuscript circa 1979.

[47] I. D. Macdonald, Commutators and their products, *Amer. Math. Monthly* **93** (1986), no. 6, 440–444.

[48] I. D. Macdonald and B. H. Neumann, On commutator laws in groups, *J. Austral. Math. Soc. Ser. A*, **45** (1988), no. 1, 95–103.

[49] I. D. Macdonald and B. H. Neumann, On commutator laws in groups. II, in *Combinatorial group theory (College Park, MD, 1988)*, Contemp. Math. **109**, 113–129, Amer. Math. Soc., Providence, RI, 1990.

[50] I. D. Macdonald and B. H. Neumann, On commutator laws in groups. III, *J. Austral. Math. Soc. Ser. A* **58** (1995), no. 1, 126–133.

[51] Desmond MacHale, Minimum counterexamples in group theory, *Math. Mag.* **54** (1981), no. 1, 23–28.

[52] G. A. Miller, The regular substitution groups whose orders is less than 48, *Quart. J. Math.* **28** (1896), 232–284.

[53] G. A. Miller, On the commutator groups, *Bull. Amer. Math. Soc.* **4** (1898), 135–139.

[54] G. A. Miller, On the commutators of a given group, *Bull. Amer. Math. Soc.* **6** (1899), 105-109.

[55] J. Neubüser, H. Pahlings and E. Cleuvers, Each sporadic finasig $G$ has a class $C$ such that $CC = G$, *Abstracts Amer. Math. Soc.* **34** (1984), 6.

[56] Nikolay Nikolov and Dan Segal, Finite index subgroups in profinite groups, *C. R. Math. Acad. Sci. Paris* **337** (2003), no. 5, 303–308.

[57] Oystein Ore, Some remarks on commutators, *Proc. Amer. Math. Soc.* **2** (1951), 307–314.

[58] Samuel Pasiencier and Hsien-chung Wang, Commutators in a semi-simple Lie group, *Proc. Amer. Math. Soc.* **13** (1962), 907–913.

[59] Rimhak Ree, Commutators in semi-simple algebraic groups, *Proc. Amer. Math. Soc.* **15** (1964), 457–460.

[60] A. H. Rhemtulla, A problem of bounded expressibility in free products, *Proc. Cambridge Philos. Soc.* **64** (1968), 573–584.

[61] D. M. Rodney, On cyclic derived subgroups, *J. London Math. Soc. (2)* **8** (1974), 642–646.

[62] D. M. Rodney, Commutators and abelian groups, *J. Austral. Math. Soc. Ser. A* **24** (1977), no. 1, 79–91.

[63] Maxwell Rosenlicht, On a result of Baer, *Proc. Amer. Math. Soc.* **13** (1962), 99–101.

[64] Joseph J. Rotman, *The theory of groups, An introduction*, Allyn and Bacon Inc., Boston, Mass., 1965.

[65] Joseph J. Rotman, *An introduction to the theory of groups*, Fourth edition, Grad. Texts in Math. **148**, Springer-Verlag, New York, 1995.

[66] I. Schur, Über die Darstellung der endlichen Gruppen durch gebrochene lineare Substitutionen, *J. Reine Angew. Math.* **127** (1904), 20–50.

[67] Elizabeth Snyder, Commutators and the commutator subgroup, Master's thesis, Binghamton University, State University of New York, June 2002.

[68] Eugene Spiegel, Calculating commutators in groups, *Math. Mag.* **49** (1976), no. 4, 192–194.

[69] R. C. Thompson, Commutators in the special and general linear groups, *Trans. Amer. Math. Soc.* **101** (1961), 16–33.

[70] R. C. Thompson, Commutators of matrices with coefficients from the field of two elements, *Duke Math. J.* **29** (1962), 367–373.

[71] R. C. Thompson, On matrix commutators, *Port. Math.* **21** (1962), 143–153.

[72] K'en–ch'eng Ts'eng and Ch'eng–hao Hsü, On the commutators of two classes of finite simple groups, *Shuxue Jinzhan* **8** (1965), 202–208.

[73] K'en–ch'eng Ts'eng and Chiung–sheng Li, On the commutators of the simple Mathieu groups, *J. China Univ. Sci. Techn.* **1** (1965), no. 1, 43–48.

[74] H. Weber, *Lehrbuch der Algebra, Vol 2*, Second edition, Braunschweig, 1899.

[75] Cheng–hao Xu, The commutators of the alternating group, *Sci. Sinica* **14** (1965), 339–342.

[76] H. J. Zassenhaus, *Lehrbuch der Gruppentheorie Band 1*, Hamburger Mathematische Einzelsehriften, Leipzig und Berlin, 1937.

# INEQUALITIES FOR THE BAER INVARIANT OF FINITE GROUPS

SAEED KAYVANFAR

Department of Mathematics, Ferdowsi University of Mashhad, P.O.Box 1159-91775, Mashhad, Iran
and
Institute for Studies in Theoretical Physics and Mathematics, P.O.Box 5746-19395, Tehran, Iran
Skayvanf@math.um.ac.ir

## Abstract

In this article we obtain inequalities for the minimal number of generators and the exponent of the Baer invariant of a finite group. An equality for the order of the Baer invariant will also be presented. These extend some main results of M. R. Jones [8] and so that of [7].

*AMS Classification:* 20E10, 20F10
*Keywords:* variety, exponent, Schur multiplicator, Baer invariant.

## 1 Introduction and Motivation

In the series of papers [6, 7], M. R. Jones has obtained some inequalities for the order, the minimal number of generators and the exponent [7, 8] of the Schur multiplicator of a finite nilpotent group. In [6] he could improve the result of J. A. Green [5] on the order of $M(G)$, when $G$ is a finite $p$-group, and in [7] he obtained an improvement of his result in [6] for such groups. Also using an interesting theorem ([7, Theorem 4.4]), he was able to sharpen the upper bound of $|M(G)|$ relative to W. Gaschütz et al. [4]. Another application of [7, Theorem 4.4] obtains the numerical inequality for the minimal number of generators and the exponent of $M(G)$, when $G$ is a finite nilpotent group. These results have been generalized in [8] using the generalization of [7, Theorem 4.4] (see [8, Theorem 2.1]).

On the other hand J. Burns et al. [2] and G. Ellis [3] have improved the bound obtained in [8] for the exponent of $M(G)$ and this bound has been sharpened by S. Kayvanfar et al. [10] for some cases.

In this note we intend to study some main results of M. R. Jones in a different way. More precisely, we want to generalize the key theorem of [8] and so [7] ([8, Theorem 2.1] and [7, Theorem 4.4]) to any Schur–Baer variety of groups and thereby some bounds for the order, exponent and minimal number of generators for the Baer invariant of a finite group will be obtained.

## 2 Preliminaries

Let $F_\infty$ be a free group of countable rank and $V$ be a nonempty subset of $F_\infty$. For an arbitrary group $G$, the subgroup of $G$ generated by all the values in $G$ of words in $V$ is called the *verbal subgroup* of $G$ determined by $V$, and is denoted by $V(G)$, i.e.,

$$V(G) = \langle v(g_1, \ldots, g_r) \mid g_i \in G,\ v \in V,\ 1 \le i \le r,\ r \in \mathbf{N} \rangle.$$

Also the *marginal subgroup* of $G$ is denoted by $V^*(G)$ and defined as follows:

$$V^*(G) = \{a \in G \mid v(g_1, \ldots, g_i a, g_{i+1}, \ldots, g_r) = v(g_1, \ldots, g_i, \ldots, g_r),\quad 1 \le i \le r,\ v \in V,\ g_i \in G,\ r \in \mathbf{N}\}.$$

It is obvious that the verbal subgroup, $V(G)$, is fully invariant and the marginal subgroup, $V^*(G)$, is characteristic in $G$. The class of all groups $G$ such that every word of $V$ has the identity value in $G$ is called the *variety* $\mathcal{V}$ with respect to $V$.

Now let $N$ be a normal subgroup of a given group $G$, and $\mathcal{V}$ be a variety of groups. Define $[NV^*G]$ to be the subgroup of $G$ generated by:

$$\{v(g_1, \ldots, g_{i-1}, g_i n, g_{i+1}, \ldots, g_r)(v(g_1, \ldots, g_{i-1}, g_i, g_{i+1}, \cdots g_r))^{-1} \mid 1 \le i \le r < \infty,\ v \in V,\ g_1, \ldots, g_r \in G,\ n \in N\}.$$

It is easily checked that $[NV^*G]$ is the smallest normal subgroup $H$ of $G$ contained in $N$ such that $N/H \subseteq V^*(G/H)$. In other words, $G/[NV^*G]$ is the largest quotient group of $G$ in which $N$ becomes marginal.

We are now ready to define the Baer invariant of a group.

Let a group $G$ be presented as the quotient of a free group $F$ by a normal subgroup $R$, and $\mathcal{V}$ be a variety of groups. Then the *Baer invariant of* $G$ with respect to the variety $\mathcal{V}$, which is denoted by $\mathcal{V}M(G)$, is defined to be

$$\mathcal{V}M(G) = \frac{R \cap V(F)}{[RV^*F]}.$$

It is known that this invariant $\mathcal{V}M(G)$ is always abelian and up to group isomorphism, independent of the choice of the free presentation of $G$ (for instance, see [11]).

In particular, if successively $\mathcal{V}$ is the variety of abelian groups $\mathcal{A}$ or nilpotent groups $\mathcal{N}_c$, say, of class at most $c$ $(c \ge 1)$, then the Baer invariant of the group $G$ will be $(R \cap [F,F])/[R,F]$, (which is isomorphic to the Schur multiplicator $M(G)$ of $G$) or $(R \cap \gamma_{c+1}(F))/[R, {}_cF]$ (denoted by $\mathcal{N}_c M(G)$), where $[R, {}_cF]$ is $[R, F, \ldots, F]$ ($F$ being repeated $c$ times), respectively. (Note that the $k^{th}$-term of the lower central series of a group $X$ is denoted by $\gamma_k(X)$).

**Definition 2.1** Let $\mathcal{V}$ be an arbitrary variety of groups defined by the set of laws $V$. Then $\mathcal{V}$ is said to be a *Schur–Baer variety* whenever, for any group $G$, if the marginal factor group, $G/V^*(G)$, is finite of order $t$, say, then the verbal subgroup, $V(G)$, is also finite and $|V(G)| \mid t^k$, for some $k \in \mathbf{N}$.

I. Schur showed in [13] that the variety of abelian groups has the Schur–Baer property. R. Baer also in [1] proved that the variety defined by outer commutator words has the same property (for the definition one may refer to [9] or [12]).

The following theorem provides an important property of Schur–Baer varieties.

**Theorem 2.2** *The following conditions on the variety $\mathcal{V}$ are equivalent:*

(i) *$\mathcal{V}$ is a Schur–Baer variety.*

(ii) *For every finite group $G$, the Baer invariant $\mathcal{V}M(G)$ is of order dividing a power of $|G|$.*

**Proof** See [11].

## 3 The Inequalities

The following theorem is one of our main objectives which gives a bound for the exponent of the Baer invariant of a group and also a numerical identity for the order of the Baer invariant of a group $G$ with respect to the verbal subgroup of $G$ relative to another variety.

**Theorem 3.1** *Let $G$ be a finite group and $G = F/R$ a presentation for $G$ as a factor group of the free group $F$. Suppose also that $\mathcal{V}$ and $\mathcal{W}$ are two Schur–Baer varieties of groups defined by the sets of laws $V$ and $W$, respectively, such that $V \supseteq W$. Then*

(i) $|W(G)| \cdot |\mathcal{V}M(G)| = |\mathcal{V}M(G/W(G))| \cdot |[(W(F)R)V^*F]/[RV^*F]|$,

(ii) $\exp(\mathcal{V}M(G)) \leq \exp(\mathcal{V}M(G/W(G))) \cdot \exp([(W(F)R)V^*F]/[RV^*F])$

*in which $\leq$ can be taken to mean "divides".*

**Proof** By the definition of the Baer invariant and the Dedekind modular law, we have;

$$\mathcal{V}M(G/W(G)) = \frac{W(F)R \cap V(F)}{[(W(F)R)V^*F]} = \frac{(R \cap V(F)) \cap W(F)}{[(W(F)R)V^*F]}.$$

Since $G$ is finite and the varieties are Schur–Baer, by Theorem 2.2 we conclude that

$$\left|\frac{R \cap V(F)}{[RV^*F]}\right| = |\mathcal{V}M(G/W(G))| \cdot \left|\frac{[(W(F)R)V^*F]}{[RV^*F]}\right|,$$

but

$$\left|\frac{R \cap V(F)/[RV^*F]}{\mathcal{V}M(G)}\right| = \left|\frac{W(F)}{R \cap W(F)}\right| = |W(F/R)| = |W(G)|,$$

It is thus readily seen that (i) holds. To prove part (ii), one can easily checked that

$$\begin{aligned}\exp(\mathcal{V}M(G)) &\leq \exp\left(\frac{(R \cap V(F))\, W(F)}{[RV^*F]}\right) \\ &\leq \exp\left(\mathcal{V}M\left(\frac{G}{W(G)}\right)\right) \cdot \exp\left(\frac{[(W(F)R)\, V^*F]}{[RV^*F]}\right),\end{aligned}$$

this finishes the proof. □

**Corollary 3.2** *Let $G$ be a finite group and $G = F/R$ be a presentation for $G$. If $\mathcal{V}$ is a Schur–Baer variety in which the set of words $V$ contains the nilpotent word $\gamma_c = [x_1, x_2, \ldots, x_c]$, then*

(i) $|\gamma_c(G)| \cdot |\mathcal{V}M(G)| = |\mathcal{V}M\left(G/\gamma_c(G)\right)| \cdot \left|\dfrac{[(\gamma_c(F)R)V^*F]}{[RV^*F]}\right|$,

(ii) $\exp(\mathcal{V}M(G)) \leq \exp(\mathcal{V}M(G/\gamma_c(G))) \cdot \exp\left(\dfrac{[(\gamma_c(F)R)V^*F]}{[RV^*F]}\right)$.

*In particular for every positive integers $n$ and $c$ which $n < c$, we have*

(i)′ $|\gamma_c(G)| \cdot |\mathcal{N}_nM(G)| = |\mathcal{N}_nM(G/\gamma_c(G))| \cdot \left|\dfrac{[\gamma_c(F)R, {}_nF]}{[R, {}_nF]}\right|$,

(ii)′ $\exp(\mathcal{N}_nM(G)) \leq \exp(\mathcal{N}_nM(G/\gamma_c(G))) \cdot \exp\left(\dfrac{[\gamma_c(F)R, {}_nF]}{[R, {}_nF]}\right)$.

To prove our other numerical inequality we need the following.

**Lemma 3.3** *Let $F$ be a free group and $N$ be a normal subgroup of $F$ which contains $\gamma_{c+1}F$, for some $c$. Then for every positive integer $n$:*

$$[\gamma_c(F)N, {}_nF, [\gamma_c(F)N, {}_nF]] \leq [N, {}_nF].$$

**Proof** Using commutator manipulations, we have

$$\begin{aligned}[\gamma_c(F)N, {}_nF, [\gamma_c(F)N, {}_nF]] &\leq [\gamma_c(F)N, {}_nF, F] \\ &\leq [N[N, F], {}_nF].\end{aligned}$$

Now it is easy to see that the last group is $[N, {}_nF]$, as required. □

Following M. R. Jones [7] we say that a finite group $X$ has *special rank* $r(X)$ if every subgroup of $X$ may be generated by $r(X)$ elements and there is at least one subgroup that can not be generated by fewer than $r(X)$ elements. The minimal number of generators of $X$ is also denoted by $d(X)$. It is clear that $d(X) \leq r(X)$ and we have equality when $X$ is abelian.

We can now apply the above lemma and definition to prove the following result.

**Theorem 3.4** *Let $G$ be a finite group with $G = F/R$ as a free presentation. Then for every positive integers $n$ and $c$ for which $n < c$, we have:*

$$d\left(\mathcal{N}_nM(G)\right) \leq d\big(\mathcal{N}_nM(G/\gamma_c(G))\big) + d\left(\frac{[\gamma_c(F)R, {}_nF]}{[R, {}_nF]}\right).$$

**Proof** By a similar argument to Theorem 3.1, we conclude that the extensions of $[\gamma_c(F)R, {}_nF]/[R, {}_nF]$ and $\mathcal{N}_nM(G)$ by the groups $\mathcal{N}_nM(G/\gamma_c(G)))$ and $\gamma_c(G)$, respectively, is $(\gamma_{n+1}(F) \cap R)\gamma_c(F)/[R, {}_nF]$. Hence we have

$$\begin{aligned}d\left(\mathcal{N}_nM(G)\right) = r\left(\mathcal{N}_nM(G)\right) &\leq r\left(\frac{(\gamma_{n+1}(F) \cap R)\gamma_c(F)}{[R, {}_nF]}\right) \\ &\leq r\left(\frac{[\gamma_c(F)R, {}_nF]}{[R, {}_nF]}\right) + r\left(\mathcal{N}_nM\left(\frac{G}{\gamma_c(G)}\right)\right).\end{aligned}$$

Now thanks to Lemma 3.3 the result follows. □

One can easily see that the result of M. R. Jones [8, Theorem 2.1] (and hence [7, Theorem 4.4]) is an immediate consequence of Corollary 3.2 (i)′ and (ii)′ and also Theorem 3.4 by taking $n = 1$. Therefore Theorems 3.1 and 3.4 are wide generalizations of [8, Theorem 2.1] and so [7, Theorem 4.4] to the variety of groups. Note also that the condition of being nilpotent is not necessary in [8, Theorem 2.1].

## References


[1] R. Baer, Endlichkeitskriterien für Kommutatorgruppen, *Math. Ann.* **124** (1952), 161–177.

[2] J. Burns and G. Ellis, Inequalities for Baer invariants of finite groups, *Canad. Math. Bull.* **41** (1998), no. 4, 385–391.

[3] G. Ellis, On the relation between upper central quotients and lower central series of a group, Max Planck Institute preprint, January 1999.

[4] W. Gaschütz, J. Neubüser and Ti Yen, Über den Multiplikator von $p$-Gruppen, *Math. Z.* **100** (1967), 93–96.

[5] J. A. Green, On the number of authomorphisms of finite groups *Proc. Roy. Soc. London, Ser A* **237** (1956), 574–581.

[6] M. R. Jones, Multiplicators of $p$-groups, *Math. Z.* **127** (1972), 165–166.

[7] M. R. Jones, Some inequalities for the multiplicator of a finite group, *Proc. Amer. Math. Soc.* **39** (1973), 450–456.

[8] M. R. Jones, Some inequalities for the multiplicator of a finite group, II, *Proc. Amer. Math. Soc.* **45** (1974), 167–172.

[9] S. Kayvanfar and M. R. R. Moghaddam, $\mathcal{V}$-perfect groups, *Indag. Mathem. (N.S.)* **8** (1997), no. 4, 537–542.

[10] S. Kayvanfar and M. A. Sanati, A bound for the exponent of the Schur multiplier of some finite $p$-groups, *Bull. Iranian Math. Soc.* **26**, no. 2, (2000), 89–95.

[11] C. R. Leedham-Green and S. McKay, Baer-invariant, isologism, varietal laws and homology, *Acta Math.* **137** (1976), 99–150.

[12] B. Mashayekhy, S. Kayvanfar and M. R. R. Moghaddam, Subgroup theorems for the Baer invariant of groups, *J. Algebra* **206** (1998), 17–32.

[13] I. Schur, Über die darstellung der endlichen gruppen durch gebrochene lineare substitutionen, *J. Reine Angew. Math.* **127** (1904), 20–50.

# AUTOMORPHISMS WITH CENTRALIZERS OF SMALL RANK[1]

EVGENY KHUKHRO* and VICTOR MAZUROV†

*School of Mathematics, Cardiff University, Cardiff, CF24 4AG, U.K.
Email: khukhro@cardiff.ac.uk

†Institute of Mathematics, Novosibirsk, 630090, Russia
Email: mazurov@math.nsc.ru

## Abstract

We obtain restrictions on the structure of a finite group $G$ with a group of automorphisms $A$ in terms of the order of $A$ and the rank of the fixed-point subgroup $C_G(A)$. When $A$ is regular, that is, $C_G(A) = 1$, there are well-known results giving in many cases the solubility of $G$ or bounds for the Fitting height. Some earlier "almost regular" results were deriving the solubility, or bounds for the Fitting height, of a subgroup of index bounded in terms of $|A|$ and $|C_G(A)|$. Now we prove rank analogues of these results: when "almost regular" in the hypothesis is interpreted as a restriction on the rank of $C_G(A)$, it is natural to seek solubility, or nilpotency, or bounds for the Fitting height of "almost" entire group modulo certain bits of bounded rank. The classification is used to prove almost solubility. For soluble groups the Hall–Higman type theorems are combined with the theory of powerful $p$-groups to obtain almost nilpotency, or bounds for the Fitting height of a normal subgroup with quotient of bounded rank. Examples are produced showing that some of our results are in a sense best-possible, while certain results on almost regular automorphism have no valid rank analogues. Several open problems are discussed, especially in the case of nilpotent $G$.

## 1 Introduction

Let $A$ be a group of automorphisms of a finite group $G$, and $C_G(A)$ its fixed point subgroup. Restricting the structure of $G$ depending on the "smallness" of $C_G(A)$ is an important direction in group theory. By Thompson's theorem [34], if $A$ is of prime order $q$ and *regular*, that is $C_G(A) = 1$, then $G$ is nilpotent; moreover, by Higman's theorem [14] then the nilpotency class of $G$ is bounded in terms of $q$ only. It follows from the classification that for any $A$ of coprime order, if $C_G(A) = 1$, then $G$ is soluble. Using the classification Fong [6] proved that if $A$ is of prime order $q$ and $|C_G(A)| = n$, then the quotient $G/S(G)$ by the soluble radical has order bounded in terms of $q$ and $n$ only (or $(q,n)$-*bounded* for short). This extends

[1]This work is supported by the Russian Foundation for Basic Research (grant no. 05-01-00797), the Presidium of the Siberian Div. of Russian Acad. of Sci. (grant no. 86-197), the Programme "Universities of Russia" (grant no. UR.04.01.028), the Programme "Developing the Scientific Potential of Higher School" (grant no. 511), and by the Grant Council of the Russian President and State Support of leading scientific schools (grant no. NSh-2069.2003.1).

the Brauer–Fowler theorem, the case of $|A| = 2$, which lies in the foundations of the classification.

When both $G$ and $A$ are soluble of coprime orders, Thompson [35] obtained bounds for the Fitting height $h(G)$ in terms of that of $C_G(A)$ and the order of $A$. This theorem is based on the techniques of representation theory developed in the Odd paper [5] and earlier in the Hall–Higman paper [11]. The Hall–Higman type theorems, often stated in terms of character theory, are a basis of numerous papers on regular and "almost regular" automorphisms, among which we mention those of Berger [2], Dade [4], Gross [10], Hartley and Isaacs [12], Kurzweil [23], Shult [32], Turull [36] (in some of Turull's other papers $A$ is not necessarily soluble).

In particular, Hartley and Meixner [13] and Pettet [30] independently proved for $A$ of prime order $q$ that the order of the Fitting quotient $G/F(G)$ is bounded in terms of $|C_G(A)|$ and $q$. In the general case where $G$ and $A$ are soluble of coprime orders Turull [36] and Hartley and Isaacs [12] proved that the order of the quotient $G/F_{2l+1}(G)$ by the $(2l+1)$-st Fitting subgroup is bounded in terms of $|A|$ and $|C_G(A)|$, where $l = l(A)$ is the number of (not necessarily distinct) primes whose product is $|A|$. The coprimeness condition is unavoidable, at least for non-nilpotent $A$, as shown by examples of Bell and Hartley [1] where $C_G(A) = 1$ but the Fitting height of $G$ is unbounded.

In the present paper we consider (mostly, finite) groups $G$ with groups of automorphisms $A$ that are almost regular in the sense that $C_G(A)$ has given rank $r$. (Recall that the rank of a finite group is the minimum number $r$ such that all of its subgroups can be generated by $r$ elements). The consequences of the restriction on the rank of $C_G(A)$ are sought as solubility, or nilpotency, or bounds for the Fitting height of certain subgroups or sections, modulo "small" pieces of rank bounded in terms of $r$ and $|A|$.

Using the classification we prove that if $(|A|, |G|) = 1$, then there is a bound for the rank of the quotient $G/S(G)$ by the soluble radical in terms of $|A|$ and the rank of $C_G(A)$ (Theorem 3.1). We also prove a similar result for the orders: if $(|A|, |G|) = 1$, then there is a bound for $|G/S(G)|$ in terms of $|C_G(A)|$ and $|A|$. If the coprimeness condition is dropped, there are examples showing that, first, the rank of $G/S(G)$ cannot be bounded in this way even for $|A| = 2$ and, second, the order of $G/S(G)$ cannot be bounded in terms of $|C_G(A)|$ and $|A|$ even for $A$ of order 4.

When both $G$ and $A$ are soluble, we obtained in [19] in a sense best-possible result for the case of $A$ of prime order $q$: there are characteristic subgroups $G > N > R$ such that $N/R$ is nilpotent, while the ranks of $G/N$ and $R$ are bounded in terms of $q$ and the rank of $C_G(A)$ (Theorems 4.1). Examples show, however, that the rank of $G/F(G)$, and even the rank of $G/F_2(G)$, cannot be bounded in this way. On the other hand, the rank of $G/F_3(G)$ is bounded (Theorem 5.2). These results for $A$ of prime order do not require the condition that the orders of $G$ and $A$ be coprime. The proofs combine the "non-modular" Hall–Higman type theorems and the theory of powerful $p$-groups. Earlier the case of $q = 2$ was settled by Shumyatsky [33].

The inverse limit argument allows us to derive similar results for locally finite locally soluble groups with an element of prime order whose centralizer has finite rank.

In the general case of soluble $A$ and $G$ it is natural to assume that $(|A|, |G|) = 1$, because of the above-mentioned examples by Bell and Hartley [1]. The aforementioned bound for the rank of $G/F_3(G)$ in the case of $A$ of prime order enables us to use a rather straightforward induction, based on Thompson's theorem [35] saying that $F(C_G(\alpha)) \leqslant F_4(G)$ for an automorphism $\alpha$ of prime order coprime to $|G|$. As a result, the rank of $G/F_{4^l-1}(G)$ is bounded in terms of the rank of $C_G(A)$ and the number $l = l(A)$ (Theorem 5.1). It remains an open question whether the index $4^l - 1$ here can be replaced by a linear function of $l$, as in the works of Turull [36] and Hartley and Isaacs [12], where the bounds for the orders were considered.

Thus, the study of the structure of $G$ with $C_G(A)$ of given rank $r$, at least when $(|A|, |G|) = 1$ and $A$ is soluble, is now largely reduced to the case where $G$ is nilpotent. Here there are more open questions than answers, even in the case of $A$ of prime order. Let $A = \langle\varphi\rangle$ be cyclic of prime order $q$. Recall that by Higman's theorem [14] the nilpotency class of a nilpotent group with a regular automorphism of prime order $q$ is $q$-bounded. By Khukhro's theorem [16, 17], if $G$ is nilpotent and $|C_G(\varphi)| = n$, then $G$ has a subgroup of $(q, n)$-bounded index whose nilpotency class is $q$-bounded. It is natural to expect that if $G$ is nilpotent and $C_G(\varphi)$ has rank $r$, then $G$ must have a subgroup of "$(q, r)$-bounded corank" whose nilpotency class is $q$-bounded, maybe a normal subgroup of $q$-bounded nilpotency class with quotient of $(q, r)$-bounded rank. (For nilpotent groups a normal subgroup of bounded rank can be eliminated at the expense of a quotient group, so there is no need for a subgroup of type $R$ as above). So far this conjecture was confirmed only for $q = 2$ by Shumyatsky [33]. Some partial results for arbitrary $q$ with bounds depending also on the derived length of $G$ were obtained by Makarenko [26] and Khukhro [18] (see Theorem 6.1).

If we move to groups of automorphisms of composite order, then even the case of $A = \langle\varphi\rangle$ cyclic and regular, $C_G(\varphi) = 1$, remains an open problem for nilpotent groups (apart from the case of $|A| = 4$ due to Kovács [20]). The conjecture is that the derived length of $G$ is bounded in terms of $|\varphi|$. This is prompted by the well-known theorem of Kreknin [21]: if a Lie ring $L$ has a regular automorphism $\varphi$ of finite order $n$ (such that the fixed-point subring is trivial, $C_L(\varphi) = 0$), then $L$ is soluble of $n$-bounded derived length (actually, $\leqslant 2^n - 2$). But the Lie ring method so far did not allow us to derive consequences for groups. For example, in the coprime situation the automorphism of the associated Lie ring $L(G)$ induced by a regular automorphism of $G$ of order $n$ remains regular, so by Kreknin's theorem $L(G)$ is soluble of $n$-bounded derived length. But the derived length of $L(G)$ may well be smaller than that of $G$ (unlike the nilpotency class, which is the same; this is why Higman's theorem is essentially about Lie rings). Recently, Khukhro and Makarenko [28] even proved that a Lie algebra with an automorphism of finite order $n$ whose fixed-point subalgebra has finite dimension $m$ is almost soluble in the sense that it has a soluble ideal of $n$-bounded derived length and $(m, n)$-bounded codimension. This result, as well as Kreknin's theorem yield analogous results for locally nilpotent torsion-free groups, due to the Mal'cev correspondence (based on the Baker–Campbell–Hausdorff formula), which preserves the derived length.

But, repeat, in general even the regular case of composite order remains on open problem for nilpotent groups.

We recall several well-known facts in § 2. In § 3 we consider the quotient by the soluble radical. In § 4 we discuss the proof of the almost nilpotency result [19] for automorphism of prime order, in particular, the combination of Hall–Higman–type results with the theory of powerful $p$-groups. Bounds for the rank of the quotient by a certain term of the Fitting series are obtained in § 5 in the soluble coprime situation. In § 6 we discuss the results and open problems for the case of nilpotent groups.

## 2 Preliminaries

Throughout the paper we shall denote by $r(G)$ the rank of a group $G$, which is defined as the minimum number $r$ such that all finitely generated subgroups of $G$ (all subgroups, of course, if $G$ is finite) can be generated by $r$ elements. Henceforth we denote the induced automorphism of an invariant subgroup, quotient group, or section by the same letter.

We now recall a few well-known facts.

**Lemma 2.1** *Let $A$ be a group of automorphisms of a finite group $H$ of coprime order $(|H|,|A|) = 1$. If $N$ is a normal $A$-invariant subgroup, then $C_{H/N}(A) = C_H(A)N/N$.* □

There are several other well-known properties of a coprime action that we shall use without special references, like the existence of $A$-invariant Sylow subgroups.

**Lemma 2.2** *If a finite abelian $p$-group $A$ admits an automorphism $\varphi$ of order $p$ whose centralizer has rank $r$, then the rank of $A$ is at most $pr$.*

**Proof** The rank of $A$ is equal to the rank of $\Omega_1(A) = \{a \in A \mid a^p = 1\}$, which can be regarded as a vector space over a field of $p$ elements. Since $0 = \varphi^p - 1 = (\varphi - 1)^p$, all eigenvalues of the linear transformation $\varphi$ are equal to 1. The number of blocks in the Jordan normal form of $\varphi$ is equal to the dimension of the centralizer $C_{\Omega_1(A)}(\varphi)$, while the dimension of each block does not exceed $p$. □

The following lemma appeared independently and simultaneously in the papers of Gorchakov [7], Merzlyakov [29], and as "P. Hall's lemma" in the paper of Roseblade [31].

**Lemma 2.3** *Let $p$ be a prime number. The rank of a $p$-group of automorphisms of a finite $p$-group of rank $r$ is bounded in terms of $r$.* □

(Although in [7], [29], and [31] $p$-groups of automorphisms of finite *abelian* $p$-groups of rank $r$ were considered, the general result can be easily derived from this special case, see, for example, [33], Lemma 4.2.)

**Lemma 2.4** *If a finite group $G$ admits an automorphism $\varphi$ of prime order $p$ with centralizer of rank $r$, then the rank of a Sylow $p$-subgroup of $G$ is bounded in terms of $p$ and $r$.*

**Proof** By including $\langle\varphi\rangle$ into a Sylow $p$-subgroup of the semidirect product $G\langle\varphi\rangle$ we find a $\varphi$-invariant Sylow $p$-subgroup $P$ of $G$. The rank of a maximal abelian normal subgroup $A$ of the semidirect product $P\langle\varphi\rangle$ does not exceed $p(r+1)$ by Lemma 2.2. Since $C_{P\langle\varphi\rangle}(A) = A$, the quotient group $PA/A$ embeds isomorphically into a Sylow $p$-subgroup of the automorphism group of $A$. By Lemma 2.3 the rank of $PA/A$ is bounded in terms of the rank of $A$. □

The following result was obtained by Kovács [20] for soluble groups on the basis of Hall–Higman type theorems and extended by Longobardi and Maj [24] using the classification.

**Lemma 2.5** (a) *If $d$ is the maximum of the ranks of the Sylow subgroups of a finite soluble group, then the rank of this group is at most $d+1$.*

(b) *If $d$ is the maximum of the ranks of the Sylow subgroups of a finite group, then the rank of this group is at most $2d$.* □

The next lemma must also be well known.

**Lemma 2.6** *A finite $p'$-group $Q$ of linear transformations of a vector space of dimension $n$ over a field of characteristic $p$ has $n$-bounded rank.*

**Proof** By Lemma 2.5 we can assume that $Q$ is a $q$-group. Choose a maximal abelian normal subgroup $A$ in $Q$; by Lemma 2.3 it suffices to bound the rank of $A$. After extension of the field, $A$ is diagonalizable, which gives the result since finite multiplicative subgroups of fields are cyclic. □

A finite $p$-group $P$ is said to be powerful if $[P,P] \leqslant P^p$ for $p \neq 2$, or $[P,P] \leqslant P^4$ for $p=2$. (Here $A^n = \langle a^n \mid a \in A\rangle$.) The following result can be found in [25].

**Lemma 2.7** *If a powerful $p$-group $P$ is generated by $d$ elements, then the rank of $P$ is at most $d$.*

Two more well-known facts.

**Lemma 2.8** *A finite soluble group of rank $r$ has $r$-bounded Fitting height.*

**Proof** For every prime $p$ the quotient $G/O_{p',p}(G)$ acts faithfully on the Frattini quotient of $O_{p',p}(G)/O_{p'}(G)$ and therefore is a linear group of dimension $\leqslant r$. By the Lie–Zassenhaus–Kolchin–Mal'cev theorem $G/O_{p',p}(G)$ has $r$-bounded derived length. Hence the same is true for $G/F(G) = G/\bigcap_p O_{p',p}(G)$. □

**Lemma 2.9** *If a finite soluble group has rank $r$ and exponent $n$, then its order is at most $n^{f(r)}$ for some $r$-bounded number $f(r)$.*

## 3 Bounding $G/S(G)$

**Theorem 3.1** *Let $A$ be a group of automorphisms of a finite group $G$ of coprime order: $(|G|, |A|) = 1$.*

(a) *If $|C_G(A)| = r$, then $|G/S(G)|$ is $(|A|, r)$-bounded.*

(b) *If $r(C_G(A)) = r$, then $r(G/S(G))$ is $(|A|, r)$-bounded.*

The proof of this theorem relies on the classification of the finite simple groups. The functions of $|A|$ and $r$ bounding the rank and the order in Theorem 3.1 can be given explicit upper estimates, although we do not write them down.

The following examples show that if the condition that $|G|$ and $|A|$ be coprime is dropped, then one cannot bound either the order or the rank of $G/S(G)$.

**Example 3.2** Let $G = PSL(2, q)$ where $q$ is a power of an odd prime and $q > 3$. Let $A$ be a group of inner automorphisms of $G$ induced by a non-cyclic subgroup of $G$ of order 4. Then $|C_G(A)| = 4$, but $S(G) = 1$ and the order of $G$ can be arbitrarily large.

**Example 3.3** For an odd prime $p$ and an arbitrary positive integer $n$, let $G = SL(2, q)$ where $q = 2^{n(p-1)}$. Then $S(G) = 1$, $|G| = q(q-1)(q+1)$, and $q - 1$ is divisible by $p$ by Fermat's Little Theorem. Let $A$ be a group of order $p$ in the group of inner automorphisms of $G$. The centralizer $C_G(A)$ is a cyclic subgroup of order $q - 1$, while the rank of a Sylow 2-subgroup of $G$ is $n(p-1)$. Thus, the rank of $G/S(G)$ can be arbitrarily large even in the case when the rank of $C_G(A)$ is equal to 1.

In the proof of Theorem 3.1 we shall need the following consequences of the classification of the finite simple groups [9]; see, for example, [3] for the description of automorphisms of the known finite simple groups. If the center of a group $G$ is trivial, we identify $G$ with the group of all inner automorphisms of $G$.

**Lemma 3.4** *Let $G$ be a finite simple non-abelian group.*

(a) $r(\mathrm{Aut}(G)/G) \leq 3$.

(b) *If $A \leq \mathrm{Aut}(G)$ and $(|G|, |A|) = 1$, then either $A = 1$, or $G = L(q)$ is a group of Lie type over a field of order $q$ and $C_G(A) \simeq L(q_0)$ where $q = q_0^{|A|}$.* □

**Lemma 3.5** (a) *Let $G = P_1 \times \cdots \times P_s$ where every $P_i$ is a finite simple non-abelian group. Then the automorphism group of $G$ permutes the factors $P_i$.*

(b) *Let $P_1, \ldots, P_s$ be pairwise non-isomorphic finite simple groups, let $Q_i = R_{i1} \times \cdots \times R_{it_i}$ where $R_{ij} \simeq P_i$ for $j = 1, \ldots, t_i$ and $i = 1, \ldots, s$, and let $G = Q_1 \times \cdots \times Q_s$. Then $\mathrm{Aut}(G) \simeq \mathrm{Aut}(Q_1) \times \cdots \times \mathrm{Aut}(Q_s)$ and $\mathrm{Aut}(Q_i)$ is isomorphic to an extension of $\mathrm{Aut}(R_{i1}) \times \cdots \times \mathrm{Aut}(R_{it_i})$ by the symmetric group of degree $t_i$.*

(c) *Let $G = P_1 \times \cdots \times P_t$ where the $P_i$ are finite simple non-abelian groups. Suppose that a group $A$ acts on $G$ so that it permutes the $P_i$ transitively, and let $A_1 = \{a \in A \mid P_1^a = P_1\}$ be the stabilizer of $P_1$ in $A$. Then $C = C_{P_1}(A_1)$ is isomorphic to a subgroup of $C_G(A)$.*

**Proof** Parts (a) and (b) are well-known. To prove (c), for every $i = 1, \ldots, t$ choose an element $a_i \in A$ such that $P_1^{a_i} = P_i$. Then $\{a_1, \ldots, a_t\}$ is a set of representatives of all the distinct right cosets of $A_1$ in $A$. For $c \in C$, denote $c_i = c^{a_i}$ and $\tau(c) = c_1 \cdots c_t$. Since $c_i \in P_i$ and $c_i \neq 1$ for $c \neq 1$, the map $\tau$ is an embedding of $C$ into $G$. On the other hand, if $a \in A$, then for every $i = 1, \ldots, t$ there exist $x_i \in A_1$ and $j_i \in \{1, \ldots, t\}$ such that $a_i a = x_i a_{j_i}$, and $\{j_1, \ldots, j_t\} = \{1, \ldots, t\}$. Hence $c_i^a = c^{a_i a} = c^{x_i a_{j_i}} = c^{a_{j_i}} = c_{j_i}$. This implies that $(\tau(c))^a = c_{j_1} \cdots c_{j_t} = \tau(c)$, so $\tau(C) \subseteq C_G(A)$. □

**Proof of Theorem 3.1** We begin with the simple groups.

**Lemma 3.6** *Theorem* 3.1 *holds if $G$ is a simple group. In addition, the order of $C_G(A)$ is even if $G$ is simple.*

**Proof** We can assume that $A$ acts on $G$ faithfully and $A \neq 1$. By Lemma 3.4(b), $G = L(q)$ is a group of Lie type and $C_G(A) = L(q_0)$ where $q = q_0^{|A|}$.

In part (a) of the theorem, both $q_0$ and the Lie rank of $G$, which is equal to the Lie rank of $L(q_0)$, are $r$-bounded. It follows that $q$ is $(r, |A|)$-bounded and hence $|G|$ is $(r, |A|)$-bounded.

Consider part (b). Since $L(q_0)$ contains an elementary abelian subgroup of order $q_0$, we have $q_0 = s^n$, where $s$ is a prime and $n$ is an $(|A|, r)$-bounded number. Since a Sylow 2-subgroup of $L(q_0)$ has $(|A|, r)$-bounded rank, the Lie rank of $G = L(q)$, which is equal to the Lie rank of $L(q_0)$, is $(|A|, r)$-bounded. Then $G$ can be embedded into $PSL_m(q)$, where $m$ is $(|A|, r)$-bounded. By Lemma 2.6 every Sylow $t$-subgroup of $G$ for $t \neq s$ has $(|A|, r)$-bounded rank. The rank of a Sylow $s$-subgroup is also $(|A|, r)$-bounded, because $q = q_0^{|A|} = s^{n|A|}$ and both the Lie rank and $n$ are $(|A|, r)$-bounded. Thus, the rank of $G$ is $(|A|, r)$-bounded by Lemma 2.5(b). □

**Lemma 3.7** *Theorem* 3.1 *holds if $G = P_1 \times \cdots \times P_t$ is a direct product of non-abelian simple groups $P_i$ and $A$ acts transitively on $\{P_1, \ldots, P_t\}$. Moreover, the order of $C_G(A)$ is even in this case.*

**Proof** Let $A_1 = \{a \in A \mid P_1^a = P_1\}$ be the stabilizer of $P_1$ in $A$. Then by Lemma 3.5(c), $C = C_{P_1}(A_1)$ is isomorphic to a subgroup of $C_G(A)$, which fact implies that $|C|$ (respectively, $r(C)$) is $(r, |A|)$-bounded. By Lemma 3.6, then $|P_1|$ (respectively, $r(P_1)$) is $(r, |A_1|)$-bounded, and $C$ is of even order. In particular, $C_G(A)$ is of even order. Since $t = |A : A_1|$ and all the $P_i$ are isomorphic, we obtain that $|G|$ (respectively, $r(G)$) is $(r, |A|)$-bounded. □

**Completion of the proof of Theorem 3.1** By Lemma 2.1 we can obviously assume that $S(G) = 1$. Let $M$ be the product of all minimal normal subgroups of $G$ (the socle). Then $M = P_1 \times \cdots \times P_t$ is a direct product of non-abelian simple groups. By Lemma 3.5(a), $A$ acts on the set $\{P_1, \ldots, P_t\}$ and $M = Q_1 \times \cdots \times Q_k$ where each $Q_i$ satisfies the conditions of Lemma 3.7 instead of $G$. By Lemma 3.7 we obtain that $|Q_i|$ (respectively, $r(Q_i)$) is $(r, |A|)$-bounded and the order of $C_{Q_i}(A)$ is even. This implies that $k$ and $t$ are $(r, |A|)$-bounded and hence $|M|$ (respectively, $r(M)$) is $(r, |A|)$-bounded.

Since $C_G(M) = 1$, the group $G$ is a subgroup of $\mathrm{Aut}(M)$. In particular, this proves part (a) of the theorem. In part (b), $t$ and $r(M)$ are $(r, |A|)$-bounded. By Lemma 3.5, the group $G$ embeds into

$$(\mathrm{Aut}(P_1) \times \cdots \times \mathrm{Aut}(P_t)) \cdot H,$$

where $H$ is a subgroup of the symmetric group of degree $t$. Since the ranks of the $\mathrm{Aut}(P_i)/P_i$ are at most 3 by Lemma 3.4(a), the rank of $G$ is $(r, |A|)$-bounded. □

## 4 Almost nilpotency in the case of $A$ of prime order

In this section $A = \langle \varphi \rangle$ is cyclic of prime order $q$. In view of Theorem 3.1 and Example 3.3 we assume the group $G$ to be soluble. However, here we shall not assume the order of $\varphi$ to be coprime to $|G|$.

### 4.1 Statement of the results

The following theorem means that $G$ is almost nilpotent, up to two bits of small rank. This result is proved in [19]; here we discuss the ideas and the technique of the proof, crucially, how the Hall–Higman type theorems work in combination with the theory of powerful $p$-groups.

**Theorem 4.1 ([19], Theorem 2)** *Suppose that a finite soluble group $G$ has an automorphism $\varphi$ of prime order $q$ with fixed point subgroup $C_G(\varphi)$ of rank $r$. Then $G$ has characteristic subgroups $G \geqslant N \geqslant R$ such that $N/R$ is nilpotent and both $G/N$ and $R$ have $(q, r)$-bounded rank.*

Examples show that, unlike in Hartley–Meixner–Pettet theorem [13, 30] for orders, one cannot get rid of the subgroup $R$ in a result of this kind; nor is it possible, of course, to get rid of the factor $G/N$.

**Example 4.2** Let $n$ be any positive integer, and let $q_1, q_2, \ldots, q_n$ be distinct primes greater than 7. For each $i = 1, \ldots, n$ the automorphism group of the elementary abelian group $E_i$ of order $q_i^6$ contains the Frobenius group $F_i$ with kernel $A_i$ of order 7 and with complement generated by the element $b_i$ of order 6 such that $A_i$ acts regularly on $E_i$. Let $B_i$ denote the semidirect product $E_iF_i$, and let $P = B_1 \times \cdots \times B_n$. Let $G$ be the subgroup of $P$ generated by all the $E_i$, $A_i$ and the element $b^2$, where $b = b_1 \cdots b_n$. The element $b^3$ induces an automorphism $\varphi$ of $G$ of order $p = 2$. Then $C_G(\varphi)$ is a group of rank 3, while the rank of the Fitting quotient of $G$, as well as the rank of any normal subgroup with nilpotent quotient, is equal to $n$.

This example (similar examples can be constructed for any prime $p$) shows that in Theorem 4.1 the series $G \geqslant N \geqslant R$ cannot in general be reduced to two subgroups. Moreover, in the next section we construct examples showing that even the rank of $G/F_2(G)$ cannot be bounded, but prove that the rank of $G/F_3(G)$ is bounded in terms of $r$ and $p$.

Theorem 4.1 can naturally be stated in terms of a finite soluble group containing an element of prime order $q$ whose centralizer has rank $r$. Moreover, the standard inverse limit argument extends this result to periodic locally finite groups. Here a group has finite rank $\leqslant r$ if every finitely generated subgroup can be generated by $r$ elements.

**Corollary 4.3 ([19], Corollary 1)** *Suppose that a periodic locally soluble group $G$ has an element $g$ of prime order $q$ whose centralizer $C_G(g)$ has finite rank $r$. Then $G$ has normal subgroups $G \geqslant N \geqslant R$ such that $N/R$ is locally nilpotent and both $G/N$ and $R$ have finite $(q,r)$-bounded rank.*

The functions of $q$ and $r$ bounding the ranks in Theorem 4.1 and Corollary 4.3 can be given explicit upper estimates, although we do not write them down. Combining Theorem 4.1 with the results in § 5 we shall see that in addition the subgroup $R$ in Theorem 4.1 can be taken to be of Fitting height 2.

In certain possible applications it may be important to have control over the primes involved in the orders of elements of the groups $G/N$ and $R$. The proofs of the main results in fact yield the corresponding corollary.

**Corollary 4.4 ([19], Corollary 2)** *Suppose that under the hypotheses of Theorem 4.1 (respectively, Corollary 4.3) the number of primes dividing the order of $C_G(\varphi)$ (the orders of elements in $C_G(\varphi)$) is (finite and equal to) $m$. Then one can ensure that the number of primes dividing the orders (of elements) of the groups $N/R$ and $R$ in the conclusion is $(q,r,m)$-bounded.*

Since the result is in some aspects a little stronger for the case where $G$ is a soluble $q'$-group of odd order (when there are no exceptional situations), we state it separately.

**Theorem 4.5 ([19], Theorem 4)** *Suppose that a finite soluble $q'$-group $G$ of odd order admits an automorphism $\varphi$ of prime order $q$ such that $C_G(\varphi)$ has rank $r$. Then $[G,\varphi]$ has a characteristic subgroup $R$ of $(q,r)$-bounded rank such that $[G,\varphi]/R$ is nilpotent.*

A certain improvement here is that in the normal series $G \geqslant [G,\varphi] \geqslant R$ with nilpotent quotient $[G,\varphi]/R$ the rank of $R$ is $(q,r)$-bounded and the rank of $G/[G,\varphi]$ is at most $r$ by Lemma 2.1.

The proof of Theorem 4.1 combines the "non-modular" Hall–Higman type theorems with the theory of powerful $p$-groups. The construction of a powerful subgroup is similar to a part of the proof in Shumyatsky's paper [33]. Known facts are used to bound the Fitting height in terms of $q$ and $r$, which enables induction. This induction is greatly facilitated by the following useful general result, which magically turns normal subgroups into characteristic ones.

**Theorem 4.6 ([19], Theorem 3)** *If a finite soluble group $G$ has normal subgroups $G \geqslant N \geqslant R$ such that $N/R$ is nilpotent and both $G/N$ and $R$ have rank $\leqslant r$, then $G$ has characteristic subgroups $G \geqslant N_1 \geqslant R_1$ such that $N_1/R_1$ is nilpotent and both $G/N_1$ and $R_1$ have $r$-bounded rank.*

## 4.2 Hall–Higman type theorems

The following is a key lemma in the "non-modular" theorems of Hall–Higman type, which appeared in many papers in various versions; this one is taken from [13].

**Lemma 4.7** *Let $T\langle\varphi\rangle$ be a semidirect product of a normal $t$-subgroup $T$ and a cyclic group $\langle\varphi\rangle$ of order $q$, where $t$ and $q$ are distinct primes. Suppose that $T = [T,\varphi] \neq 1$, and the quotient group $T/Z(T)$ is abelian of exponent $t$. Suppose that the group $T\langle\varphi\rangle$ acts faithfully and irreducibly on a vector space $V$ over an algebraically closed field whose characteristic is neither $t$ nor $q$. Then either*

(a) $C_V(\varphi) \neq 0$ *or*

(b) *the group $T$ is extraspecial, $[Z(T),\varphi] = 1$, the order $|T|$ is bounded in terms of $q$, $t = 2$, and $q = t^m + 1$ for some positive integer $m$.* □

Part (b) of this lemma is usually referred to as the exceptional case. The following lemma is a consequence of the non-exceptional part (a).

**Lemma 4.8** *Suppose that $q$ and $p$ are distinct primes. Let $H\langle\varphi\rangle$ be a semidirect product of a normal $\{2, q, p\}'$-subgroup $H$ and a cyclic group $\langle\varphi\rangle$ of order $q$. Suppose that $H = [H,\varphi] \neq 1$ and the group $H\langle\varphi\rangle$ acts faithfully on a vector space $V$ over the field $\mathbb{F}_p$ of $p$ elements. Then $C_V(\varphi) \neq 0$.*

**Proof** First we extend the ground field $\mathbb{F}_p$ to an algebraically closed field $K$; the property of $C_V(\varphi)$ to be trivial or not is not affected. A minimal $\varphi$-invariant subgroup $T$ of $H$ on which $\varphi$ acts on-trivially is known to be a $t$-subgroup for some prime $t$ satisfying the hypothesis of Lemma 4.7 (see, for example, [8], Theorem 5.3.7). Since $2 \neq t$, we can apply Lemma 4.7(a) to any faithful irreducible $KT\langle\varphi\rangle$-submodule of $V \otimes_{\mathbb{F}_p} K$. □

It is convenient to use the following result of Hartley and Isaacs [12], Theorem B, which can be regarded as a generalization of some Hall–Higman type theorems (although in our particular situation one could instead use repeatedly Lemma 4.7).

**Theorem 4.9 (Hartley–Isaacs [12])** *Let $A$ be an arbitrary finite group. Then there exists a number $\delta = \delta(A)$ depending only on $A$ with the following property. Let $A$ act on $G$, where $G$ is a finite soluble group such that $(|G|,|A|) = 1$, and let $k$ be any field of characteristic not dividing $|A|$. Let $V$ be any irreducible $kAG$-module and let $S$ be any $kA$-module that appears as a component of the restriction $V_A$. Then $\dim_k V \leqslant \delta m_S$, where $m_S$ is the multiplicity of $S$ in $V_A$.*

There are explicit estimates of the number $\delta$, which can be used for explicit estimates of the functions involved in the main results of the present paper. We shall be applying this result with $A = \langle\varphi\rangle$ to obtain upper estimates for the dimension of $V$ in terms of the dimension of $C_V(\varphi)$ (the multiplicity of the trivial $k\langle\varphi\rangle$-module) when $C_V(\varphi) \neq 0$. In the non-exceptional case Lemma 4.8 can be used to make sure that $C_V(\varphi) \neq 0$. However, special efforts are required to deal with the exceptional cases.

### 4.3 Constructing a powerful $p$-subgroup

We shall consider a model situation, the simplest case in the proof of Theorem 4.1, which, however, gives an idea how the non-modular Hall–Higman type theorems are combined with the theory of powerful $p$-groups. However, we have to state the following proposition in [19] in full, since we shall need it in the next section. We denote by $\gamma_i(X)$ the terms of the lower central series starting from $\gamma_1(X) = X$, and by $\gamma_\infty(X) = \bigcap_i \gamma_i(X)$ the nilpotent coradical. The terms of the Fitting series are denoted by $F_i(X)$ starting from $F_1(X) = F(X)$.

**Proposition 4.10 ([19], Proposition 2)** *Suppose that a finite soluble $q'$-group $G$ admits an automorphism $\varphi$ of prime order $q$ such that $C_G(\varphi)$ has rank $r$. Suppose that $G = [G,\varphi]$ and $L$ is a normal $\varphi$-invariant subgroup such that the order of $F_2(L)/F_1(L)$ is odd. Then the subgroup $\gamma_\infty(F_2(L))$ has $(q,r)$-bounded rank.*

**Proof in a special case** We consider the following simplest case: $G = L = PT$, where $P = F(G)$ is a normal $\varphi$-invariant $p$-subgroup, $T = [T,\varphi]$ is a $\varphi$-invariant $t$-subgroup, where $p$, $q$, and $t$ are distinct primes and $q \neq 2 \neq t$. It is easy to see that it suffices to bound the rank of $P_1 = [P,T]$, which is a normal $\varphi$-invariant subgroup of $G$. Note that $P_1 = [P_1,T]$ since the action is coprime. The first aim is to estimate the number of generators of $P_1$; this is achieved by applying the Hall–Higman type results.

**Lemma 4.11** *The minimum number of generators of $P_1$ is $(q,r)$-bounded.*

**Proof** We regard the Frattini quotient $V = P_1/\Phi(P_1)$ as an $\mathbb{F}_pT\langle\varphi\rangle$-module. Let

$$V = V_1 \supset V_2 \supset V_3 \supset \cdots \supset 0$$

be a composition series, so that every factor $U_i = V_i/V_{i+1}$ is an irreducible $\mathbb{F}_pT\langle\varphi\rangle$-module. By Maschke's theorem we have $[U_i,T] = U_i$ for each $i$. In particular, $T$ acts non-trivially on $U_i$. Since $[T,\varphi] = T$, by Lemma 4.8 we have $C_{U_i}(\varphi) \neq 0$. Hence by the Hartley–Isaacs theorem we can conclude that $\dim_{\mathbb{F}_p} U_i \leqslant \delta \dim_{\mathbb{F}_p} C_{U_i}(\varphi)$, where $\delta$ is a $q$-bounded number. (In the non-exceptional situation under consideration one can actually put $\delta(\langle\varphi\rangle) = q$.) Since $\dim_{\mathbb{F}_p} C_V(\varphi) = \sum_i \dim_{\mathbb{F}_p} C_{U_i}(\varphi)$, as a result we have

$$\dim_{\mathbb{F}_p} V = \sum_i \dim_{\mathbb{F}_p} U_i \leqslant \delta \dim_{\mathbb{F}_p} C_V(\varphi) \leqslant \delta r,$$

so that the number of generators of $P_1$ is $(q,r)$-bounded. □

The next goal is to show that $P_1$ has a powerful $p$-subgroup of bounded rank and "corank". The construction of a powerful subgroup is similar to a part of Shumyatsky's proof in [33]. Let $M$ be a normal $\varphi$-invariant subgroup of $G$ contained in $P_1$. Let the bar denote the images in the quotient $\overline{P}_1 = P_1/M^p$ (or $P_1/M^4$ if $p = 2$). Since $\overline{M} = M/M^p$ (or $M/M^4$) has exponent $p$ (or 4), the *order* of the centralizer of $\varphi$ in this group is at most $p^f$ (or $2^f$) for some $r$-bounded number $f = f(r)$ by Lemma 2.9. We denote by $\zeta_i(X)$ the terms of the upper central series starting from the centre $\zeta_1(X) = Z(X)$.

**Lemma 4.12** $\overline{M} \leqslant \zeta_{2f+1}(\overline{P}_1)$.

**Proof** We consider the series

$$M_1 = \overline{M} > M_2 > M_3 > \cdots > 1,$$

where

$$M_i = [\overline{M}, \underbrace{\overline{P}_1, \ldots, \overline{P}_1}_{i-1}].$$

All the $M_i$ are normal $\varphi$-invariant subgroups of $G$. Let $V_i = M_i/M_{i+1}$ be the factors of this series. Since this is a central series of $\overline{P}_1$, all the $V_i$ are elementary abelian $p$-groups and can be regarded as $\mathbb{F}_p T\langle\varphi\rangle$-modules.

Whenever $[V_i, T] \neq 0$ we have $C_{V_i}(\varphi) \neq 0$ by Lemma 4.8. Since $|C_{\overline{M}}(\varphi)| \leqslant p^f$, there can be at most $f$ factors $V_i$ with $[V_i, T] \neq 0$. Therefore for some $k \leqslant 2f + 1$ we must have both $[V_k, T] = 0$ and $[V_{k+1}, T] = 0$. In other words, we have

$$[[T, M_k], P_1] \leqslant [M_{k+1}, P_1] = M_{k+2}$$

and

$$[[M_k, P_1], T] = [M_{k+1}, T] \leqslant M_{k+2}.$$

Hence by the Three Subgroup Lemma we also have

$$[[P_1, T], M_k] = [P_1, M_k] = M_{k+1} \leqslant M_{k+2}.$$

Then, of course, $M_{k+1} = 1$, since $\overline{P}_1$ is nilpotent. This means precisely that $\overline{M} \leqslant \zeta_k(\overline{P}_1) \leqslant \zeta_{2f+1}(\overline{P}_1)$. □

**Completion of the proof of Proposition 4.10** We now put $M = \gamma_{2f+1}(P_1)$. Then

$$[\overline{M}, \overline{M}] \leqslant [\gamma_{2f+1}(\overline{P}_1), \zeta_{2f+1}(\overline{P}_1)] = 1,$$

that is, $[M, M] \leqslant M^p$ (or $[M, M] \leqslant M^4$). Thus, $M = \gamma_{2f+1}(P_1)$ is a powerful $p$-subgroup of $P_1$.

The quotient $P_1/\gamma_{2f+1}(P_1)^p$ is then nilpotent of class $4f+1$. Since $P_1$ is generated by a $(q, r)$-bounded number of elements and $P_1/\gamma_{2f+1}(P_1)^p$ is nilpotent of $(q, r)$-bounded class, the rank of $P_1/\gamma_{2f+1}(P_1)^p$ is $(q, r)$-bounded. In particular, the rank of $\gamma_{2f+1}(P_1)/\gamma_{2f+1}(P_1)^p$ is $(q, r)$-bounded too, which coincides with the rank of the powerful $p$-subgroup $\gamma_{2f+1}(P_1)$ by Lemma 2.7. As a result, the rank of $P_1$ is $(q, r)$-bounded, as required. □

## 4.4 Exceptional cases

It is the exceptional situation of Lemma 4.7(b) that obstructs extending the arguments in the proof of Proposition 4.10 and Theorem 4.5 to the general situation. Thus, in particular, we can assume that $q \neq 2$.

Simplifying the situation we just say that there is a certain reduction to the case where $G = O_{2',2}([G, \varphi])$ and $O_{2'}([G, \varphi])$ is nilpotent. Then $G$ is a semidirect product of a normal nilpotent $2'$-subgroup and a 2-group. The remaining arguments are designed to get rid of exceptional situations. Let $W$ denote a $\varphi$-invariant Sylow 2-subgroup of $H$. The idea is to "push up" the exceptionally bad pieces of $W$; it turns out that these only form a quotient of $(q, r)$-bounded rank. The remaining "good" part of $H$ then can be dealt with in the way similar to the proof of Proposition 4.10.

Let $\mathfrak{V}$ be the set of all $G\langle\varphi\rangle$-invariant sections $V$ of $O_{2'}(G)$ that are elementary abelian $p$-groups (for various primes $p$) satisfying the following two conditions:

(i) $V$ is a composition factor of $G\langle\varphi\rangle$ (that is, an irreducible $\mathbb{F}_p G\langle\varphi\rangle$-module);

(ii) $C_V(\varphi) = 1$.

(The first condition may not be essential, but it simplifies the following definition.) We set

$$K = \bigcap_{V \in \mathfrak{V}} C_G(V).$$

The Hall–Higman type Lemma 4.7 is used to show that $G/K$ has $(q, r)$-bounded rank. The arguments similar to those in the proof of the non-exceptional Proposition 4.10 can be applied to $K$. Then $\gamma_\infty(K)$ is a characteristic subgroup of $K$ of $(q, r)$-bounded rank with nilpotent quotient, which completes the proof. The definition of $K$ ensures that $C(\varphi) \neq 1$ in the sections that appear in the proof of Proposition 4.10; then Hartley–Isaacs theorem can be applied in similar fashion and the proof proceeds virtually exactly as for Proposition 4.10.

The technical Theorem 4.6 (turning normal subgroups into characteristic) greatly facilitates induction on the Fitting height. A "weak" bound for the Fitting height, that is, in terms of both $q$ and $r$, is easily obtained from known results, for example, from Lemma 2.8 and Thompson's theorem [35], see Theorem 5.5 below.

## 4.5 Completion of proofs

**Completion of proof of Theorem 4.1** In the proof we first deal with coprime case: we obtain characteristic subgroups in $O_{q'}(G)$ with the required properties; these subgroups are also characteristic in $G$. It remains to show that $G/O_{q'}(G)$ has $(q, r)$-bounded rank. Indeed, the rank of a Sylow $q$-subgroup of $G$ is $(q, r)$-bounded by Lemma 2.4. Set $Q = O_{q',q}(G)/O_{q'}(G)$. The group $G/O_{q',q}(G)$ acts faithfully by conjugation on $Q/\Phi(Q)$. The latter group is abelian of exponent $q$ and of $(q, r)$-bounded rank; hence the order of $Q/\Phi(Q)$ is $(q, r)$-bounded. As a result, the order of $G/O_{q',q}(G)$ is also $(q, r)$-bounded. Hence the rank of $G/O_{q'}(G)$ is also $(q, r)$-bounded. □

**Proof of Corollary 4.3** We apply the well-known inverse limit argument. Let $\Sigma$ be the family of all finite subgroups of $G$ containing $g$; this is a directed poset by inclusion. For each $H \in \Sigma$ let $S_H$ be the set of all pairs $(N, R)$ of normal subgroups $H > N > R$ such that $N/R$ is nilpotent and $H/N$ and $R$ have ranks $\leqslant f(q,r)$, where $f(q,r)$ is the function given by Theorem 4.1. Each $S_H$ is non-empty by Theorem 4.1 applied to $H$ and its inner automorphism induced by $g$. Clearly, each $S_H$ is finite. For any two $H_1, H_2 \in \Sigma$ such that $H_1 \geqslant H_2$ there is a map $\varphi_{H_1,H_2} : S_{H_1} \to S_{H_2}$ given by taking the intersections with $H_2$. This inverse system of finite sets has non-empty inverse limit (see, for example, Theorem 1.K.1 in [15]). The unions of the corresponding subgroups of types $N$ and $R$ over any element of the inverse limit give the required subgroups of $G$. □

## 5 Fitting height of a subgroup of bounded corank

In this section $A \leqslant \operatorname{Aut} G$ is a soluble group of automorphisms of a finite group $G$ with $C_G(A)$ of rank $r$. If the orders of $A$ and $G$ are not assumed to be coprime and $A$ is not nilpotent, there are examples constructed by Bell and Hartley [1] in which $C_G(A) = 1$ and the Fitting height of $G$ is unbounded. Therefore we assume that $(|A|, |G|) = 1$. Recall that $l = l(A)$ is the number of (not necessarily distinct) prime factors whose product is $|A|$. Our aim is to prove that for some function $f(l)$ the quotient $G/F_{f(l)}(G)$ by the $f(l)$-th Fitting subgroup has rank bounded in terms of $r$ and $|A|$. By Theorem 3.1 we can assume that $G$ is also soluble.

**Theorem 5.1** *Let $A$ be a soluble group of automorphisms of a finite group $G$ of coprime order, $(|A|, |G|) = 1$. Then*

(a) *the rank of $G/F_{4^l-1}(G)$ and*

(b) *the order of $G/F_{5\cdot(4^l-1)/3}(G)$*

*are bounded in terms of $|A|$ and the rank of $C_G(A)$.*

This result is a rank analogue, albeit with a worse function of $l$, of theorems of Turull [36] and Hartley–Isaacs [12], where under the condition that $|C_G(A)| = n$ it was proved that the order of $G/F_{2l+1}(G)$ is bounded in terms of $|A|$ and $n$. In the case of $l(A) = 1$ (when $A$ has prime order) the result holds also without the coprimeness condition and is stated as Theorem 5.2 below. We derive Theorem 5.1 from Theorem 5.2 by a rather straightforward induction on $l(A)$ based on the classical theorem of Thompson [35]. It is quite possible that the later techniques of Turull and Hartley–Isaacs could be used to significantly improve the function of $l$ in the index of the Fitting subgroup involved, say, to a linear function of $l$. But the rank result for $A$ of prime order is in a sense best possible.

**Theorem 5.2** *If a finite soluble group $G$ admits an automorphism $\varphi$ of prime order $q$ such that $C_G(\varphi)$ has rank $r$, then*

(a) *for each prime $p$ the quotient $G/O_{p',p}(G)$ has $(q,r)$-bounded rank;*

(b) *$G/F_3(G)$ has $(q,r)$-bounded rank;*

(c) *if in addition $q \nmid |G|$, then $G/F_4(G)$ has $(q,r)$-bounded order.*

Before proving the theorems, we give an example showing that Theorem 5.2(b) is best possible in the sense that the rank of $G/F_2(G)$ need not be bounded.

**Example 5.3** Let $p_i, q_i$, $i = 1, 2, \ldots, m$, be distinct primes all greater than 3 such that $q_i \mid p_i - 1$ for each $i$. Let $M_i$ be the non-abelian metacyclic semidirect product of a normal cyclic subgroup of order $p_i$ and a cyclic group of order $q_i$. Let $D_i$, $i = 1, 2, \ldots, m$, be copies of the dihedral group of order 6, so that $D_i = \langle a_i, b_i \mid a_i^3 = b_i^2 = 1;\ a_i^{b_i} = a_i^{-1} \rangle$. Let $S_i = M_i \wr D_i$ be the wreath product of $M_i$ and $D_i$ (with $D_i$ as active group), so that $S_i = B_i D_i$, where $B_i$ is the direct product of 6 copies of $M_i$. Let $G$ be the direct product $\prod_{i=1}^{m} B_i \langle a_i \rangle$ and let $\varphi$ be the automorphism of order 2 of $G$ induced by conjugation by $b_1 b_2 \cdots b_m$ in the direct product $\prod_{i=1}^{m} B_i D_i$. Then $C_G(\varphi)$ has rank at most 13 by Lemma 2.5, while $G/F_2(G)$ has rank $m$.

**Proof of Theorem 5.2** The rank of the Sylow $q$-subgroup of $G$ is $(q,r)$-bounded by Lemma 2.4. Set $Q = O_{q',q}(G)/O_{q'}(G)$. Then the order of the group $Q/\Phi(Q)$ is $(q,r)$-bounded. The group $G/O_{q',q}(G)$ acts faithfully by conjugation on $Q/\Phi(Q)$. Hence the order of $G/O_{q',q}(G)$ is $(q,r)$-bounded and therefore the rank of $G/O_{q'}(G)$ is $(q,r)$-bounded. Therefore it is sufficient to prove parts (a) and (b) of the theorem for $O_{q'}(G)$. In other words, we may assume from the outset that the order of $G$ is coprime to $q$.

(a) In Proposition 1 in [18] it was proved that the rank of the Hall $p'$-subgroup of $G/O_{p',p}(G)$ is $(q,r)$-bounded (using the Hall–Higman type Lemma 4.7). By Lemma 2.5 it remains to bound the rank of a Sylow $p$-subgroup $P$ of $G/O_{p',p}(G)$. The quotient $G/O_{p',p}(G)$ acts faithfully on the Frattini quotient of $O_{p',p}(G)/O_{p'}(G)$, which we denote by $V$. We may regard $V$ as an $\mathbb{F}_p(G/O_{p',p}(G))\langle\varphi\rangle$-module. Let $F = F(G/O_{p',p}(G))$; then $F$ is a $p'$-group and $P$ acts faithfully on $F$.

Let $F_2$ and $F_{2'}$ be the Sylow 2-subgroup and the Hall $2'$-subgroup of $F$. By Proposition 1 in [18] the rank of $F_2$ is $(q,r)$-bounded (if $p = 2$, then $F_2 = 1$). Hence the rank of $P/C_P(F_2)$ is $(q,r)$-bounded too by Lemma 2.6 applied to the action of $P/C_P(F_2)$ on the Frattini quotient of $F_2$. It remains to bound the rank of $P_1 = C_P(F_2)$.

Because actions are coprime, the group $P_1$ acts faithfully on the nilpotent $p'$-group $F_{2'}$, which in turn acts faithfully on $[V, F_{2'}]$. We set $H = [V, F_{2'}]F_{2'}P_1$. In order to apply Proposition 4.10 we need to pass to $[H,\varphi]$. Let

$$P_2 = [P_1, \varphi], \qquad W = [V, F_{2'}] \cap [H, \varphi], \qquad L = [V, F_{2'}]F_{2'} \cap [H, \varphi],$$

and let $T$ be a $P_2\langle\varphi\rangle$-invariant subgroup of $F_{2'}$ such that $L = WT$; thus, $[H,\varphi] = WTP_2$. Then $P_2$ acts faithfully on $T$, since the faithful action of $P_2$ on $F_{2'}$ is coprime and $[P_2, F_{2'}] \leqslant F_{2'} \cap [H,\varphi] = T$. Furthermore, since $T$ acts faithfully on $V$ and $[V,T] \leqslant W$, the group $T$ acts faithfully on $[W,T]$. Since $P_2$ acts faithfully on $T$, it follows that $P_2$ acts faithfully on $[W,T]$. Clearly, we only need to bound the rank of $P_2$.

By Proposition 4.10 the rank of $[W,T]$ is $(q,r)$-bounded. Hence the rank of $P_2$ is also $(q,r)$-bounded by Lemma 2.3; hence so is the rank of $P_1$ and therefore the rank of $P$.

(b) By Theorem 4.1 we may assume that $G$ has a normal subgroup $R$ of $(q,r)$-bounded rank such that $G/R$ is nilpotent.

**Lemma 5.4** *The exponent of $R/F_2(R)$ is $(q,r)$-bounded.*

**Proof** Indeed, for each prime $t$ the group $R/O_{t',t}(R)$ acts faithfully on the Frattini quotient of $O_{t',t}(R)/O_{t'}(R)$, which is an elementary abelian $t$-group of bounded rank. Thus, $R/O_{t',t}(R)$ is a soluble linear group of bounded dimension $d \leqslant d(q,r)$. After extending the ground field to an algebraically closed one we can apply the Lie–Zassenhaus–Kolchin–Mal'cev theorem: the group $R/O_{t',t}(R)$ has a normal triangularizable subgroup of index $\leqslant f(d)$. Since $O_t(R/O_{t',t}(R)) = 1$, while the field has characteristic $t$, the unitriangular part is trivial. Hence $R/O_{t',t}(R)$ has an abelian subgroup of index $\leqslant f(d)$; in particular, $R/O_{t',t}(R)$ is an extension of an abelian normal subgroup by a group of bounded exponent dividing $f(d)!$. By Remak's theorem the same is true for

$$R/\Big(\bigcap_t O_{t',t}(R)\Big) = R/F(R).$$

Hence the exponent of $R/F_2(R)$ is $(q,r)$-bounded. □

We continue the proof of Theorem 5.2(b). Consider $H = G/F_2(R)$; since $F_2(R) \leqslant F_2(G)$, it suffices to show that $H/F(H)$ has $(q,r)$-bounded rank. By Lemma 5.4 there are only $(q,r)$-boundedly many primes for that divide $|R/F_2(R)|$; we denote these primes $p_i$, $i \in I$, where $|I| \leqslant f(q,r)$. Then for any other, sufficiently large, prime $s$ we have $O_{s',s}(H) = H$. Hence $F(H) = \bigcap_p O_{p',p}(H) = \bigcap_{i\in I} O_{p_i',p_i}(H)$.

By part (a) for each prime $p$ the quotient $H/O_{p',p}(H)$ has $(q,r)$-bounded rank. By Remak's theorem the rank of

$$H/F(H) = H/\Big(\bigcap_{i\in I} O_{p_i',p_i}(H)\Big)$$

is $(q,r)$-bounded, since there are only $(q,r)$-boundedly many subgroups involved in the intersection.

(c) For a given prime number $p$ consider the factor-group $\overline{G} = G/O_{p',p}$, which has $(q,r)$-bounded rank by part (a). For every prime $t$ the group $\overline{G}/O_{t',t}(\overline{G})$ embeds isomorphically into the group of linear transformations of the space of $(q,r)$-bounded dimension, the Frattini factor-group of the group $O_{t',t}(\overline{G})/O_{t'}(\overline{G})$. Being soluble, by the Lie–Zassenhaus–Kolchin–Mal'cev Theorem the group $\overline{G}/O_{t',t}(\overline{G})$ has a triangularizable normal subgroup of $(q,r)$-bounded index. Since $\overline{G}/O_{t',t}(\overline{G})$ has no normal $t$-subgroups, the unitriangular part is trivial, so in fact $\overline{G}/O_{t',t}(\overline{G})$ has an abelian normal subgroup of $(q,r)$-bounded index. Consequently, by Remak's theorem the quotient

$$\overline{G}/F(\overline{G}) = \overline{G}/\Big(\bigcap_t O_{t',t}(\overline{G})\Big)$$

has an abelian normal subgroup whose quotient has $(q,r)$-bounded *exponent*. Thus, $G/O_{p',p}(G)$ is nilpotent-by-abelian-by-($(q,r)$-bounded exponent). By Remak's theorem then the factor-group

$$G/F(G) = G/\Big(\bigcap_p O_{p',p}(G)\Big)$$

also has this property. Thus, $G/F_3(G)$ has $(q,r)$-bounded exponent. Since in the situation considered we have $q \nmid |G|$, the centralizer of $\varphi$ in a quotient group is the image of $C_G(f)$ by Lemma 2.1. In the factor of $(q,r)$-bounded exponent the centralizer of $\varphi$, being of rank $r$, is now finite of $(q,r)$-bounded order by Lemma 2.9. By the Hartley–Meixner–Pettet theorem [13, 30] this quotient has a normal subgroup of $(q,r)$-bounded index which is nilpotent, and moreover, by Khukhro [16, 17], of $q$-bounded class. Thus, in fact, $G$ has a normal series

$$1 \leqslant G_1 \leqslant G_2 \leqslant G_3 \leqslant G_4 \leqslant G, \tag{2}$$

in which the factors $G_1$ and $G_2/G_1$ are nilpotent, the factor $G_3/G_2$ is abelian, the factor $G_4/G_3$ is nilpotent of $q$-bounded class, and the factor $G/G_4$ has $(q,r)$-bounded order. □

We recall Thompson's theorem, which will enable induction in the proof of Theorem 5.1.

**Theorem 5.5 (Thompson [35])** *Let $\alpha$ be an automorphism of prime order $q$ of a finite soluble group $G$ such that $q \nmid |G|$. Then $F(C_G(\alpha)) \leqslant F_4(G)$.*

**Corollary 5.6** *Let $B$ be a soluble group of automorphisms of a finite soluble group $G$ such that $(|G|,|B|) = 1$. Then for $l = l(B)$ we have $F_k(C_G(B)) \leqslant F_{k4^l}(G)$ for every positive integer $k$.*

**Proof** We use induction on $l(B)$. For $l(B) = 1$ we have $F(C_G(B)) \leqslant F_4(G)$ by Theorem 5.5; then the obvious induction on $k$ gives $F_k(C_G(B)) \leqslant F_{k4^l}(G)$.

For $l(B) > 1$ let $B_1$ be a normal subgroup of prime index $p$ in $B$. Then $C = C_G(B_1)$ admits the quotient $\langle\alpha\rangle = B/B_1$ of prime order $p$ as an operator group with $C_C(\alpha) = C_G(B)$. By Theorem 5.5 we have $F(C_G(B)) = F(C_C(\alpha)) \leqslant F_4(C) = F_4(C_G(B_1))$. By the induction hypothesis $F_4(C_G(B_1)) \leqslant F_{4\cdot 4^{l-1}}(G) = F_{4^l}(G)$. Thus $F(C_G(B)) \leqslant F_{4^l}(G)$. Now the obvious induction on $k$ yields the required inclusion $F_k(C_G(B)) \leqslant F_{k4^l}(G)$. □

**Proof of Theorem 5.1** (a) Induction on $l = l(A)$. For $l = 1$ the result follows from Theorem 5.2. For $l > 1$ let $A_1$ be a normal subgroup of prime index $p$ in $A$. Then $C = C_G(A_1)$ admits the quotient $\langle\varphi\rangle = A/A_1$ of prime order $q$ as an operator group with $C_C(\varphi) = C_G(A)$. Hence the rank of $C/F_3(C)$ is bounded in terms of $r = r(C_G(A))$ and $q$ by Theorem 5.2(b). By Corollary 5.6, $F_3(C) \leqslant F_{3\cdot 4^{l-1}}(G)$; thus the rank of the image of $C_G(A_1)$ in $G/F_{3\cdot 4^{l-1}}(G)$ is bounded in terms of $r$ and $q$. By the induction hypothesis applied to this quotient and its operator group $A_1$ we obtain that the rank of $G/F_{4^{l-1}-1+3\cdot 4^{l-1}}(G) = G/F_{4^l-1}$ is bounded in terms of $r$ and $q$.

(b) The proof of part (b) is quite similar, only part (c) of Theorem 5.2 should be used instead of part (b). □

## 6 Nilpotent groups

Due to Theorems 3.1 and 4.1 further studies of finite groups $G$ having an automorphism $\varphi$ of prime order $q$ with $C_G(\varphi)$ of rank $r$ are now largely reduced to the nilpotent quotient $N/R$. Because of Theorem 5.1 the study of finite groups $G$ admitting an arbitrary soluble group of automorphisms $A$ of coprime order with $C_G(A)$ of rank $r$ is also largely reduced to the case where $G$ is nilpotent.

First let $A = \langle\varphi\rangle$ be cyclic of prime order $q$. Recall that by Higman's theorem [14] the nilpotency class of a nilpotent group with a regular automorphism of prime order $q$ is $q$-bounded. By Khukhro's theorem [16, 17], if $|C_G(\varphi)| = n$, then $G$ has a subgroup of $(q, n)$-bounded index whose nilpotency class is $q$-bounded. It is natural to expect that if $G$ is nilpotent and $C_G(\varphi)$ has rank $r$, then $G$ must have a subgroup of "$(q, r)$-bounded corank" whose nilpotency class is $q$-bounded, best of all, a normal subgroup of $q$-bounded nilpotency class with quotient of $(q, r)$-bounded rank. (For nilpotent groups a normal subgroup of bounded rank can be eliminated at the expense of a quotient group by Lemma 2.3, so there is no need for a subgroup of type $R$ as in Theorem 4.1). So far this conjecture was confirmed only for $q = 2$ by Shumyatsky [33].

Here we discuss this open problem, explain why the previously used methods fail so far, and mention some partial results where a new method of group rings was used to obtain bounds depending on the derived length of the group.

The natural tool in the study of nilpotent groups is a Lie ring method. It normally consists of three stages:

1. First, a hypothesis on a group is translated into a hypothesis on a Lie ring constructed from the group in some way.
2. Then a theorem on Lie rings is proved (or used).
3. Finally, a result about the group must be recovered from the Lie ring information obtained.

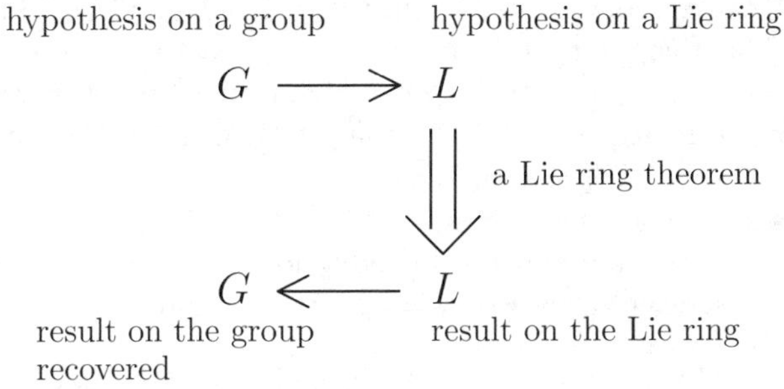


For example, in the solution of the Restricted Burnside Problem for groups of prime exponent $p$ the hypothesis on $G$ is the identity $x^p = 1$ and finite number of generators equal to $d$. The Lie ring $L$ is the associated Lie ring $L(G)$ constructed on the direct sum $\bigoplus \gamma_i(G)/\gamma_{i+1}(G)$ of the lower central factors with Lie products induced by commutators in $G$. The consequence of the hypothesis on $G$ is the $(p-1)$-Engel condition in characteristic $p$ (the Zassenhaus–Magnus–Sanov theorem)

and the same number of generators. The Lie ring theorem due to Kostrikin is that the nilpotency class of $L$ is $(p,d)$-bounded. The recovery of the information for $G$ is immediate, since the nilpotency class of $G$ is the same as that of $L$. For groups of exponent $p^k$ the scheme is similar, with both the passage to Lie rings and the Lie ring result due to Zel'manov.

In Higman's theorem on regular automorphism $\varphi$ of prime order $q$ the Lie ring $L = L(G)$ is again the associated Lie ring. Since the automorphism must be of coprime order, it induces a regular automorphism of the Lie ring: $C_L(\varphi) = 0$. It is here that we especially clearly see the advantage of passing to Lie rings as more linear objects: we can now extend the ground ring by a $q$-th root of unity $\omega$ and decompose the resulting Lie ring $\tilde{L}$ into the direct sum of the eigenspaces $L_i = \{x \in \tilde{L} \mid \varphi(x) = \omega^i x\}$. These eigenspaces behave as a $(\mathbb{Z}/q\mathbb{Z})$-grading: $[L_i, L_j] \subseteq L_{i+j \,(\mathrm{mod}\, q)}$. The Higman–Kreknin–Kostrikin theorem on Lie rings actually gives the nilpotency of $(\mathbb{Z}/q\mathbb{Z})$-graded Lie rings with $L_0 = 0$. (In the papers of Kreknin [21] and Kreknin and Kostrikin [22] a new shorter proof of Higman's theorem was found with an explicit estimate for the nilpotency class.)

In Khukhro's theorem [17] on almost regular automorphism $\varphi$ of prime order $q$, where $G$ is nilpotent and $|C_G(\varphi)| = n$, one can assume by Khukhro's result for $q$-groups [16] that $\varphi$ and $G$ have coprime orders. Then for $L = L(G)$ the induced automorphism is also almost regular: $|C_L(\varphi)| = n$. Khukhro then proves the corresponding theorem for Lie rings with such an automorphism. But here the third stage, obtaining the group consequences of the information on $L$, is far from being straightforward. Even though the Lie ring result produces a desired nilpotent subring of $(q,n)$-bounded additive index and $q$-bounded nilpotency class[2], there is no good correspondence between subgroups of $G$ and subrings of $L$. In fact, this recovery is quite difficult and occupies a good half of the paper [17].

The Lie ring theorem of Khukhro in [17] can be modified in such a way that it gives a rank analogue for Lie rings (actually, the paper [17] already contained a similar result for Lie algebras with $C_L(\varphi)$ of finite dimension $m$). In this sense the second part of the Lie ring method seems to be already in place. But, alas, here it is actually the first stage that fails: even if the prime order $q$ of the automorphism $\varphi$ is coprime to $|G|$, so that in the associated Lie ring the centralizer $C_L(\varphi)$ is the direct sum of the images of $C_G(\varphi)$ in the $\gamma_i(G)/\gamma_{i+1}(G)$, the rank of $C_L(\varphi)$ can obviously become much larger than that of $C_G(\varphi)$, being "scattered" over these direct summands. In fact, even the simplest case of $C_G(\varphi)$ cyclic remains a mystery.

As we already mentioned in the Introduction, for groups of automorphisms of composite order $n$ even the case of $A = \langle\varphi\rangle$ cyclic and regular, $C_G(\varphi) = 1$, remains

[2] Recently Makarenko [27] improved this Lie ring result proving that there is also a nilpotent *ideal* of $(q,n)$-bounded additive index and of $q$-bounded nilpotency class; for groups, though, every subgroup of index $k$ contains a normal subgroup of index $\leqslant k!$, so the existence of an ideal was not necessary in this case. On the other hand, one can also apply the same Lie ring results to Lie algebras in the Mal'cev correspondence (based on the Baker–Campbell–Hausdorff formula) with torsion-free locally nilpotent groups admitting an automorphism of prime order $q$ with "finite-dimensional" fixed-point subgroup (or, perhaps, to Lie groups in a good correspondence with a Lie algebra); then Makarenko's improvement yields also an improvement in the group-theoretic application.

an open problem for nilpotent groups (apart from the case of $|\varphi| = 4$ due to Kovács [20]); recall that the conjecture is that the derived length of $G$ is bounded in terms of $n$. Here it is the third part of the Lie ring method that does not work for the associated Lie ring. In the coprime situation the automorphism of the associated Lie ring $L(G)$ induced by $\varphi$ remains regular by Lemma 2.1. By Kreknin's theorem [21] a Lie ring with a regular automorphism of finite order $n$ is soluble of $n$-bounded derived length (actually, $\leqslant 2^n - 2$). But the derived length of $L(G)$ may well be smaller than that of $G$. Recently, Khukhro and Makarenko [28] even proved that a Lie algebra with an automorphism of finite order $n$ whose fixed-point subalgebra has finite dimension $m$ is almost soluble in the sense that it has a soluble ideal of $n$-bounded derived length and $(m, n)$-bounded codimension. This result, as well as Kreknin's theorem yield analogous results for locally nilpotent torsion-free groups, due to the Mal'cev correspondence (based on the Baker–Campbell–Hausdorff formula), which preserves the derived length. But, repeat, in general even the regular case of composite order remains on open problem for nilpotent groups.

Some progress has only been made under additional restriction for the derived length of the group. In [18] Khukhro proved the following.

**Theorem 6.1 ([18], Theorem 2)** *Let $G$ be a finite nilpotent group admitting an automorphism $\varphi$ of prime order $q$ with centralizer of rank $r$. Let $d$ be the derived length of $G$. Then the group $G$ contains a subgroup of $q$-bounded nilpotency class that is connected with $G$ by a subnormal series of $(q, r, d)$-bounded length with factors of $(q, r, d)$-bounded rank.*

It is not clear if the subgroup can be made to be normal with quotient of $(q, r, d)$-bounded rank: applying Theorem 4.6, which turns normal subgroups into characteristic ones, may upset the bound for the nilpotency class, which we prefer to keep independent of $d$ or $r$. A normal subgroup was obtained by Makarenko [26] in the case where the group $G$ is metabelian: then the $(q+1)$-th term of the lower central series of the subgroup $[G, \varphi]$ has $(q, r)$-bounded rank, which also implies the existence of a normal subgroup of $(q, r)$-bounded rank and of nilpotency class $q+1$.

The proof of the group-theoretic and Lie ring result in [17] is based on the so-called method of graded centralizers in graded Lie rings. In the proof of Theorem 6.1 analogues of graded centralizers are constructed in the associative group ring acting on an abelian group.

## References

[1] S. D. Bell and B. Hartley, A note on fixed-point-free actions of finite groups, *Quart. J. Math. Oxford Ser. (2)* **41** (1990), no. 162, 127–130.

[2] T. R. Berger, Nilpotent fixed point free automorphism groups of solvable groups, *Math. Z.* **131** (1973), 305–312.

[3] J. H. Conway, R. T. Curtis, S. P. Norton, R. A. Parker, and R. A. Wilson, *Atlas of finite groups*, Clarendon Press, Oxford, 1985.

[4] E. C. Dade, Carter subgroups and Fitting heights of finite solvable groups, *Illinois J. Math.* **13** (1969), 449–514.

[5] W. Feit and J. G. Thompson, Solvability of groups of odd order, *Pacific J. Math.* **13** (1963), 775–1029.
[6] P. Fong, On orders of finite groups and centralizers of $p$-elements, *Osaka J. Math.* **13** (1976), 483–489.
[7] Yu. M. Gorchakov, On existence of abelian subgroups of infinite ranks in locally soluble groups, *Dokl. Akad. Nauk SSSR* **146** (1964), 17–22 (Russian); English transl., *Math. USSR Doklady* **5** (1964), 591–594.
[8] D. Gorenstein, *Finite Groups*, Chelsea, New York, 1968.
[9] D. Gorenstein, *Finite simple groups. An introduction to their classification*, Plenum Publishing Corp., New York, USA, 1982).
[10] F. Gross, Solvable groups admitting a fixed-point-free automorphism of prime power order, *Proc. Amer. Math. Soc.* **17** (1966), 1440–1446.
[11] P. Hall and G. Higman, On the $p$-length of $p$-soluble groups and reduction theorems for Burnside's problem, *Proc. London Math. Soc. (3)* **6** (1956), 1–42.
[12] B. Hartley and I. M. Isaacs, On characters and fixed points of coprime operator groups, *J. Algebra* **131** (1990), 342–358.
[13] B. Hartley and T. Meixner, Finite soluble groups containing an element of prime order whose centralizer is small, *Arch. Math. (Basel)* **36** (1981), 211–213.
[14] G. Higman, Groups and rings which have automorphisms without non-trivial fixed elements, *J. London Math. Soc. (2)* **32** (1957), 321–334.
[15] O. H. Kegel and B. A. F. Wehrfritz, *Locally finite grous*, North-Holland, Amsterdam, American Elsevier, New York, 1973.
[16] E. I. Khukhro, Finite $p$-groups admitting an automorphism of order $p$ with a small number of fixed points, *Mat. Zametki* **38** (1985), 652–657 (Russian); English transl., *Math. Notes.* **38** (1986), 867–870.
[17] E. I. Khukhro, Groups and Lie rings admitting an almost regular automorphism of prime order, *Mat. Sbornik* **181** (1990), 1197–1219; English transl., *Math. USSR Sbornik* **71** (1992), 51–63.
[18] E. I. Khukhro, Finite solvable and nilpotent groups with a restriction on the rank of the centralizer of an automorphism of prime order, *Sibirsk. Mat. Zh.* **41** (2000), 451–469; English transl. in *Siberian Math. J.* **41** (2000), 373–388.
[19] E. I. Khukhro and V. D. Mazurov, Finite groups with an automorphism of prime order whose centralizer has small rank, *submitted to J. Algebra*, 2005.
[20] L. G. Kovács, On finite soluble groups, *Math. Z.* **103** (1968), 37–39.
[21] V. A. Kreknin, The solubility of Lie algebras with regular automorphisms of finite period, *Dokl. Akad. Nauk SSSR* **150** (1963), 467–469 (Russian); English transl., *Math. USSR Doklady* **4** (1963), 683–685.
[22] V. A. Kreknin and A. I. Kostrikin, Lie algebras with regular automorphisms, *Dokl. Akad. Nauk SSSR* **149** (1963), 249–251 (Russian); English transl., *Math. USSR Doklady* **4**, 355–358.
[23] H. Kurzweil, $p$-Automorphismen von auflösbaren $p'$-Gruppen, *Math. Z.* **120** (1971), 326–354.
[24] P. Longobardi and M. Maj, On the number of generators of a finite group, *Arch. Math. (Basel)* **50** (1988), 110–112.
[25] A. Lubotzky and A. Mann, Powerful $p$-groups. I: Finite groups, *J. Algebra* **105** (1987), 484–505
[26] N. Yu. Makarenko, Finite metabelian groups admitting an automorphism of prime order with a restriction on the rank of its centralizer, written communication, 1999.
[27] N. Yu. Makarenko, Ideal in a Lie ring with almost regular automorphism of prime order, *to appear in Siber. Math. J.* **46** (2005).
[28] N. Yu. Makarenko and E. I. Khukhro, Almost solubility of Lie algebras with almost

regular automorphisms, *J. Algebra* **277** (2004), 370–407.
[29] Yu. I. Merzlyakov, On locally soluble groups of finite rank, *Algebra i Logika* **3** (1964), No 2, 5–16 (Russian).
[30] M. R. Pettet, Automorphisms and Fitting factors of finite groups, *J. Algebra* **72** (1981), 404–412.
[31] J. E. Roseblade, On groups in which every subgroup is subnormal, *J. Algebra* **2** (1965), 402–412.
[32] E. E. Shult, On groups admitting fixed point free abelian operator groups, *Illinois J. Math.* **9** (1965), 701–720.
[33] P. Shumyatsky, Involutory automorphisms of finite groups and their centralizers, *Arch. Math. (Basel)* **71** (1998), 425–432.
[34] J. Thompson, Finite groups with fixed-point-free automorphisms of prime order, *Proc. Nat. Acad. Sci. U.S.A.* **45** (1959), 578–581.
[35] J. Thompson, Automorphisms of solvable groups, *J. Algebra* **1** (1964), 259–267.
[36] A. Turull, Fitting height of groups and of fixed points, *J. Algebra* **86** (1984), 555–566.

# 2-SIGNALIZERS AND NORMALIZERS OF SYLOW 2-SUBGROUPS IN FINITE SIMPLE GROUPS

ANATOLY S. KONDRATIEV*[1] and VICTOR D. MAZUROV†[2]

*Institute of Mathematics and Mechanics of UB RAS, Ekaterinburg, 620219, Russia
Email: a.s.kondratiev@imm.uran.ru

†Institute of Mathematics of SB RAS, Novosibirsk, 630090, Russia

## Abstract

We survey the recent completion of the determination of the maximal 2-signalizers and the normalizers of Sylow 2-subgroups in all finite simple groups.

## 1 Introduction

In this survey we discuss the recent completion of the determination of the maximal 2-signalizers and the normalizers of Sylow 2-subgroups in all finite simple groups (see [16, 15]).

The notion of signalizer introduced by Thompson in [21] plays an important rôle in finite group theory, in particular, in the method of signalizer functor. If $G$ is a finite group, $p$ is a prime and $P$ is a Sylow $p$-subgroup of $G$, then any $P$-invariant $p'$-subgroup of $G$ is called a *$P$-signalizer* or simply a *$p$-signalizer*. The case of 2-signalizers is of the greatest interest. In [18], the second author disproved Thompson's conjecture on the commutativity of 2-signalizers in finite simple groups. He constructed certain examples of finite simple groups with nilpotent 2-signalizers of any class of nilpotency and even with non-nilpotent 2-signalizers. In connection with announcing the classification of finite simple groups, Gorenstein [9] posed the problem of the study of the properties of signalizers in known finite simple groups. It was known that the 2-signalizers in finite simple groups of Lie type over a field of even characteristic are trivial (see [9, Thm. 4.254]) and a non-trivial 2-signalizer in the alternating group $A_n$ exists only in the case $n \equiv 3 \pmod 4$ and has order 3 (see [9, Thm. 4.255(i)]). Furthermore, Janko and Thompson in [12] described the 2-signalizers in the Ree groups ${}^2G_2(q)$.

The above-mentioned Mazurov's examples also give negative solutions of the following special questions (a) and (b) on $P$-signalizers ($P$ is a Sylow $p$-subgroup of $G$), formulated by Gorenstein in [9, Sec. 4.15.H].

(a) *Does $P$ normalize non-trivial $q$-subgroups of a finite simple group $G$ for distinct odd primes $q$?*

[1]The first author is supported by RFBR grants no. 04-01-00463, 05-01-10737.

[2]The second author is supported by RFBR grant no. 05-01-00797, the Presidium of SB RAS grant no. 86-197, the program "Universities of Russia" grant no. UR.04.01.028, the program "Developing the scientific potential of higher school" grant no. 511, the Council of the Russian President and State Support of leading scientific schools grant no. NSh-2069.2003.1.

(b) *Are any two maximal $P$-invariant $q$-subgroups ($q$ a prime) of a finite simple group $G$ conjugate by some element from $N_G(P)$ (or even from $C_G(P)$)?*

Because $N_G(P)$ acts on the set $F_G(P)$ of all maximal $P$-signalizers of the group $G$, the following problems also naturally arise:

(c) *What is the cardinality of the set $F_G(P)$?*

(d) *How many $N_G(P)$-orbits has the set $F_G(P)$?*

(e) *What $N_G(P)$-orbits in $F_G(P)$ are conjugate in $G$ (or $\mathrm{Aut}(G)$)?*

(f) *What are the isomorphism types of elements of $F_G(P)$?*

In [16] we obtained the complete classification of the maximal 2-signalizers in finite simple groups up to conjugacy, in particular, the solution of problem (f). In addition, problems (c), (d), (e) are solved for the maximal 2-signalizers in all finite simple groups except symplectic and orthogonal groups.

It is well-known that the properties of $p$-local subgroups (i.e., the normalizers of non-trivial $p$-subgroups) influence strongly the structure of a finite group. The properties of the normalizers of Sylow subgroups in finite simple groups are of particular interest. In some sence, the normalizers of Sylow $p$-subgroups are dual objects to maximal $p$-signalizers.

Yet P. Hall (see [7, Lemma 4]) proved that Sylow 2-subgroups in symmetric groups $S_n$ are self-normalizing. The normalizers of Sylow $p$-subgroups in finite simple groups of Lie type of characteristic $p$ are well-known: they are Borel subgroups (see [6]).

Glauberman and Thompson (see [11, Thm. X.8.13]) showed that if $p \geq 5$ and $G$ is a non-abelian finite simple groups then $N_G(P) \neq PC_G(P)$ for a Sylow $p$-subgroup of $G$. Recently Menegazzo and Tamburini in [19] and independently Guralnick, Malle and Navarro in [10] obtained the extension of the Glauberman–Thompson result to the prime $p = 3$. A corollary of our results (see Section 4) is that these results fail in most cases when $p = 2$.

We now focus on the case $p = 2$. Carter and Fong in [7] described the normalizers of Sylow 2-subgroups in extended classical groups over finite fields of odd characteristic. Aschbacher in [2, (6.1),(6.3)] showed that if $G$ is an alternating group or a finite simple group of Lie type of odd characteristic and $S$ is a Sylow 2-subgroup of $G$, then $N_G(S)$ can contain an element $x$ of prime order $r > 3$ only if $G \cong L_n(q)$, $U_n(q)$, $E_6(q)$ or ${}^2E_6(q)$ and $x \in C_G(S)$. In [16], the authors determined the centralizers of Sylow 2-subgroups in all finite simple groups and the normalizers of Sylow 2-subgroups in finite simple groups of exceptional Lie type over field of odd characteristic. Recently, the first author in [15] completed the determination of the normalizers of Sylow 2-subgroups in all finite simple groups.

In Section 2, we introduce some notation and describe some machinery of fundamental subgroups and maximal tori in finite simple group of Lie type of odd characteristic, which we shall need.

In Sections 3 and 4, we give the statements and some remarks on the proofs of our results on the maximal 2-signalizers and the normalizers of Sylow 2-subgroups, respectively.

## 2 Notation and preliminary results

We use basically the standart terminology and notation (see, for example, [8, 4, 6]).

Denote by $\epsilon$ the variable meaning $+$ or $-$. Let $q = p^f$ always denote a power of an odd prime $p$.

Let $I$ be one of the classical matrix groups $GL_n(q)$, $GU_n(q)$, $Sp_n(q)$, $GO_n^\delta(q)$, where $\delta$ is empty symbol for odd $n$ and $\delta = \pm$ for even $n$. Let $V$ be the vector space associated with $I$ of dimension $n$ over the field of order $q$, or $q^2$ in the case $I = GU_n(q)$, equipped (for $I \neq GL_n(q)$) with the corresponding form such that $I$ can be identified (by choosing a suitable basis in $V$) with the corresponding group $I(V) \in \{GL(V), GU(V), Sp(V), GO^\delta(V)\}$ which coincides with the group of all nondegenerate linear transformations of the space $V$ preserving (for $I \neq GL_n(q)$) the given form. The group $I$ is called the *(full) isometry group* of the space $V$. Set also

$$S = S(V) = I(V) \cap SL(V) \in \{SL(V), SU(V), Sp(V), SO^\delta(V)\}$$

and

$$\Omega = \Omega(V) = O^{p'}(S(V)).$$

Denote by

$$PI(V) \in \{PGL(V), PGU(V), PSp(V), PGO^\delta(V)\},$$

$$PS(V) \in \{PSL(V), PSU(V), PSp(V), PSO^\delta(V)\}$$

and

$$P\Omega(V) \in \{PSL(V), PSU(V), PSp(V), P\Omega^\delta(V)\}$$

the images of the groups $I(V)$, $S(V)$ and $\Omega(V)$, respectively, under the natural homomorphism of the group $I$ onto the quotient group $I/Z(I)$. For brevity, denote by $GL^\epsilon(V)$, $SL^\epsilon(V)$, $L^\epsilon(V)$, and $E_6^\epsilon(q)$, respectively, the groups $GL(V)$, $SL(V)$, $PSL(V)$, and $E_6(q)$ for $\epsilon = +$, $GU(V)$, $SU(V)$, $PSU(V)$, and ${}^2E_6(q)$ for $\epsilon = -$.

A decomposition

$$V = U_1 \oplus \cdots \oplus U_m$$

of $V$ which is an orthogonal (relative to the given form on $V$) sum (direct sum for $I = GL(V)$) of subspaces $U_1, \ldots, U_m$ is called an *orthogonal decomposition* of $V$.

Let $G = G(q)$ be a finite simple group of Lie type over the field $GF(q)$ and let $q \equiv \epsilon 1 \pmod 4$. Let $S$ be a fixed Sylow 2-subgroup of $G$. We describe some subgroups of $G$ which contain $S$.

First we describe some of the requisite notions and results in the Aschbacher's papers [1, 2] relating to fundamental subgroups of $G$. Let $G$ be distinct from $L_2(q)$ and ${}^2G_2(q)$. Let $U$ be a long root subgroup of $G$ with $U^-$ its negative root subgroup and let $K = \langle U, U^- \rangle$. Then $K \cong SL_2(q)$. Define $\Omega(G) = \{K^g \mid g \in G\}$, unless $G \cong G_2(q)$ or ${}^3D_4(q)$, where $\Omega(G) = \{K^g,\ K_0^g \mid g \in G\}$, and let $K_0$ be generated by a short root subgroup of $G$ and its negative, so that $K_0 \cong SL_2(q)$ if $G \cong G_2(q)$ and $K_0 \cong SL_2(q^3)$ if $G \cong {}^3D_4(q)$. The class of subgroups conjugate with $K$ in $G$ is called the set of *fundamental subgroups* of $G$. For a fundamental subgroup $K$ of

$G$ denote by $z(K)$ the unique involution of $K$ (such involutions are called *classical involutions* of $G$). Let

$$\Delta = Fun_G(S) = \{K^g \mid K^g \cap S \in Syl_2(K)\}.$$

Then $\Delta$ is a maximal set of pairwise commuting fundamental subgroups of $G$. Let $k = |\Delta|$ and let $\rho : N_G(\Delta) \to S(\Delta)$ be the permutation representation of degree $k$ of the group $N_G(\Delta)$ corresponding to its action on $\Delta$ by conjugation. The numbers $k$ and the permutation groups $\rho(G) = \rho(N_G(\Delta))$ were determined by Aschbacher in [2, Theorem 2] and given in Table 1.

Table 1:

| $G$ | $k$ | $\rho(G)$ |
|---|---|---|
| $L_n(q)$ | $[n/2]$ | $S_k$ |
| $PSp_{2n}(q)$ | $n$ | $S_k$ |
| $U_n(q)$ | $[n/2]$ | $S_k$ |
| $P\Omega^+_{2n}(q)$ | $2[n/2]$ | $2^{(k/2)+1-(2,n)}.S_{k/2}$ |
| $P\Omega_{2n+1}(q)$ | $2[n/2]$ | $2^{k/2}.S_{k/2}$ |
| $P\Omega^-_{2n}(q)$ | $2[(n-1)/2]$ | $2^{k/2}.S_{k/2}$ |
| $G_2(q)$ | 2 | 1 |
| ${}^3D_4(q)$ | 2 | 1 |
| $F_4(q)$ | 4 | $S_4$ |
| $E_6(q)$ | 4 | $S_4$ |
| ${}^2E_6(q)$ | 4 | $S_4$ |
| $E_7(q)$ | 7 | $L_3(2)$ |
| $E_8(q)$ | 8 | $AGL_3(2)$ |

Note that $N_G(S) \leq N_G(\Delta)$ and Sylow 2-subgroups in $\rho(G)$ are self-normalizing. Let $\Delta = \{X_1, \dots, X_k\}$, $\langle t_i \rangle = Z(X_i)$, and $S_i = S \cap X_i$ (this is a Sylow 2-subgroup of $X_i$). Let $Z = \langle t_1, \dots, t_k \rangle$. Then $Z = Z(\langle \Delta \rangle)$.

The definition of a fundamental subgroup can be extended to any quotient group by a central subgroup of the universal group $G^*$ of Lie type with $G^*/Z(G^*) \cong G$, in particular, to the classical group $\Omega = \Omega(V)$. Consider the following explicit Aschbacher's construction of the set $\Delta$ for $\Omega$.

Let $\Omega = SL^\delta(V)$ or $Sp(V)$ and $n = 2k + a$ with $0 \leq a \leq 1$. By [2, (2.1)], there exsists an orthogonal decomposition

$$V = V_0 \oplus V_1 \oplus \cdots \oplus V_k$$

with $\dim V_i = 2$ for $1 \leq i \leq k$ and $\dim V_0 = a$ such that $\Delta = \{K_i \mid 1 \leq i \leq k\}$, where $K_i = S(V_i)$ and $I(V_i)$ denotes a subgroup of $I(V)$ acting faihtfully on $V_i$ as its corresponding full isometry group and centralizing each subspace $V_j$ for $j \neq i$.

Let $\Omega = \Omega^\delta(V)$ and $n = 4k + a$ with $0 \leq a \leq 4$, $\delta = +$ if $a = 0$ and $\delta = -$ if $a = 4$. By [2, (2.2)], there exsists an orthogonal decomposition

$$V = V_0 \oplus V_1 \oplus \cdots \oplus V_k$$

of $V$ with the $V_i$ of dimension 4 and sign + for $1 \leq i \leq k$ and $\dim V_0 = a$ such that $\Delta = \{K_i,\ K_i^* \mid 1 \leq i \leq k\}$, where $K_i \circ K_i^* \cong \Omega^+(V_i)$ and $I(V_i)$ denotes a subgroup of $I(V)$ acting faihtfully on $V_i$ as its corresponding full isometry group and centralizing each subspace $V_j$ for $j \neq i$.

We now distinguish some maximal tori of $G$. There is a maximal torus $T^\epsilon$ of $G$ such that $S \leq N_G(T^\epsilon)$. This torus is defined as follows. Let $\overline{G}$ be the simple adjoint algebraic group over the algebraic closure of $GF(q)$ corresponding to $G$, i.e., there is an surjective endomorphism $\sigma$ of $\overline{G}$ such that $O^{p'}(\overline{G}_\sigma) = G$. Let $\overline{T}$ be a $\sigma$-invariant split maximal torus of $\overline{G}$, i.e., $\overline{T}$ is contained in a $\sigma$-invariant Borel subgroup of $\overline{G}$. Then $W = N_{\overline{G}}(\overline{T})/\overline{T}$ is by definition the Weyl group of $\overline{G}$. Let $w_0$ be the longest element of $W$. If $\epsilon = +$ we define $T^+ = \overline{T}_\sigma \cap G$ (the torus $T^+$ is defined in the same way in the case when $G = {}^2G_2(q)$, where $q = 3^{2n+1} > 3$, although here always $\epsilon = -$). If $\epsilon = -$ we define $T^- = ({}^{w_0}\overline{T})_\sigma \cap G$, where ${}^{w_0}\overline{T}$ is the $\sigma$-invariant maximal torus of $\overline{G}$ obtained from $\overline{T}$ by twisting by the element $w_0$ (see [20, E.II.1.3]). Set $N^\epsilon = N_G(T^\epsilon)$.

## 3 Maximal 2-signalizers in finite simple groups

In this section, we describe our results on maximal 2-signalizers in all finite simple groups.

As mentioned in the Introduction, the orders of maximal 2-signalizers in the alternating groups are known. But the proof of this fact is not published. We prove a somewhat more precise statement in the following theorem.

**Theorem 3.1** *Let $G$ be the alternating group of degree $n$, and let $S$ be a Sylow 2-subgroup of $G$. Then $G$ possesses only one maximal $S$-signalizer $F$. If $n \equiv 3$ (mod 4), then $F$ is a group of order 3, generated by a cycle of length 3; otherwise $F = 1$.*

Using Atlas [8] and Aschbacher's results [5], we prove

**Theorem 3.2** *Let $G$ be a finite sporadic simple group, and let $S$ be a Sylow 2-subgroup of $G$. Then $G$ possesses only one maximal $S$-signalizer $F$. If $G \cong M_{11}$, then $F$ is a Sylow 3-subgroup of $G$ which is isomorphic to the elementary abelian group of order 9; if $G \cong Ly$, then $F$ is a group of the order 3 generated by an element of the conjugacy class $3A$; for all remaining groups, $F = 1$.*

Up to the end of the Section, let $G = G(q)$ be a finite simple group of Lie type over the field $GF(q)$ of an odd characteristic $p$, let $q \equiv \epsilon 1 \pmod 4$, and let $S$ be a fixed Sylow 2-subgroup of $G$. We shall use the notation of Section 2.

The following general fact, which we extracted from [13, Prop. 3.3], is very useful for us.

**Proposition 3.3** *If $G$ is other than $L_2(q)$ or ${}^2G_2(q)$, then every abelian $S$-signalizer of $G$ whose order is not divisable by $p$ centralizes $Z$.*

The case $G = L_2(q)$ is very important for all cases of $G$ and is studied in detail in the following proposition.

**Proposition 3.4** *Let $G = L_2(q)$ and let $S^*$ be a Sylow 2-subgroup of $PGL_2(q)$. Then*

(a) *maximal $S$-signalizers of $G$ are conjugate under $N_G(S)$ with the cyclic subgroup $O(C_G(z))$ of order $(q-\epsilon 1)/(q-\epsilon 1)_2$, where $z$ is an involution of $Z(S)$;*

(b) *the number of maximal $S$-signalizers of $G$ is equal to 1 if $q \equiv \pm 1 \pmod 8$ or $(q-\epsilon 1) = (q-\epsilon 1)_2$ and to 3 otherwise;*

(c) *the cyclic subgroup of index 2 in the dihedral group $C_G(z)$ coincides with a maximal torus $T^\epsilon$ of order $(q-\epsilon 1)/2$ of the group $G$,*

(d) *there exists a unique maximal $S^*$-signalizer in $PGL_2(q)$, namely, $O(C_G(z))$, where $z$ is the unique involution in $Z(S^*)$*

In the case where $G$ is a classical group, we prove the following theorem.

**Theorem 3.5** (a) *If $G \cong L_n(q)$ with $n = 2^{t_1} + \cdots + 2^{t_r}$ for some integers $r$ and $t_1, \ldots, t_r$ such that $t_1 > \cdots > t_r \geq 0$, then the following holds:*

(a1) *In the case where $n = 2$, $q - \epsilon 1$ is not equal to 4 and not divisible by 8, $G$ contains exactly 3 maximal $S$-signalizers. They are conjugate under $N_G(S)$ and are isomorphic to a cyclic group of order $(q-\epsilon 1)_{2'}$;*

(a2) *In all other cases, $G$ contains exactly $r!$ maximal $S$-signalizers $F_\pi$ for $\pi \in S_r$ and the subgroup $F_\pi$ is isomorphic to an extension of $P_\pi$ by the direct product of $[n/2]$ cyclic groups of order $(q-\epsilon 1)_{2'}$, $\max(0, [n/2] - 2)$ groups of order $(q-1)_{2'}$ and a cyclic group of order $((q-1)/(q-1, n))_{2'}$, where $P_\pi$ is the group of all linear transformations of a vector space $V$ of dimension $n$ over a field of order $q$ such that $P_\pi$ stabilizes a chain of subspaces $0 = V_0 < V_1 < \cdots < V_r = V$ with $\dim(V_i/V_{i-1}) = 2^{t_{i\pi}}$ and $P_\pi$ induces in each $V_i/V_{i-1}$, $i = 1, \ldots, r$, the trivial transformation group. For $\pi$, $\sigma \in S_r$, the subgroups $F_\pi$ and $F_\sigma$ are conjugate in $G$ only if $\pi = \sigma$. Every $F_\pi$, $\pi \in S_r$, is $N_G(S)$-invariant.*

(b) *If $G \cong U_n(q)$, $n \geq 3$, then $G$ possesses only one maximal $S$-signalizer $F$ and $F$ is isomorphic to the direct product of $[n/2]$ cyclic groups of order $(q-\epsilon 1)_{2'}$, $\max(0, [n/2] - 2)$ cyclic groups of order $(q+1)_{2'}$ and a cyclic group of order $((q+1)/(q+1, n))_{2'}$.*

(c) *If $G \cong PSp_{2n}(q)$ or $G \cong O_{2n+1}(q)$, then all maximal $S$-signalizers of $G$ are conjugate in $G$ and are isomorphic to the direct product of $n$ cyclic groups of order $(q-\epsilon 1)_{2'}$. If $q - \epsilon 1$ is equal to 4 or divisible by 8, then $G$ contains only one maximal $S$-signalizer.*

(d) *If $G \cong P\Omega^+_{2n}(q)$, then all maximal $S$-signalizers of $G$ are conjugate in $G$. For even $n$, a maximal $S$-signalizer $F$ of $G$ is isomorphic to the direct product of $n$ cyclic groups of order $(q-\epsilon 1)_{2'}$. For odd $n$, $F$ is isomorphic to the direct product of $n-1$ cyclic groups of order $(q-\epsilon 1)_{2'}$ and a cyclic group of order $(q-1)_{2'}$. If $q - \epsilon 1$ is equal to 4 or divisible by 8, then $G$ contains only one maximal $S$-signalizer.*

(e) *If* $G \cong P\Omega^-_{2n}(q), n \geq 2$, *then all maximal* $S$*-signalizers of* $G$ *are conjugate in* $G$. *For even* $n$, *a maximal* $S$*-signalizer* $F$ *of* $G$ *is isomorphic to the direct product of* $n-2$ *cyclic groups of order* $(q-\epsilon 1)_{2'}$ *and a cyclic group of order* $(q^2-1)_{2'}$. *For odd* $n$, $F$ *is isomorphic to the direct product of* $n-1$ *cyclic groups of order* $(q-\epsilon 1)_{2'}$ *and a cyclic group of order* $(q+1)_{2'}$. *If* $q-\epsilon 1$ *is equal to* 4 *or divisible by* 8, *then* $G$ *contains only one maximal* $S$*-signalizer.*

The proof of Theorem 3.5 is carried out by induction on the dimension $n$ and uses Aschbacher's results [3].

In the case where $G$ is of an exceptional Lie type, we prove the following theorem.

**Theorem 3.6** (a) *If* $G$ *is not isomorphic to* ${}^2G_2(q)$ *or* $E_6(q)$, *then* $G$ *possesses only one maximal* $S$*-signalizer, namely,* $O(T^\epsilon)$.

(b) *If* $G \cong {}^2G_2(q)$, *then* $G$ *possesses exactly* 7 *maximal* $S$*-signalizers, which are conjugate in* $N_G(S)$ *with a cyclic subgroup* $O(T^\epsilon) = O(C_G(V))$ *of order* $(q+1)/4$, *where* $V$ *is a four-subgroup of* $S$.

(c) *If* $G \cong E_6(q)$, *then* $S$ *lies in exactly two parabolic maximal subgroups* $P_1$ *and* $P_2$ *of* $G$ *conjugate under a non-trivial graph automorphism of* $G$ *which fixes* $S$ *and such that* $P_i$ *is of type* $D_5$ *and* $O_p(P_i)$ *is an elementary abelian group of order* $q^{16}$; $G$ *possesses only two maximal* $S$*-signalizers, namely,* $O_p(P_1)O(T^\epsilon)$ *and* $O_p(P_2)O(T^\epsilon)$).

The orders of the maximal tori $T^\epsilon$ for the exceptional groups $G$ and for $G = L_2(q)$ were found in [20, E and G] and are given in Table 2 (see also [17, Table 3]).

Table 2:

| $G$ | $\lvert T^+\rvert$ | $\lvert T^-\rvert$ |
|---|---|---|
| $L_2(q)$ | $(q-1)/2$ | $(q+1)/2$ |
| ${}^2G_2(q)$, $q = 3^{2n+1} > 3$ | $q-1$ | $q+1$ |
| $G_2(q)$ | $(q-1)^2$ | $(q+1)^2$ |
| ${}^3D_4(q)$ | $(q-1)(q^3-1)$ | $(q+1)(q^3+1)$ |
| $F_4(q)$ | $(q-1)^4$ | $(q+1)^4$ |
| $E_6(q)$ | $\frac{(q-1)^6}{(3,q-1)}$ | $\frac{(q^2-1)^2(q+1)^2}{(3,q-1)}$ |
| ${}^2E_6(q)$ | $\frac{(q^2-1)^2(q-1)^2}{(3,q+1)}$ | $\frac{(q+1)^6}{(3,q+1)}$ |
| $E_7(q)$ | $(q-1)^7/2$ | $(q+1)^7/2$ |
| $E_8(q)$ | $(q-1)^8$ | $(q+1)^8$ |

We make some remarks on the proof of Theorem 3.6. Assume that $G$ is of exceptional Lie type. The 2-signalizers of ${}^2G_2(q)$ were described in [12]. In part (b) of Theorem 3.6, we give this description in a more precise form. Hence we can assume that $G \not\cong {}^2G_2(q)$.

At first we determine the structure of the core $\ker\rho$ of the permutation representation $\rho$ of the group $N_G(\Delta)$ on the set $\Delta$.

**Proposition 3.7** (a) $\ker\rho = C_G(Z)$.

(b) *$Z$ belongs to any $S$-invariant maximal torus $T^\epsilon$ of $G$.*

(c) $\ker\rho = \langle\Delta\rangle T^\epsilon$, *$T^\epsilon$ contains $C_G(\langle\Delta\rangle))$ and some maximal torus of each subgroup $X_i$. The structure of the subgroup $\langle\Delta\rangle C_G(\langle\Delta\rangle)$ and the quotient group $\ker\rho/(\langle\Delta\rangle C_G(\langle\Delta\rangle))$ is given in Table* 3.

(d) *The quotient group $\ker\rho/C_{\ker\rho}(X_i)$ is isomorphic to $PGL_2(q)$ or $PGL_2(q^3)$ for $1 \leq i \leq k$.*

Table 3:

| $G$ | $\langle\Delta\rangle C_G(\langle\Delta\rangle)$ | $\ker\rho/(\langle\Delta\rangle C_G(\langle\Delta\rangle))$ |
|---|---|---|
| $G_2(q)$ | $SL_2(q) \circ SL_2(q)$ | 2 |
| ${}^3D_4(q)$ | $SL_2(q) \circ SL_2(q^3)$ | 2 |
| $F_4(q)$ | $2^3.L_2(q)^4$ | 2 |
| $E_6(q)$ | $2^3.(L_2(q)^4 \times ((q-1)/2)^2/(3,\ q-1))$ | $2^3$ |
| ${}^2E_6(q)$ | $2^3.(L_2(q)^4 \times ((q+1)/2)^2/(3,\ q+1))$ | $2^3$ |
| $E_7(q)$ | $2^3.L_2(q)^7$ | $2^3$ |
| $E_8(q)$ | $2^4.L_2(q)^8$ | $2^4$ |

In Table, 3 we denote by $((q\pm1)/2)^2/(3,\ q\pm1)$ the quotient group of the group $((q\pm1)/2)^2$ by a subgroup of order $(3,\ q\pm1)$.

Proposition 3.7 implies the following useful result.

**Proposition 3.8** *There is an unique maximal $S$-signalizer (respectively $C_S(Z)$-signalizer) in the subgroups $N_G(\Delta)$ and $N^\epsilon$ (respectively $C_G(Z)$), namely, $O(T^\epsilon)$.*

The case where $G = G_2(q)$ or ${}^3D_4(q)$ is studied separately. Suppose that $G$ is not equal to $G_2(q)$ or ${}^3D_4(q)$. Let $F$ be a non-trivial maximal $S$-signalizer of $G$. Hence, for some odd prime $r$, there exists a non-trivial elementary abelian $r$-subgroup of $F$ which is normal in $FS$. If $r \neq p$, then we show that $F = O(T^\epsilon)$. If $r = p$, then by [17] or [13] we obtain that $G = E_6(q)$ and prove our theorem in this case.

## 4 Normalizers of Sylow 2-subgroups in finite simple groups

We prove the following theorem.

**Theorem 4.1** *Let $G$ be a finite simple group, and $S$ a Sylow 2-subgroup of $G$. Then $N_G(S) = S$ except the following cases:*

(a) *$G$ is a group of Lie type of characteristic 2 and $N_G(S)$ is a Borel subgroup of $G$ with $N_G(S) \neq S$;*

(b) $G \cong L_2(q)$, $3 < q \equiv \pm3 \pmod 8$ *and* $N_G(S) \cong A_4$;

(c) $G \cong L_m^\epsilon(q)$, $m \geq 3$, *q is odd and* $S \neq N_G(S) = S \times C_1 \times \cdots \times C_{t-1}$, *where the number* $t \geq 2$ *is found from the 2-adic decomposition* $m = 2^{s_1} + \cdots + 2^{s_t}$, $s_1 > \cdots > s_t \geq 0$, *and* $C_1, \ldots, C_{t-2}, C_{t-1}$ *are cyclic groups of orders* $(q-\epsilon 1)_{2'}, \ldots, (q-\epsilon 1)_{2'}$, $(q-\epsilon 1)_{2'}/(q-\epsilon 1, m)_{2'}$;

(d) $G \cong PSp_{2m}(q)$, $m \geq 2$, $q \equiv \pm 3 \pmod 8$, $C_G(S) < S$ *and the quotient group* $N_G(S)/S$ *is isomorphic to the elementary abelian 3-group of order* $3^t$, *where the number t is found from the 2-adic decomposition* $m = 2^{s_1} + \cdots + 2^{s_t}$, $s_1 > \cdots > s_t \geq 0$;

(e) $G \cong E_6^\epsilon(q)$ *and* $N_G(S) = S \times R$, *where R is a cyclic group of order* $(q-\epsilon 1)_{2'}/(3, q-\epsilon 1)$;

(f) $G \cong {}^2G_2(q)$ *or* $J_1$, $C_G(S) = S$ *and* $N_G(S) \cong 2^3.7.3 < Hol(2^3)$;

(g) $G \cong J_2$, $J_3$, *Suz or HN*, $C_G(S) < S$ *and* $|N_G(S) : S| = 3$.

Note that the conclusions of Theorem 4.1 for alternating and sporadic groups are formulated without proof in [10, Ex. 5.1].

The description of $C_G(S)$ in Theorem 4.1 is extracted from Theorems 3.1, 3.2, 3.5, 3.6. Further, Theorem 4.1 arises from the following three more general theorems 4.2–4.4.

**Theorem 4.2** *Let G be one of the finite classical groups* $GL_m(q)$, $GU_m(q)$, $Sp_{2m}(q)$, $GO_{2m+1}(q)$, *and* $GO_{2m}^\pm(q)$, *where* $m \geq 1$ *and q is a power of an odd prime p. Let T be a Sylow 2-subgroup of G and let* $T^0 = T \cap O^{p'}(G)$. *Then* $N_G(T^0) = TC_G(T)$, *except the following cases:*

(a) $G = GL_2(q)$ *or* $GU_2(q)$, $q \equiv \pm 3 \pmod 8$ *and* $N_G(T^0)/Z(G) \cong S_4$;

(b) $G = Sp_{2m}(q)$, $q \equiv \pm 3 \pmod 8$ *and the quotient group* $N_G(S)/S$ *is isomorphic to the elementary abelian 3-group of order* $3^t$, *where the number t is found from the 2-adic decomposition* $m = 2^{s_1} + \cdots + 2^{s_t}$, $s_1 > \cdots > s_t \geq 0$;

(c) $G = GO_3(q) \cong 2 \times PGL_2(q)$, $q \equiv \pm 3 \pmod 8$ *and* $N_G(T^0)/Z(G) \cong S_4$;

(d) $G = GO_2^\pm(q) \cong D_{2(q\mp 1)}$, $q \mp 1$ *is not a power of 2 and* $N_G(T^0) = G$;

(e) $G = GO_4^+(q) \cong (SL_2(q) \circ SL_2(q)).2^2$, $q \equiv \pm 3 \pmod 8$ *and* $N_G(T^0) \cong GO_4^+(3)$;

(f) $G = GO_5(q) \cong 2 \times (PSp_4(q).2)$, $q \equiv \pm 3 \pmod 8$ *and* $N_G(T^0)/Z(G) \cong GO_4^+(3)$.

The result of Carter and Fong mentioned in Introduction follows from Theorem 4.2.

We make some remarks on the proof of Theorem 4.2. We use the notation from Section 2. Let $G \cong I = I(V)$. For small $n$, the assertion of the theorem is obtained immediately. Further we shall assume that $n$ is at least 3, 4, 5 in the cases when $G = GL_n^\delta(q)$, $Sp_n(q)$, $GO_n^\delta(q)$, respectively. It is well-known that then $P\Omega(V)$ is a nonabelian simple group. We consider the set $\Delta$ of fundamental subgroups of the group $\Omega(V)$ corresponding to the orthogonal decomposition

$$V = V_0 \oplus V_1 \oplus \cdots \oplus V_k$$

as in Section 2. Set $\mathcal{D} = \{V_0, V_1, \ldots, V_k\}$. Let $I_{\mathcal{D}}$ denote the global stabilizer in $I$ of the set $\mathcal{D}$. We show that $N_I(T^0) \leq N_I(\Delta) = I_{\mathcal{D}}$. Using the results on the structure of stabilizers of orthogonal decompositions in finite classical groups (see [14]) we obtain our theorem.

By analogy with the proof of Theorem 4.2, we prove the following.

**Theorem 4.3** *Let $T$ be a Sylow 2-subgroups in $S_n$ and let $T^0 = T \cap A_n$. Then $N_{S_n}(T^0) = T$, except the following cases:*

(a) *$n = 3$ and $N_{S_3}(T^0) = S_3$;*

(b) *$n = 4$ or 5 and $N_{S_n}(T^0) \cong S_4$.*

The above-mentioned result of P. Hall follows from Theorem 4.3.

The analogy between the proofs of Theorems 4.3 and 4.2 consists in the construction of a certain partition of the set $X = \{1, 2, \ldots, n\}$ $(n \geq 5)$ and a corresponding to it set of isomorphic to $A_4$ subgroups of the alternating group $A(X)$ on $X$ which have the properties similar to the properties of the orthogonal decomposition $V = V_0 \oplus V_1 \oplus \cdots \oplus V_k$ and the corresponding set $\Delta$ of fundamental subgroups of the group $\Omega(V)$ (see Section 2), respectively.

Using Atlas [8] and Theorems 4.2 and 4.3, we easily prove the following.

**Theorem 4.4** *Let $G$ be one of the finite simple sporadic groups, $A = \mathrm{Aut}(G)$, $T$ a Sylow 2-subgroup in $A$ and let $T^0 = T \cap G$. Then $N_A(T^0) = T$, except the following cases:*

(a) *$G = A = J_1$ and $N_A(T^0) \cong 2^3.7.3 < Hol(2^3)$;*

(b) *$G = J_2$, $J_3$, $Suz$ or $HN$, $|A : G| = 2$ and $N_A(T^0)/T^0 \cong S_3$.*

## References


[1] M. Aschbacher, A characterization of Chevalley groups over fields of odd order, *Ann. Math.* **106** (1977), 353–468; Correction: *Ann. Math.* **111** (1980), 411–414.

[2] M. Aschbacher, On finite groups of Lie type and odd characteristic, *J. Algebra* **66** (1980), 400–424.

[3] M. Aschbacher, On the maximal subgroups of the finite classical groups, *Invent. Math.* **76** (1984), 469–514.

[4] M. Aschbacher, *Finite group theory*, Cambridge Univ. Press, 1986.

[5] M. Aschbacher, Overgroups of Sylow subgroups in sporadic groups, *Mem. Amer. Math. Soc.* **60**, Amer. Math. Soc., 1986.

[6] R. Carter, *Simple groups of Lie type*, Wiley, 1972.

[7] R. Carter and P. Fong, The Sylow 2-subgroups of the finite classical groups, *J. Algebra* **1** (1964), 139–151.

[8] J. H. Conway, R. T. Curtis, S. P. Norton, R. A. Parker and R. A. Wilson, *Atlas of finite groups*, Clarendon Press, 1985.

[9] D. Gorenstein, *Finite simple groups. An introduction to their classification*, Plenum Press, 1982.

[10] R. M. Guralnik, G. Malle and G. Navarro, Self-normalizing Sylow subgroups, *Proc. Amer. Math. Soc.* **132** (2004), 973–979.

[11] B. Huppert and N. Blackburn, *Finite groups III*, Springer-Verlag, 1982.

[12] Z. Janko and J. G. Thompson, On a class of finite simple groups of Ree, *J. Algebra* **4** (1966), 274–292.
[13] W. M. Kantor, Primitive permutation groups of odd degree, and an application to finite projective planes, *J. Algebra* **106** (1987), 15–45.
[14] P. B. Kleidman and M. W. Liebeck, *The subgroup structure of finite classical groups*, Cambridge Univ. Press, 1990.
[15] A. S. Kondratiev, Normalizers of Sylow 2-subgroups in finite simple groups, *Mat. Zametki*, to appear.
[16] A. S. Kondratiev and V. D. Mazurov, 2-signalizers of finite simple groups, *Algebra and Logic* **42** (2003), 333–348.
[17] M. W. Liebeck and J. Saxl, The primitive permutation groups of odd degree, *J. London Math. Soc. (2)* **31** (1985), 250–264.
[18] V. D. Mazurov, On 2-signalizers of finite groups, *Algebra i Logika* **7** (1968), 60–62.
[19] F. Menegazzo and M. C. Tamburini, A property of Sylow $p$-normalizers in simple groups, *Quaderni del Seminario Matematico di Brescia* **45/02** (2002).
[20] *Seminar on algebraic groups and related finite groups* (A. Borel et al., eds.), Springer-Verlag, 1970.
[21] J. G. Thompson, 2-signalizers of finite groups, *Pacific J. Math.* **14** (1964), 364–364.

# ON PROPERTIES OF ABNORMAL AND PRONORMAL SUBGROUPS IN SOME INFINITE GROUPS[1]

LEONID A. KURDACHENKO*, JAVIER OTAL† and IGOR YA. SUBBOTIN§

*Department of Algebra, National Dnipropetrovsk University,
Vul. Naukova 13. Dnipropetrovsk 50, Ukraine 49050
Email: lkurdachenko@hotmail.com

†Department of Mathematics, University of Zaragoza,
Pedro Cerbuna 12. 50009 Zaragoza, Spain
Email: e-mail@unizar.es

§Mathematics Department, National University, Inglewood CA 90301, USA
Email: isubboti@nu.edu

Abnormal and pronormal subgroups have appeared in the process of investigation of some important subgroups of finite (soluble) groups such as Sylow subgroups, Hall subgroups, system normalizers, and Carter subgroups. Let $H$ be a subgroup of a group $G$. We recall that a subgroup $H$ is *abnormal* in $G$ if $g \in \langle H, H^g \rangle$ for each element $g \in G$; and a subgroup $H$ is *pronormal* in $G$ if for each element $g \in G$, $H$ and $H^g$ are conjugate in $\langle H, H^g \rangle$. Pronormal subgroups have been introduced by P. Hall in his lectures in Cambridge; he also introduced abnormal subgroups in his paper [6], whereas the term *abnormal* comes from R. Carter [2]. These subgroups and their generalizations have shown to be very useful in the finite group theory. It appears to be logical to employ such fruitful concepts in infinite groups. However, in some classes of infinite groups these mentioned subgroups gain such properties that they cannot posses in the finite case. For example, it is well-known that every finite $p$-group has no proper abnormal subgroups. Nevertheless, A. Yu. Olshanskii has constructed a series of impressive examples of infinite finitely generated $p$-groups saturated with abnormal subgroups. Concretely, for a large enough prime $p$ there exists an infinite $p$-group $G$ whose all proper subgroups have prime order $p$ [18, Theorem 28.1]. In particular, every proper non-identity subgroup of $G$ is maximal, and being non-normal, is abnormal. In general, we quite frequently observe that the situation in infinite groups is significantly different from the situation in the corresponding finite case. Therefore, the first task is to find such classes of infinite groups for which a given notion is effective at the same level as in the finite case. It is logical to focus our search on infinite but near to finite groups; that is, on the groups with finiteness conditions.

The aim of this survey is to discuss the main accomplishments in extending some fundamental results concerning topics related to pronormal subgroups on some classes of infinite groups. We cannot find the unique universal class in which such a transfer is possible. Moreover, very frequently this extending can be realized for some distinct and weakly related classes of infinite groups. The selection of these classes is a very subjective process. However, as we will show below, in each particular case the choice of such a class was not only convenient, but mostly logical.

[1]Supported by Proyecto MTM2004-04842 of Dirección General de Investigación MEC (Spain)

## 1 Pronormal, Abnormal and Carter Subgroups in Infinite Groups

We begin with extension of the characterizations of abnormal and pronormal subgroups which have been obtained for finite soluble groups a long time ago. In a certain sense abnormal subgroups are some antipodes to normal subgroups. Thus, in finite soluble groups abnormality is tightly connected to self-normalizing. For example, D. Taunt has shown that *a subgroup $H$ of a finite soluble group $G$ is abnormal if and only if every intermediate subgroup for $H$ coincides with its normalizer in $G$*; that is, such a subgroup is self-normalizing (see, for example, [25, 9.2.11]). Recall that a subgroup $S$ is said to be *an intermediate subgroup for $H$* if $H \leq S$. The following theorem extends this result on the radical groups.

**Theorem 1.1 (L. A. Kurdachenko, I. Ya. Subbotin [12])** *Let $G$ be a radical group and let $H$ be a subgroup of $G$. Then $H$ is abnormal in $G$ if and only if every intermediate subgroup for $H$ is self-normalizing.*

The following results are straightforward consequences of this theorem.

**Corollary 1.2 (F. de Giovanni, G. Vincenzi [5])** *Let $G$ be a hyperabelian group and let $H$ be a subgroup of $G$. Then $H$ is abnormal in $G$ if and only if every intermediate subgroup for $H$ is self-normalizing.*

**Corollary 1.3** *Let $G$ be a soluble group and let $H$ be a subgroup of $G$. Then $H$ is abnormal in $G$ if and only if every intermediate subgroup for $H$ is self-normalizing.*

As normality and pronormality, the abnormality is not a transitive property (the symmetric group $S_4$ can serve as an example). The groups (finite and infinite), in which normality is transitive and groups in which pronormality is transitive [10] are studied well enough. The case of the transitivity of abnormality is very different. One cannot count here on an explicit description. In a finite metanilpotent group, abnormality is transitive [28]. For infinite groups the most general result known up till now is the following theorem.

**Theorem 1.4 (L. A. Kurdachenko and I. Ya. Subbotin [12])** *Let $G$ be a group and suppose that $A$ is a normal subgroup of $G$ such that $G/A$ has no proper abnormal subgroups. If $A$ satisfies the normalizer condition, then abnormality is transitive in $G$.*

We recall that a group $G$ is said to be a group with *the normalizer condition* ($G$ is *an $N$-group*) if $H \neq N_G(H)$ for every proper subgroup $H$. Note that the class of $N$-groups is a proper subclass of the class of locally nilpotent groups.

We also recall that a subgroup $H$ of a group $G$ is said to have *the Frattini property* if given two intermediate subgroup $K$ and $L$ for $H$ such that $K$ is normal in $L$ we obtain then $L \leq N_G(H)K$ (in this case it is also said that $H$ is *weakly pronormal* in $G$). T. A. Peng in his paper [19] has characterized pronormal subgroups in finite soluble groups. He has proved that *a subgroup $H$ of a finite soluble group $G$ is pronormal if and only if $H$ is weakly pronormal.*

Let $\mathfrak{X}$ be a class of groups. A group $G$ is said to be *a hyper-$\mathfrak{X}$-group* if $G$ has an ascending series of normal subgroups whose factors are $\mathfrak{X}$-groups. Peng's characterization of pronormal subgroups is extended in the following way.

**Theorem 1.5 (L. A. Kurdachenko, J. Otal, I. Ya. Subbotin [8])** *Let $G$ be a hyper-$N$-group. Then a subgroup $H$ of $G$ is pronormal in $G$ if and only if $H$ is weakly pronormal in $G$.*

This result has two immediate corollaries.

**Corollary 1.6 (F. de Giovanni, G. Vincenzi [5])** *Let $G$ be a hyperabelian group and let $H$ be a subgroup of $G$. Then $H$ is pronormal in $G$ if and only if $H$ is weakly pronormal in $G$.*

**Corollary 1.7** *Let $G$ be a soluble group and let $H$ be a subgroup of $G$. Then $H$ is pronormal in $G$ if and only if $H$ is weakly pronormal in $G$.*

T. A. Peng [20] also proved that the finite soluble groups whose subgroups are pronormal are exactly the finite soluble groups whose subgroups satisfy the transitivity for normality. Extending this result to infinite groups N. F. Kuzennyi and I. Ya. Subbotin obtained the following result.

**Theorem 1.8 (N. F. Kuzennyi, I. Ya. Subbotin [14])** *Suppose that $G$ is a locally soluble group or a periodic locally graded group. Then the following conditions are equivalent.*

1. *Every cyclic subgroup of $G$ are pronormal in $G$.*
2. *$G$ is a soluble group in which all subgroups are groups with transitivity of normality.*

N. F. Kuzennyi and I. Ya. Subbotin also completely described *non-periodic locally soluble groups and periodic locally graded groups in which all subgroups are pronormal* [13], *locally graded periodic groups in which all primary subgroups are pronormal* [16], and *infinite locally soluble groups in which all infinite subgroups are pronormal* [14]. They also have shown that in the infinite case, *the class of groups with all pronormal subgroups is a proper subclass of the class of the groups with the transitivity of normality; and moreover, it is also a proper subclass of the class of groups in which all primary subgroups are pronormal.* However, *the pronormality condition for all subgroups could be weakened to pronormality for only abelian subgroups* (N. F. Kuzennyi, I. Ya. Subbotin [17]).

An important type of abnormal subgroups are the Carter subgroups. In the finite group theory these subgroups have been introduced by R. Carter [2] as the self-normalizing nilpotent subgroups. These subgroups are very tightly connected to abnormality. In their original definition one can find no specifications related to finite groups. Some attempts of extending the notion of Carter subgroups on infinite groups has been already done by S. E. Stonehewer [29, 30], A. D. Gardiner, B. Hartley and M. J. Tomkinson [4], and M. R. Dixon [3]. We want to extend this very useful notion on the following class of nilpotent-by-hypercentral (not necessary periodic) groups. Note that we may define a Carter subgroup in a finite metanilpotent group as a minimal abnormal subgroup.

*Let $\mathfrak{X}$ be a class of groups. A group $G$ is said to be an artinian-by-$\mathfrak{X}$-group if $G$ has a normal subgroup $H$ such that $G/H$ belongs to $\mathfrak{X}$ and $H$ satisfies Min-$G$.*

The groups of this kind have been handled by D. J. S. Robinson [22, 23, 24, 26, 27] and D. I. Zaitsev [31, 32, 33, 34, 35] in their series of papers dedicated to existence of complements to the $\mathfrak{X}$-residual (when this one is abelian) for some classes $\mathfrak{X}$ of groups, such as hypercentral groups, locally nilpotent groups, hypercyclic groups, locally supersoluble groups, and hyperfinite groups.

The logical first step here is consideration of artinian-by-hypercentral groups whose locally nilpotent residual is nilpotent. For these groups, the following results hold.

**Theorem 1.9 (L. A. Kurdachenko, I. Ya. Subbotin [12])** *Let $G$ be an artinian–by–hypercentral group and suppose that its locally nilpotent residual $K$ is nilpotent. Then*

1. *$G$ has a minimal abnormal subgroup $L$. Moreover, $L$ is maximal hypercentral subgroup and it includes the upper hypercenter of $G$. In particular, $G = KL$.*
2. *Two minimal abnormal subgroups of $G$ are conjugate.*

**Theorem 1.10 (L. A. Kurdachenko, I. Ya. Subbotin [12])** *Let $G$ be an artinian-by-hypercentral group and suppose that its locally nilpotent residual $K$ is nilpotent. Then*

1. *$G$ has a hypercentral abnormal subgroup $L$. Moreover, $L$ is a maximal hypercentral subgroup and it includes the upper hypercenter of $G$. In particular, $G = KL$; and*
2. *Two hypercentral abnormal subgroups of $G$ are conjugate.*

Thus, given an artinian-by-hypercentral group with a nilpotent hypercentral residual, *a subgroup $L$ is called a Carter subgroup of a group $G$ if $H$ is a hypercentral abnormal subgroup of $G$ or, equivalently, if $H$ is a minimal abnormal subgroup of $G$.*

A Carter subgroup of a finite soluble group can be characterized as *a covering subgroup for the formation of nilpotent groups.* In the paper [12] this definition was extended on the class of artinian-by-hypercentral groups with a nilpotent locally nilpotent residual.

## 2 Pronormal Subgroups and Generalized Nilpotency

The following well-known characterizations of finite nilpotent groups are tightly connected to abnormal and pronormal subgroups.

*A finite group $G$ is nilpotent if and only if $G$ does not include a proper abnormal subgroup.*

*A finite group $G$ is nilpotent if and only if every pronormal subgroup of $G$ is normal.*

Note that since the normalizer of a pronormal subgroup is abnormal *the absence of abnormal subgroups is equivalent to the normality of all pronormal subgroups.* These above characterizations can be extended on finitely generated hyper-(abelian-by-finite) groups (and, in particular, on soluble-by-finite groups). This is a consequence of the following fundamental result due to D. J. S. Robinson [21]: *If $G$ is a finitely generated hyper-(abelian-by-finite) group and every finite factor-group of $G$ is nilpotent, then $G$ itself is nilpotent.*

For infinite groups we have the following simple result [15]: *Let $G$ be a locally nilpotent group. Then $G$ has no proper abnormal subgroups and every pronormal subgroup of $G$ is normal.* However, we do not know whether or not the converse of this result holds.

L. A. Kurdachenko, J. Otal, I. Ya. Subbotin [7] introduced the following wide extension of the class of soluble minimax groups. *Let $G$ be a group, $A$ a normal subgroup of $G$. We say that $A$ satisfies the condition Max-$G$ (respectively Min-$G$) if $A$ satisfies the maximal (respectively the minimal) condition for $G$-invariant subgroups. A group $G$ is said to be a generalized minimax group, if it has a finite series of normal subgroups*

$$\langle 1 \rangle = H_0 \leq H_1 \leq \cdots \leq H_n = G,$$

*every factor of which is abelian and either satisfies Max-$G$ or Min-$G$.*

Every soluble minimax group is obviously generalized minimax. However, the class of generalized minimax groups is significantly wider than the class of soluble minimax groups.

In the paper [7], the following generalization of the famous nilpotency criterion was obtained.

**Theorem 2.1 (L. A. Kurdachenko, J. Otal, I. Ya. Subbotin [7])** *Let $G$ be a generalized minimax group. If every pronormal subgroup of $G$ is normal, then $G$ is hypercentral.*

Let $G$ be a group. Then the set

$$FC(G) = \{x \in G \mid x^G \text{ is finite}\}$$

is a characteristic subgroup of $G$, which is called *the $FC$-center of $G$*. Note that a group $G$ is an $FC$-group if and only if $G = FC(G)$. Starting from the $FC$-center, we construct *the upper $FC$-central series of a group $G$*

$$\langle 1 \rangle = C_0 \leq C_1 \leq \cdots \leq C_\alpha \leq C_{\alpha+1} \leq \cdots C_\gamma$$

where $C_1 = FC(G)$, $C_{\alpha+1}/C_\alpha = FC(G/C_\alpha)$ for all $\alpha < \gamma$, and $FC(G/C_\gamma) = \langle 1 \rangle$.

The term $C_\alpha$ is called *the $\alpha$-$FC$-hypercenter of $G$*, while the last term $C_\gamma$ of this series is called *the upper $FC$-hypercenter of $G$*. If $C_\gamma = G$, then the group $G$ is called *$FC$-hypercentral*, and, if $\gamma$ is finite, then $G$ is called *$FC$-nilpotent.*

**Theorem 2.2 (L. A. Kurdachenko, A. Russo, G. Vincenzi [9])** *Let $G$ be a group whose pronormal subgroups are normal. Then every $FC$-hypercenter of $G$ having finite number is hypercentral.*

**Theorem 2.3 (L. A. Kurdachenko, A. Russo, G. Vincenzi [9])** *Let $G$ be an $FC$-nilpotent group. If all pronormal subgroups in $G$ are normal, then $G$ is hypercentral.*

**Theorem 2.4 (L. A. Kurdachenko, A. Russo, G. Vincenzi [9])** *Let $G$ be a group whose pronormal subgroups are normal. Suppose that $H$ is a $FC$-hypercenter of $G$ having finite number. If $C$ is a normal subgroup of $G$ such that $C \geq H$ and $C/H$ is hypercentral, then $C$ is hypercentral.*

For periodic groups the above results were obtained by L. A. Kurdachenko and I. Ya. Subbotin [11].

The next results connect the conditions of nilpotency with descendant subgroups. Recall that a subgroup $H$ is called *descendant* if there exists a descending chain

$$G = G_0 \geq G_1 \geq \cdots G_\alpha \geq G_{\alpha+1} \geq \cdots G_\gamma = H$$

where $G_{\alpha+1}$ is normal in $G_\alpha$ for each $\alpha < \gamma$. Descendant subgroups are a very wide generalization of subnormal subgroups. There are many useful properties of the groups with all subnormal subgroups have been obtained, but our knowledge about the groups with all descendant subgroups are very limited.

**Theorem 2.5 (L. A. Kurdachenko and I. Ya. Subbotin [11])** *Let $G$ be a group, every subgroup of which is descendant. If $G$ is $FC$-hypercentral, then $G$ is hypercentral.*

**Theorem 2.6 (L. A. Kurdachenko, J. Otal and I. Ya. Subbotin [8])** *Let $G$ be a generalized minimax group. Then every subgroup of $G$ is descendant if and only if $G$ is nilpotent.*

**Theorem 2.7 (A. Ballester-Bolinches and T. Pedraza [1])** *Let $G$ be a group, every subgroup of which is descendant. If $G$ is a radical group with Chernikov Sylow $p$-subgroups for all primes $p$, then $G$ is hypercentral and the center of $G$ includes the divisible part of $G$.*

## References


[1] A. Ballester-Bolinches and T. Pedraza, On a class of generalized nilpotent groups, *J. Algebra* **248** (2002), 219–229.

[2] R. W. Carter, Nilpotent self-normalizing subgroups of soluble groups, *Math. Z.* **75** (1961), 136–139.

[3] M. R. Dixon, A conjugacy theorem for Carter subgroups in groups with min-$p$ for all $p$, *Lecture Notes in Math.* **848** (1981), 161–168.

[4] A. D. Gardiner, B. Hartley and M. J. Tomkinson, Saturated formations and Sylow structure in locally finite groups, *J. Algebra* **17** (1971), 177–211.

[5] F. De Giovanni and G. Vincenzi, Some topics in the theory of pronormal subgroups of groups, in *Topics in Infinite Groups*, Quaderni di Matematica **8** (2001), 175–202.
[6] P. Hall, On the system normalizers of a soluble group, *Proc. London Math. Soc.* **43** (1937), 507–528.
[7] L. A. Kurdachenko, J. Otal and I. Ya. Subbotin, On some criteria of nilpotency, *Comm. Algebra* **30** (2002), 3755–3776.
[8] L. A. Kurdachenko, J. Otal and I. Ya. Subbotin, Abnormal, pronormal, contranormal and Carter subgroups in some generalized minimax groups, *Comm. Algebra*, to appear.
[9] L. A. Kurdachenko, A. Russo and G. Vincenzi, Groups without the proper abnormal subgroups, *J. Group Theory*, to appear.
[10] L. A. Kurdachenko and I. Ya. Subbotin, On transitivity of pronormality, *Comment. Matemat. Univ. Caroline* **43** (2002), 583–594.
[11] L. A. Kurdachenko and I. Ya. Subbotin, Pronormality, contranormality and generalized nilpotency in infinite groups, *Pub. Mat.* **47** (2003), 389–414.
[12] L. A. Kurdachenko and I. Ya. Subbotin, Abnormal subgroups and Carter subgroups in some infinite groups, *Algebra and Discrete Mathematics*, Number **1**, (2005), 63–77.
[13] N. F. Kuzennyi and I. Ya. Subbotin, Groups in which all subgroups are pronormal, *Ukrainian Mat. Zh.* **39** (1987), 325–329; English translation: *Ukrainian Math. J.* **39** (1987), 251–254.
[14] N. F. Kuzennyi and I. Ya. Subbotin, Locally soluble groups in which all infinite subgroups are pronormal, *Izv. Vyssh. Ucheb. Zaved., Mat.* **11** (1988), 77–79; English translation: *Soviet Math.* **11** (1988), 126–131.
[15] N. F. Kuzennyi and I. Ya. Subbotin, New characterization of locally nilpotent $\overline{IH}$-groups, *Ukrainian Math. J.* **40** (1988), 274–277.
[16] N. F. Kuzennyi and I. Ya. Subbotin, Groups with pronormal primary subgroups, *Ukrainian Mat. Zh.* **41** (1989), 323–327; English translation: *Ukrainian Math. J.* **41** (1989), 286–289.
[17] N. F. Kuzennyi and I. Ya. Subbotin, On groups with fan subgroups, *Contemporary Mathematics* **131** Part 1 (1992), 383–388.
[18] A. Yu. Olshanskii, *Geometry of defining relations in groups,* Kluwer Acad. Publ., Dordrecht, 1991.
[19] T. A. Peng, Finite groups with pronormal subgroups, *Proc. Amer. Math. Soc.* **20** (1969), 232–234.
[20] T. A. Peng, Pronormality in finite groups, *J. London Math. Soc.* **3** (1971), 301–306.
[21] D. J. S. Robinson, A theorem on finitely generated hyperabelian groups, *Invent. Math.* **10** (1970), 38–43.
[22] D. J. S. Robinson, The vanishing of certain homology and cohomology group, *J. Pure Applied Algebra* **7** (1976), 145–167.
[23] D. J. S. Robinson, On the homology of hypercentral groups, *Arch. Math.* **32** (1979), 223–226.
[24] D. J. S. Robinson, Applications of cohomology to the theory of groups, *London Math. Soc. Lecture Notes Series* **71** (1982), 46–80.
[25] D. J. S. Robinson, *A course in the theory of groups*, Springer, New York, 1982.
[26] D. J. S. Robinson, Cohomology of locally nilpotent groups, *J. Pure Applied Algebra* **48** (1987), 281–300.
[27] D. J. S. Robinson, Homology and cohomology of locally supersoluble groups, *Math. Proc. Cambridge Philos. Soc.* **102** (1987), 233–250.
[28] J. S. Rose, Nilpotent subgroups of finite soluble groups, *Math. Z.* **106** (1968), 250–251.
[29] S. E. Stonehewer, Abnormal subgroup of a class of periodic locally soluble groups, *Proc. London Math. Soc.* **14** (1964), 520–536.
[30] S. E. Stonehewer, Locally soluble $FC$-groups, *Arch. Math.* **16** (1965), 158–177.

[31] D. I. Zaitsev, The hypercyclic extensions of abelian groups, in *The groups defined by the properties of systems of subgroups, Math. Inst. Kiev 1979*, 16–37.

[32] D. I. Zaitsev, On splitting of extensions of the abelian groups, in *The investigations of groups with the prescribed properties of systems of subgroups, Math. Inst. Kiev 1981*, 14–25.

[33] D. I. Zaitsev, The splitting extensions of abelian groups, in *The structure of groups and the properties of its subgroups, Math. Inst. Kiev 1986*, 22–31.

[34] D. I. Zaitsev, The hyperfinite extensions of abelian groups, in *The investigations of groups with the restrictions on subgroups, Math. Inst. Kiev 1988*, 17–26.

[35] D. I. Zaitsev, On locally supersoluble extensions of abelian groups, *Ukrainian Math. J.* **42** (1990), 908–912.

# $P$-LOCALIZING GROUP EXTENSIONS

KARL LORENSEN

Mathematics Department, Pennsylvania State University, Altoona College,
3000 Ivyside Park, Altoona, PA 16601-3760, USA
Email: kql3@psu.edu

## Abstract

We examine the effect of the $P$-localization functor on three types of group extensions: extensions that give rise to a nilpotent action on the kernel, extensions with a nilpotent kernel and a torsion quotient, and extensions with a finite kernel.

## 1 Introduction

Assume $P$ is a family of primes. A group $G$ is said to be *$P$-local* if the function $x \mapsto x^q$ from $G$ to $G$ is bijective for any prime $q$ in the complement of $P$. As shown in [16], any group $G$ can be mapped canonically into a unique $P$-local group $G_P$; moreover, the assignment $G \mapsto G_P$ defines a functor from the category of groups to the category of $P$-local groups. This functor is called the *$P$-localization functor*, and it plays an important role in homotopy theory (see [5] and [12]). For nilpotent groups the properties of the $P$-localization functor are well understood (see [1], [9], and [12]); however, its properties outside this subcategory remain largely a mystery.

One avenue to a more complete understanding of this functor is to determine its effect on short exact sequences. This is the aim of three papers, one by C. Casacuberta and M. Castellet [4] and two by the author [13, 14]. The first of these examines the effect of $P$-localization on group extensions with a nilpotent kernel and a torsion quotient, the second looks at extensions with a finite kernel, and the third investigates extensions that give rise to a nilpotent action on the kernel.

In the present paper we present a survey, with some elaborations, of the results in [4], [13], and [14]. Section 2 discusses the results in [14], treating extensions of the form $1 \to N \xrightarrow{\iota} G \xrightarrow{\epsilon} Q \to 1$, where the action of $G$ on $N$ is nilpotent. For such extensions the sequence $N_P \xrightarrow{\iota_P} G_P \xrightarrow{\epsilon_P} Q_P \to 1$ is always exact. Moreover, if $Q$ satisfies a certain pair of homological conditions, the sequence $1 \to N_P \xrightarrow{\iota_P} G_P \xrightarrow{\epsilon_P} Q_P \to 1$ is exact.

Section 3 is devoted to the results on extensions with nilpotent kernel and torsion quotient from [4]. The principal contribution of that paper is to show that, if $1 \to N \to G \to Q \to 1$ is a group extension in which $N$ is nilpotent of class $c$ and $Q$ is torsion, then there is an exact sequence

$$1 \to (K/\gamma_{c+1}K)_P \to G_P \to Q_P \to 1,$$

where $K$ is the inverse image in $G$ of the kernel of the $P$-localization map $l_P : Q \to Q_P$ and $\gamma_{c+1}K$ is the $(c+1)$st term of the lower central series of $K$

(in which $\gamma_1 K = K$).

In Section 4 we discuss the contents of [13]. Here, as opposed to the previous sections, we provide proofs of most of our assertions, simplifying the proof of the main theorem from [13] and presenting the results in light of some recent advances in homotopy theory. This section treats extensions of the form $1 \to N \xrightarrow{\iota} G \xrightarrow{\epsilon} Q \to 1$, where $N$ is finite. In this case, as above with extensions with nilpotent action, we have that the sequence $N_P \xrightarrow{\iota_P} G_P \xrightarrow{\epsilon_P} Q_P \to 1$ is exact. Furthermore, if $Q$ belongs to a particular class of groups, denoted $\mathfrak{U}_P$, then we can provide an explicit description of $\operatorname{Ker} \iota_P$ by employing the left action of $G$ on $N_P$ that is induced by the action of $G$ on $N$ by conjugation. We establish that $\operatorname{Ker} \iota_P$ is the subgroup of $N_P$ generated by all the elements of the form $(g \cdot a)a^{-1}$ such that $a \in N_P$ and $g$ is an element of $G$ whose image in $\operatorname{Aut}(N)$ has order relatively prime to all the primes in $P$.

The class $\mathfrak{U}_P$ referred to above is the class of all groups whose classifying space has a $P$-localization that is also aspherical. Such groups were first studied by C. Casacuberta and G. Peschke [5], who proved that both nilpotent and free groups are contained in $\mathfrak{U}_P$ for every family $P$. In [7] A. Descheemaeker and W. Malfait discuss some virtually nilpotent groups that are contained in $\mathfrak{U}_P$, demonstrating, for example, that the five groups that occur as the fundamental groups of non-orientable flat manifolds of dimension less than four are all in $\mathfrak{U}_P$ for every set $P$. We discuss the class $\mathfrak{U}_P$ in the final section of the paper (Section 5), showing how to construct new groups in $\mathfrak{U}_P$ by forming certain types of extensions involving groups in $\mathfrak{U}_P$.

**Definitions and notation** Throughout this paper, $G$ will represent a group. We will use the following definition of a commutator: $[g, h] = ghg^{-1}h^{-1}$ for $g,\ h \in G$. Commutators of weight $n \geq 3$ are defined inductively by $[g_1, \cdots, g_n] = [[g_1, \cdots, g_{n-1}], g_n]$ for $g_1, \cdots, g_n \in G$. The subgroup of $G$ generated by all the commutators of weight $n$ is denoted $\gamma_n G$. We will let $Z(G)$ represent the center of $G$.

In this paper $P$ will always represent a set of prime positive integers and $P'$ the set of all positive primes outside of $P$. A *$P$-number* is an integer all of whose prime divisors are in $P$. The group $G$ is a *$P$-group* if each element of $G$ has a finite order that is a $P$-number. The group $G$ is *$P$-local* if the function $g \mapsto g^n$ from $G$ to $G$ is a bijection for every $P'$-number $n$.

The *$P$-localization functor* $()_P$ is a functor from the category of groups to the category of $P$-local groups such that there is a natural transformation $l_P : G \to G_P$ with the following universal property: for any group homomorphism $\phi : G \to H$ with $H$ $P$-local, there exists a unique map $\psi : G_P \to H$ with $\psi l_P = \phi$. Any group homomorphism $\phi : G \to K$ obeying the same universal property as $l_P : G \to G_P$ is a *$P$-equivalence*; moreover, if $K$ is $P$-local, then $\phi$ is said to *$P$-localize* $G$.

Let $\phi : G \to H$ be a homomorphism of nilpotent groups. The map $\phi$ is *$P$-injective* if $\operatorname{Ker} \phi$ is a $P'$-group and *$P$-surjective* if $\operatorname{Coker} \phi$ is a $P'$-group. If $\phi$ is both $P$-injective and $P$-surjective, then it is *$P$-bijective*. Moreover, $\phi$ is $P$-bijective if and only if $\phi$ is a $P$-equivalence.

As described in [11], any group extension $1 \to N \to G \to Q \to 1$ embeds into a diagram of the form

$$\begin{array}{ccccccccc} 1 & \longrightarrow & N & \longrightarrow & G & \longrightarrow & Q & \longrightarrow & 1 \\ & & \downarrow{\scriptstyle l_P} & & \downarrow{\scriptstyle l_{(P)}} & & \| & & \\ 1 & \longrightarrow & N_P & \longrightarrow & G_{(P)} & \longrightarrow & Q & \longrightarrow & 1, \end{array}$$

where the bottom row is exact. The group $G_{(P)}$ is called the *relative $P$-localization of $G$*. Moreover, as shown in [4, Proposition 1.3], the map $l_{(P)} : G \to G_{(P)}$ is a $P$-equivalence.

For the proofs of our results in Section 4 we require some terminology and notation regarding group extensions. Any extension $1 \to N \to G \to Q \to 1$ gives rise to a natural homomorphism $\chi : Q \to \mathrm{Out}(N)$ called the *coupling* of the extension. The set of all extensions of $N$ by $Q$ with a given coupling $\chi : Q \to \mathrm{Out}(N)$ is denoted by $\mathcal{E}(Q, N, \chi)$. An extension $1 \to N \to G \to Q \to 1$ also gives rise to a left action of $Q$ on $Z(N)$; we will regard $Z(N)$ as a left $Q$-module under this action.

## 2 Extensions with a nilpotent action on the kernel

In this section we investigate the effect of $P$-localization on a group extension that gives rise to a nilpotent action on the kernel. We observe, first of all, that $P$-localization is right exact on such extensions, and then we examine when full exactness is preserved. The proofs of the results in this section may be found in [14].

We begin by recalling the definition of a nilpotent action.

**Definitions and notation** The term *$G$-group* will be used for a left $G$-group. If $N$ is a $G$-group, then the function $G \times N \to N$ will be denoted by $(g, a) \mapsto g \cdot a$. When $N$ is a normal subgroup of $G$, we will assume that the action of $G$ on $N$ arises from conjugation; that is, $g \cdot a = gag^{-1}$ for every $a \in N$ and $g \in G$.

Let $N$ be a $G$-group. We define the *lower central $G$-series*

$$\cdots \subseteq \gamma_3^G N \subseteq \gamma_2^G N \subseteq \gamma_1^G N$$

of $N$ as follows: $\gamma_1^G N = N$; $\gamma_i^G N = \langle a(g \cdot b)a^{-1}b^{-1} \mid a \in N,\ b \in \gamma_{i-1}^G N,\ g \in G\rangle$ for $i > 1$.

We let $Z^G N = \{a \in N \mid b(g \cdot a)b^{-1} = a \ \text{ for all } \ g \in G,\ b \in N\}$. The *upper central $G$-series*

$$Z_1^G N \subseteq Z_2^G N \subseteq Z_3^G N \subseteq \cdots$$

of $N$ is defined as follows: $Z_1^G N = Z^G N$; $Z_i^G N / Z_{i-1}^G N = Z^G(N/Z_{i-1}^G N)$.

It is an elementary fact that, if $N$ is a $G$-group, then $Z_n^G N = N$ if and only if $\gamma_G^{n+1} N = 1$.

We say that the action of $G$ on $N$ is *nilpotent* if there is a nonnegative integer $c$ such that $Z_c^G N = N$. The smallest such integer $c$ is called the *nilpotency class* of the action and denoted $\mathrm{nil}_G N$.

In [11, Theorem 2.8] and [11, Corollary 2.9], P. Hilton establishes the following two properties of nilpotent $G$-groups.

**Lemma 2.1 (Hilton)** *Assume $N$ is a nilpotent $G$-group. Then the following two statements hold.*

(i) *For any positive integer $n$, the map $\gamma_n^G N \to \gamma_n^G N_P$ induced by $l_P : N \to N_P$ $P$-localizes.*

(ii) *If the action of $G$ on $N$ is nilpotent, then the action of $G$ on $N_P$ is also nilpotent and $\mathrm{nil}_G N_P \leq \mathrm{nil}_G N$.*

For extensions that give rise to a nilpotent action on the kernel the $P$-localization functor is always right exact.

**Theorem 2.2** *Assume $1 \to N \xrightarrow{\iota} G \xrightarrow{\epsilon} Q \to 1$ is a short exact sequence such that the action of $G$ on $N$ is nilpotent. Then the sequence $N_P \xrightarrow{\iota_P} G_P \xrightarrow{\epsilon_P} Q_P \to 1$ is exact. Furthermore, the action of $G_P$ on $\mathrm{Im}\,\iota_P$ is nilpotent with $\mathrm{nil}_{G_P}(\mathrm{Im}\,\iota_P) \leq \mathrm{nil}_G N$.*

That full exactness is not always preserved after $P$-localizing an extension with nilpotent action as illustrated by the example of the central extension $1 \to \mathbb{Z}/2 \to SL_2(3) \to PSL_2(3) \to 1$ (see [13, p. 204]). If this extension is localized at $\{2\}$, then the injectivity of the map $\mathbb{Z}/2 \to SL_2(3)$ is not preserved since the $\{2\}$-localization of $SL_2(3)$ is trivial. However, full exactness is retained after $P$-localizing an extension with nilpotent action in some important cases. This holds, for instance, if the extension splits.

**Theorem 2.3** *Assume $1 \to N \xrightarrow{\iota} G \xrightarrow{\epsilon} Q \to 1$ is a split group extension in which the action of $Q$ on $N$ is nilpotent. Then the sequence $1 \to N_P \xrightarrow{\iota_P} G_P \xrightarrow{\epsilon_P} Q_P \to 1$ is exact.*

As the following two theorems show, we may also guarantee that $P$-localization preserves exactness in an extension with nilpotent action by placing certain homological conditions on the quotient.

**Theorem 2.4** *Assume $0 \to A \xrightarrow{\iota} G \xrightarrow{\epsilon} Q \to 1$ is a central group extension such that the map $H_2 Q \to H_2 Q_P$ induced by $l_P : Q \to Q_P$ is $P$-bijective. Then the sequence $0 \to A_P \xrightarrow{\iota_P} G_P \xrightarrow{\epsilon_P} Q_P \to 1$ is exact.*

**Theorem 2.5** *Assume $1 \to N \xrightarrow{\iota} G \xrightarrow{\epsilon} Q \to 1$ is a group extension in which the action of $G$ on $N$ is nilpotent. Suppose, further, that $Q$ satisfies the following two properties:*

(i) *The map $H_2 Q \to H_2 Q_P$ induced by $l_P : Q \to Q_P$ is $P$-bijective.*

(ii) *The map $H_3 Q \to H_3 Q_P$ induced by $l_P : Q \to Q_P$ is $P$-surjective.*

*Then the sequence $1 \to N_P \xrightarrow{\iota_P} G_P \xrightarrow{\epsilon_P} Q_P \to 1$ is exact.*

As observed in [14], it is impossible to dispense with the second condition on $Q$ in the above theorem.

In [14] a number of examples of groups that satisfy conditions (i) and (ii) of Theorem 2.5 are cited. The most important source of examples of such groups is the class $\mathfrak{U}_P$ discussed in Section 5 of the present paper. This class includes all nilpotent groups, all free groups, and all the groups that occur as fundamental groups of non-orientable flat manifolds of dimension less than four. In addition to the groups in $\mathfrak{U}_P$, there are also other groups that satisfy conditions (i) and (ii) of Theorem 2.5. For instance, all the dihedral groups—both finite and infinite—manifest these properties, as well as any finite simple group with a trivial Schur multiplier. Furthermore, the class of groups satisfying these two conditions is closed under finite direct products, direct limits, and central extensions.

We conclude this section by applying Theorem 2.5 to calculate a $P$-localization of a particular group.

**Example 2.6** Let $G$ be the group with the presentation

$$\begin{aligned} G = \langle a, b, c, \alpha, \beta \mid & [a,b] = c^2;\ [a,c] = [b,c] = 1; \\ & [\alpha, a] = [\alpha, b] = [\alpha, c] = [\beta, a] = [\beta, c] = 1; \\ & \beta b \beta^{-1} = ab;\ \beta\alpha\beta^{-1} = \alpha^{-1}c\rangle. \end{aligned}$$

Let $N$ be the subgroup of $G$ generated by $a$, $b$ and $c$. Then

$$N \cong \begin{pmatrix} 1 & \mathbb{Z} & \mathbb{Z} \\ 0 & 1 & 2\mathbb{Z} \\ 0 & 0 & 1 \end{pmatrix}.$$

Moreover, $N$ is normal, and $G$ acts nilpotently on $N$ with $\mathrm{nil}_G N = 3$. Also, $G/N \cong \Pi = \langle \bar{\alpha}, \bar{\beta} \mid \bar{\beta}\bar{\alpha}\bar{\beta}^{-1} = \bar{\alpha}^{-1}\rangle$, the fundamental group of the Klein bottle, a group that satisfies both homological conditions in Theorem 2.5 (see Proposition 5.4). Assume $P$ is a set of odd primes. Then, since $(\bar{\beta}\bar{\alpha}\bar{\beta}^{-1})^2 = \bar{\alpha}^2$, $\bar{\alpha}$ is in the kernel of $l_P : \Pi \to \Pi_P$, implying that $\Pi_P \cong \mathbb{Z}_P$. Applying Theorem 2.5 yields a split short exact sequence

$$1 \to \begin{pmatrix} 1 & \mathbb{Z}_P & \mathbb{Z}_P \\ 0 & 1 & \mathbb{Z}_P \\ 0 & 0 & 1 \end{pmatrix} \to G_P \to \mathbb{Z}_P \to 1,$$

in which the quotient $\mathbb{Z}_P$ acts on the kernel in the following way:

$$x \cdot \begin{pmatrix} 1 & s & t \\ 0 & 1 & 2u \\ 0 & 0 & 1 \end{pmatrix} = \begin{pmatrix} 1 & s + xu & t + xu(u-1) \\ 0 & 1 & 2u \\ 0 & 0 & 1 \end{pmatrix}$$

for all $x$, $s$, $t$, $u$ and $v$ in $\mathbb{Z}_P$.

## 3 Extensions with nilpotent kernel and torsion quotient

In this section we examine the effect of $P$-localization on a group extension with nilpotent kernel and torsion quotient. Combining [4, Theorem 2.1], [14, Theorem 5.2], and [14, Theorem 5.3], we have the following result.

**Theorem 3.1** *Assume $1 \to N \xrightarrow{\iota} G \xrightarrow{\epsilon} Q \to 1$ is a group extension with $N$ nilpotent of class $c$ and $Q$ torsion. Let $S$ be the kernel of $l_P : Q \to Q_P$, and set $K = \epsilon^{-1}(S)$. Then the following four statements hold.*

(i) *There is an exact sequence*

$$1 \to (K/\gamma_{c+1}K)_P \to G_P \xrightarrow{\epsilon_P} Q_P \to 1.$$

(ii) *If $H_2S$ is a $P'$-group, then the sequence*

$$1 \to \gamma^K_{c+1}N_P \xrightarrow{\subseteq} N_P \xrightarrow{\iota_P} G_P \xrightarrow{\epsilon_P} Q_P \to 1 \tag{3.1}$$

*is exact.*

(iii) *If $K/\gamma^K_{c+1}N$ splits over $N/\gamma^K_{c+1}N$, then the sequence (3.1) is exact.*

Below we apply the above theorem to a particular group extension.

**Example 3.2** Let $G$ be the group with presentation

$$\begin{aligned} G = \langle a, b, c, d, \alpha, \beta, \gamma \mid\ & [a,b] = [a,c] = [a,d] = 1; \\ & [b,d] = [c,d] = 1;\ \alpha a \alpha^{-1} = a; \\ & \alpha b \alpha^{-1} = b^{-1};\ \alpha c \alpha^{-1} = c^{-1}; \\ & \alpha d \alpha^{-1} = d;\ \beta a \beta^{-1} = a; \\ & \beta b \beta^{-1} = b;\ \beta c \beta^{-1} = c^{-1}; \\ & \beta d \beta^{-1} = d^{-1};\ \gamma a \gamma^{-1} = a; \\ & \gamma b \gamma^{-1} = c;\ \gamma c \gamma^{-1} = d; \\ & \gamma d \gamma^{-1} = b^{-1};\ \alpha^2 = d;\ \beta^2 = b; \\ & \gamma^6 = a;\ \beta\gamma = b\gamma\alpha; \\ & \alpha\beta = b^{-1}c^{-1}d\beta\alpha;\ \alpha\gamma = bd\gamma\alpha\beta \rangle \end{aligned}$$

This group is the fundamental group of a four dimensional flat manifold of the type 25/01/01/010 discussed in [2]; it fits into an extension of the form $1 \to \mathbb{Z}^4 \to G \to A_4 \times \mathbb{Z}/2 \to 1$. We will localize $G$ at the set of primes $P$. First suppose that $2, 3 \in P$. Then, by Theorem 3.1(ii) (or by Corollary 4.4 in the next section), $G_P$ fits into an extension $1 \to \mathbb{Z}^4_P \to G_P \to A_4 \times \mathbb{Z}/2 \to 1$.

Now assume $2 \notin P$ and $3 \in P$. Then the kernel $S$ of the map that $P$-localizes $A_4 \times Z/2$ is isomorphic to $(\mathbb{Z}/2)^3$. Moreover, the preimage $K$ of $S$ in $G$ is generated by $a$, $b$, $c$, $d$, $\alpha$, $\beta$ and $\gamma^3$. Thus $N/\gamma^K_2 N \cong \mathbb{Z} \times \mathbb{Z}/2 \times \mathbb{Z}/2 \times \mathbb{Z}/2$, where the image of $a$ corresponds to the generator of the first factor in the direct product. Therefore, by Theorem 3.1(ii), $G_P$ fits into an exact sequence $1 \to \mathbb{Z}_P \to G_P \to \mathbb{Z}/3 \to 1$. Moreover, in view of the relations $\gamma^6 = a$ and $\gamma a = a\gamma$, we have $G_P \cong \mathbb{Z}_P$.

Suppose $2 \notin P$ and $3 \notin P$. Then the kernel of the map that $P$-localizes $A_4 \times \mathbb{Z}/2$ is $A_4 \times \mathbb{Z}/2$. Moreover, $H_2(A_4 \times \mathbb{Z}/2) = \mathbb{Z}/2$. Thus, since $N/\gamma^G_2 N \cong \mathbb{Z} \times \mathbb{Z}/2$, we have, by Theorem 3.1(ii), that $G_P \cong \mathbb{Z}_P$.

Finally, consider the case where $2 \in P$ and $3 \notin P$. Here the kernel of the map that $P$-localizes the quotient is $A_4$, so that its inverse image $K$ in $G$ is generated by $a$, $b$, $c$, $d$ $\alpha$, $\beta$ and $\gamma^2$. A simple calculation reveals that $K/\gamma_2 K = \mathbb{Z}$, where $K/\gamma_2 K$ is generated by the image of $\gamma^2$. Therefore, by Theorem 3.1(i), $G_P$ fits into an extension of the form $1 \to \mathbb{Z}_P \to G_P \to \mathbb{Z}/2 \to 1$. It follows, then, that $G_P \cong \mathbb{Z}_P$

## 4 Group extensions with a finite kernel

In this section we discuss the effect of $P$-localization on a group extension with finite kernel. In this endeavor, the following notion will be important.

**Definition** Let $N$ be a $G$-group. The action of $G$ on $N$ is $P$-*local* if the function $a \mapsto a(g{\cdot}a)\cdots(g^{n-1}{\cdot}a)$ from $N$ to $N$ is a bijection for every $g \in G$ and $P'$-number $n$.

The relevance of the above notion to group extensions is elucidated by the following proposition.

**Proposition 4.1** *Assume $N$ is a normal subgroup of $G$.*

(i) *If $G$ is a $P$-local group, then $G/N$ is a $P$-local group if and only if the action of $G$ on $N$ is $P$-local.*

(ii) *If $G/N$ is a $P$-local group and the action of $G$ on $N$ is $P$-local, then $G$ is a $P$-local group.*

**Proof** The results follow quite easily from the fact that

$$(ag)^n = a(g \cdot a) \cdots (g^{n-1} \cdot a) g^n$$

for all $g \in G$ and $a \in N$. □

In [13] the author identifies the following class of $P$-local actions.

**Proposition 4.2** *Let $N$ be a $G$-group and $\omega : G \to \mathrm{Aut}(N)$ the homomorphism arising from the action of $G$ on $N$. Assume*

(i) *$N$ is $P$-local, and*

(ii) *for each $\alpha \in \mathrm{Im}\,\omega$, there is a $P$-number $n$ such that $\alpha^n \in \mathrm{Inn}(N)$.*

*Then the action of $G$ on $N$ is $P$-local.*

The above proposition has the following corollary.

**Corollary 4.3** *Let $1 \to N \to G \to Q \to 1$ be a group extension with coupling $\chi : Q \to \mathrm{Out}(N)$. If $Q$ is $P$-local and $\mathrm{Im}\,\chi$ is a $P$-group, then there is an exact sequence $1 \to N_P \to G_P \to Q \to 1$.*

**Proof** Form the relative $P$-localization $1 \to N_P \to G_{(P)} \to Q \to 1$. By Proposition 4.2 and Proposition 4.1, $G_{(P)}$ is $P$-local. Moreover, the map $l_{(P)} : G \to G_{(P)}$ is a $P$-equivalence. Therefore, $G_{(P)} \cong G_P$, proving the corollary. □

In [8] Garcia proves the following special case of the above corollary.

**Corollary 4.4 (Garcia)** *Let $1 \to N \to G \to Q \to 1$ be a group extension in which $Q$ is a $P$-group. Then there is an exact sequence $1 \to N_P \to G_P \to Q \to 1$.*

In this section our primary interest is in $P$-local actions on finite groups. These actions are characterized by

**Proposition 4.5 ([13, Proposition 2.8])** *Assume $N$ is a finite $G$-group. Then the action of $G$ on $N$ is $P$-local if and only if the following two conditions hold:*
(i) *$N$ is a $P$-group;*
(ii) *the image of $G$ in $\mathrm{Aut}(N)$ is a $P$-group.*

The above proposition provides us with the following property.

**Corollary 4.6** *If $N$ is a finite group on which $G$ acts $P$-locally, then $G$ acts $P$-locally on every $G$-subgroup and every $G$-quotient group of $N$.*

Next we establish that any $P$-local action of $G$ on a finite group $N$ induces an action of $G_P$ on $N$.

**Lemma 4.7** *Assume $G$ acts $P$-locally on a finite group $N$. Let $\omega : G \to \mathrm{Aut}(N)$ be the homomorphism arising from the action of $G$ on $N$. Then there exists a unique map $\omega^P : G_P \to \mathrm{Aut}(N)$ such that the diagram*

$$\begin{array}{ccc} G & \xrightarrow{\omega} & \mathrm{Aut}(N) \\ {\scriptstyle \iota_P}\downarrow & \nearrow{\scriptstyle \omega^P} & \\ G_P & & \end{array}$$

*commutes. Moreover, $\mathrm{Im}\,\omega^P = \mathrm{Im}\,\omega$.*

**Proof** By Proposition 4.5, $\mathrm{Im}\,\omega$ is a $P$-group. An appeal to the "existence part" of the universal property of $P$-localization provides us with a map $\omega^P$ that makes the diagram commute and whose image is the same as $\mathrm{Im}\,\omega$. The uniqueness of $\omega^P$ follows from the "uniqueness part" of the universal property of $P$-localization and the fact that any finite homomorphic image of a $P$-local group must be a $P$-group. □

Given a finite $G$-group $N$, we now consider the maximal $G$-quotient of $N$ on which $G$ acts $P$-locally.

**Definition** Assume that $N$ is a finite $G$-group. We define $\Gamma^G_{P'}N$ to be the subgroup of $N$ generated by all the elements of the form $(g \cdot a)a^{-1}$ such that $a \in N$, $g \in G$, and the order of the image of $g$ in $\mathrm{Aut}(N)$ is a $P'$-number. The subgroup of $N$ generated by all the elements with $P'$-orders is denoted by $T_{P'}N$. Let $T^G_{P'}N$ be the subgroup of $N$ generated by $\Gamma^G_{P'}N$ and the subgroup $T_{P'}N$.

It is easy to see that $\Gamma^G_{P'}N$ and $T^G_{P'}N$ are both $G$-subgroups of $N$. The $G$-subgroup $T^G_{P'}N$ possesses the following two properties.

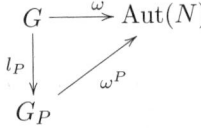

**Lemma 4.8** *Assume $N$ is a finite $G$-group. Then the $G$-subgroup $T^G_{P'}N$ is the minimal $G$-subgroup of $N$ such that the action of $G$ on $N/T^G_{P'}N$ is $P$-local.*

**Proof** This follows at once from Proposition 4.5. □

**Lemma 4.9** *If $N$ is a finite normal subgroup of $G$, then, for any group homomorphism $\phi : G \to H$ where $H$ is $P$-local, $T^G_{P'}N$ is contained in* $\operatorname{Ker}\phi$.

**Proof** That $T_{P'}N \subseteq \operatorname{Ker}\phi$ is plain. To show that, in addition, $\Gamma^G_{P'}N \subseteq \operatorname{Ker}\phi$, let $a \in N$ and $g \in G$ such that the order $n$ of the image of $g$ in $\operatorname{Aut}(N)$ is a $P'$-number. Then $[\phi(a)\phi(g^{-1})\phi(a^{-1})]^n = [\phi(g^{-1})]^n$, so that $\phi(a)\phi(g^{-1})\phi(a^{-1}) = \phi(g^{-1})$. Hence $(g \cdot a)a^{-1} \in \operatorname{Ker}\phi$. □

Now we turn our attention to group extensions with finite kernel. First we establish the following two elementary properties of finite subgroups.

**Lemma 4.10** *Let $H$ be a finite subgroup of a $P$-local group $G$. If $g \in G$ such that $g^n \in N_G(H)$ for some $P'$-number $n$, then $g \in N_G(H)$.*

**Proof** Since $H$ is finite, there exists $m \in \mathbb{Z}$ such that $g^m h g^{-m} = h$ for all $h \in H$. Hence $(hgh^{-1})^m = g^m$ for all $h \in G$. This means that, for each $h \in H$, $(hgh^{-1})^k = g^k$, where $k$ is the largest $P$-number dividing $m$. It follows that $g^k \in C_G(H) \subseteq N_G(H)$. Therefore, since $n$ and $k$ are relatively prime, $g \in N_G(H)$. □

**Lemma 4.11** *Let $N$ be a finite normal subgroup of $G$. Then the image of $N$ under $l_P : G \to G_P$ is normal in $G_P$.*

**Proof** As described in [16], we can view $G_P$ as the direct limit of

$$G = G_1 \to G_2 \to G_3 \to \cdots,$$

where $G_i$ is obtained from $G_{i-1}$ by adjoining any missing $n$th roots for all $P'$-numbers $n$ and dividing out all the elements of the form $gh^{-1}$ such that $g^n = h^n$ in $G_{i-1}$ for some $P'$-number $n$. Hence $G_i$ is generated by elements that have $P'$-powers in the image of $G_{i-1}$. Therefore, by Lemma 4.10, if the image of $N$ in $G_P$ is normalized by the image of $G_{i-1}$ in $G_P$, then the image of $N$ must also be normalized by the image of $G_i$. It follows, then, by induction, that the image of $N$ in $G_P$ is normalized by the image of $G_i$ for all $i$. Thus the image of $N$ in $G_P$ is normal. □

Now we are prepared to consider the $P$-localization of a group extension with finite kernel.

**Theorem 4.12** *Assume $1 \to N \xrightarrow{\iota} G \xrightarrow{\epsilon} Q \to 1$ is a group extension in which $N$ is finite. Then the sequence*

$$N_P \xrightarrow{\iota_P} G_P \xrightarrow{\epsilon_P} Q_P \to 1$$

*is exact. Moreover, $\Gamma^G_{P'}N_P \subseteq$* $\operatorname{Ker}\iota_P$.

**Proof** By Lemma 4.9, $\Gamma^G_{P'}N_P \subseteq \operatorname{Ker}\iota_P$. Let $\hat{N}$ be the image of $\iota_P : N_P \to G_P$. The kernel of the epimorphism $N \to \hat{N}$ induced by $l_P : G \to G_P$ is normal in $G$, which means that the action of $G$ on $N$ by conjugation induces an action of $G$ on $\hat{N}$. Moreover, since $\hat{N}$ is a $G$-homomorphic image of the $G$-group $N/T^G_{P'}N$ with $P$-local action, $G$ must also act $P$-locally on $\hat{N}$ by Corollary 4.6. Let $\omega$ be the homomorphism $G \to \operatorname{Aut}(\hat{N})$ arising from the action of $G$ on $\hat{N}$. By Lemma 4.11, we have $\hat{N} \trianglelefteq G_P$, making $\hat{N}$ a $G_P$-group. Let $\omega^P$ be the homomorphism $G_P \to \operatorname{Aut}(\hat{N})$ that arises from the action of $G_P$ on $\hat{N}$. The two maps $\omega$ and $\omega^P$ make the diagram

$$\begin{array}{ccc} G & \xrightarrow{\ \omega\ } & \operatorname{Aut}(\hat{N}) \\ {\scriptstyle l_P}\downarrow & \nearrow{\scriptstyle \omega^P} & \\ G_P & & \end{array}$$

commute. It follows from Lemma 4.7 that $\operatorname{Im}\omega = \operatorname{Im}\omega^P$. Therefore, the action of $G_P$ on $\hat{N}$ is also $P$-local. Hence, by Proposition 4.1, $R = G_P/\hat{N}$ is a $P$-local group. Let $\phi : Q \to R$ be the map induced by $l_P : G \to G_P$. We claim that $\phi$ $P$-localizes $Q$, which will yield the desired result. To show this, let $\psi : Q \to S$ be a homomorphism such that $S$ is $P$-local. The universal property of $l_P : G \to G_P$ provides a unique map $\theta : G_P \to S$ such that $\theta l_P = \psi\epsilon$. The map $\theta$, then, induces a unique homomorphism $\eta : R \to S$ such that $\eta\phi = \psi$. Thus $\phi$ $P$-localizes $Q$. □

In the case of a split extension, we have that $\Gamma^G_{P'}N_P = \operatorname{Ker}\iota_P$ under the hypotheses of Theorem 4.12.

**Theorem 4.13** *Assume* $1 \to N \xrightarrow{\iota} G \xrightarrow{\epsilon} Q \to 1$ *is a split group extension in which* $N$ *is finite. Then the sequence*

$$1 \to \Gamma^G_{P'}N_P \xrightarrow{\subseteq} N_P \xrightarrow{\iota_P} G_P \xrightarrow{\epsilon_P} Q_P \to 1$$

*is exact.*

**Proof** Set $\bar{N} = N/T^G_{P'}N \cong N_P/\Gamma^G_{P'}N_P$, and let $\omega : Q \to \operatorname{Aut}(\bar{N})$ be the homomorphism arising from the action of $G$ on $\bar{N}$. By Lemma 4.7, there exists a map $\omega^P : Q_P \to \operatorname{Aut}(\bar{N})$ that makes the diagram

$$\begin{array}{ccc} Q & \xrightarrow{\ \omega\ } & \operatorname{Aut}(\bar{N}) \\ {\scriptstyle l_P}\downarrow & \nearrow{\scriptstyle \omega^P} & \\ Q_P & & \end{array}$$

commute. Therefore, the canonical map $N \to \bar{N}$ and the $P$-localizing map $Q \to Q_P$ induce a map $G \to \bar{N} \rtimes_{\omega^P} Q_P$. Moreover, by Lemma 4.7 and Proposition 4.5, the action of $Q_P$ on $\bar{N}$ is $P$-local. Thus, by Proposition 4.1, the group $\bar{N} \rtimes_{\omega^P} Q_P$ is $P$-local. This implies that the intersection of the kernel of $l_P : G \to G_P$ with $N$ must be exactly $T^G_{P'}N$, which, in view of Theorem 4.12, suffices to prove the theorem. □

Our next aim is to show that, under the hypotheses of Theorem 4.12, $\Gamma^G_{P'}N_P = \operatorname{Ker}\iota_P$ if $Q$ is in a special class of groups, denoted $\mathfrak{U}_P$. This class is defined as follows.

**Definition** The class $\mathfrak{U}_P$ is the class of groups $G$ such that the space $(K(G,1))_P$ is aspherical. Here $(K(G,1))_P$ represents the $P$-localization of the Eilenberg–Mac Lane space $K(G,1)$ in the sense discussed in Section 3 of [5].

In [5] Casacuberta and Peschke provide two algebraic characterizations of the class $\mathfrak{U}_P$.

**Proposition 4.14 (Casacuberta and Peschke)** *The following statements are equivalent.*

(i) *The group $G$ is in $\mathfrak{U}_P$.*

(ii) *For any abelian group $A$ upon which $G_P$ acts $P$-locally, the map $H_n(G,A) \to H_n(G_P,A)$ induced by $l_P : G \to G_P$ is an isomorphism for all $n \geq 0$.*

(iii) *For any abelian group $A$ upon which $G_P$ acts $P$-locally, the map $H^n(G_P,A) \to H^n(G,A)$ induced by $l_P : G \to G_P$ is an isomorphism for all $n \geq 0$.*

In the succeeding section we discuss examples of groups in the class $\mathfrak{U}_P$. These examples include all nilpotent groups, all free groups, as well as the fundamental groups of all the nonorientable flat manifolds of dimension $\leq 3$.

Below in Theorem 4.16 we show that, under the hypotheses of Theorem 4.12, $\Gamma^G_{P'}N_P = \operatorname{Ker}\iota_P$ if $Q$ is in the class $\mathfrak{U}_P$. The proof of this result is based on the following lemma which is established using S. Mac Lane's [15] theory of group extensions with a nonabelian kernel.

**Lemma 4.15** *Assume $1 \to N \to G \to Q \to 1$ is a group extension such that $N$ is finite and the action of $G$ on $N$ is $P$-local. Let $\chi : Q \to \operatorname{Out}(N)$ be the coupling of this extension. Then the following statements hold.*

(i) *There exists a unique homomorphism $\chi^P : Q_P \to \operatorname{Out}(N)$ such that the diagram*

$$\begin{array}{ccc} Q & \xrightarrow{\ \chi\ } & \operatorname{Out}(N) \\ {\scriptstyle l_P}\big\downarrow & \nearrow{\scriptstyle \chi^P} & \\ Q_P & & \end{array}$$

*commutes. Moreover,* $\operatorname{Im}\chi^P = \operatorname{Im}\chi$.

(ii) *If $Q$ is in the class $\mathfrak{U}_P$, then there exists an extension of $N$ by $Q_P$ with coupling $\chi^P : Q_P \to$* Out(N).

(iii) *If $Q$ is in the class $\mathfrak{U}_P$, then $l_P : Q \to Q_P$ induces a bijection $\mathcal{E}(Q_P, N, \chi^P) \to \mathcal{E}(Q, N, \chi)$.*

**Proof**

(i) This can be deduced similarly to Lemma 4.7.

(ii) Following Mac Lane [15], we can associate an element $\zeta_P$ of $H^3(Q_P, Z(N))$ to the map $\chi^P : Q_P \to \mathrm{Out}(N)$ such that the existence of an extension of $N$ by $Q$ with coupling $\chi^P$ is equivalent to $\zeta_P$ being trivial. Since $Q$ acts $P$-locally on $Z(N)$, the map $H^3(Q_P, Z(N)) \to H^3(Q, Z(N))$ induced by $l_P : Q \to Q_P$ is an isomorphism. Let $\zeta$ be the image of $\zeta_P$ under this map. Then $\zeta = 0$ because we have an extension of $N$ by $Q$ with coupling $\chi$. Thus $\zeta_P = 0$, and the conclusion follows.

(iii) According to [15], the cohomology group $H^2(Q, Z(N))$ acts freely and transitively on $\mathcal{E}(Q, N, \chi)$, and $H^2(Q_P, Z(N))$ does the same on $\mathcal{E}(Q_P, N, \chi^P)$. Moreover, if the equivalence class $E_P \in \mathcal{E}(Q_P, N, \chi^P)$ is mapped to $E \in \mathcal{E}(Q, N, \chi)$ by the function $\mathcal{E}(Q_P, N, \chi^P) \to \mathcal{E}(Q, N, \chi)$ induced by $l_P : Q \to Q_P$, then, for any $\xi_P \in H^2(Q_P, Z(N))$, $\xi_P \cdot E_P$ is mapped to $\xi \cdot E$, where $\xi$ is the image of $\xi_P$ in $H^2(Q, Z(N))$ under the map induced by $l_P : Q \to Q_P$. However, the homomorphism $H^2(Q_P, Z(N)) \to H^2(Q, Z(N))$ is an isomorphism. Therefore, the function $\mathcal{E}(Q_P, N, \chi^P) \to \mathcal{E}(Q, N, \chi)$ must be a bijection. □

Armed with the above lemma, we are ready to prove Theorem 4.16.

**Theorem 4.16** *Assume* $1 \to N \xrightarrow{\iota} G \xrightarrow{\epsilon} Q \to 1$ *is a group extension in which* $N$ *is finite and* $Q$ *is in the class* $\mathfrak{U}_P$. *Then the sequence*

$$1 \to \Gamma^G_{P'} N_P \xrightarrow{\subseteq} N_P \xrightarrow{\iota_P} G_P \xrightarrow{\epsilon_P} Q_P \to 1$$

*is exact.*

**Proof** Set $\bar{N} = N/T^G_{P'}N \cong N_P/\Gamma^G_{P'}N_P$ and $\bar{G} = G/T^G_{P'}N$. Then the canonical map $G \to \bar{G}$ is a $P$-equivalence by Lemma 4.9. Let $\chi : Q \to \mathrm{Out}(\bar{N})$ be the coupling of $1 \to \bar{N} \to \bar{G} \to Q \to 1$, and let $\chi^P : Q_P \to \mathrm{Out}(\bar{N})$ be the map described in Lemma 4.15(i). By parts (ii) and (iii) of Lemma 4.15, the set $\mathcal{E}(Q_P, \bar{N}, \chi_P)$ is nonempty, and the map $\mathcal{E}(Q_P, \bar{N}, \chi_P) \to \mathcal{E}(Q, \bar{N}, \chi)$ is surjective. Therefore, there is a commutative diagram

$$\begin{array}{ccccccccc}
1 & \longrightarrow & \bar{N} & \longrightarrow & \bar{G} & \longrightarrow & Q & \longrightarrow & 1 \\
 & & \| & & \downarrow & & \downarrow & & \\
1 & \longrightarrow & \bar{N} & \longrightarrow & G_* & \longrightarrow & Q_P & \longrightarrow & 1
\end{array}$$

in which the bottom row is a group extension with coupling $\chi_P$. Also, in view of the second assertion in Lemma 4.15(i), we have that $G_*$ acts $P$-locally on $\bar{N}$, which guarantees that $G_*$ is a $P$-local group. It follows, then, that $\bar{N}$ injects into $G_P$, proving the theorem. □

Theorem 4.16 can be employed to calculate the $P$-localization of certain groups that possess a finite normal subgroup. We illustrate this with the following example.

**Example 4.17** Let $G$ be the group with the presentation

$$G = \langle a, b, c, \alpha, \beta \mid a^3 = b^6 = c^2 = 1;\ [a, b] = c;\\ [b, c] = 1;\ [\alpha, a] = [\alpha, b] = [\alpha, c] = 1;\\ [\beta, a] = [\beta, b] = [\beta, c] = 1;\\ \beta\alpha\beta^{-1} = \alpha^{-1}b^2\rangle.$$

Then $G$ fits into an extension $1 \to N \to G \to \Pi \to 1$, where $\Pi$ is the fundamental group of the Klein bottle and $N = A_4 \times \mathbb{Z}/3$. Note that, by Proposition 5.4 in the next section, $\Pi$ is in $\mathfrak{U}_P$. Assume $P$ is a set of odd primes. Then, as observed in Example 2.6, the map $\Pi \to \mathbb{Z}_P$ taking the image of $\alpha$ to 0 and the image of $\beta$ to 1 $P$-localizes $\Pi$. If $3 \in P$, then $N_P/\Gamma^G_{P'}N_P \cong \mathbb{Z}/3 \times \mathbb{Z}/3$, so that, by Theorem 4.16, there is an exact sequence $1 \to \mathbb{Z}/3 \times \mathbb{Z}/3 \to G_P \to \mathbb{Z}_P \to 1$. Moreover, if $3 \notin P$, then $N_P/\Gamma^G_{P'}N_P = 1$, which means that $G_P \cong \mathbb{Z}_P$.

## 5 The class of groups whose classifying space has an aspherical $P$-localization

We conclude this paper by discussing the class of groups $\mathfrak{U}_P$ which played such a prominent role in Theorem 4.16. First we observe that this class contains all the nilpotent groups and all the free groups.

**Proposition 5.1 (Casacuberta and Peschke)** *All nilpotent groups are in $\mathfrak{U}_P$.*

**Proposition 5.2 (Casacuberta and Peschke)** *All free groups are in $\mathfrak{U}_P$.*

Both of the above propositions are proved in [5]. The latter result is established using G. Baumslag's [1] construction of the $P$-localization of a free group.

In [4, Theorem 1.5] Casacuberta shows that the finite groups that are in $\mathfrak{U}_P$ are exactly the $P$-nilpotent groups.

**Proposition 5.3 (Casacuberta)** *Let $G$ be a finite group. Then $G$ is in $\mathfrak{U}_P$ if and only if $G$ is $P$-nilpotent.*

Recall that a finite group is *$P$-nilpotent* if the subgroup of $G$ generated by all the elements with $P'$-order is a $P'$-group.

Another set of examples of groups in $\mathfrak{U}_P$ is provided by

**Proposition 5.4 (Descheemaeker and Malfait [7])** *The five groups that occur as fundamental groups of nonorientable flat manifolds of dimension $\leq 3$ are in $\mathfrak{U}_P$.*

In dimension two, the only nonorientable flat manifold is the Klein bottle. However, in dimension three, there are four such groups; for a list of them see [2] or [7].

The class $\mathfrak{U}_P$ is closed under certain types of operations. For example, below we show that $\mathfrak{U}_P$ is closed under finite direct products and direct limits.

**Proposition 5.5** *If $G$ and $H$ are in $\mathfrak{V}_P$, then $G \times H$ is in $\mathfrak{V}_P$.*

**Proof** First we remark that $(H \times G)_P = H_P \times G_P$. Now let $A$ be an abelian group on which $H_P \times G_P$ acts $P$-locally. To establish that $H_n(H \times G, A) \cong H_n(H_P \times G_P, A)$, we will employ the Lyndon–Hochschild–Serre homology spectral sequences associated with the extensions $1 \to H \to H \times G \to G \to 1$ and $1 \to H_P \to H_P \times G_P \to G_P \to 1$. Since $H$ is in $\mathfrak{U}_P$, the map $H_q(H, A) \to H_q(H_P, A)$ induced by $l_P : H \to H_P$ is an isomorphism for every $q \geq 0$. Moreover, since $G$ acts $P$-locally on $A$, the action of $G$ on $H_q(H, A)$ is also $P$-local. Consequently, since $G$ is in $\mathfrak{U}_P$, the map $H_p(G, H_q(H, A)) \to H_p(G_P, H_q(H_P, A))$ induced by $l_P : H \times G \to H_P \times G_P$ is an isomorphism for all $p,\ q \geq 0$. Hence all the maps induced by $l_P : H \times G \to H_P \times G_P$ between corresponding terms of the spectral sequences are isomorphisms. Therefore, the map $H_n(H \times G, A) \to H_n(H_P \times G_P, A)$ is an isomorphism for all $n \geq 0$. □

**Proposition 5.6** *The class $\mathfrak{U}_P$ is closed under direct limits.*

**Proof** This follows immediately from the fact that both the homology functor and the $P$-localization functor commute with direct limits. □

Next we show that $\mathfrak{U}_P$ is closed under central extensions.

**Proposition 5.7** *Assume $0 \to A \xrightarrow{\iota} G \xrightarrow{\epsilon} Q \to 1$ is a central group extension Then $G$ is in $\mathfrak{V}_P$ if $Q$ is in $\mathfrak{U}_P$.*

**Proof** By Theorem 2.8, the sequence $0 \to A_P \xrightarrow{\iota_P} G_P \xrightarrow{\epsilon_P} Q_P \to 1$ is exact. Let $B$ be a $P$-local abelian group upon which $G_P$ acts $P$-locally. We will employ the Lyndon–Hochschild–Serre homology spectral sequences associated with the extensions $0 \to A \to G \to Q \to 1$ and $0 \to A_P \to G_P \to Q_P \to 1$. The conclusion will follow if we can show that the map $H_p(Q, H_q(A, B)) \to H_p(Q_P, H_q(A_P, B))$ induced by $l_P : G \to G_P$ is an isomorphism for all $p,\ q \geq 0$. To see this, notice first that the map $H_p(Q, H_q(A, B)) \to H_p(Q, H_q(A_P, B))$ is an isomorphism for all $p,\ q \geq 0$ since $A$ is in $\mathfrak{U}_P$. Moreover, since $Q$ acts trivially on $A_P$, the action of $Q$ on $H_q(A_P, B)$ must be $P$-local. Thus the map $H_p(Q, H_q(A_P, B)) \to H_p(Q_P, H_q(A_P, B))$ is an isomorphism for all $p,\ q \geq 0$, proving the proposition. □

Below we show that the class $\mathfrak{U}_P$ is also closed under certain types of extensions by $P$-local groups.

**Proposition 5.8** *Assume $1 \to N \to G \to Q \to 1$ is a group extension with coupling $\chi : Q \to \mathrm{Out}(N)$. Then $G$ is in $\mathfrak{V}_P$ if each of the following statements is true:*

(i) *$Q$ is $P$-local;*

(ii) *$N$ is in $\mathfrak{U}_P$;*

(iii) $\mathrm{Im}\,\chi$ *is a $P$-group.*

**Proof** By Corollary 4.3, we have an exact sequence $1 \to N_P \to G_P \to Q \to 1$. Let $A$ be an abelian group upon which $G_P$ acts $P$-locally. We invoke the Lyndon–Hochschild–Serre homology spectral sequences associated with the extensions $1 \to N \to G \to Q \to 1$ and $1 \to N_P \to G_P \to Q \to 1$. The result will follow if the map $H_p(Q, H_q(N, A)) \to H_p(Q, H_q(N_P, A))$ induced by $l_P : G \to G_P$ is an isomorphism for all $p,\ q \geq 0$. However, since $N$ is in $\mathfrak{U}_P$, the map $H_q(N, A) \to H_q(N_P, A)$ is an isomorphism for all $q \geq 0$, so that the result follows. □

In conclusion, we invoke our results on extensions with a finite kernel to point out yet another way that we can construct examples of groups in $\mathfrak{U}_P$.

**Proposition 5.9** *Let $1 \to N \to G \to Q \to 1$ be a group extension with coupling $\chi : Q \to \mathrm{Out}(N)$. Then $G$ is in $\mathfrak{U}_P$ if all of the following conditions are satisfied:*

(i) *$Q$ is in $\mathfrak{U}_P$;*

(ii) *$N$ is a finite group in $\mathfrak{U}_P$;*

(iii) *$\mathrm{Im}\,\chi$ is a $P$-group.*

**Proof** By Theorem 4.16, we have an exact sequence $1 \to N_P \to G_P \to Q_P \to 1$. Let $A$ be an abelian group on which $G_P$ acts $P$-locally. Consider the Lyndon–Hochschild–Serre cohomology spectral sequences associated with the extensions $1 \to N \to G \to Q \to 1$ and $1 \to N_P \to G_P \to Q_P \to 1$. The conclusion will follow if we can show that the map $H_p(Q, H_q(N, A)) \to H_p(Q_P, H_q(N_P, A))$ induced by $l_P : G \to G_P$ is an isomorphism for all $p,\ q \geq 0$. Since $N$ is in $\mathfrak{V}_P$, the map $H_p(Q, H_q(N, A)) \to H_p(Q, H_q(N_P, A))$ is an isomorphism for all $p,\ q \geq 0$. Moreover, for each $g \in G$, there exists a $P$-number $n$ such that $g^n \cdot a = a$ for all $a$ in $N_P$, implying that the the action of $\langle g^n \rangle$ on $H_q(N_P, A)$ is $P$-local for all $q \geq 0$. Consequently, as a result of Lemma 5.10 below, $Q$ acts $P$-locally on $H_q(N_P, A)$ for all $q \geq 0$. Therefore, the map $H_p(Q, H_q(N_P, A)) \to H_p(Q_P, H_q(N_P, A))$ is an isomorphism, proving the proposition. □

**Lemma 5.10** *Let $N$ be a $G$-group. If $g \in G$ such that $\langle g^n \rangle$ acts $P$-locally on $N$ for some $P$-number $n$, then $\langle g \rangle$ acts $P$-locally on $N$.*

**Proof** Let $H = N \rtimes \langle g \rangle$ and $K = N \rtimes \langle g^n \rangle$. Then $K$ is $P$-local by Proposition 4.1, and $H/K$ is a $P$-group. Hence $H$ is $P$-local by Corollary 4.3, which implies that $\langle g \rangle$ acts $P$-locally on $N$. □

## References


[1] G. Baumslag, Some aspects of groups with unique roots, *Acta. Math.* **104** (1960), 217–303.

[2] H. Brown, R. Bülow, J. Neubüser, H. Wondratsscheck, and H. Zassenhaus, *Crystallographic Groups of Four Dimensional Space*, Wiley, 1978.

[3] C. Casacuberta, The behavior of homology in the localization of finite groups, *Canad. Math. Bulletin* **3** (1991), 311–320.

[4] C. Casacuberta and M. Castellet, Localization methods in the study of homology of virtually nilpotent groups, *Math. Proc. Cambridge Philos. Soc.* **112** (1992), 551–564.

[5] C. Casacuberta and G. Peschke, Localizing with respect to self-maps of the circle, *Trans. Amer. Math. Soc.* **339** (1993), 117–140.
[6] K. Dekimpe. *Almost-Bieberbach Groups: Affine and Polynomial Structures*, Lecture Notes in Math. **1639**, Springer-Verlag, 1996.
[7] A. Descheemaeker and W. Malfait, Preserving asphericity of virtually nilpotent spaces under $P$-localization, *Topology* **42** (2003), 1143–1154.
[8] A. García Rodicio, On the relationship between the localization of groups and that of relative groups, *Rend. Circ. Mat. Palermo* **II**, Ser. 40, 1 (1991) 122–127.
[9] P. Hilton, Localization and cohomology of nilpotent groups, *Math. Z.* **132** (1973), 263–286.
[10] P. Hilton, Nilpotent actions on nilpotent groups, in *Algebra and Logic*, Lecture Notes in Math. **450**, 174–196, Springer-Verlag, 1975.
[11] P. Hilton, Relative nilpotent groups, in *Categorical Aspects of Topology and Analysis*, Lecture Notes in Math. **915**, 136–147, Springer-Verlag, 1982.
[12] P. Hilton, G. Mislin, and J. Roitberg, *Localization of Nilpotent Groups and Spaces*, North Holland Math. Studies **15**, North-Holland, 1975.
[13] K. Lorensen, $P$-localizing group extensions with a finite kernel, *Math. Proc. Cambridge Philos. Soc.* **139** (2005), 193–206.
[14] K. Lorensen, $P$-localizing group extensions with a nilpotent action on the kernel, *Comm. Algebra* (to appear).
[15] S. Mac Lane, *Homology*, Springer-Verlag, 1975.
[16] P. Ribenboim, Torsion et localisation de groupes arbitraires, in *Séminaire d'Algèbre Paul Dubreil*, Lecture Notes in Math. **740**, 444–456, Springer-Verlag, 1979.

# ON THE $n$-COVERS OF EXCEPTIONAL GROUPS OF LIE TYPE

MARIA SILVIA LUCIDO

Dipartimento di Matematica e Informatica, Università di Udine,
via delle Scienze 200, I-33100 Udine, Italy
Email: mslucido@dimi.uniud.it

## Abstract

A group is *n-coverable* if it is the union of $n$ conjugacy classes of proper subgroups. We prove that the Suzuki groups and the Ree groups are respectively 3- and 4-coverable.

## 1 Introduction

Let $G$ be a group and let $H_1, \ldots, H_n$ be proper subgroups of $G$. For a subgroup $H$ of $G$, let $[H]$ denote the conjugacy class of $H$. Then $\{H_1, \ldots, H_n\}$ is called an *n-covering* of $G$, if every element of $G$ is $G$-conjugate to an element of $H_i$, $i = 1, \ldots, n$, i.e.,

$$G = \bigcup_{i=1}^{n} [H_i].$$

If $G$ has an $n$-covering, then it has an $m$-covering for any $m \geq n$. We say that $G$ is $n$-coverable if it has an $n$-covering, but not an $(n-1)$-covering.

There is a connection between the factorization of an integer polynomial with the $n$-coverings of its Galois group over the rationals, as follows (see [1]). Let $f(x) \in \mathbb{Z}[x]$ with $f(x)$ the product of $l$ irreducible factors, none of which is linear. Suppose that $\Omega$ is the set of all roots of $f(x)$ and $G$ is the Galois group of $f(x)$ over $\mathbb{Q}$. Choose an element $\omega_i$, $i = 1, \ldots, l$ in each orbit of $G$ on $\Omega$ and let $U_i$ the stabilizer of $\omega_i$. If $f(x)$ has roots modulo $p$ for all primes $p$, then $G$ is covered by the conjugates of $U_1, \ldots, U_l$. This proves that, in this case, $l \geq 2$.

The first question is then: does there exist 2-coverable groups? For the alternating and symmetric groups, there exists a complete answer. In [2] it was proved that $Alt(n)$ is 2-coverable if and only if $4 \leq n \leq 8$ and $Sym(n)$ is 2-coverable if and only if $3 \leq n \leq 6$. Similar statements hold for the general linear groups. In [3], it was proved that the linear groups $GL_n(q)$, $SL_n(q)$ and the projective groups $PGL_n(q)$, $PSL_n(q)$ are 2-coverable if and only if $2 \leq n \leq 4$. In general if $G$ is a group of Lie type of sufficiently high rank, then $G$ is not 2-coverable, except for the symplectic groups defined over a field of even characteristic (see [4]). In fact in [5] it was proved that the symplectic groups $Sp_{2l}(2^f)$ have a 2-cover $\{M_1, M_2\}$ with $M_1 \cong O_{2l}^{+}(2^f)$ and $M_2 \cong O_{2l}^{-}(2^f)$, thus showing that examples exist in arbitrary high rank. Moreover there are groups of low Lie rank which are not 2-coverable. In fact we prove that the two families of groups ${}^2B_2(q)$ and ${}^2G_2(q)$ of minimal Lie rank 1 are not 2-coverable, showing that the Suzuki groups are 3-coverable and the Ree groups ${}^2G_2(2)$ are 4-coverable.

Observe that the notion of $n$-cover extends that of a $(**)$-cover defined in the 2001 edition of these Proceedings [8].

## 2 Some small rank examples

We prove that the Suzuki and Ree groups are respectively 3- and 4-coverable.

**Lemma 2.1** *Let $G = Sz(q) \leq GL(4, q)$ be a Suzuki group, where $q = 2^{2m+1}$, $m$ a positive integer. Then $G$ is 3-coverable.*

**Proof** Let $G = Sz(q) \leq GL(4, q)$ be a Suzuki group, where $q = 2^{2m+1}$, $m$ a positive integer and let $r = 2^{m+1}$. Then $|G| = q^2(q-1)(q^2+1)$ and $(q-r+1)(q+r+1) = q^2+1$. Let $U$ be the subgroup of the lower unitriangular matrices of $G$: then $U$ is a Sylow 2-subgroup of exponent 4 and of order $q^2$. Let $H$ be the subgroup of the diagonal matrices of $G$. Then $H$ is isomorphic to the multiplicative group of the field, and has therefore order $q-1$, it is a $\pi(q-1)$-Hall subgroup of $G$ and it normalizes $U$. By Theorem 3.10, chapter XI of [6], the set $\Psi = \{U^g, H^g, T_1^g, T_2^g\}$ is a partition of $G$, where $T_1$ is a cyclic maximal torus of order $q+r+1$ and $T_2$ is a cyclic maximal torus of order $q-r+1$. The only maximal subgroup of $G$ containing $T_i$ is $N_i = N_G(T_i) = T_i\langle t_i \rangle$, with $t_i$ an element of order 4 and $|N_i : T_i| = 4$, for $i = 1, 2$ (see [9]). Therefore, if we consider an $n$-covering $\mathcal{A}$ of $G$, then $\mathcal{A}$ must contain (a conjugate of) $N_i$, $i = 1, 2$. Since for example the elements of order dividing $q-1$ are not contained in $[N_1] \cup [N_2]$, then $G$ is not 2-coverable.

Let $B = N_G(U) = UH$, then $G = [B] \cup [N_1] \cup [N_2]$, therefore $G$ is 3-coverable. □

**Lemma 2.2** *Let $G = Ree(q)$ be a Ree group, where $q = 3^{2m+1}$, $m$ a positive integer. Then $G$ is 4-coverable.*

**Proof** Let $G = Ree(q)$ be a Ree group, where $q = 3^{2m+1}$, $m$ a positive integer and let $r = 3^{m+1}$. Then $|G| = q^3(q-1)(q^3+1)$ and $(q-r+1)(q+r+1) = q^2-q+1$. The conjugacy classes of the Ree groups are well known, see for example [10]. We recall here the conjugacy classes of the elements not dividing $q^3$ or $q^2-q+1$.

| $x$ class representative | order of $x$ | $\lvert C_G(x)\rvert$ |
|---|---|---|
| $R^a$ | $(q-1)/2a$ | $q-1$ |
| $S^b$ | $(q+1)/4b$ | $q+1$ |
| $JT$ | 6 | $2q$ |
| $JT^{-1}$ | 6 | $2q$ |
| $JR^a$ | $(q-1)/a$ | $q-1$ |
| $JS^b$ | $(q+1)/2b$ | $q+1$ |
| $J$ | 2 | $q(q-1)(q+1)$ |

with $1 \leq a \leq (q-3)/4$, $1 \leq b \leq (q-3)/8$. Here $J$ is an element of order 2, whose centralizer is $C = \langle J \rangle \times L$, with $L \cong PSL(2,q)$ and therefore has order $q(q-1)(q+1)$. Moreover $R$, $S$ and $T$ are elements of $L$ such that $|R| = (q-1)/2$ (odd), $|S| = (q+1)/4$ (odd) and $|T| = 3$.

Let $T_i$ be a cyclic maximal torus of order $q+(-1)^i r+1$, $i = 1, 2$. By [7], Theorem C, the only maximal subgroup of $G$ containing $T_i$ is $N_i = N_G(T_i) = T_i\langle t_i \rangle$, with $t_i$ an element of order 6 and $|N_i : T_i| = 6$, for $i = 1, 2$

Let now $\mathcal{A}$ be an $n$-covering of $G$ with maximal components. Since $T_i$ is cyclic and $N_i$ is the only maximal subgroup containing it, a conjugate of $N_1$ and $N_2$ must be among the components of $\mathcal{A}$.

Let $x$ be an element of order 9. By [7], Theorem C, the only maximal subgroups of $G$ containing $x$ are those of type $B = N_G(U)$, where $U$ is a Sylow 3-subgroup of $G$ and those of type $F = {}^2G_2(q_0)$, where $q = q_0^f$, $f$ a prime number. Then either (a conjugate of) $B$ or $F$ must be contained in $\mathcal{A}$.

We now consider the elements of order $(q+1)/2$. These can be contained in the conjugates of the maximal subgroups of type $C$ or $M = (2^2 \times D_{(q+1)/2}) : 3$ (by [7], Theorem C). Then either (a conjugate of) $C$ or $M$ must be contained in $\mathcal{A}$ and at least 4 classes of subgroups are needed to cover $G$. In fact they are enough: we prove that

$$G = [B] \cup [N_1] \cup [N_2] \cup [C].$$

We observe that $B$ contains a Sylow 3-subgroup.

The elements $R^a$, $S^b$, $JT$, $JT^{-1}$, $JR^a$, $JR^b$, $J$ are all contained in $C_G(J) = C$. Therefore $\mathcal{A}$ is a 4-covering and $G$ is 4-coverable. □

## References


[1] R. Brandl, D. Bubboloni and I. Hupp, Polynomials with roots mod $p$ for all primes $p$, *J. Group Theory* **4** (2001), 233–239.

[2] D. Bubboloni, Coverings of the symmetric and alternating groups, *Dipartimento di matematica "U. Dini" — Universita' di Firenze* **7**, (1998).

[3] D. Bubboloni and M. S. Lucido, Coverings of linear groups, *Comm. Algebra* **30** (2002), no. 5, 2143–2159.

[4] D. Bubboloni, M. S. Lucido and T. Weigel, Coverings of classical groups, in preparation.

[5] R. H. Dye, Interrelations of symplectic and orthogonal groups in characteristic two, *J. Algebra* **59** (1979), 202–221.

[6] B. Huppert and N.Blackburn, *Finite Groups III*, Springer-Verlag, Berlin–Heidelberg–New York, 1982.

[7] P. B. Kleidman, The maximal subgroups of the Chevalley groups $G_2(q)$, with $q$ odd, the Ree groups ${}^2G_2(q^2)$ and their automorphism groups, *J. Algebra* **117** (1988), 30–71.

[8] M. S. Lucido, On the covers of finite groups, in *Groups St Andrews 2001 in Oxford, Vol. II*, London Math. Soc. Lecture Note Ser. **305**, 395–399, Cambridge Univ. Press, Cambridge, 2003.

[9] M. Suzuki, On a class of doubly transitive groups, *Ann. Math.* **75** (1962), 105–145.

[10] H. N. Ward, On Ree's series of simple groups, *Trans. Amer. Math. Soc.* **121** (1966), 62–89.

# POSITIVELY DISCRIMINATING GROUPS

O. MACEDOŃSKA

Institute of Mathematics, Silesian University of Technology,
ul. Kaszubska 23, 44-100 Gliwice, Poland
Email: Olga.Macedonska@polsl.pl

## Abstract

A group is positively discriminating if any finite subset of positive equations $u = v$, which are not laws in $G$, can be simultaneously falsified in $G$. All known groups which are not positively discriminating satisfy positive laws. The question whether every group without positive laws must be positively discriminating is open. We give an affirmative answer to this question for the class of locally graded groups.

*AMS Classification:* 20E10 (primary), 20M07 (secondary).

An equation in a group is an expression of the form $u = v$, where $u = u(x_1, \dots, x_n)$, $v = v(x_1, \dots, x_n)$ are different words ($v$ may be the empty word 1) in the free group $F$, freely generated by $x_1, x_2, \dots$. If $n = 2$, the equation is called *binary* and we use $x, y$ instead of $x_1, x_2$. The equation is called *positive* if $u$ and $v$ are written *without the inverses* of the $x_i$'s. A positive equation is called *balanced* if the exponent sum of $x_i$ is the same in $u$ and $v$ for each fixed $i$. A balanced equation $u = v$ is of *degree* $n$ if the $x$-length of $u$ and $v$ is equal to $n$. We say that the $n$-tuple of elements $g_1, \dots, g_n$ in $G$ satisfies the equation $u = v$, if under the substitution $x_i \to g_i$ we get the equality $u(g_1, \dots, g_n) = v(g_1, \dots, g_n)$. If equality does not hold, the $n$-tuple *falsifies* the equations. The equation $u = v$ is *a law* in a group $G$ if every $n$-tuple of elements in $G$ satisfies this equation. The equation is *a non-law* in $G$ if it is not a law in $G$, so that there is an $n$-tuple which falsifies the equation.

Let $\mathfrak{V}$ be a finite set of equations. Since the equations need not be cancelled, we can assume that for some $n$ all the equations in $\mathfrak{V}$ are written in $n$ variables. If there is an $n$-tuple in a group $G$ which falsifies each equation in $\mathfrak{V}$, we say that $\mathfrak{V}$ can be *simultaneously* falsified in the group $G$. For example, in the symmetric group $S_3$ the set of two equations $\{xy = y,\ xy^2 = x\}$ can be simultaneously falsified by a pair of elements $a, d \in S_3$, of orders $2, 3$, respectively: $ad \neq d,\ ad^2 \neq a$. However the set $\{xy^2 = x,\ xy^3 = x\}$ can not be simultaneously falsified, because either $y^2$ or $y^3$ has image 1 in $S_3$ and hence each pair satisfies at least one equation.

**More examples** The following binary sets $\mathfrak{V}$ of non-laws can not be simultaneously falsified in the indicated group. Each pair of elements satisfies some equation in $\mathfrak{V}$:

1. The quaternion group $Q_8$: $\mathfrak{V} = \{\, xy = yx,\ x^3y = yx \,\}$.
2. The cyclic group $C_3$: $\mathfrak{V} = \{\, xy = 1,\ x^2y = 1,\ xy = x,\ xy = y \,\}$.

We recall that a group $G$ is *discriminating* (see [10] 17.12, 17.23) if any finite subset $\mathfrak{V}$ of non-laws in $G$ can be simultaneously falsified in $G$. In the terminology of [1] this means that $G$ discriminates the free group in $var\, G$. We consider here only subsets $\mathfrak{V}$ of positive equations, and of binary equations, and we speak of positively or binary discriminating groups, respectively.

**Definition** A group $G$ is called *positively discriminating* if for any finite subset $\mathfrak{V}$ of positive equations $u = v$, which are non-laws in $G$, there exist elements $g_1, \ldots, g_n$ in $G$, such that $u(g_1, \ldots, g_n) \neq v(g_1, \ldots, g_n)$ for all $u = v$ in $\mathfrak{V}$ *simultaneously.*

A consequence of the definition is the following:

**Proposition 1** *If $G$ contains a free non-cyclic subsemigroup, then $G$ is positively discriminating. Every discriminating group is positively discriminating; however the converse is not true.*

**Proof** The first statement follows because each free non-cyclic subsemigroup contains a free subsemigroup of infinite rank, where every subset of positive equations can be falsified simultaneously.

By definition, every discriminating group is positively discriminating. To see that the converse is not true, take the group $G = A_5 \times F/F''$, where $A_5$ is the alternating group and $F/F''$ a free metabelian group of rank $> 1$. The group $G$ is positively discriminating, because, by [9], $F/F''$ contains a free non-cyclic subsemigroup. However $G$ is not discriminating, because the set of the commutator non-laws $\mathfrak{V} = \{\, [[x^d, y], [x, y]] = 1\, ;\ d|60,\ 0 < d < 60 \,\}$, cannot be simultaneously falsified (every pair of elements in $G$ satisfies at least one of them). □

**Proposition 2 (cf. [10] 17.32)** *No finite group is positively discriminating.*

**Proof** If $|G| = n$, we take $\mathfrak{V}$ to consist of the $n(n+1)$ nontrivial equations $x_i = x_j,\ \ i, j = 1, 2, \ldots, n+1$. Since there are more variables then elements in $G$, the pigeon-hole principle implies that the equations in $\mathfrak{V}$ cannot be falsified in $G$ simultaneously. □

All known groups that are not positively discriminating, e.g., finite groups, satisfy positive laws, and the known groups without positive laws, e.g., free soluble groups ([10] 32.23), [9], are positively discriminating. So the natural question arises:

**Question** *Must a group without positive laws be positively discriminating?*

We give an affirmative answer in a large class of groups. First we note that in the class of groups which do not satisfy any positive laws, the definition of positively discriminating group can be restricted to any binary equations; such a group is positively discriminating if and only if it is binary positively discriminating.

**Theorem 1** *A group $G$ which does not satisfy any positive laws is positively discriminating if and only if for any finite subset $\mathfrak{V}$ of positive binary equations $u(x, y) = v(x, y)$, there exist elements $g, h$ in $G$, such that $u(g, h) \neq v(g, h)$ for all equations in $\mathfrak{V}$ simultaneously.*

**Proof** The "only if" part is clear, because if any finite subset of positive equations can be simultaneously falsified in $G$, then the same is true for any finite subset of binary positive equations.

Conversely, let $G$ be a binary positively discriminating group, so any finite subset of *binary* positive equations *can be* simultaneously falsified in $G$. Assume that $G$ is not positively discriminating; then there exists a finite subset $\mathfrak{V}$ of positive equations on $n > 2$ variables, which cannot be simultaneously falsified in $G$. Let $\alpha$ map $x_i$ to $xy^i$; then the subset $\mathfrak{V}$ determines a correspomding subset $\mathfrak{V}^\alpha$ of non-trivial binary positive equations.

Since $\mathfrak{V}$ cannot be simultaneously falsified, every $n$-tuple of elements in $G$ satisfies at least one equation in $\mathfrak{V}$. So for every pair $g, h \in G$, the $n$-tuple $gh, gh^2, \ldots, gh^n$ satisfies some equation $u(x_1, \ldots, x_n) = v(x_1, \ldots, x_n)$ in $\mathfrak{V}$, that is the equality $u(gh, gh^2, \ldots, gh^n) = v(gh, gh^2, \ldots, gh^n)$ holds. This means that the pair $g, h$ satisfies the binary equation $u(xy, xy^2, \ldots, xy^n) = v(xy, xy^2, \ldots, xy^n)$ in $\mathfrak{V}^\alpha$. So the set $\mathfrak{V}^\alpha$ of binary positive equations cannot be simultaneously falsified in $G$, a contradiction. □

We need the following two technical lemmas.

**Lemma 1** *Let $G$ be finitely generated and not contain any free non-cyclic subsemigroups. If $G/N$ is nilpotent-by-finite, then $N$ is finitely generated.*

**Proof** By assumption, $G/N$ contains a nilpotent normal subgroup $H/N$ of finite index, hence $H$ and $H/N$ are finitely generated. Then by ([10] 31.12), there is a finite normal series with cyclic factors $H = N_0 \triangleright N_1 \triangleright \cdots \triangleright N_m = N$. We know that $N_0$ is finitely generated and assume inductively that $N_i$ is finitely generated. Since by assumption $G$ does not contain free non-cyclic subsemigroups, and the group $N_i/N_{i+1}$ is cyclic, it follows from ([5], Lemmas 5 and 1) that $N_{i+1}$ is finitely generated, which completes the induction, and proves that $N$ is finitely generated. □

**Lemma 2** *A finitely generated group, which is finite-by-nilpotent-by-finite, is nilpotent-by-finite.*

**Proof** It suffices to show that a finite-by-nilpotent group is nilpotent-by-finite. Let $G$ be a finitely generated group and let $N$ be a finite normal subgroup such that $G/N$ is nilpotent of class $c$. Then $\gamma_{c+1}(G) \subseteq N$. Moreover, since $G$ is finitely generated and $N$ is finite, the centralizer $C$ of $N$ in $G$ is a normal subgroup of finite index in $G$. Hence $\gamma_{c+2}(C) = [\gamma_{c+1}(C), C] \subseteq [N, C] = 1$, so $C$ is a nilpotent normal subgroup of finite index in $G$, which means that $G$ is nilpotent-by-finite, as required. □

We show that the question whether a group without positive laws must be positively discriminating, has an affirmative answer in the large class of so called locally graded groups, introduced in 1970 by Černikov. This class was defined to avoid groups with finitely generated infinite simple sections, such as infinite Burnside groups and Ol'shanskii–Tarski monsters.

A group $G$ is called *locally graded* if every nontrivial finitely generated subgroup in $G$ has a proper subgroup of finite index.

**Theorem 2** *Every locally graded group without positive laws is positively discriminating.*

**Proof** In view of Proposition 1, it suffices to consider only locally graded groups without free non-cyclic subsemigroups which do not satisfy positive laws. Let $G$ be such a group. We show that the assumption that $G$ is not positively discriminating leads to a contradiction.

If $G$ is not positively discriminating then by Theorem 1, there is a finite subset $\mathfrak{V}$ of binary positive equations, such that every pair $g, h \in G$ satisfies some of these equations. We can assume that $\mathfrak{V}$ consists of balanced binary positive equations $u_i(x, y) = v_i(x, y)$, because if elements satisfy an equation $u = v$, then they satisfy the balanced equation $uv = vu$. If elements satisfy an equation $u = 1$, then they also satisfy the balanced equation $ux = xu$. Since the equations need not be cancelled, we can assume all of them of the same degree $n$, say.

Thus for any two-element set $S = \{g, h\}$ in $G$, some words $u(g, h)$ and $v(g, h)$ in $S^n$ are equal. Hence we have $|S^n| < 2^n$ which means that $G$ is an $(n, 2)$-collapsing group [11]. Now there are two possibilities. If $G$ is locally a residually finite group, then by ([7], Theorem 3) $G$ satisfies a positive law, which contradicts the assumption.

If $G$ is not locally residually finite, then it contains a finitely generated subgroup which is not residually finite. We assume that $G$ itself is finitely generated and the intersection of all subgroups of finite index in $G$, denoted by $N$, is nontrivial. Since $G/N$ is finitely generated, residually finite and collapsing, it must be nilpotent-by-finite by [11].

Since $G$ does not contain free non-cyclic subsemigroups, we may apply Lemma 1 to get that $N$ is finitely generated. As a subgroup of a locally graded group, $N$ must contain a proper subgroup of finite index. Then by ([6] p.196), $N$ contains a proper characteristic subgroup $K \subsetneqq N$ of finite index in $N$, which is normal in $G$. So $N/K$ is finite, $(G/K)/(N/K) \cong G/N$ is nilpotent-by-finite and hence $G/K$ is finite-by-nilpotent-by-finite. Then by Lemma 2, $G/K$ is nilpotent-by-finite and by [4], $G/K$ is residually finite. So the intersection of all normal subgroups of finite index in $G$ is in $K$. That is $N \subseteq K$, which together with $K \subsetneqq N$, gives the required contradiction. □

**Corollary** *A locally graded group which is not positively discriminating, must be nilpotent-by-locally finite of finite exponent.*

**Proof** If a locally graded group $G$ is not positively discriminating, then by Theorem 2, $G$ must satisfy a positive law. By ([2] Theorem B, corrected in [3]), the locally graded group, which satisfies a positive law, must be nilpotent-by-locally finite of finite exponent. □

We show the region of known positively discriminating groups on the Groupland-map, introduced in [8]. It shows the mutual relations of different properties of groups. For example, the left half of the picture contains groups without free non-cyclic subsemigroups, and the right half groups containing free non-cyclic subsemigroups. There are three disjoint regions of groups with positive laws, non-positive laws and without laws. The locally graded groups are in the biggest inner ellipse.

By Proposition 1, the right half, and by Theorem 2 part of the left half of Groupland consist of positively discriminating groups. These regions of positively discriminating groups are shaded grey.

**GROUPLAND**

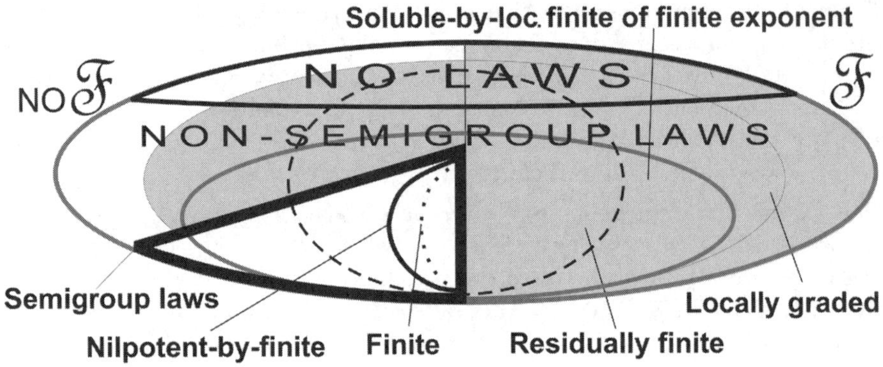


More details on Groupland can be found via http://www.google.pl

## References


[1] G. Baumslag, A. G. Myasnikov and V. N. Remeslennikov, Discriminating and co-discriminating groups, *J. Group Theory* **3** (2000), 467–479.

[2] Robert G. Burns, Olga Macedońska and Yuri Medvedev, Groups satisfying semigroup laws, and nilpotent-by-Burnside varieties, *J. Algebra* **195** (1997), 510–525.

[3] Robert G. Burns and Yuri Medvedev, Group laws implying virtual nilpotence, *J. Austral. Math. Soc.* **74** (2003), 295–312.

[4] P. Hall, On the finiteness of certain soluble groups, *Proc. London Math. Soc.* **9** (1959), 595–622.

[5] Yangkok Kim and Akbar H. Rhemtulla, On locally graded groups, in *Groups—Korea '94 (Pusan)*, 189–197, de Gruyter, Berlin, 1995.

[6] R. C. Lyndon and P. E. Schupp, *Combinatorial Group Theory*, Springer-Verlag Berlin, Heidelberg, New York, 1977.

[7] O. Macedońska, Collapsing groups and positive laws, *Comm. Algebra* **28** (2000), 3661–3666.

[8] O. Macedońska, Groupland, in *Groups St Andrews 2001 in Oxford, Vol. II*, London Math. Soc. Lecture Note Ser. **305**, 400–404, Cambridge Univ. Press, Cambridge, 2003.

[9] A. I. Malcev, Nilpotent semigroups, *Ivanov. Gos. Ped. Inst. Uc. Zap. Fiz. Mat. Nauk* **4** (1953), 107–111.

[10] Hanna Neumann, *Varieties of groups*, Springer-Verlag, Berlin, Heidelberg, New York, 1967.

[11] J. F. Semple and A. Shalev, Combinatorial conditions in residually finite groups I, *J. Algebra* **157** (1993), 43–50.

# AUTOMORPHISM GROUPS OF SOME CHEMICAL GRAPHS

G. A. MOGHANI*, A. R. ASHRAFI†, S. NAGHDI* and M. R. ADMADI†

*Color Control and Color Reproduction Department, Iran Color Research Center (ICRC), Tehran, Iran
Email: moghani@icrc.ac.ir

†Department of Mathematics, Faculty of Science, University of Kashan, Kashan, Iran
Email: ashrafi@kashanu.ac.ir

## Abstract

An Euclidean graph associated with a molecule is defined by a weighted graph with adjacency matrix $M = [d_{ij}]$, where for $i \neq j$, $d_{ij}$ is the Euclidean distance between the nuclei $i$ and $j$. In this matrix $d_{ii}$ can be taken as zero if all the nuclei are equivalent. Otherwise, one may introduce different weights for distinct nuclei. In this paper we present some MATLAB and GAP programs and use them to find the automorphism group of the Bis Benzene Chromium(0) with $D_{6d}$ point group symmetry and the big fullerene $C_{80}$.

*AMS Subject Classification:* 92E10.
*Keywords:* Euclidean graph, symmetry, Bis Benzene Chromium(0), fullerene.

## 1 Introduction

Let $G = (V, E)$ be a simple graph. $G$ is called a weighted graph if each edge $e$ is assigned a non-negative number $w(e)$, called the weight of $e$. The Euclidean graph of a molecule is a complete weighted graph in which the edges are weighted by the Euclidean distances of vertices.

An automorphism of a weighted graph $G$ is a permutation $g$ of the vertex set of $G$ with the property that, (i) for any vertices $u$ and $v$, $g(u)$ and $g(v)$ are adjacent if and only if $u$ is adjacent to $v$; (ii) for every edge $e$, $w(g(e)) = w(e)$. The set of all automorphisms of a weighted graph $G$, with the operation of composition of permutations, is a permutation group on $V(G)$, denoted $\mathrm{Aut}(G)$.

By symmetry we mean the automorphism group symmetry of a graph. The symmetry of a graph, also called a topological symmetry, accounts only for the bond relations between atoms, and does not fully determine molecular geometry. The symmetry of a graph does not need to be the same as the molecular point group symmetry. However, it does represent the maximal symmetry which the geometrical realization of a given topological structure may possess.

In [21, 20], it was shown by Randić that a graph can be depicted in different ways such that its point group symmetry or three dimensional perception may differ, but the underlying connectivity symmetry is still the same as characterized by the automorphism group of the graph. However, the molecular symmetry depends on the coordinates of the various nuclei which relate directly to their three dimensional geometry. Although the symmetry as perceived in graph theory by the automorphism group of the graph and the molecular group are quite different, it was shown by Balasubramanian [8] that the two symmetries are connected.

In this paper we consider only weighted graphs. The motivation for this study is outlined in [5–11] and the reader is encouraged to consult these papers for background material as well as for basic computational techniques. Our notation is standard and taken mainly from [12, 18, 19] and [22].

## 2 Computational Details

In this section we first describe some notation, which will be kept throughout. Let $G$ be a group and $N$ be a subgroup of $G$. Suppose $X$ is a set. The set of all permutations on $X$, denoted by $S_X$, is a group which is called the symmetric group on $X$. In the case that, $X = \{1, 2, \cdots, n\}$, we denote $S_X$ by $S_n$ or $\mathrm{Sym}(n)$.

The last years have seen a rapid spread of interest in the understanding, design and even implementation of group theoretical algorithms. These are gradually becoming accepted both as standard tools for a working group theoretician, like certain methods of proof, and as worthwhile objects of study, like connections between notions expressed in theorems.

Our computations of the symmetry properties of molecules were carried out with the use of GAP [13]. In this paper, we use freely GAP functions and the reader is encouraged to consult the manual of GAP and [1, 4].

Consider the equation $(P_\sigma)^t A P_\sigma = A$, where $A$ is the adjacency matrix of the weighted graph $G$. Suppose $\mathrm{Aut}(G) = \{\sigma_1, \sigma_2, \cdots, \sigma_m\}$. The matrix $S_G = [s_{ij}]$, where $s_{ij} = \sigma_i(j)$ is called a solution matrix for $G$. Clearly, for computing the automorphism group of $G$, it is enough to calculate a solution matrix for $G$. The second author in [2] proved a result that is useful for computing symmetry of molecules. Using this result, we presented a MATLAB program [3] for computing a solution matrix for the automorphism group of Euclidean graphs.

Our program in [3], needs the Cartesian coordinates of the atoms to determine the Euclidean distances in the molecule under consideration. If we calculate these distances by HyperChem, Gaussian 98 or another software, then for computing the symmetry of molecule under consideration, it is enough to delete the first eight lines of that program and load the distance matrix of the molecule.

## 3 Results and Discussion

In this section, we apply our program to compute the automorphism group of Euclidean graph of Bis Benzene Chromium(0) and fullerene $C_{80}$, the Cartesian coordinates of them were computed by Hyper Chem 97. It is useful to mention that in our program the accuracy is very important. Our calculations on the symmetry of some fullerenes show that, if we change the accuracy then the automorphism group will be changed.

We now calculate the symmetry of Bis Benzene Chromium(0). Let $G$ be the automorphism group of Euclidean graph of the mentioned molecule with $D_{6d}$ symmetry point group. Consider this molecule, Figure 1, to illustrate the Euclidean graphs and their automorphism group. We don't have to work with exact Euclidean distances since a mapping of weights into a set of integers suffices as long as different weights are identified with different integers. To illustrate let us use a Euclidean edge weighting for Bis Benzene Chromium(0) obtained from Table 1 and our program. Suppose $E$ is the $13 \times 13$ matrix defined by Euclidean distances. By Table 1, the matrix $E$ is as follows:

$$E = \begin{bmatrix} 0 & 1 & 1 & 2 & 3 & 4 & 3 & 5 & 7 & 8 & 10 & 9 & 6 \\ 1 & 0 & 3 & 2 & 1 & 3 & 4 & 6 & 5 & 7 & 8 & 10 & 9 \\ 1 & 3 & 0 & 2 & 4 & 3 & 1 & 7 & 8 & 10 & 9 & 6 & 5 \\ 2 & 2 & 2 & 0 & 2 & 2 & 2 & 2 & 2 & 2 & 2 & 2 & 2 \\ 3 & 1 & 4 & 2 & 0 & 1 & 3 & 9 & 6 & 5 & 7 & 8 & 10 \\ 4 & 3 & 3 & 2 & 1 & 0 & 1 & 10 & 9 & 6 & 5 & 7 & 8 \\ 3 & 4 & 1 & 2 & 3 & 1 & 0 & 8 & 10 & 9 & 6 & 5 & 7 \\ 5 & 6 & 7 & 2 & 9 & 10 & 8 & 0 & 1 & 3 & 4 & 3 & 1 \\ 7 & 5 & 8 & 2 & 6 & 9 & 10 & 1 & 0 & 1 & 3 & 4 & 3 \\ 8 & 7 & 10 & 2 & 5 & 6 & 9 & 3 & 1 & 0 & 1 & 3 & 4 \\ 10 & 8 & 9 & 2 & 7 & 5 & 6 & 4 & 3 & 1 & 0 & 1 & 3 \\ 9 & 10 & 6 & 2 & 8 & 7 & 5 & 3 & 4 & 3 & 1 & 0 & 1 \\ 6 & 9 & 5 & 2 & 10 & 8 & 7 & 1 & 3 & 4 & 3 & 1 & 0 \end{bmatrix}$$

Not all 13! permutations of the vertices of the molecule belong to the automorphism group of its weighted graph since the weights of all the edges are not the same. For example, the permutation $(1, 2, 3, 4, 5, 6, 7)$ does not belong to the automorphism group since the resulting graph does not preserve connectivity. Let $X$ denote the set of all solutions of matrix equation $P^t AP = A$. Set $Y = \{\alpha \in S_{13} \mid P_\alpha \in X\}$. Then $Y$ is the automorphism group of Euclidean graph of Bis Benzen Chromium(0). We now apply our MATLAB program to find a solution matrix for this group. After running this program, we can see that $G$ has order 8. Using the solution matrix of Bis Benzen Chromium(0) and a simple GAP program, we can find the structure of the automorphism group $G$ of the Euclidean graph of Bis Benzen Chromium(0).

$$\begin{aligned} G = \{&(1)(2)(3)(4)(5)(6)(7)(8)(9)(10), (2,3)(4,7)(5,9)(6,8)(11,12), \\ &(1,4,10,7)(2,5,11,8)(3,6,12,9), (1,4)(2,6)(3,5)(7,10)(8,12)(9,11), \\ &(1,7,10,4)(2,8,11,5)(3,9,12,6), (1,7)(2,9)(3,8)(4,10)(5,12)(6,11), \\ &(1,10)(2,11)(3,12)(4,7)(5,8)(6,9), (1,10)(2,12)(3,11)(5,6)(8,9)\}. \end{aligned}$$

Next we calculate the symmetry of fullerene $C_{80}$. Fullerenes are molecules in the form of polyhedral closed cages made up entirely of $n$ three-coordinate carbon atoms and having 12 pentagonal and $(n/2 - 10)$ hexagonal faces, where $n$ is even and greater or equal 20, with the exception of $n = 22$. Hence, the fullerene $C_{80}$ $(n = 80)$ has but 12 pentagons and 30 hexagons. Let $G$ be the automorphism group of Euclidean graph of $C_{80}$ fullerene with $I_h$ symmetry point group. Consider this molecule, Figure 3, to illustrate the Euclidean graphs and their automorphism group. To illustrate let us use a Euclidean edge weighting for fullerenes $C_{80}$

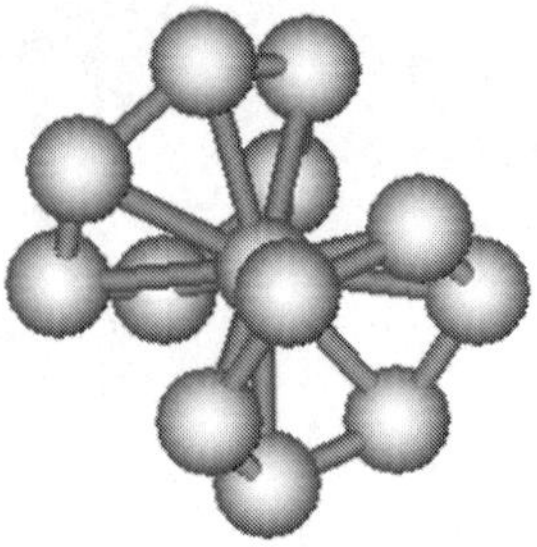

Figure 1. The Structure of Bis Benzene Chromium(0) with $D_{6d}$ Point Group.

| $a_{ij}$ | 1 | 2 | 3 | 4 | 5 | 6 | 7 | 8 | 9 | 10 | 11 | 12 | 13 |
|---|---|---|---|---|---|---|---|---|---|---|---|---|---|
| 1 | 0.00 | 1.46 | 1.46 | 1.91 | 2.52 | 2.91 | 2.52 | 3.17 | 3.72 | 3.75 | 3.25 | 2.60 | 2.56 |
| 2 | 1.46 | 0.00 | 2.52 | 1.91 | 1.46 | 2.52 | 2.91 | 2.55 | 3.17 | 3.72 | 3.75 | 3.25 | 2.60 |
| 3 | 1.46 | 2.52 | 0.00 | 1.91 | 2.91 | 2.52 | 1.46 | 3.72 | 3.75 | 3.25 | 2.60 | 2.55 | 3.17 |
| 4 | 1.91 | 1.91 | 1.91 | 0.00 | 1.91 | 1.91 | 1.91 | 1.91 | 1.91 | 1.91 | 1.91 | 1.91 | 1.91 |
| 5 | 2.52 | 1.46 | 2.91 | 1.91 | 0.00 | 1.46 | 2.52 | 2.60 | 2.56 | 3.17 | 3.72 | 3.75 | 3.25 |
| 6 | 2.91 | 2.52 | 2.52 | 1.91 | 1.46 | 0.00 | 1.46 | 3.25 | 2.60 | 2.56 | 3.17 | 3.72 | 3.75 |
| 7 | 2.52 | 2.91 | 1.46 | 1.91 | 2.52 | 1.46 | 0.00 | 3.75 | 3.25 | 2.60 | 2.56 | 3.17 | 3.72 |
| 8 | 3.17 | 2.55 | 3.72 | 1.91 | 2.60 | 3.25 | 1.46 | 0.00 | 1.46 | 2.52 | 2.91 | 2.52 | 1.46 |
| 9 | 3.72 | 3.17 | 3.75 | 1.91 | 2.56 | 2.60 | 3.25 | 1.46 | 0.00 | 1.46 | 2.52 | 2.91 | 2.52 |
| 10 | 3.75 | 3.72 | 3.25 | 1.91 | 3.17 | 2.56 | 2.60 | 2.52 | 1.46 | 0.00 | 1.46 | 2.52 | 2.91 |
| 11 | 3.25 | 3.75 | 2.60 | 1.91 | 3.72 | 3.17 | 2.56 | 2.91 | 2.52 | 1.46 | 0.00 | 1.46 | 2.52 |
| 12 | 2.60 | 3.25 | 2.55 | 1.91 | 3.75 | 3.72 | 3.17 | 2.52 | 2.91 | 2.52 | 1.46 | 0.00 | 1.46 |
| 13 | 2.56 | 2.60 | 3.17 | 1.91 | 3.25 | 3.75 | 3.72 | 1.46 | 2.52 | 2.91 | 2.52 | 1.46 | 0.00 |

Table 1. Distance matrix (angstroms) of $C_r(C_6H_6)_2$.

obtained from Table 2. Suppose $A$ is the $80 \times 80$ matrix defined by Euclidean distances and $X$ is the set of all solutions of matrix equation $P^tAP = A$. Set $Y = \{\alpha \in S_{80} \mid P_\alpha \in X\}$. Then $Y$ is the automorphism group of Euclidean graph of $C_{80}$. We now apply our MATLAB program to find a solution matrix for this group. After running this program, we can see that $G$ has order 120. Using the solution matrix of $C_{80}$ and a simple GAP program, we can find the structure of the automorphism group $G$ of Euclidean graph of $C_{80}$.

Our GAP program is as follows:

**A GAP Program for Computing the Structure of the Automorphism Group of the Euclidean Graphs of $C_{80}$**

```
G := Group(X);Size(G);
R := NormalSubgroups(G);
I:=Intersection(R[2],R[3]);
GeneratorsOfGroup(R[2]);
GeneratorsOfGroup(R[3]);
IsSimple(R[2]);
```

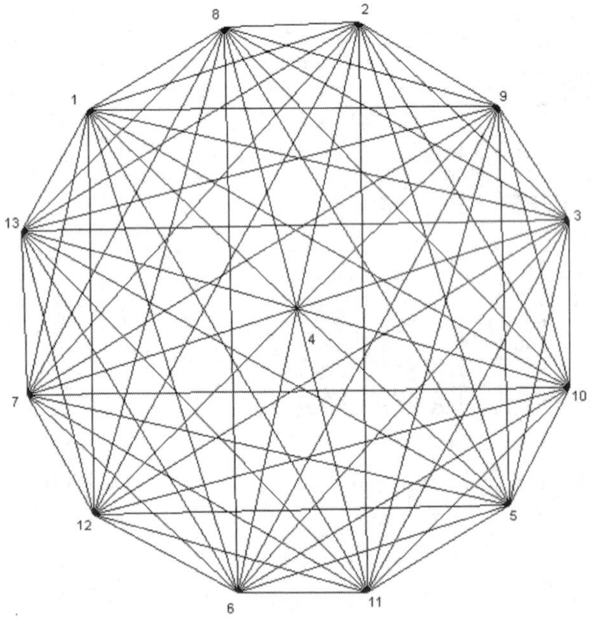


Figure 2. The Euclidean graph of Bis Benzen Chromium(0).

Using this GAP program two proper non-trivial normal subgroups $N = R[2]$ and $M = R[3]$ are obtained which intersect trivially. Therefore, $G$ is isomorphic to the direct product $Z_2 \times A_5$, where $Z_2$ is a cyclic group of order 2 and $A_5$ is the unique simple group of order 60. We now consider the following permutations:

$$\begin{aligned} A_1 = {} & (2,3,4)(5,9,8)(6,10,7)(11,13,12)(14,19,17)(15,18,16)(20,26,23) \\ & (21,27,24)(22,28,25)(29,37,33)(30,40,34)(31,39,35)(32,38,36) \\ & (41,49,45)(42,52,46)(43,51,47)(44,50,48)(53,59,56)(54,60,57) \\ & (55,61,58)(62,64,63)(65,70,68)(66,69,67)(71,75,74)(72,76,73) \\ & (77,78,79), \end{aligned}$$

$$\begin{aligned} A_2 = {} & (1,11,21,58,64,80,62,54,25,13)(2,20,30,70,75,77,53,42,19,9) \\ & (3,5,29,48,76,78,71,41,36,10)(4,6,16,47,59,79,72,67,35,26) \\ & (7,38,31,69,66,73,50,43,18,15)(8,14,17,46,49,74,65,68,34,37) \\ & (12,28,22,57,60,63,61,55,24,27)(23,39,32,45,52,56,51,44,33,40), \end{aligned}$$

$$\begin{aligned} B_1 = {} & (1,80)(2,77)(3,78)(4,79)(5,71)(6,72)(7,73)(8,74)(9,75)(10,76)(11,62) \\ & (12,63)(13,64)(14,65)(15,66)(16,67)(17,68)(18,69)(19,70)(20,53)(21,54) \\ & (22,55)(23,56)(24,57)(25,58)(26,59)(27,60)(28,61)(29,41)(30,42)(31,43) \\ & (32,44)(33,45)(34,46)(35,47)(36,48)(37,49)(38,50)(39,51)(40,52), \end{aligned}$$

$$\begin{aligned} C_1 = {} & (1,80)(2,77)(3,79)(4,78)(5,73)(6,74)(7,71)(8,72)(9,76)(10,75)(11,63) \\ & (12,62)(13,64)(14,69)(15,70)(16,68)(17,67)(18,65)(19,66)(20,56)(21,55) \\ & (22,54)(23,53)(24,61)(25,60)(26,59)(27,58)(28,57)(29,44)(30,43)(31,42) \\ & (32,41)(33,50)(34,51)(35,52)(36,49)(37,48)(38,45)(39,46)(40,47), \end{aligned}$$

$$\begin{aligned} C_2 = {} & (1,60,54)(2,66,42)(3,49,67)(4,52,41)(5,75,43)(6,59,68)(7,65,36) \\ & (8,51,35)(9,31,71)(10,32,53)(11,64,55)(12,61,25)(13,22,62)(14,48,73) \\ & (15,30,77)(16,78,37)(17,72,26)(18,33,34)(19,23,50)(20,76,44)(21,80,27) \\ & (28,58,63)(29,79,40)(38,70,56)(39,47,74)(45,46,69). \end{aligned}$$

By this program, $\{A_1, A_2\}$ is a generating set for $G$. Also, $N = \langle B_1 \rangle$ and $M = \langle C_1, C_2 \rangle$ are normal subgroups of $G$. Since $M$ is a simple group of order 60 and $A_5$ is the unique simple group of this order, $M \cong A_5$. Therefore, the automorphism group of the Euclidean graph of fullerene $C_{80}$ has order 120 and is isomorphic to $Z_2 \times A_5$.

## 4 Conclusions

Suppose T is a complete weighted graph and

$$\mathrm{Supp}(T) = |\{w(e) \mid e \text{ is an edge of } T\}|.$$

If Supp($T$) is large enough, for example greater than $|V(T)|$, then our MATLAB program is fast for computing the symmetry of the graph $T$. In particular, our program is suitable for computing symmetry of fullerenes. We applied our programs for computing symmetries of all molecules in Fullerene Gallery presented by Mitsuho Yoshida (for details see http://www.cochem2.tutkie.tut.ac.jp/Fuller/higher/higherE.html) with running time less than 0.1 s for GAP program and less than 40s for MATLAB program. The maximum running time is obtained in the case of fullerene with $I_h$ symmetry group.

We also mentioned that, our calculations with GAP and calculations done by Balasubramanian [5–11], Hao–Xu [14], Ivanov [16, 17], show that the automorphism group of the Euclidean graph of every molecule is trivial or has an even number of elements.

**Acknowledgement.** The authors would like to thank Professor M. Klin for some constructive discussion.

| No. | Cartesian Coordinates | No. | Cartesian Coordinates |
|---|---|---|---|
| C(1) | (-1.408246,-0.234747,3.734826) | C(21) | (0.613091,3.477989,1.876630) |
| C(2) | (-1.425582,1.154427,3.569457) | C(22) | (-2.050478,3.323037,0.862601) |
| C(3) | (-0.162489,-0.844639,3.919980) | C(23) | (-3.653731,1.488449,0.738862) |
| C(4) | (-2.393305,-0.973526,3.071041 ) | C(24) | (-3.919410,-0.787758,-0.100455 ) |
| C(5) | (-0.238983,1.897231,3.528427) | C(25) | (-2.411613,-3.174201,0.318601) |
| C(6) | (1.024065,-0.101706,3.879941) | C(26) | (-0.412541,-3.640735,1.637571) |
| C(7) | (-2.353758,1.774150,2.724498) | C(27) | (1.897532,-2.925157,1.959053) |
| C(8) | (-3.321435,-0.353931,2.225092) | C(28) | (3.053305,-0.383763,2.554027) |
| C(9) | ( 0.106700,-2.119051,3.406902) | C(29) | (1.876115,2.898533,2.046535) |
| C(10) | (-2.124107,-2.248931,2.558092) | C(30) | (0.291670,3.957380,0.602553) |
| C(11) | (1.009633,1.278402,3.652287) | C(31) | (-1.015560,3.881907,0.105503) |
| C(12) | (-3.299504,1.028367,2.011964) | C(32) | (-3.040886,2.614334,0.174489) |
| C(13) | (-0.859299,-2.832393,2.689232) | C(33) | (-3.958296,0.597117,-0.297560) |
| C(14) | (2.026729,-0.917566,3.340570) | C(34) | (-3.497917,-1.579070,-1.174670) |
| C(15) | (1.459771,-2.164267,3.048561) | C(35) | (-2.758644,-2.750313,-0.969673) |
| C(16) | (-0.433729,2.975636,2.658166) | C(36) | (-1.174199,-3.809160,0.474310) |
| C(17) | (-1.740913,2.900035,2.160125) | C(37) | (0.940530,-3.685950,1.279230) |
| C(18) | (-3.626000,-1.245263,1.188670) | C(38) | (3.073480,1.004397,2.378416) |
| C(19) | (-2.885774,-2.416364,1.394702) | C(39) | (3.532814,-1.171676,1.502252) |
| C(20) | (2.069816,1.820242,2.917743) | C(40) | (2.965856,-2.418377,1.210244) |
| C(41) | (-1.875657,-2.898698,-2.046089) | C(61) | (-3.052847,0.383598,-2.553580) |
| C(42) | (-0.292211,-3.957559,-0.602151) | C(62) | (-1.010174,-1.278581,-3.651885) |
| C(43) | (1.015019,-3.882087,-0.105101) | C(63) | (3.299962,-1.028532,-2.011518) |
| C(44) | (3.040391,-2.614642,-0.175078) | C(64) | (0.859757,2.832228,-2.688785) |
| C(45) | (3.957755,-0.597297,0.297962) | C(65) | (-2.026272,0.917400,-3.340123) |
| C(46) | (3.498375,1.578905,1.175116) | C(66) | (-1.459313,2.164101,-3.048115) |
| C(47) | (2.758149,2.750005,0.969084) | C(67) | (0.434187,-2.975801,-2.657719) |
| C(48) | (1.173704,3.808852,-0.474898) | C(68) | (1.740418,-2.900343,-2.160714) |
| C(49) | (-0.941033,3.686634,-1.279947) | C(69) | (3.625505,1.244956,-1.189259) |
| C(50) | (-3.072976,-1.004690,-2.378961) | C(70) | (2.885233,2.416184,-1.394300) |
| C(51) | (-3.533355,1.171497,-1.501851) | C(71) | (0.239478,-1.896533,-3.529100) |
| C(52) | (-2.966397,2.418198,-1.209842) | C(72) | (-1.023615,0.102533,-3.879624) |
| C(53) | (-2.070365,-1.819430,-2.917470) | C(73) | (2.354216,-1.774315,-2.724052) |
| C(54) | (-0.613594,-3.477305,-1.877348) | C(74) | (3.321939,0.353638,-2.225636) |
| C(55) | (2.049983,-3.323345,-0.863189) | C(75) | (-0.106251,2.119878,-3.406584) |
| C(56) | (3.654189,-1.488614,-0.738416) | C(76) | (2.124565,2.248765,-2.557646) |
| C(57) | (3.919914,0.787464,0.099911) | C(77) | (1.426086,-1.154720,-3.570001) |
| C(58) | (2.411072,3.174021,-0.318199) | C(78) | (0.162993,0.844345,-3.920524) |
| C(59) | (0.412038,3.641419,-1.638288) | C(79) | (2.393755,0.974353,-3.070724) |
| C(60) | (-1.898073,2.924978,-1.958651) | C(80) | (1.408742,0.235445,-3.735499) |

Table 2. Cartesian Coordinates of $C_{80}$ Molecule(Angstroms)

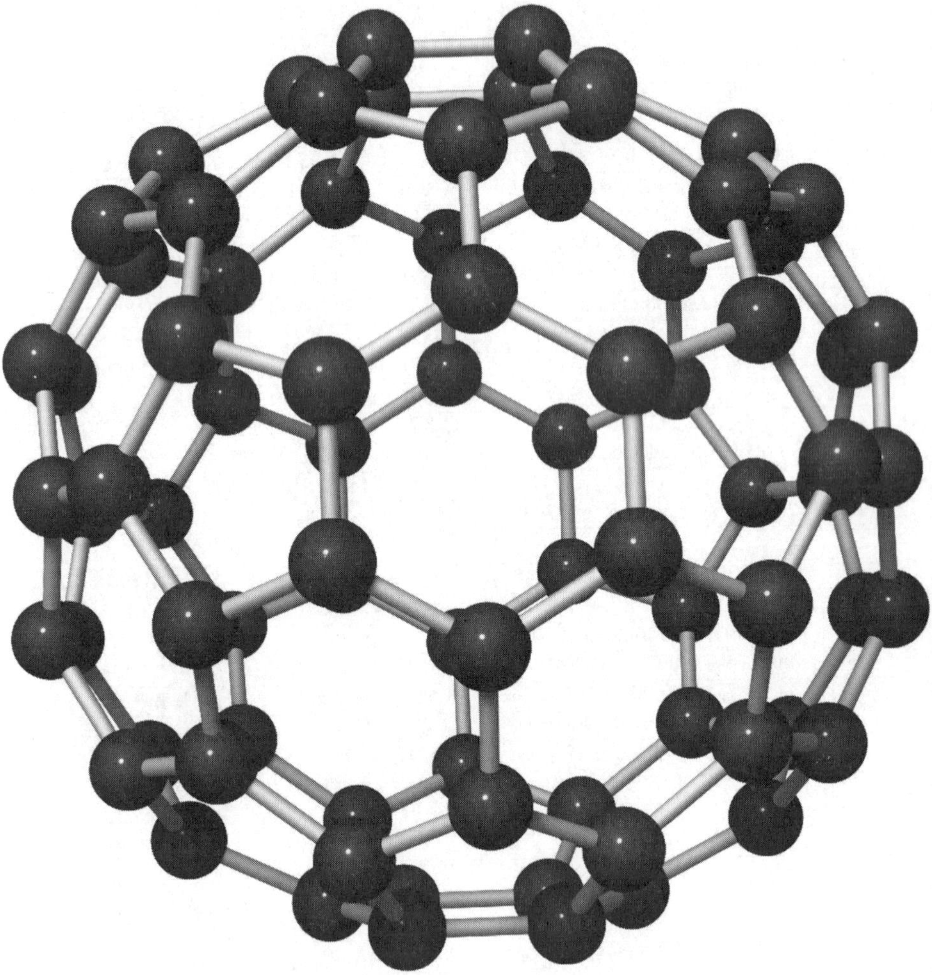

Figure 3. The Big Fullerene $C_{80} - I_h$.

## References


[1] A. R. Ashrafi, On Non-Rigid Group Theory For Some Molecules, *MATCH Commun. Math. Comput. Chem.* **53** (2005), 161–174.

[2] A. R. Ashrafi, On symmetry properties of molecules, *Chem. Phys. Letters* **406** (2005), 75–80.

[3] A. R. Ashrafi and M. R. Ahmadi, New Computer Program to Calculate the Symmetry of Molecules, *Cent. Eur. J. Chem.* **3** (2005), no. 4, 647–657.

[4] A. R. Ashrafi and M. Hamadanian, The Full Non-Rigid Group Theory for Tetraamino-Platinum(II), *Croatica Chem Acta* **76** (2003), 299–303.

[5] K. Balasubramanian, The Symmetry Groups of Non-rigid Molecules as Generalized Wreath Products and Their Representations, *J. Chem. Phys.* **72** (1980), 665–677.

[6] K. Balasubramanian, The Symmetry Groups of Chemical Graphs, *Internat. J. Quantum Chem.* **21** (1982), 411–418.

[7] K. Balasubramanian, Applications of Combinatorics and Graph Theory to Spectroscopy and Quantum Chemistry, *Chem. Rev.* **85** (1985), 599–618.

[8] K. Balasubramanian, Graph-Theoretical Perception of Molecular Symmetry, *Chem. Phys. Letters* **232** (1995), 415–423.

[9] K. Balasubramanian, Non-rigid Group Theory, Tunneling Splitting and Nuclear Spin Statistics of Water Pentamer, (H2O)5, *J. Phys. Chem.* **108** (2004), 5527–5536.

[10] K. Balasubramanian, Group Theoretical Analysis of Vibrational Modes and Rovibronic Levels of extended aromatic C48N12 Azafullerene, *Chem. Phys. Letters* **391** (2004), 64–68.

[11] K. Balasubramanian, Nuclear Spin Statistics of extended aromatic $C_{48}N_{12}$ Azafullerene, *Chem. Phys. Letters* **391** (2004), 69–74.

[12] G. S. Ezra, *Symmetry Properties of Molecules*, Lecture Notes in Chemistry **28**, Springer, Berlin–Heidelberg, 1982.

[13] The GAP Team, *GAP, Groups, Algorithms and Programming*, Lehrstuhl De fúr Mathematik, RWTH, Aachen, 1995.

[14] J.-F. Hao and L. Xu, The study on automorphism group of ESESOC, *Computers and Chemistry* **26** (2002), 119–123.

[15] D. J. Higham and N. J. Higham, *MATLAB Guide*, Society for Industrial and Applied Mathematics (SIAM), Philadelphia, PA, 2000.

[16] J. Ivanov, Molecular symmetry perception, *J. Chem. Inf. Comput. Sci.* **44** (2004), 596–600.

[17] J. Ivanov and G. Schürmann, Simple Algorithms for Determining the Molecular Symmetry, *J. Chem. Inf. Comput. Sci.* **39** (1999), 728–737.

[18] H. C. Longuet-Higgins, The symmetry groups of non-rigid molecules, *Mol. Phys.* **6** (1963), 445–460.

[19] G. A. Moghani, A. R. Ashrafi and M. Hamadanian, Symmetry properties of tetraammine platinum (II) with $C_{2v}$ and $C_{4v}$ point groups, *J. Zhejiang Univ. SCI.* **6B(3)** (2005), 222–226.

[20] M. Randić, On the recognition of identical graphs representing molecular topology, *J. Chem. Phys.* **60** (1974), 3920–3928.

[21] M. Randić, On discerning symmetry properties of graphs, *Chem. Phys. Letters* **42** (1976), 283–287.

[22] N. Trinajstić, *Chemical Graph Theory*, CRC Press, Boca Raton, FL., 1992.

# ON $c$-NORMAL SUBGROUPS OF SOME CLASSES OF FINITE GROUPS

Z. MOSTAGHIM

Department of Mathematics, Iran University of Science and Technology, Tehran, Iran
Email: mostaghim@iust.ac.ir

## Abstract

A subgroup $H$ is called $c$-normal in a group $G$ if there exists a normal subgroup $N$ of $G$ such that $HN = G$ and $H \cap N \leq H_G$, where $H_G =: \mathrm{Core}(H)$ is the maximal normal subgroup of $G$ which is contained in $H$. We obtain the $c$-normal subgroups in symmetric and dihedral groups. Also we find the number of $c$-normal subgroups of order 2 in symmetric groups. We conclude by giving a program in GAP for finding $c$-normal subgroups.

*AMS Classification:* 20D25.
*Keywords:* $c$-normal, symmetric, dihedral.

## 1 Introduction

The relationship between the properties of maximal subgroups of a finite group $G$ and the structure of $G$ has been studied extensively. The normality of subgroups in a finite group plays an important role in the study of finite groups. It is well known that a finite group $G$ is nilpotent if and only if every maximal subgroup of $G$ is normal in $G$.

In [2] Wang introduced the concept of $c$-normality of a finite group. He used the $c$-normality of a maximal subgroup to give some conditions for the solvability and supersolvability of a finite group. For example, he showed that $G$ is solvable if and only if $M$ is $c$-normal in $G$ for every maximal subgroup $M$ of $G$.

In this paper, we obtain the $c$-normal subgroups in symmetric and dihedral groups, and also we find the number of $c$-normal subgroups of order 2 in symmetric groups. In the end, we state a program in GAP for finding $c$-normal subgroups.

All groups considered in this paper are finite.

## 2 Basic Definitions and Elementary Properties.

In this section, we give two definitions and some properties that are needed in this paper.

**Definition 2.1** **([2])** Let $G$ be a group. We call a subgroup $H$ *$c$-normal* in $G$ if there exists a normal subgroup $N$ of $G$ such that $HN = G$ and $H \cap N \leq H_G$.

It is clear that a normal subgroup of $G$ is a $c$-normal subgroup of $G$ but the converse is not true.

**Definition 2.2** ([2]) We call a group $G$ *c-simple* if $G$ has no $c$-normal subgroup except the identity group 1 and $G$.

We can easily show that $G$ is $c$-simple if and only if $G$ is simple.

**Lemma 2.3** ([2]) *Let $G$ be a group. Then*

1. *if $H$ is normal in $G$, then $H$ is c-normal in $G$;*
2. *$G$ is c-simple if and only if $G$ is simple;*
3. *if $H$ is c-normal in $G$, $H \leq K \leq G$, then $H$ is c-normal in $K$;*
4. *Let $K \lhd G$ and $K \leq H$. Then $H$ is c-normal in $G$ if and only if $H/K$ is c-normal in $G/K$.*

## 3 Main Results

In this section we obtain the $c$-normal subgroups in symmetric and dihedral groups.

**Theorem 3.1** *Let $H$ be a non-normal subgroup of $S_n$ for $n \geq 5$. Then $H$ is c-normal in $S_n$ if and only if $H = \langle \tau \rangle$ where $\tau$ is an odd permutation of order 2.*

**Proof** Let $N$ be a normal subgroup of $G = S_n$ such that $HN = G$ and $H \cap N \leq H_G$. Since $H$ is a non-normal subgroup of $G$, then $N = A_n$. So $[H : H \cap N] = 2$ and $[H : H_G] = 2$. So we have $H \cap N = H_G$ and therefore $H \cap N \lhd A_n$. Therefore $H \cap N = 1$ and $|H| = 2$. Now we have $H = \langle \tau \rangle$ where $\tau$ is an odd permutation of order 2. □

**Remark** It is easy to see that all the subgroups of $S_3$ are $c$-normal and thus for $n = 3$ the above theorem is true. In $S_4$ all the subgroups except the subgroups $H = \langle \tau \rangle$ where $\tau$ is an even permutation of order two or $\tau$ is a cycle of order 4 are $c$-normal.

**Corollary 3.2** *The number of c-normal subgroups of order 2 in $S_n$ for $n \geq 5$ is*

$$\sum_{r=1}^{[n/2]} \frac{n(n-1)\cdots(n-(2r-1))}{2^r r!}$$

*where $r$ is odd.*

**Theorem 3.3** *Let $G = D_{2n} = \langle a, b \mid a^n = 1,\ b^2 = 1,\ bab = a^{-1} \rangle$ and $n$ be odd. Then all the subgroups of $G$ are c-normal.*

**Proof** Let $H$ be a subgroup of $G$. We consider two cases.

*Case 1:* Let $H$ have no element of order 2. Then $H = \langle a^j \rangle$ for $0 \leq j \leq n-1$ and $H \lhd G$. Thus by Lemma 2.3, $H$ is $c$-normal in $G$.

*Case 2:* Let $2 \mid |H|$.

1. Let $a^j \notin H$ for $0 < j \leq n-1$ so we have $|H| = 2$ . Now let $N = \langle a \rangle$, $N \lhd G$ and $G = HN$, $H \cap N = 1$ so $H$ is $c$-normal in $G$.
2. There exists $i > 0$ such that $a^i \in H$. Now let $m = \min\{i \mid i > 0,\ a^i \in H\}$. Then $|H| = 2l$, $n = lt$. It is easy to see that $G$ has a cyclic subgroup $N$ of order $nt/(n,m)$. Since $N \leq \langle a \rangle$, $N \lhd G$ and $HN = G$. Also we have $H \cap N = \langle a^m \rangle$ and so $H \cap N \lhd G$ and $H \cap N \leq H_G$. Then $H$ is $c$-normal in $G$.

□

## 4 GAP Program

In this section we use GAP [1] and give a program for finding $c$-normal subgroups. By using this program we can find all $c$-normal subgroups of a finite group with two generators. With a few changes in this program we can find a program for finding $c$-normal subgroups in a finite group with any number of generators and relations.

```
F:=FreeGroup("a","b");
a:=GeneratorsOfGroup(F)[1];
b:=GeneratorsOfGroup(F)[2];
Read("r");
G:=F/r;
n:=Order(G);
z:=LowIndexSubgroupsFpGroup(G,TrivialSubgroup(G),n);
s:=[];
        for i in [1..Size(z)] do
        t:=ConjugacyClassSubgroups(G,z[i]);
            for j in [1..Size(t)] do
            Add(s,t[j]);
            od;
        od;
cnorm:=[];
N:=[];
H:=[];
        for i in [1..Size(s)] do
        vi:=IsNormal(G,s[i]);
        if vi=true then Add(N,s[i]);fi;
        if vi=false then Add(H,s[i]);fi;
        od;
for y in [1..Size(H)] do
l:=0;
m:=0;
h:=false;
   while (m=0 or h=false)  and  l<=Size(N)  do
   l:=l+1;
```

```
        eH:=Elements(H[y]);
        eN:=Elements(N[1]);
        HN:=[];
            for i in [1..Order(H[y])] do
                  for j in [1..Order(N[1])] do
                  u:=eH[i]*eN[j];
                  AddSet(HN,u);
                  od;
            od;
        h:=IsSubgroup(Core(G,H[y]),Intersection(H[y],N[1]));
        if HN=G then m:=1;fi;
            od;
    if HN=G and h=true then Add(cnorm,H[y]);fi;
    od;
```

## References


[1] The GAP Group, Aachen, St Andrews, *GAP – Groups, Algorithms and Programming, Version 4.4*, 2004.

[2] Y. Wang, c-normality of groups and its properties, *J. Algebra* **180** (1995), 954–965.

# FONG CHARACTERS AND THEIR FIELDS OF VALUES

LUCIA SANUS[1]

Departament d'Àlgebra, Facultat de Matemàtiques, Universitat de València,
46100 Burjassot (València), Spain
Email: lucia.sanus@uv.es

Suppose that $G$ is a finite $p$-solvable group and let $H$ be a $p$-complement of $G$. Let $\varphi \in \mathrm{IBr}(G)$ be an irreducible Brauer character of $G$. P. Fong showed that there exists an irreducible constituent $\alpha \in \mathrm{Irr}(H)$ of $\varphi_H$ such that $\alpha^G = \Phi_\varphi$, the projective indecomposable character. We say that $\alpha$ is a *Fong character* associated with $\varphi$.

In this note we are concerned about the fields of values of Fong characters. As usual, if $\alpha$ is a complex-valued function on some set, we write $\mathbb{Q}(\alpha)$ to denote the subfield of $\mathbb{C}$ generated by all of the values of $\alpha$. If $\varphi \in \mathrm{IBr}(G)$ and $\alpha$ is a Fong character of $\varphi$, then it is easy to show that $\mathbb{Q}(\varphi) \subseteq \mathbb{Q}(\alpha)$ . But in fact, as we recently proved in [2], we have the following.

**Theorem 0.1** *Let $G$ be a $p$-solvable group, let $\varphi \in \mathrm{IBr}(G)$, and let $H$ be a $p$-complement subgroup of $G$. Then there exists a Fong character $\alpha \in \mathrm{Irr}(H)$ associated with $\varphi$ such that*

$$|\mathbb{Q}(\alpha) : \mathbb{Q}(\varphi)| \quad \text{divides} \quad \varphi(1)_p\,.$$

Not every Fong character associated with $\varphi$ satisfies the condition in Theorem 0.1. Moreover, different Fong characters associated with the same element of $\mathrm{IBr}(G)$ can have different fields of values. We provide some examples in [2].

We can obtain the following consequence.

**Corollary 0.2** *Let $G$ be a $p$-solvable group, let $\varphi \in \mathrm{IBr}(G)$ be real-valued, and let $H$ be a $p$-complement subgroup of $G$. If $p \neq 2$, then there exists a real-valued Fong character $\alpha \in \mathrm{Irr}(H)$ associated with $\varphi$.*

The conclusion of this corollary can fail if $p = 2$. It is enough to consider $G = S_3$, the symmetric group of degree 3. Even if $p = 2$, we can still ensure the existence of real Fong characters if the corresponding Brauer characters have odd degree.

**Proposition 0.3** *Let $G$ be a $p$-solvable group and let $\varphi \in \mathrm{IBr}(G)$ be real-valued of odd degree. Suppose that $H$ is a $p$-complement for $G$. Then there exists a real-valued Fong character $\alpha \in \mathrm{Irr}(H)$ associated with $\varphi$.*

**Proof** If $\varphi(1)$ is a $p'$-number, then $\varphi_H \in \mathrm{Irr}(H)$ is a real-valued Fong character associated with $\varphi$ and the proposition is proved. Thus, we may assume that $\varphi$ does not have $p'$-degree. By Proposition (3.4) of [1] we can choose $N \lhd G$ such that

[1]Research partially supported by Ministerio de Educación y Ciencia and Generalitat Valenciana.

the irreducible constituents of $\varphi_N$ have $p'$-degree and are not $G$-invariant. We may choose one of these constituents $\theta$ such that $H \cap I_G(\theta)$ is a $p$-complement for $I = I_G(\theta)$. Let $T = \{g \in G \,|\, \theta^g = \theta \text{ or } \theta^g = \bar{\theta}\}$. Notice that $I \triangleleft T$ and $T/I$ is a 2-group. Since $|G : I|$ is odd, it follows that $T = I$. Hence, $\theta$ is real-valued. Let $\mu \in \mathrm{IBr}(I \,|\, \theta)$ be the Clifford correspondent of $\varphi$ over $\theta$. We have that $\bar{\mu} \in \mathrm{IBr}(I \,|\, \theta)$ and $\varphi = \mu^G = \bar{\mu}^G = \overline{\mu^G} = \bar{\varphi}$. By uniqueness of the Clifford correspondent, it follows that $\mu$ is real-valued of odd degree. By induction, we have that there exists a real-valued Fong character $\alpha_1 \in \mathrm{Irr}(H)$ associated with $\mu$. Now, by Proposition (3.3) of [1] we know that $\alpha = (\alpha_1)^H$ is a Fong character associated with $\varphi$. Moreover,

$$\alpha = (\alpha_1)^H = (\overline{\alpha_1})^H = \overline{(\alpha_1)^H} = \bar{\alpha}\,,$$

as desired. □

In general, it is not true that if $\varphi \in \mathrm{IBr}(G)$ is real, then every Fong character associated with $\varphi$ is real. Also, if we have two Fong characters satisfying Theorem A it is not true that these characters are conjugate by an element of $\mathbf{N}_G(H)$, as the following example shows.

**Example 0.4** Let $G = B \rtimes A$ where $B$ is the direct product of 6 copies of the cyclic group of order 3 and the alternating group $A = A_4$ is acting in a transitive permutation representation of degree 6. Let $\mu$ be a non-trivial irreducible character of the cyclic group of order 3. Let $\lambda = \mu \times 1 \times 1 \times 1 \times 1 \times 1 \in \mathrm{Irr}(B)$. It follows that $|G : I_G(\lambda)| = 6$. Take $p = 2$. It follows that there is $\chi \in \mathrm{Irr}(G)$ lying over $\lambda$ such that $\chi^0 \in \mathrm{IBr}(G)$, where $\chi^0$ is the restriction of $\chi$ to $p$-regular elements. Let $H$ be a 2-complement for $G$. Since $H$ is a maximal subgroup of $G$, we have that $\mathbf{N}_G(H) = H$. Moreover, $\chi_H = \alpha_1 + \alpha_2$ where $\alpha_1$ and $\alpha_2$ are Fong characters associated with $\chi^0$. We deduce that $\mathbb{Q}(\alpha_1) = \mathbb{Q}(\alpha_2)$. Then $\alpha_1$ and $\alpha_2$ satisfy Theorem A but these are not conjugate by an element of $\mathbf{N}_G(H)$.

## References

[1] I. M. Isaacs, Fong characters in $\pi$-separable groups, *J. Algebra* **99** (1986), 89–107.
[2] I. M. Isaacs, G. Navarro and L. Sanus, Fields of values of Fong characters, *Arch. Math.* (accepted)

# ARITHMETICAL PROPERTIES OF FINITE GROUPS

W. J. SHI[1]

Department of Mathematics, Southwest China University, Chongqing 400715 China
and
School of Mathematics, Suzhou University, Suzhou 215006, Jiangsu, P. R. China
E-mail: wjshi@suda.edu.cn

## Abstract

Let $G$ be a finite group and $Ch_i(G)$ some quantitative sets. In this paper we study the influence of $Ch_i(G)$ to the structure of $G$. We present a survey of author and his colleagues' recent works.

*AMS Classification:* 20D60; 20D05; 20D06; 20D08; 20D25; 20E45; 20C15
*Keywords:* characterizable group, element orders, finite simple group; conjugacy class; irreducible character

Let $G$ be a finite group and $Ch(G)$ be one of the following sets:

(a) $Ch_1(G) = |G|$, that is, the order of $G$;

(b) $Ch_2(G) = \pi_e(G) = \{o(g) \mid g \in G\}$, that is, the set of element orders of $G$;

(c) $Ch_3(G) = cs(G) = \{|g^G| \mid g \in G\}$, that is, the set of conjugacy class sizes of $G$;

(d) $Ch_4(G) = cd(G)$, that is, the set of irreducible character degrees of $G$.

Our aim is to study the structure of $G$ under certain arithmetical hypotheses of $Ch_i(G)$, $i = 1$, 2, 3 or 4. Further to the above quantitative sets, we may define $Ch_5(G)$ to be the set of the maximal subgroup orders of $G$ (see [26]), $Ch_6(G)$ to be the set of Sylow normalizer orders of $G$ (see [3]), and other quantitative sets (for example, see [1]). In this paper we discuss the cases of $Ch_i(G)$, $i = 1, 2, 3$ or 4, especially for the cases of $i = 1$, 2.

**Question A** *If $Ch(G)$ is fixed, what can we say about the structure of $G$?*

For the set $Ch_i(G)$, $i = 2$, 3 or 4, we can define a graph $\Gamma_i(G)$ as follows: Its vertices are the primes dividing the numbers in $Ch_i(G)$; and two distinct vertices $p, q$ are connected if $pq \mid m$ holds for $m \in Ch_i(G)$.

**Question B** *If we know the information of graph $\Gamma_i(G)$, $i = 2$, 3 or 4, what can we say about the structure of $G$?*

In our characterization using the element orders, the graph $\Gamma_2(G)$ (prime graph) and the Gruenberg–Kegel theorem on groups with disconnected prime graphs (see [47]) plays an important role.

[1]The author gratefully acknowledges the support of K. C. Wong education foundation, Hong Kong.

For any $a \in Ch_i(G)$, $i = 2$, 3 or 4, for example, $a \in Ch_2(G)$, we define $M_2(a)$, the multiplicity of $a$ in $G$ as the number of elements of order $a$. Also, $DC_{3,2}(G) = |Ch_3'(G)| - |Ch_2(G)|$, the difference of conjugacy classes number $|Ch_3'(G)|$ and same order classes number in $G$. Furthermore, $QC_{1,3}(G) = |G|/|Ch_3'(G)|$, the quotient of $|G|$ and the number of conjugacy classes.

**Question C** *If $Ch(G)$ and $M(a)$, for all $a \in G$, are known what can we say about the structure of $G$? If $DC(G)$ is "small," what can we say about the structure of $G$? If $QC(G)$ is known what can we say about the structure of $G$?*

For example, $DC_{3,2}(G) = 0$, that is, any elements of a finite group $G$ with same order are conjugate. This is Syskin's problem, and some group theorists have proved that $G \cong 1$, $Z_2$, or $S_3$ using the classification theorem of finite simple groups (see [12], [50]). In [11] we classified all finite groups of $DC_{3,2}(G) = 1$, and studied such finite groups in which elements of the same order outside the center are conjugate (see [32]).

**Question D** *Let $b_i(G) = \max\{Ch_i(G)\}$, $i = 2$, 3 or 4. If some information of $b_i(G)$ is known, what can we say about the structure of $G$?*

For Problem A we know there exists some very well known results for $Ch_1(G) = |G|$, for example, **Sylow's theorem**, the **odd order theorem**, and **Burnside's $p^aq^b$ theorem**, etc. (see [6], [13] and the important bibliography listed in [14]).

> **Burnside's $p^aq^b$ theorem** implies that if $G$ is a nonabelian simple group, then $|\pi(G)| \geq 3$.

A finite group $G$ is called a $K_n$-group if $|\pi(G)| = n$. There are only eight simple $K_3$-groups (see[17]). In [34] we classified all simple $K_4$-groups, but we do not know whether the number of $K_4$-groups is finite or not. This problem depends on the solutions of some special Diophantine equations (also see [5]).

Twenty years ago, we studied such finite groups in which every non-identity element has prime order, i.e., finite EPO groups and got an interesting result: $G \cong A_5$ if and only if $\pi_e(G) = \{1, 2, 3, 5\}$ (that is, we may characterize the integral property only using the local property, see [42]).

After this, we developed such characterizations for all finite simple groups using the "two orders", and for the finite nonsolvable groups only using the "set of elements orders" (or "spectrum"). That is, we researched the following characterizations of two kinds.

(1) Characterizing all finite simple groups unitization using only the two sets $Ch_1(G)$ and $Ch_2(G)$.

We have finished the above works except $B_n$, $C_n$ and $D_n$ ($n$ even) (see[35], [36], [37], [38], [39], [7] and [48]).

(2) Many finite simple groups are characterizable using only the set $Ch_2(G)$.

The most recent version of the latter characterization is presented in Table 1 of [30].

Let $|G| = n$ and $f(n)$ denote the numbers of $G$, pairwise non-isomorphic, such that $|G| = n$. It is easy to see that $f(n) = 1$ if and only if $(n, \varphi(n)) = 1$, where $\varphi(n)$ is a Euler function of $n$. Also the solutions for $f(n) = 2, 3$, and 4 are found (see [46]).

The following question is posed: For any integer $k$, is there a solution for $f(n) = k$?

Now we consider substituting $|G|$ by the set $\pi_e(G)$ where $\pi_e(G)$ is a set of some positive integers. Similarly, for a set $\Gamma$ of positive integers, let $h(\Gamma)$ be the number of isomorphic classes of finite group $G$ such that $\pi_e(G) = \Gamma$. If $\Gamma = \pi_e(G)$, then we have $h(\pi_e(G)) \geq 1$. If $\Gamma = \{1, 2, 3, 5\}$, then $h(\Gamma) = 1$. Conversely, which groups $G$ satisfy $h(\pi_e(G)) = 1$? Such groups are called **characterizable groups** or **recognizable groups**.

Regarding characterizable groups, summarizing many scholars' work, we have the following results:

**Theorem 1** *The following groups are characterizable groups:*

(a) *Alternating groups $A_n$, where $n = 5$, 16, $p$, $p+1$, $p+2$, and $p \geq 7$ is a prime; Symmetric groups $S_n$, where $n = 7$, 9, 11, 12, 13, 14, 19, 20, 23, 24.*

(b) *Simple groups of Lie type $L_2(q)$, $q \neq 9$, $L_3(2^m)$, $L_n(2^m)$, $n = 2^k > 8$, $F_4(2^m)$, $U_3(2^m)$, $m \geq 2$, the series of simple groups of Suzuki–Ree type $Sz(2^{2m+1})$, ${}^2G_2(3^{2m+1})$, ${}^2F_4(2^{2m+1})$; $S_4(3^{2m+1})$ $(m > 0)$, $G_2(3^m)$; $L_3(7)$, $L_4(3)$, $L_5(3)$, $L_5(2)$, $L_6(2)$, $L_7(2)$, $L_8(2)$, $U_3(9)$, $U_3(11)$, $U_4(3)$, $U_6(2)$, $S_6(3)$, $O_8^-(2)$, $O_{10}^-(2)$, ${}^3D_4(2)$, $G_2(4)$, $G_2(5)$, $F_4(2)$, ${}^2F_4(2)'$, ${}^2E_6(2)$; and the non-solvable groups $PGL_2(p^m)$, $m > 1$, $p^m \neq 9$, $L_2(9).2_3$ $(\cong M_{10})$, $L_3(4).2_1$.*

(c) *All sporadic simple groups except $J_2$.*

The work [24] extended the characterization of the above some groups from finite groups to periodic groups. For the case of $h(\pi_e(G)) = \infty$ we have the next result.

**Theorem 2** *If all the minimal normal subgroups of $G$ are elementary (especially $G$ is solvable), or $G$ is one of the following: $A_6$, $A_{10}$, $L_3(3)$, $U_3(3)$, $U_3(5)$, $U_3(7)$, $U_4(2)$, $U_5(2)$; $J_2$; $S_4(q)$ $(q \neq 3^{2m+1}$ and $m > 0)$, then $h(\pi_e(G)) = \infty$.*

In the case of $k$-recognized groups, we have found infinite pairs of 2-recognizable groups as follows:

(a) $L_3(q)$, $L_3(q)\langle\theta\rangle$, where $\theta$ is a graph automorphism of $L_3(q)$ of order 2, $q = 5$, 29, 41, or $q \equiv \pm 2 \pmod 5$ and $(6, (q-1)/2) = 2$ (see [28], [31] and some references listed in [31]);

(b) $L_3(9)$, $L_3(9).2_1$ (see [9]);

(c) $S_6(2)$, $O_8^+(2)$ (see [29] and [40]);

(d) $O_7(3)$, $O_8^+(3)$ (see [40]);

(e) $L_6(3)$, $L_6(3)\langle\theta\rangle$, where $\theta$ is a graph automorphism of $L_6(3)$ of order 2 (see [43]);

(f) $U_4(5)$, $U_4(5)\langle\gamma\rangle$, where $\gamma$ is a graph automorphism of $U_4(5)$ of order 2 (see [43]).

For any $r > 0$, $h(\pi_e(L_3(7^{3^r}))) = r+1$, and these $r+1$ groups are $L_3(7^{3^r})\langle\rho\rangle$, where $\rho$ is a field automorphism of $L_3(7^{3^r})$, $k = 0, 1, 2, \ldots, r$ (see [49]).

**Problem 1** *For the cases of $B_n(q)$ and $C_n(q)$, $q$ odd, we have $|B_n(q)| = |C_n(q)|$, and $B_n(q)$ is not isomorphic to $C_n(q)$. How to distinguish them using the set $\pi_e(G)$?*

**Problem 2** *Find new characterizable simple groups within a particular class of finite simple groups.*

Considering the independent number of the prime graph A. V. Vasil'ev and E. P. Vdovin find recently a new approach which makes possible to study the case of a finite simple group with the connected prime graph (see [45], also see [44] and [16] for its application).

**Problem 3** *Determine whether or not there exist two section-free finite groups $G_1$ and $G_2$ such that $\pi_e(G_1) = \pi_e(G_2)$ and $h(\pi_e(G_1))$ is finite?*

**Problem 4** *For a enough large positive integer $n$, is the alternating group $A_n$ characterizable?*

**Problem 5** *For any $n \geq 3$, does $h(L_n(2)) = 1$ (see [15])?*

In 1987, when the author communicated the first characterization (that is, characterizing all finite simple groups unitization using only the two sets $|G|$ and $\pi_e(G)$) to Prof. J. G. Thompson, he put forward the following questions and conjectures in his letters:

For any finite group $G$ and any integer $d > 0$, let $G(d) = \{x \in G \mid x^d = 1\}$. Two finite groups $G_1$ and $G_2$ are of **the same order type** if and only if $|G_1(d)| = |G_2(d)|$ for every $d$. (That is, $Ch_2(G_1) = Ch_2(G_2)$ and $M_2(G_1) = M_2(G_2)$.)

**Problem 6 (J. G. Thompson)** *Suppose $G_1$ and $G_2$ are groups of the same order type. If $G_1$ is solvable, is $G_2$ necessarily solvable?*

The problem that the solvability of groups in which the number of elements whose orders are largest are given, induced by Thompson's problem, interested many Chinese group theorists. They proved the following results:

**Theorem 3** *Let $G$ be a finite group and $b_2'(G)$ be the number of elements of maximal order in $G$. If $b_2'(G)$ = odd, 32, $2p$, $4p$, $6p$, $8p$, $10p$, $2p^2$, $2p^3$, $2pq$ ($p, q$ are primes), or $b_2'(G) = \varphi(k)$, where $\varphi(k)$ is the Euler function of maximal order $k$, then $G$ is solvable except $G \cong S_5$.*

**Corollary** *For the above cases, Thompson's problem has an affirmative answer.*

**Problem 7 (J. G. Thompson's conjecture)** *Let $G$ and $H$ be two finite groups with $Ch_3(G) = Ch_3(H)$ ($cs(G) = cs(H)$). If $H$ is a nonabelian simple group and $Z(G) = 1$, then $G \cong H$.*

G. Y. Chen has introduced the concept of **order components** and proved that Thompson's conjecture holds if $H$ is a finite simple group with at least three prime graph components (see [8]).

Now we consider the case of "small" difference number $DC(G)$.

**Theorem 4** *Let $G$ be a finite group. Then $DC_{3,2}(G) = 1$ (i.e., $G$ have one and only one same order class containing two conjugacy classes of $G$) if and only if $G \cong A_5$, $L_2(7)$, $S_5$, $S_4$, $A_4$, $D_{10}$, $Z_3$, $Z_4$, $\mathrm{Hol}(Z_5)$ or $[Z_3]Z_4$ (see [11]).*

**Theorem 5** *Let $G$ be a finite group.*

(a) *If $G$ is nonabelian, then $QC_{1,3}(G) \geq 8/5$, and $QC_{1,3}(G) = 8/5$ if and only if $G = P \times A$, where $A$ is abelian with odd order and $P$ is a specific nonabelian 2-group.*

(b) *If $Z(G) = 1$, then $QC_{1,3}(G) \geq 2$, and $QC_{1,3}(G) = 2$ if and only if $G \cong S_3$.*

(c) *If $G$ is nonabelian simple, then $QC_{1,3}(G) \geq 12$, and $QC_{1,3}(G) = 12$ if and only if $G \cong A_5$ (see [41]).*

Some papers improved the above result (see [51] and [10]).

For the case of $Ch_3(G)$ and $Ch_4(G)$, we may pose the similar problems. For example, **which positive integers set can become $Ch_3(G)$ or $Ch_4(G)$ for some groups $G$?** Also, we may define the corresponding graphs. In [25], M. L. Lewis presented an integral overview. The following results are just relate to the author's joint works.

**Theorem 6** (a) *([20]) If $G$ is a finite group with $|cs(G)| = 2$, then (1) $G = P \times A$ with $P$ a $p$-group and $N$ Abelian. (2) $P/Z(P)$ has exponent $p$. Also, if $dl(P) \leq 2$, then $c(P) \leq 3$.*

(b) *([4]) Let $G$ be a finite group. Then $|cd(G)| = 2$ if and only if one of the following is true: (1) $G$ possesses a normal and Abelian subgroup $N$ with $|G : N| = q$, $q$ is a prime. (2) $G/Z(G)$ is a Frobenius group with kernel $(G' \times Z(G))/Z(G)$ and a cyclic complement. (3) $G = P \times A$, where $A$ is Abelian and $P$ is a $p$-group with $|cd(P)| = 2$.*

For the case (a) it is proved that the nilpotency class of $G$ is at most 3 (see [19]).

**Theorem 7** (a) *([21, 22, 23]) If $|cs(G)| \leq 3$, then $G$ is solvable. If $G$ is simple with $|cs(G)| = 4$, then $G \cong PSL(2, 2^m)$ (and conversely). If $G$ is simple with $|cs(G)| = 5$, then $G \cong PSL(2, q)$, where $q$ is an odd prime power greater than 5 (and conversely).*

(b) *([18]) If $|cd(G)| = k \leq 3$, then $G$ is solvable. If $G$ is solvable and $|cd(G)| = k \leq 4$, then $dl(G) \leq k$.*

(c) *([33]) If $cd(G) = \{1, m, n, mn\}$, then $G$ is solvable and one of the following is true: (1) $dl(G) \leq 3$. (2) $cd(G) = \{1, 3, 13, 39\}$. (3) $cd(G) = \{1, p^a, p^b, p^{a+b}\}$.*

In [27], the authors classified nonsolvable groups with four irreducible character degrees. Furthermore, it is proved that if $cs(G) = \{1, m, n, mn\}$ and $(m, n) = 1$, then $G$ is solvable (see [2]).

**Acknowledgement** The author would like to thank Dr. G. H. Qian, Dr. A. Moretó and the referee for their great help.

## References

[1] S. Abe, A characterization of some finite simple groups by orders of their solvable subgroups, *Hokkaido Math. J.* **31** (2002), 349–361.

[2] A. Beltrán and M. J. Felipe, Some class size conditions implying solvability of finite groups, to appear.

[3] J. X. Bi, On the group with the same orders of Sylow normalizers as the finite projective special unitary group, *Sci. in China (Ser. A)* **47** (2004), 801–811.

[4] M. Bianchi, A. Gillio Berta Mauri, M. Herzog, G.H. Qian and W.J. Shi, Characterization of non-nilpotent groups with two irreducible character degrees, *J. Algebra* **284** (2005), 326–332.

[5] Y. Bugeaud, Z. Cao and M. Mignotte, On simple $K_4$-groups, *J. Algebra* **241** (2001), 658–668.

[6] W. Burnside, On groups of order $p^a q^b$, *Proc. London Math. Soc.* **2** (1904), 388–392.

[7] H. P. Cao and W. J. Shi, Pure quantitative characterization of finite projective special unitary groups, *Sci. in China (Ser. A)* **32** (2002), 761–772.

[8] G. Y. Chen, On Thompson's conjecture, *J. Algebra* **185** (1996), 184–193.

[9] N. Chigira and W. J. Shi, More on the set of element orders of finite groups, *Northeast. Math. J.* **12** (1996), 257–260.

[10] X. L. Du, On the quotient of the group order and the number of conjugacy classes, *J. Southwest China Normal University* **29** (2004), 159–162 (Chinese).

[11] X. L. Du and W. J. Shi, Finite groups with conjugacy classes number one greater than its same order classes number, *Comm. Algebra*, to appear.

[12] W. Feit and G. M. Seitz, On finite rational groups and related topics, *Illinois J. Math.* **33** (1989), 103–131.

[13] W. Feit and J. G. Thompson, Solvablility of groups of odd order, *Pacific J. Math.* **13** (1963), 775–1029.

[14] D. Gorenstein, *Finite Simple Groups*, Plenum Press, New York and London, 1982.

[15] M. A. Grechkoseeva, M. S. Lucido, M. D. Mazurov, A. R. Moghaddamfar and A. V. Vasil'ev, On recognition of the projective special linear groups over the binary field, to appear.

[16] M. A. Grechkoseeva, W. J. Shi and A. V. Vasil'ev, Recognition by spectrum of $L_{16}(2^m)$, to appear.

[17] M. Herzog, On finite simple groups of order divisible by three primes only, *J. Algebra* **10** (1968), 383–388.

[18] I. M. Isaacs, *Character Theory of Finite Groups*, Academic Press, New York 202–206, 1976.

[19] K. Ishikawa, On finite $p$-groups which have only two conjugacy lengths, *Israel J. Math.* **129** (2002), 119–123.

[20] N. Itô, On finite groups with given conjugate types I, *Nagoya Math. J.* **6** (1953), 17–28.

[21] N. Itô, On finite groups with given conjugate types II, *Osaka J. Math.* **7** (1970), 231–251.

[22] N. Itô, On finite groups with given conjugate types III, *Math. Z.* **117** (1970), 267–271.

[23] N. Itô, Simple groups of conjugate type rank 4, *J. Algebra* **20** (1972), 226–249.
[24] A. S. Kondrat'ev and M. D. Mazurov, On the recognition of the finite simple groups $L_2(2^m)$ in the class of all groups, *Siberian Math. J.* **40** (1999), 62–64 (Russian).
[25] M. L. Lewis, An overview of graphs associated with character degrees and conjugacy class sizes in finite groups, to appear.
[26] X. H. Li, A characterization of the finite simple groups, *J. Algebra* **245** (2001), 620–649.
[27] G. Mallea and A. Moretó, Nonsolvable groups with few character degrees, *J. Algebra* **294** (2005), 117–126.
[28] V. D. Mazurov, The set of orders of elements in a finite group, *Algebra and Logic* **33** (1994), 49–55.
[29] V. D. Mazurov, Characterization of finite groups by sets of orders of their elements, *Algebra and Logic* **36** (1997), 23–32.
[30] V. D. Mazurov, Characterizations of groups by arithmetic properties, Proceedings of the International Conference on Algebra, *Algebra Colloquium* **11** (2004), 129–140.
[31] A. R. Moghaddamfar and M. R. Darafsheh, A family of finite simple groups which are 2-recognizable by their elements order, *Comm. Algebra* **32** (2004), 4507–4513.
[32] G. H. Qian, W. J. Shi and X. Z. You, Conjugacy classes outside a normal subgroup, to appear.
[33] G. H. Qian and W. J. Shi, A note on character degrees of finite groups, *J. Group Theory* **7** (2004), 187–196.
[34] W. J. Shi, On simple $K_4$-groups, *Science Bulletin* **36** (1991), 1281–1283 (Chinese).
[35] W. J. Shi, A new characterization of the sporadic simple groups, in *Group Theory, Proc. Singapore Group Theory Conf. 1987*, Walter de Gruyter Berlin-New York, 1989, 531–540.
[36] W. J. Shi, The pure quantitative characterization of finite simple groups (I), *Progress in Natural Science* **4** (1994), 316–326.
[37] W. J. Shi and J. X. Bi, A characteristic property for each finite projective special linear group, *Lecture Notes Math.* **1456**, Springer, 1990, 171–180.
[38] W. J. Shi and J. X. Bi, A characterization of Suzuki–Ree groups, *Science in China (Ser. A)*, **34** (1991), 14–19.
[39] W. J. Shi and J. X. Bi, A new characterization of the alternating groups, *Southeast Asian Bull. Math.* **16** (1992), 81–90.
[40] W. J. Shi and C. Y. Tang, A characterization of some orthogonal groups, *Progress in Natural Science* **7** (1997), 155–162.
[41] W. J. Shi and Y. R. Xiao, Quotient of the group order and the number of conjugacy classes, *Southeast Asian Bull. Math.* **22** (1998), 301–305.
[42] W. J. Shi and W. Z. Yang, A new characterization of $A_5$ and the finite groups in which every non-identity element has prime order, *J. Southwest Teachers College* **9** (1984), 36–40 (Chinese).
[43] A. V. Vasil'ev, On recognition of all finite nonabelian simple groups with orders having prime divisor at most 13, *Siberian Math. J.* **46** (2005), 246–253.
[44] A. V. Vasil'ev and M. A. Grechkoseeva, On recognition by spectrum of finite simple linear groups over fields of characteristic 2, *Siberian Math. J.* **46** (2005), 593–600.
[45] A. V. Vasil'ev and E. P. Vdovin, An adjacency criterion for two vertices of the prime graph of a finite simple group, *Algebra and Logic*, to appear. (See also Preprint No. 152, Sobolev Institute of Mathematics, Novosibirsk, 2005).
[46] G. M. Wei, Enumeration formular for the number of finite groups and their application, *Southeast Asian Bull. Math.* **22** (1998), 93–102.
[47] J. S. Williams, Prime graph components of finite groups, *J. Algebra,* **69** (1981), 487–513.

[48] M. C. Xu and W. J. Shi, Pure quantitative characterization of finite simple groups $^2D_n(q)$ and $D_l(q)$ ($l$ odd), *Algebra Colloq.* **10** (2003), 427-443.
[49] A. V. Zavarnitsine, Recognition of the simple groups $L_3(q)$ by element orders, *J. Group Theory* **7** (2004), 81–97.
[50] J. P. Zhang, On Syskin's problem for finite groups, *Sci. in China (Ser. A)*, **18**:2 (1988), 124–128.
[51] X. G. Zhong, On the group order and the number of conjugacy classes, *Journal of Mathematical Study (China)*, **34** (2001), 356–359.

# ON PREFRATTINI SUBGROUPS OF FINITE GROUPS: A SURVEY

X. SOLER-ESCRIVÀ

Departament de Matemàtica Aplicada, Universitat d'Alacant, Campus de Sant Vicen
Ap. Correus 99 – 03080 Alacant, Spain
Email: xaro.soler@ua.es

## 1 Introduction

In 1962 Gaschütz introduced a characteristic conjugacy class of subgroups of a finit soluble group, called the *prefrattini subgroups* (see [11]). These subgroups appea as intersection of some maximal subgroups of the group and they are interestin because they localize some particular information of the normal structure of th whole group. Ever since, the Gaschütz's original idea has been widely investigate by many authors and it exists a varied range of generalizations. Most of thes extensions have been developed in the universe of all finite groups, in spite c the fact that there are some works about prefrattini subgroups in some classes c infinite groups (see for example [16]). Our aim in this paper is to do a survey c this topic in the finite universe.

In section 2 we go through the main extensions of prefrattini subgroups in th soluble universe. We deal with generalizations in the general universe of all finit groups in section 3.

Every group considered in the sequel will be finite.

## 2 Prefrattini subgroups in the soluble universe

We begin with the original definition of prefrattini subgroups, given by Gaschütz

**Definition 2.1 ([11])** Let $G$ be a soluble group and $\Sigma$ a Hall system of $G$. Th *prefrattini subgroup* of $G$ associated to $\Sigma$, denoted as $W(\Sigma)$, is the intersection o all maximal subgroups of $G$ such that $\Sigma$ reduces into them,

$$W(\Sigma) = \bigcap\{M <\cdot G \ : \ \Sigma \text{ reduces into } M\}.$$

Given that a soluble group acts transitively by conjugation on its set of Ha systems, Gaschütz obtained that the prefrattini subgroups form a characteristi conjugacy class of subgroups. The intersection of all prefrattini subgroups of a so luble group is the Frattini subgroup of the group. Moreover, a prefrattini subgrou of a soluble group has the cover and avoidance properties. In fact, a prefrattin subgroup of a soluble group covers the Frattini chief factors of the group and avoid the complemented ones.

In the rest of this section, we will refer to any extension of the prefrattini sub groups of Gaschütz with the expression *subgroups of prefrattini type*. In 196

Hawkes made the first generalization. To do so, he used the theory of formations. For a subgroup $T$ of a group $G$, let us denote $T_G$ the core of $T$ in $G$, that is, the largest normal subgroup of $G$ contained in $T$. Let $\mathfrak{F}$ be a saturated formation. A maximal subgroup $M$ of a group $G$ is said to be $\mathfrak{F}$-*normal* if the quotient $G/M_G$ belongs to $\mathfrak{F}$, and $\mathfrak{F}$-*abnormal* otherwise. This definition allows the author to distinguish maximal subgroups according to $\mathfrak{F}$.

**Definition 2.2 ([12])** Let $\mathfrak{F}$ be a saturated formation, $G$ a soluble group and $\Sigma$ a Hall system of $G$. The $\mathfrak{F}$-*prefrattini subgroup* of $G$ associated to $\Sigma$, denoted as $W_{\mathfrak{F}}(\Sigma)$, is

$$W_{\mathfrak{F}}(\Sigma) = \bigcap\{M <\cdot\, G \ : M \text{ is } \mathfrak{F}\text{-abnormal and } \Sigma \text{ reduces into } M\}.$$

Obviously, $W(\Sigma) \leq W_{\mathfrak{F}}(\Sigma)$. Hawkes proved that the $\mathfrak{F}$-prefrattini subgroups form a characteristic conjugacy class of subgroups having the cover and avoidance property. A complemented chief factor $H/K$ of a group $G$ is said to be $\mathfrak{F}$-*central* if $H/K$ is complemented by an $\mathfrak{F}$-normal maximal subgroup of $G$. $H/K$ is said to be $\mathfrak{F}$-*eccentric* if it is complemented by an $\mathfrak{F}$-abnormal maximal subgroup of $G$. An $\mathfrak{F}$-prefrattini subgroup avoids each complemented $\mathfrak{F}$-eccentric chief factor of the group and covers the rest. Moreover, if $\mathfrak{F} = \{1\}$, then the $\mathfrak{F}$-prefrattini subgroups of a group are exactly the prefrattini subgroups defined by Gaschütz.

Saturated formations are contained in a considerably larger family of classes called Schunck classes. In the soluble universe, the prefrattini subgroups associated to Schunck classes were defined by Förster in [10]. Förster's approach is based on the concept of a crown introduced by Gaschütz in [11]. Förster distinguished the crowns of a soluble group according to a Schunck class. Thus, the prefrattini subgroups of Förster appear as intersection of the complements of certain crowns into which a fixed Hall system reduces. They extend the prefrattini subgroups defined by Hawkes and form a characteristic conjugacy class with the cover and avoidance property.

As we will see in the sequel, subgroups of prefrattini type permute with certain relevant subgroups of the group. The first result on this line appears in [12].

**Theorem 2.3 ([12])** *Let $\mathfrak{F}$ be a saturated formation and $D$ the $\mathfrak{F}$-normalizer of a soluble group $G$ associated to a Hall system $\Sigma$ of $G$. Then $DW(\Sigma) = W_{\mathfrak{F}}(\Sigma)$.*

In 1970, always in the soluble universe, Makan obtained another factorization of prefrattini subgroups in [14]. For this, the author defined a new family of prefrattini subgroups.

**Definition 2.4 ([14])** Let $\mathfrak{L}$ be a Fischer class and $V$ an $\mathfrak{L}$-injector of a soluble group $G$. Let $\Sigma$ be a Hall system of $G$ reducing into $V$. The $V$-*prefrattini subgroup* of $G$ associated to $\Sigma$, $W_V(\Sigma)$, is

$$W_V(\Sigma) = \bigcap\{M <\cdot\, G \ : V \leq M \text{ and } \Sigma \text{ reduces into } M\}.$$

Makan proved that the $V$-prefrattini subgroups of $G$ form a characteristic conjugacy class which have the cover and avoidance property. We say that a chief factor of a group is *partially $V$-complemented* if it is a complemented chief factor and at least one of its complements contains $V$. The above defined subgroup $W_V(\Sigma)$ avoids all partially $V$-complemented chief factors and covers the rest ([14]). If $\mathfrak{L}$ is trivial or, equivalently, $V = 1$, then $W_V(\Sigma) = W(\Sigma)$ and we recover the prefrattini subgroups defined by Gaschütz. The factorization obtained by Makan is the following.

**Theorem 2.5 ([14])** *Let $G$, $\Sigma$ and $V$ be as in Definition 2.4. Then $VW(\Sigma) = W_V(\Sigma)$.*

Two years later, Chambers remarked in [5] that Makan's construction depended only on the fact that the $\mathfrak{L}$-injector $V$, for a Fischer class $\mathfrak{L}$, is in particular a normally embedded subgroup and not on the fact that $V$ is an $\mathfrak{L}$-injector. Therefore the factorization stated in Theorem 2.5 is true whenever $V$ is a normally embedded subgroup of $G$ into which the Hall system $\Sigma$ reduces. In addition, Chambers obtained that normally embedded subgroups also permute with the prefrattini subgroups defined by Hawkes.

**Theorem 2.6 ([5])** *Let $\mathfrak{F}$ be a saturated formation and $V$ a normally embedded subgroup of a soluble group $G$. Let us consider a Hall system $\Sigma$ of $G$ which reduces into $V$ and the $\mathfrak{F}$-prefrattini subgroup $W_{\mathfrak{F}}(\Sigma)$ of $G$ associated to $\Sigma$.*

*Then $VW_{\mathfrak{F}}(\Sigma)$ is a subgroup of $G$, which avoids all partially $V$-complemented $\mathfrak{F}$-eccentric chief factors of $G$ and covers the rest.*

To obtain Theorem 2.6, Chambers first proved that if $V$ is a normally embedded subgroup of a soluble group $G$ into which a Hall system $\Sigma$ of $G$ reduces and $D$ is the $\mathfrak{F}$-normalizer of $G$ associated to $\Sigma$, for a saturated formation $\mathfrak{F}$, then $DV$ is a subgroup of $G$, which avoids all $\mathfrak{F}$-eccentric $V$-avoided chief factors of $G$ and covers the rest (see [5, Theorem 3]). With this result and Hawkes factorization (Theorem 2.3), Chambers showed Theorem 2.6.

At this point, given a Hall system $\Sigma$ of a soluble group $G$, there are three subgroups of $G$ associated to $\Sigma$ which are pairwise permutable: a normally embedded subgroup $V$, the $\mathfrak{F}$-normalizer $D$ of $G$ corresponding to a saturated formation $\mathfrak{F}$ and the prefrattini subgroup $W(\Sigma)$. In [15] Makan considered the lattice $\mathcal{L}(V, D, W(\Sigma))$ generated by these three subgroups and proved the following.

**Theorem 2.7 ([15])** *The lattice $\mathcal{L}(V, D, W(\Sigma))$ is a distributive lattice composed of pairwise permutable subgroups with the cover and avoidance property. Moreover $\Sigma$ reduces into each subgroup of $G$ in $\mathcal{L}(V, D, W(\Sigma))$.*

We finish this section with another interesting generalization of prefrattini subgroups in the soluble universe. It was performed by Kurzweil in [13]. He defined the $H$-prefrattini subgroups of a soluble group $G$, where $H$ is a proper subgroup of $G$. In this case, the choice of maximal subgroups is given by considering those maximal subgroups which contain $H$ and complement some chief factor in a fixed chief series of $G$.

**Definition 2.8** **([13])** Let $H$ be a proper subgroup of a soluble group $G$. Consider $1 = N_0 < \cdots < N_n = G$ a chief series of $G$ and, for each $i$, denote by $\mathcal{M}_i$ the set of all maximal subgroups of $G$ which complement the chief factor $N_i/N_{i-1}$ and contain $H$. Let $I$ be the set of suffixes $i$ for which $\mathcal{M}_i$ is nonempty. We choose $M_i$ in $\mathcal{M}_i$, for $i \in I$.

The intersection $\bigcap_{i \in I} M_i$ is an *$H$-prefrattini subgroup* of $G$.

In [13] it is proved that this definition is independent of the particular chief series used. The $H$-prefrattini subgroups form a conjugate class of subgroups with the cover and avoidance properties. When $H = 1$, these subgroups coincide with the prefrattini subgroups of Gaschütz. If $H$ is an $\mathfrak{F}$-normalizer, for a saturated formation $\mathfrak{F}$, then we obtain the $\mathfrak{F}$-prefrattini subgroups of Hawkes.

## 3 Prefrattini subgroups of finite groups

The extension of prefrattini subgroups to the general finite, non-necessarily soluble, universe was an important challenge successfully solved in [2] and [3]. These generalizations, done with no use of the classical Hall theory of soluble groups, bring new light to non-arithmetical properties of maximal subgroups. It is surprising to realize that, in the general context, although subgroups of prefrattini type lose their cover and avoidance properties, they keep their excellent permutability properties, as we will see in the sequel.

The first step to develop the theory of prefrattini subgroups in the universe of all finite groups was to define something playing the role of Hall systems in order to choose maximal subgroups. This task was tackled by Ballester-Bolinches and Ezquerro in [2], where the authors defined something they called *systems of maximal subgroups*. These are sets of maximal subgroups which exist in each finite group. If the group is soluble, then each system of maximal subgroups $\mathcal{S}$ may be associated to a Hall system of the group, so that $\mathcal{S}$ is exactly the set of maximal subgroups into which the Hall system reduces. In general, the systems of maximal subgroups are not conjugate; in fact, conjugacy characterizes solubility.

The existence of systems of maximal subgroups allows Ballester-Bolinches and Ezquerro to define the prefrattini subgroups associated to a Schunck class in each finite group (see [2]). This definition includes the classical ones due to Gaschütz, Hawkes and Förster in the soluble universe.

Four years later, the same authors presented in [3] a definition of subgroups of prefrattini type in the general universe of all finite groups. This definition covers all previous types of prefrattini subgroups, including the Kurzweil ones, in a more unified setting. To do so, the authors used some ideas due to Tomkinson. In [16] Tomkinson extended the results of Gaschütz and Hawkes to a class $\mathcal{U}$ of locally finite groups with a satisfactory Sylow structure. The intersection of $\mathcal{U}$ with the class of all finite groups is just the class of all finite soluble groups.

Based in Tomkinson's ideas, in [3] was introduced the concept of a *weakly-solid* (or simply *w-solid*) set of maximal subgroups. In this way, the prefrattini subgroups appear attached to both a w-solid set $\boldsymbol{X}$ of maximal subgroups and a system $\mathcal{S}$ of maximal subgroups.

**Definition 3.1** (**[3]**) Let $G$ be a group, $\boldsymbol{X}$ a w-solid set of maximal subgroups $G$ and $\mathcal{S}$ a system of maximal subgroups of $G$. Suppose that $\boldsymbol{X} \cap \mathcal{S} \neq \emptyset$ and for the subgroup

$$W(G, \boldsymbol{X}, \mathcal{S}) = \bigcap\{M <\cdot G \ : \ M \in \boldsymbol{X} \cap \mathcal{S}\}.$$

We say that $W(G, \boldsymbol{X}, \mathcal{S})$ is the $\boldsymbol{X}$*-prefrattini subgroup* of $G$ associated to $\mathcal{S}$. $\boldsymbol{X} \cap \mathcal{S} = \emptyset$ we put $W(G, \boldsymbol{X}, \mathcal{S}) = G$.

We will say that a subgroup $W$ is a *subgroup of prefrattini type* of a group $G$ $W$ is the $\boldsymbol{X}$-prefrattini subgroup of $G$ associated to $\mathcal{S}$, for some system of maxim subgroups and some w-solid set $\boldsymbol{X}$ of maximal subgroups of $G$.

When $\boldsymbol{X}$ is the set of all maximal subgroups of a soluble group, the $\boldsymbol{X}$-prefratti subgroups are just the prefrattini subgroups introduced by Gaschütz.

Given a w-solid set $\boldsymbol{X}$ of maximal subgroups of a group $G$ and a subgroup of $G$, $\boldsymbol{X}_H$ denotes the set of all maximal subgroups of $G$ which are in $\boldsymbol{X}$ ar contain $H$. The set $\boldsymbol{X}_H$ is also w-solid (see [3]). If $G$ is soluble and $\boldsymbol{X}$ is the s of all maximal subgroups of $G$, then the $\boldsymbol{X}_H$-prefrattini subgroups of $G$ are th $H$-prefrattini subgroups studied by Kurzweil.

Given a Schunck class $\mathfrak{H}$ and a w-solid set $\boldsymbol{X}$ of maximal subgroups of a group ( $\boldsymbol{X}^a_{\mathfrak{H}}$ denotes the set of all $\mathfrak{H}$-abnormal maximal subgroups of $G$ which are in $\boldsymbol{X}$ In [3] it is proved that $\boldsymbol{X}^a_{\mathfrak{H}}$ is w-solid whenever so $\boldsymbol{X}$ is. If $\boldsymbol{X}$ is the set of all ma imal subgroups of $G$, then the $\boldsymbol{X}^a_{\mathfrak{H}}$-prefrattini subgroups of $G$ are the $\mathfrak{H}$-prefratti subgroups studied in [2].

In general, the $\boldsymbol{X}$-prefrattini subgroups are not conjugate and they do not hav the cover and avoidance properties. In fact, if $\boldsymbol{X}$ is the set of all maximal subgrou of a group $G$, then conjugacy of the $\boldsymbol{X}$-prefrattini subgroups characterizes solubilit (see [2, Theorem 4.3]). The consideration of primitive non-soluble groups, whos core-free maximal subgroups are neither conjugate nor $CAP$-subgroups, cause these properties to fail in the general non-soluble universe.

In a group $G$, the subgroup $W(G, \boldsymbol{X}, \mathcal{S})$ covers all $\boldsymbol{X}$-Frattini chief factors of ( and avoids the $\boldsymbol{X} \cap \mathcal{S}$-complemented chief factors of $G$. Moreover, if $H/K$ is a abelian $\boldsymbol{X}$-complemented chief factor and does not possess an $\boldsymbol{X}$-complement in $\mathcal{S}$ then $W(G, \boldsymbol{X}, \mathcal{S})$ covers $H/K$.

Once prefrattini subgroups were defined in any finite group, the natural que tion arising was to obtain a factorization similar to the one proved by Hawkes i the soluble universe (Theorem 2.3). In the general universe of finite groups, th $\mathfrak{H}$-normalizers were defined in [1] for any Schunck class $\mathfrak{H}$ of the form $\mathfrak{H} = \mathrm{E}_\Phi\mathfrak{F}$ where $\mathfrak{F}$ is a formation. In [2], the authors established a correspondence betwee $\mathfrak{H}$-normalizers and systems of maximal subgroups. Then, a factorization simila to the Hawkes one was possible. It was obtained in [2] for prefrattini subgroup associated to the set of all maximal subgroups. Later, in [3] it is proved that thi factorization can be obtained for any subgroup of prefrattini type.

**Theorem 3.2** (**[3]**) *Let $G$ be a group and $\mathfrak{H}$ a Schunck class of the form $\mathfrak{H}$ = $\mathrm{E}_\Phi\mathfrak{F}$, where $\mathfrak{F}$ is a formation. Consider a system of maximal subgroups $\mathcal{S}$ of $G$*

*Then, if* $\boldsymbol{X}$ *is a w-solid set of maximal subgroups of* $G$, *we have* $DW(G, \boldsymbol{X}, \mathcal{S}) = W(G, \boldsymbol{X}^a_{\mathfrak{H}}, \mathcal{S})$, *where* $D$ *is an* $\mathfrak{H}$*-normalizer of* $G$ *associated with* $\mathcal{S}$.

After this factorization, it seems natural to wonder if it possible to extend Makan and Chambers factorizations (Theorems 2.5 and 2.6) to the general universe of all finite groups. The proofs of these results that appear in [14] and [5] depend heavily on the structure of normally embedded subgroups of finite soluble groups. In fact, these proofs are based on the cover and avoidance properties of normally embedded subgroups and prefrattini subgroups of a soluble group. Since prefrattini subgroups of a finite group do not have these properties in general, new techniques are needed in order to prove similar results. In addition, to obtain an extension of the lattice of Theorem 2.7, it would be interesting that the new subgroup playing the role of a normally embedded subgroup permutes with the normalizers associated to a saturated formation $\mathfrak{F}$. The suitable subgroup for this task is a hypercentrally embedded subgroup. A subgroup $T$ of a group $G$ is said to be *hypercentrally embedded* in $G$ if every chief factor of $G$ between the core and the normal closure of $T$ in $G$ is a central chief factor. In [4] Carocca and Maier characterized hypercentrally embedded subgroups of a finite group as those subgroups which permute with every pronormal subgroup of the group. In particular, hypercentrally embedded subgroups always permute with maximal subgroups. In [8] appears the following factorization theorem.

**Theorem 3.3 ([8])** *Let* $G$ *be a group,* $\boldsymbol{X}$ *a w-solid set of maximal subgroups of* $G$ *and* $\mathcal{S}$ *a system of maximal subgroups of* $G$. *If* $T$ *is a hypercentrally embedded subgroup of* $G$, *then* $TW(G, \boldsymbol{X}, \mathcal{S}) = W(G, \boldsymbol{X}_T, \mathcal{S})$.

Moreover, in the same paper it is proved that hypercentrally embedded subgroups also permute with $\mathfrak{F}$-normalizers, for any saturated formation $\mathfrak{F}$.

Notice that a hypercentrally embedded subgroup is in particular an $S$-permutable subgroup, that is a subgroup which permutes with all Sylow subgroups of the group. It is worth remarking that factorizations with subgroups of prefrattini type and with $\mathfrak{F}$-normalizers are no longer valid if we use $S$-permutable subgroups instead of hypercentrally embedded subgroups (see [8, Example 2]).

Accordingly, given a system of maximal subgroups, there are three subgroups which are pairwise permutable: a hypercentrally embedded subgroup, a subgroup of prefrattini type and an $\mathfrak{F}$-normalizer, for a saturated formation $\mathfrak{F}$. At this point, the natural question is to analyze the lattice generated by these three subgroups, in order to obtain a result which extends Theorem 2.7. Papers [8] and [9] deal with this matter.

A subgroup $V$ of a group $G$ is said to be *local-hypercentrally embedded* in $G$ if each Sylow subgroup of $V$ is also a Sylow subgroup of a hypercentrally embedded subgroup of $G$. In the soluble universe, Theorem 3.3 and also the factorization with $\mathfrak{F}$-normalizers can be extended to factorizations involving local-hypercentrally embedded subgroups (see [8]). In addition, in [9] we can find the following result.

**Theorem 3.4 ([9])** *Let* $\mathfrak{F}$ *be a saturated formation and* $G$ *a soluble group. Let* $\Sigma$ *be a Hall system of* $G$ *and* $\mathcal{S}$ *a system of maximal subgroups of* $G$ *associated to* $\Sigma$.

*Consider the $\mathfrak{F}$-normalizer $D$ of $G$ associated with $\Sigma$, the $\boldsymbol{X}$-prefrattini subgroup $W(G, \boldsymbol{X}, \mathcal{S})$ associated with $\mathcal{S}$ and a local-hypercentrally embedded subgroup $V$ of $G$ such that $\Sigma$ reduces into $V$.*

*It follows that the lattice $\mathcal{L}(V, D, W(G, \boldsymbol{X}, \mathcal{S}))$ generated by $V$, $D$ and $W(G, \boldsymbol{X}, \mathcal{S})$ is a distributive lattice of pairwise permutable subgroups with the cover and avoidance properties.*

## References


[1] A. Ballester-Bolinches, $\mathfrak{H}$-Normalizers and local definitions of saturated formations of finite groups, *Israel J. Math.* **67** (1989), 312–326.

[2] A. Ballester-Bolinches and L. M. Ezquerro, On maximal subgroups of finite groups, *Comm. Algebra* **19** (1991), no. 8, 2373–2394.

[3] A. Ballester-Bolinches and L. M. Ezquerro, The Jordan–Hölder theorem and prefrattini subgroups of finite groups, *Glasgow Math. J.* **37** (1995), 265–277.

[4] A. Carocca and R. Maier, Hypercentral embedding and pronormality, *Arch. Math.* **71** (1998), 433–436.

[5] G. Chambers, On $\mathfrak{F}$-prefrattini subgroups, *Canad. Math. Bull.* **15** (1972), no. 3, 345–348.

[6] K. Doerk and T. O. Hawkes, *Finite soluble groups*, De Gruyter, 1992.

[7] L. M. Ezquerro and X. Soler-Escrivà, Some permutability properties related to $\mathfrak{F}$-hypercentrally embedded subgroups of finite groups, *J. Algebra* **264** (2003), 279–295.

[8] L. M. Ezquerro and X. Soler-Escrivà, Some new permutability properties of hypercentrally embedded subgroups of finite groups, *J. Austral. Math. Soc.* **79** (2005), 1–13.

[9] L. M. Ezquerro and X. Soler-Escrivà, On certain lattices of subgroups of finite soluble groups, preprint.

[10] P. Förster, Prefrattini groups, *J. Austral. Math. Soc. (Series A)* **34** (1983), 234–247.

[11] W. Gaschütz, Praefrattinigruppen, *Arch. Math.* **13** (1962), 418–426.

[12] T. O. Hawkes, Analogues of Prefrattini subgroups, in *Proc. Internat. Conf. Theory of Groups*, Austral. Nat. Univ. Canberra (1965), 145–150, Gordon and Breach Science, 1967.

[13] H. Kurzweil, Die Praefrattinigruppe im Intervall eines Untergruppen-verbandes, *Arch. Math.* **53** (1989), 235–244.

[14] A. Makan, Another characteristic conjugacy class of subgroups of finite soluble groups, *J. Austral. Math. Soc.* **11** (1970), 395–400.

[15] A. Makan, On certain sublattices of the lattice of subgroups generated by the prefrattini subgroups, the injectors and the formation subgroups, *Canad. J. Math.* **25** (1973), 862–869.

[16] M. J. Tomkinson, Prefrattini subgroups and cover-avoidance properties in $\mathcal{U}$-groups, *Canad. J. Math.* **27** (1975), 837–851.

# FRATTINI EXTENSIONS AND CLASS FIELD THEORY

TH. WEIGEL

Università di Milano-Bicocca, U5-3067, Via R.Cozzi, 53 20125 Milano, Italy
Email: thomas.weigel@unimib.it

## Abstract

A. Brumer has shown that every profinite group of strict cohomological $p$-dimension 2 possesses a class field theory — the tautological class field theory. In particular, this result also applies to the universal $p$-Frattini extension $\tilde{G}_p$ of a finite group $G$. We use this fact in order to establish a class field theory for every $p$-Frattini extension $\pi\colon \tilde{G} \to G$ (Thm. A). The role of the class field module will be played by the $p$-Frattini module. The universal norms of this class field theory will carry important information about the $p$-Frattini extension $\pi\colon \tilde{G} \to G$. A detailed analysis will lead to a characterization of finite groups $G$ which have a $p$-Frattini extension $\pi\colon \tilde{G} \to G$ in which $\tilde{G}$ is a weakly-orientable $p$-Poincaré duality group of dimension 2 (Thm. B).

In section §5 we characterize the $p$-Frattini extensions $\pi_{A_1}: Sl_2(\mathbb{Z}_p) \to Sl_2(\mathbb{F}_p)$, $p \neq 2, 3, 5$, by some kind of localization technique. This answers a question posed by M. D. Fried and M. Jarden (Thm. C). It is quite likely that such an approach might also be successful for the characterization of the $p$-Frattini extensions $\pi_D\colon X_D(\mathbb{Z}_p) \to X_D(\mathbb{F}_p)$, where $X_D$ is the simple simply-connected split $\mathbb{Z}$-Chevalley group scheme with Dynkin diagram $D$.

*AMS Classification:* Primary 20E18

## 1 Introduction

Let $G$ be a finite group and let $p$ be a prime number. An extension of $G$ by a pro-$p$ group $A$

$$1 \longrightarrow A \stackrel{\iota}{\longrightarrow} \tilde{G} \stackrel{\pi}{\longrightarrow} G \longrightarrow 1 \tag{1.1}$$

is called a *$p$-Frattini extension*, if $\mathrm{im}(\iota)$ is contained in the Frattini subgroup of $\tilde{G}$. The study of $p$-Frattini extensions of finite groups has a long history. W. Gaschütz (cf. [8]) showed that every finite group $G$ has a universal elementary $p$-abelian Frattini extension $\pi_{/p}\colon \tilde{G}_{/p} \to G$ whose kernel — considered as (left) $\mathbb{F}_p[G]$-module — is isomorphic to $\Omega_2(G, \mathbb{F}_p)$, where $\Omega_k(G, \_) = \Omega^{-k}(G, \_)$ denotes the $k^{\text{th}}$-Heller translate in the category ${}_G\mathrm{mod}_p$ of finitely generated (left) $\mathbb{F}_p[G]$-modules. Based on this result J. Cossey, L. G. Kovács and O. H. Kegel [3] showed the existence of a universal $p$-Frattini cover $\pi_p\colon \tilde{G}_p \to G$. As the universal $p$-Frattini cover coincides with the minimal projective cover (cf. [6, Prop. 20.33]), K. Gruenberg's theorem [7] implies that $\tilde{G}_p$ is of cohomological $p$-dimension less or equal to 1, i.e, $cd_p(\tilde{G}_p) \leq 1$. In particular, $\ker(\pi_p)$ is a finitely generated free pro-$p$ group (cf. [12, §I.4.2, Cor. 2]).

If $p$ divides the order of $G$, the profinite group $\tilde{G}_p$ is of strict cohomologica $p$-dimension 2. For these groups A. Brumer [2] showed the existence of a *tautologica class field theory*. The goal of this paper is to use this tautological class field theor for the group $\tilde{G}_p$ in order to obtain new result on $p$-Frattini extensions.

The most efficient way to establish a class field theory is to use the theor of *cohomological Mackey functors*. A. Dress introduced this notion in [4]. Th exposition given by P. Webb in [15] will be particularly useful for our purpose, an therefore we will follow it closely as far as possible.

The following theorem can be seen as a "structure theorem for $p$-Frattini ex tensions", which combines W. Gaschütz theorem with the fact that the inflatio mapping $H^1(\pi, S)$ is bijective for a $p$-Frattini extension $\pi$ as in (1.1) and an irre ducible (left) $\mathbb{F}_p[G]$-module $S$ [17, Prop. 3.1]. Its proof can be found in section 3. (cf. Thm. 3.1, Cor. 3.2).

**Theorem A** *Let $G$ be a finite group, let $p$ be a prime number and let $\pi\colon \tilde{G} \to$ ( be a $p$-Frattini extension. Let $\mathcal{F}(\tilde{G})$ be the set of all open normal subgroups of ( being contained in* $\ker(\pi)$. *Then there exists a $p$-class field theory* $(\mathbf{C}, \gamma)$ *for* $(\tilde{G}, \mathcal{F}$ *i.e.,*

(i) $\mathbf{C}$ *is a cohomological $\mathcal{F}(\tilde{G})$-Mackey functor of type $H^0$ (this is a short for to say that it has Galois descent),*

(ii) $\mathbf{C}_U = \Omega_2(\tilde{G}/U, \mathbb{Z}_p)$ *for all $U \in \mathcal{F}(\tilde{G})$,*

(iii) $\gamma\colon \mathbf{C} \to \mathbf{Ab}^p$ *is a surjective morphism of cohomological $\mathcal{F}(\tilde{G})$-Mackey func tors, where $\mathbf{Ab}^p$ denotes the cohomological $\mathcal{F}(\tilde{G})$-Mackey functor of maxim $p$-abelian quotients (cf. §3.1),*

(iv) *for all $U, V \in \mathcal{F}(\tilde{G})$, $V \leq U$, $\gamma$ induces an isomorphism*

$$\mathbf{C}_U/\operatorname{im}(N^{\mathbf{C}}_{V,U}) \simeq (U/V)^{ab}_p, \tag{1.2}$$

(v) *let $U, V, W \in \mathcal{F}(\tilde{G})$, $V, W \leq U$, such that $U/V$ and $U/W$ are abelian* $p$ *groups. Then $\operatorname{im}(N^{\mathbf{C}}_{V,U}) = \operatorname{im}(N^{\mathbf{C}}_{W,U})$ implies $V = W$.*

The class field theory $(\mathbf{C}, \gamma)$ has also two further properties one would usu ally require from a class field theory: (vi) There exists a canonical class $c$ $\mathbf{nat}^2(\mathfrak{X}(\mathbb{Z}_p), \mathbf{C})$, (vii) $H^1(\tilde{G}/V, \mathbf{C}_V) = H^1(U/V, \mathbf{C}_V) = 0$ for all $U, V \in \mathcal{F}(\tilde{G}$ $V \leq U$ (cf. Rem. 3.3). However, this will not be of importance for our purpose.

The kernel of $\gamma$ will be called *the universal norms* (of $\mathbf{C}$). Its analysis will finall enable us to characterize finite groups $G$ possessing a $p$-Frattini cover $\pi\colon \tilde{G} \to G$ i which $\tilde{G}$ is a weakly-orientable profinite $p$-Poincaré duality group of dimension (cf. Cor. 4.6). Here we call a profinite $p$-Poincaré duality group $\tilde{G}$ of dimension *weakly-orientable*, if $H^d(\tilde{G}, \mathbb{F}_p[\![\tilde{G}]\!]) \simeq \mathbb{F}_p$ is the trivial module.

**Theorem B** *Let $G$ be a finite group, and let $p$ be a prime number. Then th following are equivalent:*

(i) *There exist a $p$-Frattini extension $\pi\colon \tilde{G} \to G$, where $\tilde{G}$ is a profinite weakl orientable $p$-Poincaré duality group of dimension* 2.

(ii) *There exists an injective map*

$$\alpha\colon \Omega^1(G,\mathbb{F}_p) \longrightarrow \Omega_2(G,\mathbb{F}_p) \tag{1.3}$$

*which is not an isomorphism.*

**Remark 1.1** Theorem B raises the following two questions: (1) For which finite groups $G$ and prime numbers $p$ does there exist an injective but not surjective map $\alpha\colon \Omega^1(G,\mathbb{F}_p) \longrightarrow \Omega_2(G,\mathbb{F}_p)$? (2) Provided such a mapping exists, how many isomorphism types of $p$-Frattini covers $\pi\colon \tilde{G} \to G$ exist, where $\tilde{G}$ is a weakly-orientable $p$-Poincaré duality group of dimension 2?

Unfortunately, we cannot say anything about the second question. Explicit computations using the work of K. Erdmann [5] show that for $q \equiv 3 \mod 4$, such a mapping $\alpha$ exists for $G := PSl_2(q)$ and $p = 2$ (cf. [17], [18]). However, it seems a very difficult problem to characterize or classify the tuples $(G,p)$ for which such a mapping exists.

Let $\mathfrak{S}_p(G)$ denote the set of isomorphism types of irreducible (left) $\mathbb{F}_p[G]$-modules, and let $\Delta \subseteq \mathfrak{S}_p(G)$ be a subset of $\mathfrak{S}_p(G)$. For short we call a $p$-Frattini extension $\pi\colon \tilde{G} \to G$ a *$\Delta$-Frattini extension*, if the isomorphism type of every $G$-composition factor of $\ker(\pi)$ is contained in $\Delta$. From the existence of the universal $p$-Frattini extension one deduces easily the existence of a universal $\Delta$-Frattini extension $\pi_\Delta\colon \tilde{G}_\Delta \to G$ (cf. §5.2). Obviously, $\tilde{G}_{\mathfrak{S}_p(G)}$ coincides with $\tilde{G}_p$, and $\tilde{G}_\emptyset$ coincides with $G$ itself. For our purpose it will be useful that the universal $\Delta$-Frattini extension can be characterized by vanishing of second degree cohomology in a similar way as it is known for the universal $p$-Frattini extension (cf. Prop. 5.1).

It is well-known that for $p \neq 3$, the extension

$$\pi_{A_1}\colon Sl_2(\mathbb{Z}_p) \longrightarrow Sl_2(\mathbb{F}_p) \tag{1.4}$$

is indeed a $p$-Frattini extension (cf. [16]). However, it remained an open problem to characterize the extension $\pi_{A_1}$ among all $p$-Frattini extension (cf. [6, Problem 20.40]).

For $p \neq 2,3$, M. Lazard's theorem implies that $Sl_2(\mathbb{Z}_p)$ is an orientable $p$-Poincaré duality group of dimension 3 (cf. [13]). From this fact we will deduces the following characterization:

**Theorem C** *Let $p$ be a prime different from 2, 3 and 5. Let $M_k$, $k = 0,\dots,p-1$, denote the simple $\mathbb{F}_p[Sl_2(\mathbb{F}_p)]$-module of weight $k$ and $\mathbb{F}_p$-dimension $k+1$. Then for every subset $\Delta \subset \mathfrak{S}_p(Sl_2(\mathbb{F}_p))$ satisfying*

(i) $[M_2] \in \Delta$,

(ii) $[M_{p-3}] \notin \Delta$,

*the universal $\Delta$-Frattini extension $\pi_\Delta$ of $Sl_2(\mathbb{F}_p)$ coincides with $\pi_{A_1}$, i.e., one has an isomorphism*

$$\phi\colon \widetilde{Sl}_2(\mathbb{F}_p)_\Delta \longrightarrow Sl_2(\mathbb{Z}_p) \tag{1.5}$$

*satisfying $\pi_{A_1} \circ \phi = \pi_\Delta$.*

For a given Dynkin diagram $D$ let $X_D$ be the simple simply-connected $\mathbb{Z}$-Ch valley group scheme associated to $D$. It has been proved in [16] that apart fro finitely many (more or less explicitly known) values of $(D, p)$,

$$\pi_D \colon X_D(\mathbb{Z}_p) \longrightarrow X_D(\mathbb{F}_p) \qquad (1$$

is a $p$-Frattini extension. Therefore, one wonders whether one can charact ize $X_D(\mathbb{Z}_p)$ in a similar fashion as $Sl_2(\mathbb{Z}_p)$ answering the problem raised in Prob. 20.40] in a wider context:

**Question 1.2** *Assume that $p$ is large with respect to the Coxeter number of Let $\mathfrak{L}_D(\mathbb{F}_p)$ denote the $\mathbb{F}_p$-Chevalley Lie algebra associated to $D$ considered (left) $\mathbb{F}_p[X_D(\mathbb{F}_p)]$-module and put $\Delta_D := \{[\mathfrak{L}_D(\mathbb{F}_p)]\}$. Are $\pi_D$ and $\pi_{\Delta_D}$ isom phic $p$-Frattini covers?*

**Remark 1.3** Proposition 5.1 shows that Question 1.2 is equivalent to the questi whether

$$H^2(X_D(\mathbb{Z}_p), \mathfrak{L}_D(\mathbb{F}_p)) = 0. \qquad (1.$$

## 2 Cohomological Mackey functors

### 2.1 Profinite modules of profinite groups

Let $p$ be a prime number, and let $\hat{G}$ be a profinite group. The *completed $\mathbb{Z}_p$-gro algebra* of $\hat{G}$ is given by

$$\mathbb{Z}_p[\![\hat{G}]\!] := \varprojlim_U \mathbb{Z}_p[\hat{G}/U], \qquad (2.$$

where the inverse system is running over all open normal subgroups of $\hat{G}$. By ${}_{\hat{G}}\mathbf{pr}$ we denote the abelian category the objects of which are abelian pro-$p$ groups wi continuous left $\hat{G}$-action. The morphisms from $M$ to $N$, $M, N \in ob({}_{\hat{G}}\mathbf{prf}_p)$, a defined to be the continuous morphisms of profinite groups commuting with t action of $\hat{G}$. The abelian group of morphisms from $M$ to $N$ will be denoted b $\mathbf{Hom}_{\hat{G}}(M, N)$. This category can be identified with the full subcategory of t category of topological left $\mathbb{Z}_p[\![\hat{G}]\!]$-modules, the objects of which are also abeli pro-$p$ groups. It is well-known that ${}_{\hat{G}}\mathbf{prf}_p$ has enough projectives, and in particul minimal projective covers. If $\hat{G}$ is the trivial group, then ${}_{\hat{G}}\mathbf{prf}_p$ coincides with t category of abelian pro-$p$ groups, which we will denote by $\mathbf{prf}_p$.

By ${}_{\hat{G}}\mathbf{prf}_{/p}$ we denote the abelian category the objects of which are profini $\mathbb{F}_p$-vector spaces with continuous left $\hat{G}$-action. It is a full subcategory of ${}_{\hat{G}}\mathbf{prf}$ and objects can be considered as topological modules for the *completed $\mathbb{F}_p$-gro algebra*

$$\mathbb{F}_p[\![\hat{G}]\!] := \varprojlim_U \mathbb{F}_p[\hat{G}/U]. \qquad (2.$$

For further details the reader may wish to consult [2], [11] or [13].

## 2.2 Cohomological Mackey functors

There are several equivalent ways to define a cohomological Mackey functor. Here we will follow more or less the approach chosen by P. Webb (cf. [15, §2]).

Let $\hat{G}$ be a profinite group and let $\mathcal{N}$ be a set of open normal subgroups of $\hat{G}$. For short we call $\mathcal{N}$ a *normal Mackey system* if $\mathcal{N}$ is closed with respect to products and intersections and if $\bigcap_{U\in\mathcal{N}} U = 1$.

Let $\mathcal{N}$ be a normal Mackey system of the profinite group $\hat{G}$. A *cohomological $\mathcal{N}$-Mackey functor* $\mathbf{X}$ with coefficients in $\mathbf{prf}_p$ is a collection $(\mathbf{X}_U)_{U\in\mathcal{N}}$ of $\hat{G}$-modules $\mathbf{X}_U \in ob({}_{\hat{G}/U}\mathbf{prf}_p)$, together with two series of mappings $i^{\mathbf{X}}_{U,V}$ and $N^{\mathbf{X}}_{V,U}$ for $U, V \in \mathcal{N},\ V \leq U$, where

$$\begin{aligned} i^{\mathbf{X}}_{U,V} &\in \mathbf{Hom}_{\hat{G}/V}(\mathbf{X}_U, \mathbf{X}_V), \\ N^{\mathbf{X}}_{V,U} &\in \mathbf{Hom}_{\hat{G}/V}(\mathbf{X}_V, \mathbf{X}_U), \end{aligned} \tag{2.3}$$

and which satisfy the following relations:

$$i^{\mathbf{X}}_{U,U} = N^{\mathbf{X}}_{U,U} = id_{\mathbf{X}_U} \qquad \text{for all } U \in \mathcal{N}, \tag{2.4}$$

$$i^{\mathbf{X}}_{U,W} = i^{\mathbf{X}}_{V,W} \circ i^{\mathbf{X}}_{U,V} \qquad \text{for all } U, V, W \in \mathcal{N},\ U \leq V \leq W, \tag{2.5}$$

$$N^{\mathbf{X}}_{W,U} = N^{\mathbf{X}}_{V,U} \circ N^{\mathbf{X}}_{W,V} \qquad \text{for all } U, V, W \in \mathcal{N},\ U \leq V \leq W, \tag{2.6}$$

$$i^{\mathbf{X}}_{UV,V} \circ N^{\mathbf{X}}_{U,UV} = N^{\mathbf{X}}_{U\cap V,V} \circ i^{\mathbf{X}}_{U,U\cap V} \qquad \text{for all } U, V \in \mathcal{N}, \tag{2.7}$$

$$i^{\mathbf{X}}_{U,V} \circ N^{\mathbf{X}}_{V,U} = \sum_{x\in U/V} x \qquad \text{for all } U, V \in \mathcal{N},\ U \leq V, \tag{2.8}$$

$$N^{\mathbf{X}}_{V,U} \circ i^{\mathbf{X}}_{U,V} = |U : V|.id_{\mathbf{X}_U} \qquad \text{for all } U, V \in \mathcal{N},\ U \leq V. \tag{2.9}$$

The notation we have chosen is closer related to number theory than the one introduced in [15]. One can easily verify that the role of $I^U_V$ in [15] is played by $N^{\mathbf{X}}_{V,U}$, and $i^{\mathbf{X}}_{U,V}$ plays the role of $R^U_V$. Our axioms (2.3) and (2.4)–(2.6) are obviously equivalent to the axioms (0)–(5) in [15, §2]. The axioms (2.7) and (2.8) are reformulating axiom (6) in [15], as we assumed that all open subgroups of $\hat{G}$ under consideration are normal in $\hat{G}$. Axiom (2.9) characterizes cohomological Mackey functors among all Mackey functors (cf. [15, §7]).

By $\mathfrak{CM}_{\mathcal{N}}(\hat{G}, \mathbf{prf}_p)$ we denote the category of cohomological $\mathcal{N}$-Mackey functors of $\hat{G}$ with coefficients in $\mathbf{prf}_p$. A morphism between cohomological $\mathcal{N}$-Mackey functors $\eta\colon \mathbf{X} \to \mathbf{Y}$ is a sequence of mappings $(\eta_U)_{U\in\mathcal{N}}$, $\eta_U \in \mathbf{Hom}_{\hat{G}/U}(\mathbf{X}_U, \mathbf{Y}_U)$, for which the diagrams

$$\begin{array}{ccc} \mathbf{X}_U & \xrightarrow{\eta_U} & \mathbf{Y}_U \\ {\scriptstyle i^{\mathbf{X}}_{U,V}}\downarrow & & \downarrow{\scriptstyle i^{\mathbf{Y}}_{U,V}} \\ \mathbf{X}_V & \xrightarrow{\eta_V} & \mathbf{Y}_V \end{array} \qquad \begin{array}{ccc} \mathbf{X}_U & \xrightarrow{\eta_U} & \mathbf{Y}_U \\ {\scriptstyle N^{\mathbf{X}}_{V,U}}\uparrow & & \uparrow{\scriptstyle N^{\mathbf{Y}}_{V,U}} \\ \mathbf{X}_V & \xrightarrow{\eta_V} & \mathbf{Y}_V \end{array} \tag{2.10}$$

commute for all $U, V \in \mathcal{N}$, $V \leq U$. By $\mathbf{nat}(\mathbf{X}, \mathbf{Y})$ we denote the abelian group of morphisms of cohomological $\mathcal{N}$-Mackey functors from $\mathbf{X}$ to $\mathbf{Y}$.

Using the interpretation of $\mathfrak{CM}_{\mathcal{N}}(\hat{G}, \mathbf{prf}_p)$ as the category of additive $\mathbb{Z}_p$-linear functors from the category of $\hat{G}$-permutation modules of discrete $\hat{G}$-sets with isotropy group being contained in $\mathcal{N}$ to the category $\mathbf{prf}_p$ of abelian pro-$p$ groups (cf. [15, Prop.7.2]), one sees easily that $\mathfrak{CM}_{\mathcal{N}}(\hat{G}, \mathbf{prf}_p)$ is an abelian category. Kernels and cokernels are defined in the obvious way.

## 2.3 From cohomological Mackey functors to $\hat{G}$-modules and vice versa

Taking the inverse limit over the norm maps $N_{V,U}$ defines a covariant left exact functor

$$\begin{aligned} m : \mathfrak{CM}_{\mathcal{N}}(\hat{G}, \mathbf{prf}_p) &\longrightarrow {}_{\hat{G}}\mathbf{prf}_p, \\ m(\mathbf{X}) := \varprojlim\nolimits_{U \in \mathcal{N}} \mathbf{X}_U, &\quad \text{for } \mathbf{X} \in ob(\mathfrak{CM}_{\mathcal{N}}(\hat{G}, \mathbf{prf}_p)). \end{aligned} \tag{2.11}$$

In case $\mathcal{N}$ contains a countable basis of neighbourhoods of $1 \in \hat{G}$, $\varprojlim^1$ vanishes, since all modules $\mathbf{X}_U$ are compact. Hence in this case $m$ is exact.

Let $M \in ob({}_{\hat{G}}\mathbf{prf}_p)$ be an abelian pro-$p$ group with continuous left $\hat{G}$-action. For an open normal subgroup $U \in \mathcal{N}$ we denote by

$$M_U := \mathbb{Z}_p[\hat{G}/U]\hat{\otimes}_{\hat{G}} M = M/cl(\langle\, (1-u).M \mid u \in U \,\rangle) \tag{2.12}$$

the $U$-coinvariants of $M$. Here $\hat{\otimes}$ denotes the pro-$p$ tensor product as defined by A. Brumer (cf. [2, §2]), and $cl$ denotes the closure operation. The assignment $\mathfrak{X}(M)$ which assigns $U \in \mathcal{N}$ the $U$-coinvariants $\mathfrak{X}(M)_U := M_U$ together with the natural map $N^{\mathfrak{X}(M)}_{V,U} : M_V \to M_U$, $V \leq U$, and the mapping $i^{\mathfrak{X}(M)}_{U,V} : M_U \to M_V$, $V \leq U$,

$$i^{\mathfrak{X}(M)}_{U,V}(m + cl(\langle\, (1-u).M \mid u \in U \,\rangle)) := \sum_{x \in V/U} x.m + cl(\langle\, (1-v).M \mid v \in V \,\rangle), \tag{2.13}$$

defines a cohomological $\mathcal{N}$-Mackey functor $\mathfrak{X}(M) \in ob(\mathfrak{CM}_{\mathcal{N}}(\hat{G}, \mathbf{prf}_p))$. It induces a covariant additive right exact functor

$$\mathfrak{X}(\_) : {}_{\hat{G}}\mathbf{prf}_p \longrightarrow \mathfrak{CM}_{\mathcal{N}}(\hat{G}, \mathbf{prf}_p), \tag{2.14}$$

which will be in general not exact. As we will see in the next subsection, the cohomological $\mathcal{N}$-Mackey functors obtained this way have a particular property which characterizes them.

## 2.4 Cohomology and homology of cohomological $\mathcal{N}$-Mackey functors

Let $\mathbf{X}$ be a cohomological $\mathcal{N}$-Mackey functor for $\hat{G}$ with coefficients in $\mathbf{prf}_p$. For short we call $\mathbf{X}$ *i-injective*, if all maps $i^{\mathbf{X}}_{U,V}$, $U, V \in \mathcal{N}$, $V \leq U$, are injective. Similarly, $\mathbf{X}$ is called *N-surjective*, if $N^{\mathbf{X}}_{V,U}$ is surjective for all $U, V \in \mathcal{N}$, $V \leq U$.

Assume that $\mathbf{X} \in ob(\mathfrak{CM}_{\mathcal{N}}(\hat{G}, \mathbf{prf}_p))$ is $i$-injective. Then we call $\mathbf{X}$ of *type* $H^0$ if

$$\operatorname{im}(i^{\mathbf{X}}_{U,V}) = \mathbf{X}_V^{U/V} \tag{2.15}$$

for all $U, V \in \mathcal{N}$, $V \leq U$. Here $\mathbf{X}_V^{U/V}$ denotes the abelian group of $U/V$-fixed points on $\mathbf{X}_V$. Cohomological $\mathcal{N}$-Mackey functors of type $H^0$ are sometimes also called to have *Galois descent.* The $N$-surjective cohomological $\mathcal{N}$-Mackey functor is called of *type* $H_0$ if

$$\ker(N_{V,U}^{\mathbf{X}}) = \sum_{x \in U/V} (x-1).\mathbf{X}_V \tag{2.16}$$

for all $U, V \in \mathcal{N}$, $V \leq U$. From this definition it is straightforward that a cohomological $\mathcal{N}$-Mackey functor is of type $H_0$ if and only if it is isomorphic to a functor $\mathfrak{X}(M)$ for some $M \in ob({}_{\hat{G}}\mathbf{prf}_p)$. The cohomological $\mathcal{N}$-Mackey functors being oy type $H_0$ are sometimes also called to have *Galois codescent.*

It is possible to interpret the definitions of being of type $H_0$ or of type $H^0$ in a more general homological context. For a cohomological $\mathcal{N}$-Mackey functor $\mathbf{X}$ we define for $U, V \in \mathcal{N}$, $V \leq U$,

$$\mathbf{k}^0(U/V, \mathbf{X}) := \ker(i_{U,V}^{\mathbf{X}}), \qquad \mathbf{k}^1(U/V, \mathbf{X}) := \mathbf{X}_V^{U/V} / \operatorname{im}(i_{U,V}^{\mathbf{X}}), \tag{2.17}$$

$$\mathbf{c}_0(U/V, \mathbf{X}) := \operatorname{coker}(N_{V,U}^{\mathbf{X}}), \qquad \mathbf{c}_1(U/V, \mathbf{X}) := \ker(N_{U,V}^{\mathbf{X}}) / \sum_{x \in U/V} (x-1)\mathbf{X}_V. \tag{2.18}$$

Let $0 \to \mathbf{X} \to \mathbf{Y} \to \mathbf{Z} \to 0$ be a short exact sequence of cohomological $\mathcal{N}$-Mackey functors. Then the snake lemma implies that one has exact sequences

$$\begin{aligned} 0 \to \mathbf{k}^0(U/V, \mathbf{X}) &\to \mathbf{k}^0(U/V, \mathbf{Y}) \to \mathbf{k}^0(U/V, \mathbf{Z}) \ldots \\ \to \mathbf{k}^1(U/V, \mathbf{X}) &\to \mathbf{k}^1(U/V, \mathbf{Y}) \to \mathbf{k}^1(U/V, \mathbf{Z}), \end{aligned} \tag{2.19}$$

$$\begin{aligned} \mathbf{c}_1(U/V, \mathbf{X}) &\to \mathbf{c}_1(U/V, \mathbf{Y}) \to \mathbf{c}_1(U/V, \mathbf{Z}) \to \ldots \\ \mathbf{c}_0(U/V, \mathbf{X}) &\to \mathbf{c}_0(U/V, \mathbf{Y}) \to \mathbf{c}_0(U/V, \mathbf{Z}) \to 0. \end{aligned} \tag{2.20}$$

One can therefore think of $\mathbf{k}^{0/1}(U/V, \_)$ as the 0- and 1-dimensional *section cohomology* of cohomological $\mathcal{N}$-Mackey functors, and of $\mathbf{c}_{0/1}(U/V, \_)$ as the 0- and 1-dimensional *section homology* of cohomological $\mathcal{N}$-Mackey functors. It is possible to extend these functors to cohomological and homological functors, respectively. Since we will not make use of the higher derived functors we omit a detailed discussion here. However, we would like to remark, that these functors are not unrelated.

**Proposition 2.1** *Let* $\mathbf{X} \in \mathfrak{CM}_{\mathcal{N}}(\hat{G}, \mathbf{prf}_p)$ *be a cohomological* $\mathcal{N}$*-Mackey functor and let* $U, V \in \mathcal{N}$, $V \leq U$. *Then one has an exact sequence of* $\hat{G}/U$*-modules*

$$\begin{aligned} 0 \longrightarrow \mathbf{c}_1(U/V, \mathbf{X}) &\xrightarrow{\alpha_1} \hat{H}^{-1}(U/V, \mathbf{X}_V) \xrightarrow{\alpha_2} \mathbf{k}^0(U/V, \mathbf{X}) \xrightarrow{\alpha_3} \ldots \\ \mathbf{c}_0(U/V, \mathbf{X}) &\xrightarrow{\alpha_4} \hat{H}^0(U/V, \mathbf{X}_V) \xrightarrow{\alpha_5} \mathbf{k}^1(U/V, \mathbf{X}) \longrightarrow 0, \end{aligned} \tag{2.21}$$

*where* $\hat{H}^\bullet(U/V, \_)$ *denotes Tate cohomology.*

**Proof** The mapping $\alpha_1 \colon \mathbf{c}_1(U/V, \mathbf{X}) \to \hat{H}^{-1}(U/V, \mathbf{X}_V)$ is clearly injective. Since $\alpha_2$ is induced by the norm map $N_{V,U}^{\mathbf{X}}$, one has

$$\ker(\alpha_2) = \ker(N_{V,U}^{\mathbf{X}}) / \sum_{x \in U/V} (x-1)\mathbf{X}_V = \operatorname{im}(\alpha_1). \tag{2.22}$$

Furthermore, by axiom (2.9)

$$\ker(\alpha_3) = \ker(i^{\mathbf{X}}_{U,V}) \cap \operatorname{im}(N_{V,U}) = N_{V,U}\Big(\ker\Big(\sum_{x\in U/V} x\Big)\Big) = \operatorname{im}(\alpha_2). \tag{2.23}$$

The mapping $\alpha_4$ is induced by $i^{\mathbf{X}}_{U,V}$. Hence

$$\ker(\alpha_4) = \big(\ker(i^{\mathbf{X}}_{U,V}) + \operatorname{im}(N^{\mathbf{X}}_{V,U})\big) / \operatorname{im}(N^{\mathbf{X}}_{V,U}) = \operatorname{im}(\alpha_3). \tag{2.24}$$

The mapping $\alpha_5$ is the canonical map and thus surjective. Furthermore,

$$\ker(\alpha_5) = \operatorname{im}(i^{\mathbf{X}}_{U,V})/(\textstyle\sum_{x\in U/V} x).\mathbf{X}_V = \operatorname{im}(\alpha_4). \tag{2.25}$$

This yields the claim. □

**Remark 2.2** Let $\hat{G}$ be a finite cyclic group and let $\mathcal{N} := \{1, \hat{G}\}$. Using an alternative approach for the definition of $\mathbf{c}_\bullet(\hat{G}, \_)$ and $\mathbf{k}^\bullet(\hat{G}, \_)$ one sees that there exist connecting homomorphisms making the sequence

$$(\mathbf{k}^0(\hat{G}, \_), \mathbf{k}^1(\hat{G}, \_), \mathbf{c}_1(\hat{G}, \_), \mathbf{c}_0(\hat{G}, \_)) \tag{2.26}$$

a (co)homological functor. Let $M \in ob({}_{\hat{G}}\mathbf{prf}_p)$ be a finitely generated $\mathbb{Z}_p[\hat{G}]$-module. Then (2.21) says that the Herbrand quotient (cf. [10, Kap. IV, §7])

$$h(\hat{G}, M) := \frac{|\hat{H}^0(\hat{G}, M)|}{|\hat{H}^{-1}(\hat{G}, M)|} \tag{2.27}$$

can be interpreted as a kind of multiplicative Euler characteristic, i.e., one has

$$h(\hat{G}, M) = \frac{|\mathbf{c}_0(\hat{G}, \mathfrak{X}(M))| \cdot |\mathbf{k}^1(\hat{G}, \mathfrak{X}(M))|}{|\mathbf{c}_1(\hat{G}, \mathfrak{X}(M))| \cdot |\mathbf{k}^0(\hat{G}, \mathfrak{X}(M))|} =: \chi(\mathfrak{X}(M)). \tag{2.28}$$

For short we say that a cohomological $\mathcal{N}$-Mackey functor $\mathbf{X}$ is *cohomologically trivial*, if $\mathbf{X}$ is of type $H^0$ and $H_0$. From Proposition 2.1 follows that such a functor satisfies

$$\hat{H}^{-1}(U/V, \mathbf{X}_V) = \hat{H}^0(U/V, \mathbf{X}_V) = 0 \tag{2.29}$$

for all $U, V \in \mathcal{N}$, $V \leq U$.

**Proposition 2.3** *Let $P \in ob({}_{\hat{G}}\mathbf{prf}_p)$ be projective. Then for $V \in \mathcal{N}$, $\mathfrak{X}(P)$ (cf. 2.3) is a projective $\mathbb{Z}_p[\hat{G}/V]$-module. In particular, $\mathfrak{X}(P)$ is a cohomological trivial cohomological $\mathcal{N}$-Mackey functor.*

**Proof** The first statement follows from the fact that deflation from ${}_{\hat{G}}\mathbf{prf}_p$ to ${}_{\hat{G}/V}\mathbf{prf}_p$ is mapping projectives to projectives. Since restriction to closed subgroups is mapping projectives to projectives, it suffices to prove the second claim for $U = \hat{G}$. Since $\mathfrak{X}(P)$ is of type $H_0$, $\mathbf{c}_{0/1}(\hat{G}/V, \mathfrak{X}(P)) = 0$. As $P_V \in ob({}_{\hat{G}/V}\mathbf{prf}_p)$ is projective, $\hat{H}^{-1}(\hat{G}/V, P_V) = \hat{H}^0(\hat{G}/V, P_V) = 0$. Hence Proposition 2.1 yields the claim.

## 3 Class field theories

Throughout this section let $\hat{G}$ be a profinite group, and let $p$ be a prime number. We also assume that $\mathcal{N}$ is a normal Mackey system for $\hat{G}$.

For a finite group $G$ we denote by $\mathfrak{S}_p(G)$ the set of isomorphism types of irreducible (left) $\mathbb{F}_p[G]$-modules. For an irreducible $\mathbb{F}_p[G]$-module $S$ we use the symbol $[S] \in \mathfrak{S}_p(G)$ to denote its isomorphism type.

### 3.1 The cohomological Mackey functors $\mathbf{Ab}^p$ and $\mathbf{Ab}^{/p}$

For $U \in \mathcal{N}$, let

$$\mathbf{Ab}^p_U := U^{ab}_p = U/cl([U,U])/O_{p'}(U/cl([U,U]) \tag{3.1}$$

denote the largest continuous homomorphic image of $U$ which is an abelian pro-$p$ group. Here $[\_,\_]$ stands for the commutator subgroup, and $cl$ denotes the closure operation. Then for $U, V \in \mathcal{N}$, $V \le U$, one has a canonical map $N^{\mathbf{Ab}^p}_{V,U} : V^{ab}_p \to U^{ab}_p$. This map together with the *transfer map* (cf. [10, p.312])

$$i^{\mathbf{Ab}^p}_{U,V} : = tr^U_V : U^{ab}_p \to V^{ab}_p \tag{3.2}$$

makes $\mathbf{Ab}^p \in ob(\mathfrak{CM}_{\mathcal{N}}(\hat{G}, \mathbf{prf}_p))$ a cohomological $\mathcal{N}$-Mackey functor. By $\mathbf{Ab}^{/p}$ we denote its reduction modulo $p$, i.e., for $U \in \mathcal{N}$ one has

$$\mathbf{Ab}^{/p}_U : = U^{ab}_{/p} = \mathbf{Ab}^p_U / p.\mathbf{Ab}^p_U, \tag{3.3}$$

and the maps $i^{\mathbf{Ab}^{/p}}_{U,V}$ and $N^{\mathbf{Ab}^{/p}}_{V,U}$, $U, V \in \mathcal{N}$, $V \le U$, are the maps induced from $i^{\mathbf{Ab}^p}_{U,V}$ and $N^{\mathbf{Ab}^p}_{V,U}$, respectively. It is obviously a cohomological $\mathcal{N}$-Mackey functor.

### 3.2 Weak $p$-class field theories

We define a *weak $p$-class field theory* $(\mathbf{X}, \eta)$ (for $(\hat{G}, \mathcal{N})$) to be a cohomological $\mathcal{N}$-Mackey functor $\mathbf{X} \in ob(\mathfrak{CM}_{\mathcal{N}}(\hat{G}, \mathbf{prf}_p))$, together with a surjective morphism $\eta \colon \mathbf{X} \to \mathbf{Ab}^p$ of cohomological $\mathcal{N}$-Mackey functors with the following properties:

(i) $\mathbf{X}$ is of type $H^0$,

(ii) $\mathbf{c}_0(U/V, \eta) \colon \mathbf{c}_0(U/V, \mathbf{X}) \to (U/V)^{ab}_p$ is an isomorphism for all $U, V \in \mathcal{N}$, $V \le U$.

The property (i) implies that $\mathbf{k}^{0/1}(U/V, \mathbf{X}) = 0$ for all $U, V \in \mathcal{N}$, $V \le U$. In particular, one has an isomorphism $\mathbf{c}_0(U/V, \mathbf{X}) = \hat{H}^0(U/V, \mathbf{X}_V)$. The property (ii) is one of the properties one would expect from a $p$-class field theory. However, in order to state the other property, one has also to require some structure on the normal Mackey system $\mathcal{N}$.

### 3.3 $p$-Class field theories

For short we call a normal Mackey system *$p$-closed* if it satisfies the following property: Assume that $W$ is an open normal subgroup of $\hat{G}$ which is contained in an open normal subgroup in $U \in \mathcal{N}$ such that $U/W$ is a finite abelian $p$-grou Then $W$ is also contained in $\mathcal{N}$.

Let $\mathcal{N}$ be a $p$-closed normal Mackey system of $\hat{G}$. Then we call the weak $p$-cla field theory $(\mathbf{X}, \eta)$ a *$p$-class field theory* if it satisfies additionally the followin property:

(iii) Let $U \in \mathcal{N}$ and let $V, W \leq U$ be open and normal in $\hat{G}$, such that $U/V$ an $U/W$ are finite abelian $p$-groups. Assume that $\mathrm{im}(N^{\mathbf{X}}_{V,U}) = \mathrm{im}(N^{\mathbf{X}}_{W,U})$. The $V = W$.

In a similar fashion one defines a */p-class field theory*: Let $\mathcal{N}$ be a $p$-closed no mal Mackey system of $\hat{G}$. A cohomological $\mathcal{N}$-Mackey functor $\mathbf{X}$ together with surjective morphism of $\mathcal{N}$-Mackey functors $\eta\colon \mathbf{X} \to \mathbf{Ab}^{/p}$ is called a /$p$-class fiel theory, if the following properties hold:

(i) $\mathbf{X}$ is of type $H^0$,

(ii) $\mathbf{c}_0(U/V, \eta)\colon \mathbf{c}_0(U/V, \mathbf{X}) \to (U/V)^{ab}_{/p}$ is an isomorphism for all $U, V \in \mathcal{N}$ $V \leq U$,

(iii) Let $U \in \mathcal{N}$ and let $V, W \leq U$ be open and normal in $\hat{G}$, such that $U/V$ and $U/W$ are finite elementary abelian $p$-groups. Assume that $\mathrm{im}(N^{\mathbf{X}}_{V,U}) = \mathrm{im}(N^{\mathbf{X}}_{W,U})$. Then $V = W$.

### 3.4 The $p$-Frattini class field theory and the /$p$-Frattini class field theory

Let $G$ be a finite group, and let $\pi_p\colon \tilde{G}_p \to G$ denote its universal $p$-Frattini cove We are considering the normal Mackey system

$$\mathcal{F} := \{\, U \leq \ker(\pi_p) \mid U \text{ open and normal in } \tilde{G}_p \,\}. \tag{3.4}$$

As $\ker(\pi_p)$ is a pro-$p$ group, it is obviously $p$-closed.

Let

$$0 \longrightarrow P_1 \xrightarrow{\delta} P_0 \xrightarrow{\varepsilon} \mathbb{Z}_p \longrightarrow 0 \tag{3.5}$$

be a minimal projective resolution of the trivial $\mathbb{Z}_p[\![\tilde{G}_p]\!]$-module $\mathbb{Z}_p$ in ${}_{\tilde{G}_p}\mathbf{prf}_p$. I particular, $\varepsilon\colon P_0 \to \mathbb{Z}_p$ and $\delta'\colon P_1 \to \ker(\varepsilon)$ are minimal projective covers in ${}_{\tilde{G}_p}\mathbf{prf}$

Let $\mathfrak{S}_p(G)$ denote the set of isomorphism types of irreducible $\mathbb{F}_p[G]$-module and let $\tau_S\colon P_S \to S$ denote a minimal projective cover in ${}_{\tilde{G}_p}\mathbf{prf}_p$, $[S] \in \mathfrak{S}_p(G)$. A (3.5) is minimal, one has isomorphisms

$$\mathbf{Hom}_{\tilde{G}_p}(P_1, S) \simeq H^1(\tilde{G}_p, S) \tag{3.6}$$

for all $[S] \in \mathfrak{S}_p(G)$. In particular, $P_1 \simeq \coprod_{[S]\in\mathfrak{S}_p(G)} P_S^{\mu_S}$, where

$$\mu_S\colon = \frac{\dim_{\mathbb{F}_p}(H^1(\tilde{G}_p, S))}{\dim_{\mathbb{F}_p}(\mathrm{End}_G(S))}. \tag{3.7}$$

Let $U \in \mathcal{F}$. As $\_{}_U$ is right exact, one has an exact sequence

$$(P_1)_U \xrightarrow{\delta_U} (P_0)_U \xrightarrow{\varepsilon_U} \mathbb{Z}_p \longrightarrow 0. \tag{3.8}$$

As $\tilde{G}_p \to \tilde{G}_p/U$ is a $p$-Frattini extension, inflation induces isomorphisms

$$H^1(\tilde{G}_p, S) \simeq H^1(\tilde{G}_p/U, S) \tag{3.9}$$

for all $[S] \in \mathfrak{S}_p(G)$ (cf. [17, Prop. 3.1]). This yields that

$$H^1(\tilde{G}_p/U, S) \simeq \mathbf{Hom}_{\tilde{G}_p/U}((P_1)_U, S) \tag{3.10}$$

for all $[S] \in \mathfrak{S}_p(G)$, and from this one concludes easily that (3.8) is a partial minimal projective resolution. In particular, $\ker(\delta_U) = \Omega_2(\tilde{G}_p/U, \mathbb{Z}_p)$.

Let $\boldsymbol{\Omega}_2 := \ker(\mathfrak{X}(\delta))$. Then one has an exact sequence of cohomological $\mathcal{F}$-Mackey functors

$$0 \longrightarrow \boldsymbol{\Omega}_2 \longrightarrow \mathfrak{X}(P_1) \xrightarrow{\mathfrak{X}(\delta)} \mathfrak{X}(P_0) \xrightarrow{\mathfrak{X}(\varepsilon)} \mathfrak{X}(\mathbb{Z}_p) \longrightarrow 0, \tag{3.11}$$

and $\boldsymbol{\Omega}_{2,U} = \Omega_2(\tilde{G}_p/U, \mathbb{Z}_p)$.

From the Eckmann–Shapiro lemma for $\mathbf{Tor}_\bullet$ (cf. [13, Lemma 3.3.4]), and the canonical isomorphism $\mathbf{H}_1(U, \mathbb{Z}_p) \simeq U_p^{ab} = \mathbf{Ab}_U^p$, where $\mathbf{H}_\bullet$ denotes homology as defined by A. Brumer (cf. [2, §2]), one obtains an isomorphism

$$\eta \colon \boldsymbol{\Omega}_2 \longrightarrow \mathbf{Ab}^p \tag{3.12}$$

of cohomological $\mathcal{F}$-Mackey functors.

By $\boldsymbol{\Omega}_2^{/p}$ we denote the reduction mod $p$ of $\boldsymbol{\Omega}_2$, i.e., one has a short exact sequence in $\mathfrak{CM}_{\mathcal{F}}(\tilde{G}_p, \mathbf{prf}_p)$

$$0 \longrightarrow \boldsymbol{\Omega}_2 \xrightarrow{p.id} \boldsymbol{\Omega}_2 \longrightarrow \boldsymbol{\Omega}_2^{/p} \longrightarrow 0. \tag{3.13}$$

By $\eta^{/p} \colon \boldsymbol{\Omega}_2^{/p} \to \mathbf{Ab}^{/p}$ we denote the induced isomorphism.

**Theorem 3.1** *Let $G$ be a finite group, $\pi_p \colon \tilde{G}_p \to G$ its universal $p$-Frattini cover, and let $\mathcal{F}$ be given as in* (3.4).

(a) *The tuple $(\boldsymbol{\Omega}_2, \eta)$ is a $p$-class field theory for $(\tilde{G}_p, \mathcal{F})$.*

(b) *The tuple $(\boldsymbol{\Omega}_2^{/p}, \eta^{/p})$ is a $/p$-class field theory for $(\tilde{G}_p, \mathcal{F})$.*

We call $(\boldsymbol{\Omega}_2, \eta)$ the *$p$-Frattini class field theory* for $(\tilde{G}_p, \mathcal{F})$, and $(\boldsymbol{\Omega}_2^{/p}, \eta^{/p})$ the *$/p$-Frattini class field theory* for $(\tilde{G}_p, \mathcal{F})$.

**Proof** (a) One has to verify the axioms (i)–(iii). Axiom (ii) is obviously satisfied. Consider the short exact sequence

$$0 \longrightarrow \boldsymbol{\Omega}_2 \xrightarrow{\iota} \mathfrak{X}(P_1) \longrightarrow \operatorname{coker}(\iota) \longrightarrow 0. \tag{3.14}$$

Since $\operatorname{coker}(\iota)$ is a cohomological $\mathcal{F}$-subMackey functor of $\mathfrak{X}(P_0)$, $\mathbf{k}^0(\operatorname{coker}(\iota)) = 0$ (cf. (2.19), Prop. 2.3). The long exact sequence (2.19) applied to (3.14) and the cohomological triviality of $\mathfrak{X}(P_0)$ and $\mathfrak{X}(P_1)$ yields that $\boldsymbol{\Omega}_2$ is of type $H^0$. Hence axiom (i) is satisfied. It remains to verify (iii). We may assume that $p$ divides the order of the finite group $G$, since otherwise $\boldsymbol{\Omega}_2 = 0$, and there is nothing to prove. In this case $\tilde{G}_p$ is of cohomological $p$-dimension 1, and thus of strict cohomological $p$-dimension 2 (cf. [12, §I.3.2]). In particular, by Brumer's theorem (cf. [2], [10, Kap. IV, §6, Aufg. 6]) $\hat{G}$ possesses a *tautological class field theory*. Let $(\mathfrak{H}, \rho)$ denote its restriction to the Mackey system $\mathcal{F}$, i.e., $\mathfrak{H}_U = \mathbf{Ab}^p_U$ and $\rho_U$ is the identity on $\mathbf{Ab}^p_U$. In particular $(\mathfrak{H}, \rho)$ and $(\mathbf{\Omega}_2, \eta)$ essentially coincide, i.e., one has a commutative diagram in $\mathfrak{CM}_{\mathcal{F}}(\tilde{G}_p, \mathbf{prf}_p)$

$$\begin{array}{ccc} \mathbf{\Omega}_2 & \xrightarrow{\eta} & \mathfrak{H} \\ {\scriptstyle\eta}\downarrow & & \| \\ \mathbf{Ab}^p & = & \mathbf{Ab}^p \end{array} \tag{3.15}$$

The property (iii) is well-known for $(\mathfrak{H}, \rho)$ (cf. [10, Kap. IV, Thm. 6.7]). Thus it also holds for $(\mathbf{\Omega}_2, \eta)$.

(b) It suffices to prove that $\mathbf{\Omega}_2^{/p}$ is of type $H^0$. The axiom (ii) is obvious, and axiom (iii) follows from axiom (iii) for $(\mathbf{\Omega}_2, \eta)$.

Let $\mathfrak{X}(P_{0/1})^{/p}$ denote the reduction mod $p$ of $\mathfrak{X}(P_0)$ and $\mathfrak{X}(P_1)$, respectively. Then one has a short exact sequence

$$0 \longrightarrow \mathbf{\Omega}_2^{/p} \xrightarrow{\iota^{/p}} \mathfrak{X}(P_1)^{/p} \longrightarrow \operatorname{coker}(\iota^{/p}) \longrightarrow 0, \tag{3.16}$$

and $\operatorname{coker}(\iota^{/p})$ is a cohomological $\mathcal{F}$-sub Mackey functor of $\mathfrak{X}(P_0)^{/p}$. From Proposition 2.1 one concludes that $\mathfrak{X}(P_0)^{/p}$ and $\mathfrak{X}(P_1)^{/p}$ are cohomologically trivial. Hence the long exact sequence (2.19) yields the claim. □

Let $\pi \colon \tilde{G} \to G$ be any $p$-Frattini extension, finite or infinite. By universality, there exists a mapping $\tau \colon \tilde{G}_p \to \tilde{G}$, such that $\pi_p = \pi \circ \tau$. Since $\pi$ is a $p$-Frattini extension, $\tau$ is surjective. For short we put $N := \ker(\tau)$.

The morphism $\tau$ induces a canonical bijection of sets $\tau_* \colon \mathcal{F}_N \to \mathcal{F}(\tilde{G})$, where $\mathcal{F}$ is given as in (3.4) and

$$\begin{aligned} \mathcal{F}_N &:= \{\, U \in \mathcal{F} \mid N \leq U \,\}, \\ \mathcal{F}(\tilde{G}) &:= \{\, U' \leq \ker(\pi) \mid U' \text{ open and normal in } \tilde{G} \,\}. \end{aligned} \tag{3.17}$$

Let $\mathbf{C} \in ob(\mathfrak{CM}_{\mathcal{F}(\tilde{G})}(\tilde{G}, \mathbf{prf}_p))$ denote the cohomological $\mathcal{F}(\tilde{G})$-Mackey functor given by

$$\mathbf{C}_U := \mathbf{\Omega}_{2,\tau_*^{-1}(U)}, \quad U \in \mathcal{F}(\tilde{G}) \tag{3.18}$$

equipped with the obvious maps $i^{\mathbf{C}}_{U,V}$, $N^{\mathbf{C}}_{V,U}$, $U, V \in \mathcal{F}(\tilde{G})$, $V \leq U$. Let $\gamma \colon \mathbf{C} \to \mathbf{Ab}^p$ denote the morphism of $\mathcal{F}(\tilde{G})$-Mackey functors induced by $\eta$. In particular $\gamma$ is surjective, but if $\tilde{G}$ does not coincide with the universal $p$-Frattini cover, $\gamma$ will not be an isomorphism.

Similarly, we define the reduction mod $p$ $\mathbf{C}^{/p}$ of $\mathbf{C}$, i.e., one has

$$\mathbf{C}^{/p}_U : = \mathbf{\Omega}^{/p}_{2,\tau_*^{-1}(U)}, \quad U \in \mathcal{F}(\tilde{G}), \tag{3.19}$$

and by $\gamma^{/p} \colon \mathbf{C}^{/p} \to \mathbf{Ab}^{/p}$ we denote the surjective morphism induced by $\eta^{/p}$. Again apart from the case $\tilde{G} \simeq \tilde{G}_p$, $\gamma^{/p}$ will not be surjective. From Theorem 3.1 one concludes:

**Corollary 3.2** *Let $G$ be a finite group, and let $\pi\colon \tilde{G} \to G$ be any p-Frattini extension. Then*

(a) *The tuple $(\mathbf{C}, \gamma)$ is a p-class field theory for $(\tilde{G}, \mathcal{F}(\tilde{G}))$.*

(b) *The tuple $(\mathbf{C}^{/p}, \gamma^{/p})$ is a /p-class field theory for $(\tilde{G}, \mathcal{F}(\tilde{G}))$.*

**Remark 3.3** The definition of a $p$ or a $/p$-class field theory we have given here is very much adapted to our main purpose, which is to prove Theorem B. Nevertheless, $(\mathbf{\Omega}_2, \eta)$ satisfies all class field theory axioms, which are usually required in number theory, i.e., using Tate cohomology one sees easily that for all $U, V \in \mathcal{F},\ V \leq U$,

$$H^1(U/V, \mathbf{\Omega}_{2,V}) = H^1(\tilde{G}_p/V, \mathbf{\Omega}_{2,V}) = 0. \tag{3.20}$$

Moreover, (3.11) defines a *canonical class* $c \in \mathbf{nat}^2(\mathfrak{X}(\mathbb{Z}_p), \mathbf{\Omega}_2)$, where $\mathbf{nat}^\bullet(\_,\_)$ denote the derived functors of $\mathbf{nat}(\_,\_)$ (cf. [9, Chap. XII]). This also applies to the $p$-class field theory $(\mathbf{C}, \gamma)$ defined for any $p$-Frattini cover $\pi\colon \tilde{G} \to G$. However, as the reader might verify by himself, (3.20) does not hold for the $/p$-class field theories $(\mathbf{\Omega}_2^{/p}, \eta^{/p})$ or $(\mathbf{C}^{/p}, \gamma^{/p})$. Nevertheless, as we will see in the next section, these are the class field theories which are easiest to deal with.

## 4 $p$-Poincaré duality groups of dimension 2 as $p$-Frattini extensions

Throughout this section we assume that $G$ is a finite group, and that $\pi\colon \tilde{G} \to G$ is a $p$-Frattini extension. By

$$\begin{gathered} P_1 \xrightarrow{\delta} P_0 \xrightarrow{\varepsilon} \mathbb{Z}_p, \\ Q_1 \xrightarrow{\delta/p} P_0 \xrightarrow{\varepsilon/p} \mathbb{F}_p \end{gathered} \tag{4.1}$$

we denote partial minimal projective resolutions in ${}_{\tilde{G}}\mathbf{prf}_p$ and ${}_{\tilde{G}}\mathbf{prf}_{/p}$, respectively.

### 4.1 Universal norms

Let $\pi\colon \tilde{G} \to G$ be a $p$-Frattini extension, and let $(\mathbf{C}, \gamma)$ denote its $p$-Frattini class field theory. We call the cohomological $\mathcal{F}(\tilde{G})$-Mackey functor $\mathfrak{N} := \ker(\gamma)$ the *universal norms* of $(\mathbf{C}, \gamma)$. Similarly, $\mathfrak{N}^{/p} := \ker(\gamma^{/p})$ will be called the *universal norms* of $(\mathbf{C}^{/p}, \gamma^{/p})$. One has:

**Proposition 4.1** *Let $\pi\colon \tilde{G} \to G$ be a p-Frattini extension. Then:*

(a) *$\mathfrak{N}$ is $N$-surjective. Let $P_1 \xrightarrow{\delta} P_0 \longrightarrow \mathbb{Z}_p$ be a partial minimal projective resolution of $\mathbb{Z}_p$ in ${}_{\tilde{G}}\mathbf{prf}_p$. Then* $\ker(\delta) \simeq m(\mathfrak{N})$.

(b) *$\mathfrak{N}^{/p}$ is $N$-surjective. Let $Q_1 \xrightarrow{\delta} Q_0 \longrightarrow \mathbb{F}_p$ be a partial minimal projective resolution of $\mathbb{F}_p$ in ${}_{\tilde{G}}\mathbf{prf}_{/p}$. Then* $\ker(\delta) \simeq m(\mathfrak{N}^{/p})$.

**Proof** (a) For simplicity let us assume that $\iota\colon \mathfrak{N} \to \mathbf{C}$ is given by inclusion. Let $\{U_k\}_{k\in\mathbb{N}} \subseteq \mathcal{F}(\tilde{G})$ be a linearly ordered basis of neighbourhoods of $1 \in \tilde{G}$. We have to show that for $x \in \bigcap_{m \geq n} \mathrm{im}(N^{\mathbf{C}}_{U_m,U_n})$, there exists a sequence $(y_k)_{k\in\mathbb{N}_0}$, $y_k \in \mathbf{C}_{U_{n+k}}$, such that $y_0 = x$ and $y_k = N_{U_{n+k+1},U_{n+k}}(y_{k+1})$.

Let $Z := \prod_{k\in\mathbb{N}_0} \mathbf{C}_{U_{n+k}}$. Then $Z$ is compact by Tychonoff's theorem. Let

$$Z_{x,r} := \{\, (z_k)_{k\in\mathbb{N}_0} \in Z \mid z_0 = x,\ N_{U_{k+1},U_k}(z_{k+1}) = z_k \text{ for all } k \leq r \,\}. \tag{4.2}$$

Then $Z_{x,r+1} \subseteq Z_{x,r}$ and all sets $Z_{x,r}$ are closed. By definition, any finite intersection of sets $Z_{x,r}$ is non-empty. Hence $Z_{x,\infty} := \bigcap_{r\in\mathbb{N}} Z_{x,r}$ is non-empty. Any element $(y_k)_{k\in\mathbb{N}_0} \in Z_{x,\infty}$ will have the desired property.

By construction, $\ker(\mathfrak{X}(\delta)) = \mathbf{C}$. Moreover, one has a short exact sequence of $\mathcal{F}(\tilde{G})$-Mackey functors $0 \to \mathfrak{N} \to \mathbf{C} \to \mathbf{Ab}^p \to 0$. Obviously, $m(\mathbf{Ab}^p) = 0$. Thus the claim follows from the exactness of $m$. The assertion (b) follows by a similar argument. □

## 4.2 Weakly oriented $p$-Poincaré duality groups

Let $\hat{G}$ be a profinite group of cohomological $p$-dimension $d$, $d \in \mathbb{N}$. Then $\hat{G}$ is called a *$p$-Poincaré duality group of dimension $d$*, if

(i) for every finite discrete left $\hat{G}$-module of $p$-power order $X$ and for all $k \in \mathbb{N}_0$ one has
$$|H^k(\hat{G}, X)| < \infty, \tag{4.3}$$

(ii) the $p$-dualizing module $\mathbb{I}_{\hat{G},p}$ of $\hat{G}$ is isomorphic to $\mathbb{Q}_p/\mathbb{Z}_p$ as abelian group,

(iii) for every finite discrete left $\hat{G}$-module of $p$-power order $X$, cup-product induces a non-degenerate pairing
$$H^k(\hat{G}, X') \times H^{d-k}(\hat{G}, X) \overset{H^d(ev_X)\circ(.\cup.)}{\longrightarrow} H^d(\hat{G}, \mathbb{I}_{\hat{G},p}) \overset{i}{\longrightarrow} \mathbb{Q}_p/\mathbb{Z}_p, \tag{4.4}$$
where $X' := \mathrm{Hom}(X, \mathbb{I}_{\hat{G},p})$, $ev_X \colon X' \times X \to \mathbb{I}_{\hat{G},p}$ is the evaluation map and $i$ is given as in [12, §I.3.5].

The $p$-Poincaré duality group $\hat{G}$ of dimension $d$ is called *orientable*, if $\mathbb{I}_{\hat{G},p}$ is a trivial $\hat{G}$-module, and *weakly-orientable*, if the socle of $\mathbb{I}_{\hat{G},p}$ is a trivial $\hat{G}$-module, i.e., $\mathrm{soc}(\mathbb{I}_{\hat{G},p}) \simeq \mathbb{F}_p$.

One can characterize these groups by continuous cochain cohomology as introduced by J. Tate (cf. [14]) with coefficients in $\mathbb{F}_p[\![\hat{G}]\!]$ as follows:

**Proposition 4.2** *Let $\hat{G}$ be a profinite group of cohomological $p$-dimension $d$, $d \in \mathbb{N}$, and assume* (4.3) *holds for every finite discrete left $\hat{G}$-module of $p$-power order $X$. Then the following are equivalent:*

*(i) $\hat{G}$ is a weakly-orientable $p$-Poincaré duality group of dimension $d$,*

*(ii)*
$$\mathbf{H}^k(\hat{G}, \mathbb{F}_p[\![\hat{G}]\!]) = \begin{cases} \mathbb{F}_p & \text{for } k = d, \\ 0 & \text{for } k \neq d, \end{cases} \tag{4.5}$$
*where $\mathbb{F}_p$ denotes the trivial $\hat{G}$-module and $\mathbf{H}^\bullet$ denotes continuous cochain cohomology.*

**Proof** The implication (i) $\Rightarrow$ (ii) is implicitly already contained in a letter from J. Tate to J-P. Serre (cf. [12, App. 1]) Here one should only note that the second property of a Poincaré duality group ensures that $\mathbf{H}^k(\tilde{G}, \mathbb{F}_p[\![\tilde{G}]\!])^* = E_k(\mathbb{F}_p)$.

Note that property (4.5) already implies that (4.4) holds for all finite $\mathbb{F}_p$-vector spaces which are discrete $\hat{G}$-modules. Then the same argument used in the proof of [12, Prop. I.32]) shows that (4.4) holds for all finite discrete $\hat{G}$-modules of $p$-power order. □

### 4.3 Cohomological Mackey functors for $p$-Frattini extensions

Let $\mathbf{X}$ be a cohomological $\mathcal{F}(\tilde{G})$-Mackey functor, such that $\mathbf{X}_U$ are finitely generated $\mathbb{F}_p[\tilde{G}/U]$-modules for all $U \in \mathcal{F}(\tilde{G})$. Then applying $\mathrm{Hom}_{\tilde{G}}(\_, \mathbb{F}_p)$ and changing the role of $i$ and $N$ defines a new cohomological $\mathcal{F}(\tilde{G})$-Mackey functor which we denote by $\mathbf{X}^*$. The functor $^*$ is obviously contravariant and exact.

For short put $\mathbf{S}(\mathbb{F}_p) := \mathfrak{X}(\mathbb{F}_p)$, $\mathbf{T}(\mathbb{F}_p) := \mathbf{S}(\mathbb{F}_p)^*$. Then $\mathbf{S}(\mathbb{F}_p)$ is a cohomological $\mathcal{F}(\tilde{G})$-Mackey functor with all mapping $N_{V,U}^{\mathbf{S}(\mathbb{F}_p)}$ bijective, and $\mathbf{T}(\mathbb{F}_p)$ is a $\mathcal{F}(\tilde{G})$-Mackey functor with all mapping $i_{U,V}^{\mathbf{T}(\mathbb{F}_p)}$ bijective, $U, V \in \mathcal{F}(\tilde{G})$, $V \leq U$.

Thus one has an exact sequence of cohomological $\mathcal{F}(\tilde{G})$-Mackey functors

$$0 \longrightarrow \mathbf{T}(\mathbb{F}_p) \stackrel{\mathfrak{X}(\varepsilon^{/p})^*}{\longrightarrow} \mathfrak{X}(Q_0)^* \stackrel{\mathfrak{X}(\delta^{/p})^*}{\longrightarrow} \mathfrak{X}(Q_1)^*. \tag{4.6}$$

We put

$$\begin{aligned} \Omega^1(\tilde{G}/\_, \mathbb{F}_p) &:= \ker(\mathfrak{X}(\delta^{/p})^*), \\ \Omega^2(\tilde{G}/\_, \mathbb{F}_p) &:= \mathrm{coker}(\mathfrak{X}(\delta^{/p})^*). \end{aligned} \tag{4.7}$$

It is an easy exercise to show that $\Omega^1(\tilde{G}/\_, \mathbb{F}_p)$ is $i$-injective and $N$-surjective, and that $\Omega^2(\tilde{G}/\_, \mathbb{F}_p)$ is of type $H_0$.

### 4.4 Extending injective maps $\Omega^1(G, \mathbb{F}_p) \to \Omega_2(G, \mathbb{F}_p)$

The first step in proving Theorem B is establishing the following proposition:

**Proposition 4.3** *Let $G$ be a finite group, and let $\alpha\colon \Omega^1(G, \mathbb{F}_p) \to \Omega^2(G, \mathbb{F}_p)$ be a mapping of $\mathbb{F}_p[G]$-modules. Then there exists a closed normal subgroup $N$, $N \leq \ker(\pi_p)$ of the universal p-Frattini extension $\tilde{G}_p$, $\tilde{G} := \tilde{G}_p/N$, and a map of cohomological $\mathcal{F}(\tilde{G})$-Mackey functors*

$$\boldsymbol{\alpha}\colon \mathbf{\Omega}^1(\tilde{G}/\_, \mathbb{F}_p) \longrightarrow \mathbf{C}^{/p}, \tag{4.8}$$

*satisfying* $\mathrm{im}(\boldsymbol{\alpha}) = \mathfrak{N}^{/p}$ *and* $\boldsymbol{\alpha}_{\ker(\pi_p)} = \iota_{\ker(\pi_p)} : \alpha$, *where* $\iota\colon \mathfrak{N}^{/p} \to \mathbf{C}^{/p}$ *denotes the canonical map.*

*Moreover, if $\alpha$ is injective, $\boldsymbol{\alpha}$ is injective.*

**Proof** Put $V_0 := \ker(\pi_p)$ and $\boldsymbol{\alpha}_0 := \alpha\colon \Omega^1(G, \mathbb{F}_p) \to \Omega_2(G, \mathbb{F}_p)$. Assume we have constructed open normal subgroups $V_0, \ldots, V_{k-1}$ and injective morphisms

$$\boldsymbol{\alpha}_{V_i} : \Omega^1(\tilde{G}_p/V_i) \longrightarrow \Omega_2(\tilde{G}_p/V_i, \mathbb{F}_p), \tag{4.9}$$

$i = 0, \dots, k-1$, such that the diagrams

$$\begin{array}{ccc} \Omega^1(\tilde{G}_p/V_{i-1}, \mathbb{F}_p) & \xrightarrow{\boldsymbol{\alpha}_{V_{i-1}}} & \Omega_2(\tilde{G}_p/V_{i-1}, \mathbb{F}_p) \\ {\scriptstyle i^{\Omega^1}_{V_{i-1},V_i}}\downarrow & & \downarrow{\scriptstyle i^{\Omega_2}_{V_{i-1},V_i}} \\ \Omega^1(\tilde{G}_p/V_i, \mathbb{F}_p) & \xrightarrow{\boldsymbol{\alpha}_{V_i}} & \Omega_2(\tilde{G}_p/V_i, \mathbb{F}_p) \end{array} \tag{4.10}$$

$$\begin{array}{ccc} \Omega^1(\tilde{G}_p/V_{i-1}, \mathbb{F}_p) & \xrightarrow{\boldsymbol{\alpha}_{V_{i-1}}} & \Omega_2(\tilde{G}_p/V_{i-1}, \mathbb{F}_p) \\ {\scriptstyle N^{\Omega^1}_{V_i,V_{i-1}}}\uparrow & & \uparrow{\scriptstyle N^{\Omega_2}_{V_i,V_{i-1}}} \\ \Omega^1(\tilde{G}_p/V_i, \mathbb{F}_p) & \xrightarrow{\boldsymbol{\alpha}_{V_i}} & \Omega_2(\tilde{G}_p/V_i, \mathbb{F}_p) \end{array} \tag{4.11}$$

commute, $i = 1, \dots, k-1$. In the first step we construct $V_k$ and a mapping

$$\boldsymbol{\alpha}_{V_k} \colon \Omega^1(\tilde{G}_p/V_k, \mathbb{F}_p) \to \Omega_2(\tilde{G}_p/V_k, \mathbb{F}_p) \tag{4.12}$$

such the diagrams (4.10) and (4.11) commute for $(k-1, k)$.

Let $V_k \leq \ker(\pi_p)$ be the unique open normal subgroup such that $V_{k-1}/V_k$ is elementary $p$-abelian, and $\mathrm{im}(\boldsymbol{\alpha}_{V_{k-1}}) = \mathrm{im}(N^{\Omega_2}_{V_k,V_{k-1}})$. The uniqueness is guaranteed by axiom (iii) of a $/p$-class field theory. Since $(Q_0)^*_{V_k}$ is a projective $\mathbb{F}_p[\tilde{G}_p/V_k]$-module, there exists a mapping $\alpha' \colon (Q_0)^*_{V_k} \to \Omega_2(\tilde{G}_p/V_k, \mathbb{F}_p)$ making the diagram

$$\begin{array}{ccc} \Omega^1(\tilde{G}_p/V_{k-1}, \mathbb{F}_p) & \xrightarrow{\boldsymbol{\alpha}_{V_{k-1}}} & \Omega_2(\tilde{G}_p/V_{k-1}, \mathbb{F}_p) \\ {\scriptstyle N}\uparrow & & \uparrow{\scriptstyle N^{\Omega_2}_{V_k,V_{k-1}}} \\ (Q_0)^*_{V_k} & \xrightarrow{\alpha'} & \Omega_2(\tilde{G}_p/V_k, \mathbb{F}_p) \end{array} \tag{4.13}$$

commute, where $N \colon (Q_0)^*_{V_k} \to \Omega^1(\tilde{G}_p/V_{k-1}, \mathbb{F}_p)$ is the canonical map. Since the $\mathbb{F}_p[\tilde{G}_p/V_k]$-module $\Omega_2(\tilde{G}_p/V_k, \mathbb{F}_p)$ is directly indecomposable, and as $(Q_0)^*_{V_k}$ is also injective, $\alpha'$ cannot be injective. Hence $\alpha'$ factors through a mapping

$$\boldsymbol{\alpha}_{V_k} \colon \Omega^1(\tilde{G}_p/V_k, \mathbb{F}_p) \to \Omega_2(\tilde{G}_p/V_k, \mathbb{F}_p). \tag{4.14}$$

for which diagram (4.11) commutes for $(k-1, k)$.

Let $x \in \Omega^1(\tilde{G}_p/V_{k-1}, \mathbb{F}_p)$. As $\Omega^1(\tilde{G}_p/\_, \mathbb{F}_p)$ is $N$-surjective, there exists $y \in \Omega^1(\tilde{G}_p/V_k, \mathbb{F}_p)$ such that $N^{\Omega^1}_{V_k,V_{k-1}}(y) = x$. Thus

$$\begin{aligned} i^{\Omega_2}_{V_{k-1},V_k}(\boldsymbol{\alpha}_{V_{k-1}}(x)) &= i^{\Omega_2}_{V_{k-1},V_k}(\boldsymbol{\alpha}_{V_{k-1}}(N^{\Omega^1}_{V_k,V_{k-1}}(y))), \\ &= i^{\Omega_2}_{V_{k-1},V_k}(N^{\Omega_2}_{V_k,V_{k-1}}(\boldsymbol{\alpha}_{V_k}(y))) = N_{V_{k-1}/V_k}(\boldsymbol{\alpha}_{V_k}(y)), \end{aligned} \tag{4.15}$$

where $N_{V_{k-1}/V_k} := \sum_{g \in V_{k-1}/V_k} g$. On the other hand

$$\begin{aligned} \boldsymbol{\alpha}_{V_k}(i^{\Omega^1}_{V_{k-1},V_k}(x)) &= \boldsymbol{\alpha}_{V_k}(i^{\Omega^1}_{V_{k-1},V_k}(N^{\Omega^1}_{V_k,V_{k-1}}(y))) \\ &= \boldsymbol{\alpha}_{V_k}(N_{V_{k-1}/V_k}(y)) = N_{V_{k-1}/V_k}(\boldsymbol{\alpha}_{V_k}(y)), \end{aligned} \tag{4.16}$$

i.e., the diagram (4.10) commutes for $(k-1,k)$ aswell.

Since $i^{\Omega^1}_{V_{k-1},V_k}\colon \operatorname{soc}(\Omega^1(\tilde{G}_p/V_{k-1},\mathbb{F}_p) \to \operatorname{soc}(\Omega^1(\tilde{G}_p/V_k,\mathbb{F}_p)$ is bijective, and as $\mathbf{C}^{/p}$ is of type $H^0$, $\boldsymbol{\alpha}_{V_k}$ is injective provided $\boldsymbol{\alpha}_{V_{k-1}}$ is injective.

Let $N := \bigcap_{k\in\mathbb{N}_0} V_k$. Then $\{V_k/N\}_{k\in\mathbb{N}_0}$ is a basis of open neighbourhoods of $1 \in \tilde{G}_p/N$.

Let $V \in \mathcal{F}_N := \{U \in \mathcal{F} \mid N \le U\}$. Then there exist $k \in \mathbb{N}_0$ such that $V_k \le V$. Since $\Omega^1(\tilde{G}_p/\_,\mathbb{F}_p)$ and $\Omega_2(\tilde{G}_p/\_,\mathbb{F}_p)$ are $i$-injective cohomological $\mathcal{F}$-Mackey functors, there exists a unique mapping

$$\boldsymbol{\alpha}_V\colon \Omega^1(\tilde{G}_p/V,\mathbb{F}_p) \longrightarrow \Omega_2(\tilde{G}_p/V,\mathbb{F}_p) \tag{4.17}$$

making the diagram

$$\begin{array}{ccc} \Omega^1(\tilde{G}_p/V,\mathbb{F}_p) & \xrightarrow{\boldsymbol{\alpha}_V} & \Omega_2(\tilde{G}_p/V,\mathbb{F}_p) \\ {\scriptstyle i^{\Omega^1}_{V,V_k}}\big\downarrow & & \big\downarrow{\scriptstyle i^{\Omega_2}_{V,V_k}} \\ \Omega^1(\tilde{G}_p/V_k,\mathbb{F}_p) & \xrightarrow{\boldsymbol{\alpha}_{V_k}} & \Omega_2(\tilde{G}_p/V_k,\mathbb{F}_p) \end{array} \tag{4.18}$$

commute. It is easy to check that for all $U,V \in \mathcal{F}_N$, $V \le U$, the diagram

$$\begin{array}{ccc} \Omega^1(\tilde{G}_p/U,\mathbb{F}_p) & \xrightarrow{\boldsymbol{\alpha}_U} & \Omega_2(\tilde{G}_p/U,\mathbb{F}_p) \\ {\scriptstyle i^{\Omega^1}_{U,V}}\big\downarrow & & \big\downarrow{\scriptstyle i^{\Omega_2}_{U,V}} \\ \Omega^1(\tilde{G}_p/V,\mathbb{F}_p) & \xrightarrow{\boldsymbol{\alpha}_V} & \Omega_2(\tilde{G}_p/V,\mathbb{F}_p) \end{array} \tag{4.19}$$

commutes. Note that $\Omega_2(\tilde{G}/\_,\mathbb{F}_p)$ is $i$-injective, and that for $x \in \Omega^1(\tilde{G}_p/V,\mathbb{F}_p)$

$$i^{\Omega_2}_{U,V}(\boldsymbol{\alpha}_U(N^{\Omega^1}_{V,U}(x))) = \boldsymbol{\alpha}_V(i^{\Omega^1}_{U,V}(N^{\Omega^1}_{V,U}(x))) = \boldsymbol{\alpha}_V(N_{U/V}(x)), \tag{4.20}$$

$$i^{\Omega_2}_{U,V}(N^{\Omega_2}_{V,U}(\boldsymbol{\alpha}_V(x))) = N_{V/U}(\boldsymbol{\alpha}_V(x)) = \boldsymbol{\alpha}_V(N_{U/V}(x)). \tag{4.21}$$

Hence the diagram

$$\begin{array}{ccc} \Omega^1(\tilde{G}_p/U,\mathbb{F}_p) & \xrightarrow{\boldsymbol{\alpha}_U} & \Omega_2(\tilde{G}_p/U,\mathbb{F}_p) \\ {\scriptstyle N^{\Omega^1}_{V,U}}\big\uparrow & & \big\uparrow{\scriptstyle N^{\Omega_2}_{V,U}} \\ \Omega^1(\tilde{G}_p/V,\mathbb{F}_p) & \xrightarrow{\boldsymbol{\alpha}_V} & \Omega_2(\tilde{G}_p/V,\mathbb{F}_p) \end{array} \tag{4.22}$$

commutes as well showing that

$$\boldsymbol{\alpha}\colon \Omega^1(\tilde{G}_p/\_,\mathbb{F}_p) \longrightarrow \Omega_2(\tilde{G}_p/\_,\mathbb{F}_p) \tag{4.23}$$

is a morphism of cohomological $\mathcal{F}(\tilde{G}_p/N)$-Mackey functors. By construction, one has $\operatorname{im}(\boldsymbol{\alpha}) = \mathfrak{N}^{/p}$. Moreover, if $\alpha$ is injective, then the construction shows that $\boldsymbol{\alpha}$ is also injective. This yields the claim. □

### 4.5 $\Omega^1$-relator $p$-Frattini extensions

Let $\pi\colon \tilde{G} \to G$ be a $p$-Frattini extension of $G$, and let $(\mathbf{C}^{/p}, \gamma^{/p})$ denote its $/p$-Frattini class field theory. We call $\pi$ an *$\Omega^1$-relator $p$-Frattini extension*, if there exists a map

$$\boldsymbol{\alpha}\colon \Omega^1(\tilde{G}/\_, \mathbb{F}_p) \to \mathbf{C}^{/p} \tag{4.24}$$

of cohomological $\mathcal{F}(\tilde{G})$-Mackey functors with $\mathrm{im}(\alpha) = \mathfrak{N}^{/p}$. If necessary we include the mapping $\boldsymbol{\alpha}$ in the notation, i.e., we write $(\pi, \alpha)$ for a $\Omega^1$-relator $p$-Frattini extension.

For the universal $p$-Frattini extension $\pi_p\colon \tilde{G}_p \to G$ one has $\mathfrak{N}^{/p} = 0$, and thus $\pi_p$ is a $\Omega^1$-relator $p$-Frattini extension.

From Proposition 4.3 one concludes that one can also construct such a $p$-Frattini extension starting from a map $\alpha\colon \Omega^1(G, \mathbb{F}_p) \to \Omega_2(G, \mathbb{F}_p)$.

Another source of examples arises in the context of modular towers. The starting point in the study of modular towers is a fixed surjective morphism $\phi\colon \hat{G} \to G$ where $\hat{G}$ is a certain profinite orientable $p$-Poincaré duality group of dimension 2 onto a finite group $G$. A modular tower consists of all open normal subgroups $U$ in $\hat{G}$ contained in $\ker(\phi)$ such that the induced map $\phi_U\colon \tilde{G}/U \to G$ is a $p$-Frattini extension (cf. [1]). The 'limit groups' of a modular tower correspond to a closed normal subgroup $A \leq \ker(\phi)$ such that $\phi_A\colon \hat{G}/A \to G$ is a maximal $p$-Frattini extension $\phi$ can factor through. In particular, $(\phi_A, \pi_A)$, $\pi_A\colon \tilde{G} \to \hat{G}/A$ the canonical projection, is a *maximal $p$-Frattini quotient* of $\phi$ (cf. [17]). These $p$-Frattini extensions have the following property.

**Proposition 4.4** *Let $\phi\colon \hat{G} \to G$ be a surjective map of the profinite weakly-orientable $p$-Poincaré duality group $\hat{G}$ of dimension 2 onto the finite group $G$. Then for every maximal $p$-Frattini quotient $(\pi, \beta)$, $\pi\colon \mathrm{im}(\beta) \to G$ is a $\Omega^1$-relator $p$-Frattini extension of $G$.*

**Proof** Let $B := \mathrm{im}(\beta)$, and let

$$Q_1 \xrightarrow{\delta/p} Q_0 \longrightarrow \mathbb{F}_p \tag{4.25}$$

be a partial minimal projective resolution in ${}_B\mathbf{prf}_{/p}$. Put $M := \ker(\delta)$. By [17, Prop. 3.4], one has a surjective map $\alpha\colon Q_0 \to M$. Since $\mathfrak{N}^{/p}$ is norm surjective (cf. Prop. 4.1(b)), one has a surjective map of cohomological $\mathcal{F}(B)$-Mackey functors

$$\rho\colon \mathfrak{X}(Q_0) \longrightarrow \mathfrak{X}(M) \longrightarrow \mathfrak{N}^{/p}. \tag{4.26}$$

Since $\mathfrak{N}^{/p}$ is a $\mathcal{F}(B)$-sub Mackey functor of $\mathbf{C}$, and as $(Q_0)_U$ is an injective $\mathbb{F}_p[B/U]$-module, $\rho_U\colon (Q_0)_U \to \mathfrak{N}^{/p}_U \leq \Omega_2(B/U, \mathbb{F}_p)$ cannot be injective, i.e, $\mathrm{soc}((Q_0)_U) \leq \ker(\rho_U)$. Hence $\rho$ induces a surjective mapping

$$\rho_*\colon \Omega^1(B/\_, \mathbb{F}_p) \longrightarrow \mathfrak{N}^{/p} \tag{4.27}$$

of cohomological $\mathcal{F}(B)$-Mackey functors and this yields the claim. □

In order to finish the proof of Theorem B, we establish the following theorem:

**Theorem 4.5** *Let* $(\pi, \boldsymbol{\alpha})$, $\pi\colon \tilde{G} \to G$, *be a* $\Omega^1$*-relator p-Frattini extension. Assume further that* $\boldsymbol{\alpha}$ *is injective, and that* $\boldsymbol{\alpha}_{\ker(\pi)}$ *is not an isomorphism. Then* $\tilde{G}$ *is a weakly-orientable p-Poincaré duality group of dimension* 2.

**Proof** Note that $\dim_{\mathbb{F}_p}(\Omega_2(G, \mathbb{F}_p)) > \dim_{\mathbb{F}_p}(\Omega^1(G, \mathbb{F}_p))$ implies that $\tilde{G}$ is infinite (cf. [17, Prop. 3.5]). It suffices to prove that $\mathbf{H}^k(\tilde{G}, \mathbb{F}_p[\![\tilde{G}]\!]) = 0$ for $k \neq 2$, and $\mathbf{H}^2(\tilde{G}, \mathbb{F}_p[\![\tilde{G}]\!]) \simeq \mathbb{F}_p$. As before $\mathbf{H}^\bullet$ denotes continuous cochain cohomology.

By definition, one has exact sequences of cohomological $\mathcal{F}(\tilde{G})$-Mackey functors

$$0 \longrightarrow \mathbf{T}(\mathbb{F}_p) \longrightarrow \mathfrak{X}(Q_0) \longrightarrow \Omega^1(\tilde{G}/\_, \mathbb{F}_p) \longrightarrow 0, \tag{4.28}$$

$$0 \longrightarrow \Omega^1(\tilde{G}/\_, \mathbb{F}_p) \longrightarrow \Omega_2(\tilde{G}/\_, \mathbb{F}_p) \longrightarrow \mathbf{Ab}^{/p} \longrightarrow 0, \tag{4.29}$$

$$0 \longrightarrow \Omega_2(\tilde{G}/\_, \mathbb{F}_p) \longrightarrow \mathfrak{X}(Q_1) \longrightarrow \mathfrak{X}(Q_0) \longrightarrow \mathbf{S}(\mathbb{F}_p) \longrightarrow 0. \tag{4.30}$$

As $\tilde{G}$ is infinite $m(\mathbf{T}(\mathbb{F}_p)) = m(\mathbf{Ab}^{/p}) = 0$. Thus applying the functor $m$ yields that one has a minimal projective resolution

$$0 \longrightarrow Q_0 \longrightarrow Q_1 \longrightarrow Q_0 \longrightarrow \mathbb{F}_p \longrightarrow 0 \tag{4.31}$$

of $\mathbb{F}_p$ in ${}_{\tilde{G}}\mathbf{prf}_{/p}$. Hence $\tilde{G}$ is of cohomological $p$-dimension 2.

In his letter to J-P. Serre (cf. [12, App. 1]), J. Tate described how one can compute the Pontryagin dual of the cohomology groups $\mathbf{H}^k(\tilde{G}, \mathbb{F}_p[\![\tilde{G}]\!])$. Translated to our situation we obtain

$$\begin{aligned} \mathbf{H}^2(\tilde{G}, \mathbb{F}_p[\![\tilde{G}]\!])^* &= \varinjlim_U \mathbf{H}_2(U, \mathbb{F}_p), \\ \mathbf{H}^1(\tilde{G}, \mathbb{F}_p[\![\tilde{G}]\!])^* &= \varinjlim_U \mathbf{H}_1(U, \mathbb{F}_p). \end{aligned} \tag{4.32}$$

Since $\tilde{G}$ is infinite, $\mathbf{H}^0(\tilde{G}, \mathbb{F}_p[\![\tilde{G}]\!]) = 0$. From the exact sequences (4.28) it follows that one has an isomorphism of $\mathcal{F}(\tilde{G})$-Mackey functors $\mathbf{H}_2(\_, \mathbb{F}_p) \simeq \mathbf{T}(\mathbb{F}_q)$. This yields $\mathbf{H}^2(\tilde{G}, \mathbb{F}_p[\![\tilde{G}]\!]) \simeq \mathbb{F}_p$.

Let $\boldsymbol{\alpha}^*\colon \Omega^2(\tilde{G}/\_, \mathbb{F}_p) \longrightarrow \Omega_1(\tilde{G}/\_, \mathbb{F}_p)$ be the Pontryagin dual of $\boldsymbol{\alpha}$. Then by (4.32), $\mathbf{H}^1(\tilde{G}, \mathbb{F}_p[\![\tilde{G}]\!]) \simeq m(\ker(\boldsymbol{\alpha}^*))$. Moreover, $\boldsymbol{\alpha}^*$ is surjective. Since for all $U \in \mathcal{F}(\tilde{G})$, one has an isomorphism

$$hd(\boldsymbol{\alpha}_U^*)\colon hd(\Omega_2(\tilde{G}/U, \mathbb{F}_p)) \longrightarrow hd(\Omega_1(\tilde{G}/U, \mathbb{F}_p)), \tag{4.33}$$

where $hd(\_)$ denotes the head of a module, one obtains a commutative diagram

$$\begin{array}{ccccccccc} 0 & \longrightarrow & \Omega^1(\tilde{G}/\_, \mathbb{F}_p) & \longrightarrow & \mathfrak{X}(Q_1)^* & \longrightarrow & \Omega^2(\tilde{G}/\_, \mathbb{F}_p) & \longrightarrow & 0 \\ & & {\scriptstyle\rho}\downarrow & & {\scriptstyle\sigma}\downarrow & & {\scriptstyle\boldsymbol{\alpha}^*}\downarrow & & \\ 0 & \longrightarrow & \Omega_2(\tilde{G}/\_, \mathbb{F}_p) & \longrightarrow & \mathfrak{X}(Q_1)^* & \longrightarrow & \Omega_1(\tilde{G}/\_, \mathbb{F}_p) & \longrightarrow & 0. \end{array} \tag{4.34}$$

By (4.33), $\sigma$ is an isomorphism. So by the snake lemma, $\rho$ is injective, and one has an isomorphism $\operatorname{coker}(\rho) = \ker(\boldsymbol{\alpha}^*)$. Since $\Omega^1(\tilde{G}/\_, \mathbb{F}_p)$ is $N$-surjective, all elements in $\mathrm{im}(\sigma)$ are universal norms. Hence by dimension arguments, $\mathrm{im}(\rho) = \mathrm{im}(\boldsymbol{\alpha})$ and this yields

$$m(\ker(\boldsymbol{\alpha}^*)) \simeq m(\mathrm{coker}(\rho)) \simeq m(\mathbf{Ab}^{/p}) = 0. \tag{4.35}$$

This yields the claim. □

**Corollary 4.6** *Let $G$ be a finite group and let $p$ be a prime number. Then the following are equivalent:*

(i) *There exists a $p$-Frattini extension $\pi\colon \tilde{G} \to G$ with $\tilde{G}$ a profinite weakly-orientable $p$-Poincaré duality group of dimension 2.*

(ii) *There exists an injection $\alpha\colon \Omega^1(G, \mathbb{F}_p) \to \Omega_2(\mathbb{F}_p)$ which is not an isomorphism.*

**Proof** This is a direct consequence of [17, Thm. 4.1] and Theorem 4.5. □

**Remark 4.7** (a) Let $p = 2$ and let $G = PSl_2(q)$, $q \equiv 3 \mod 4$. The explicit description of the projective indecomposable $\mathbb{F}_2[G]$-modules obtained by K. Erdmann [5] shows that in this case one has an injection $\alpha\colon \Omega^1(G, \mathbb{F}_p) \to \Omega_2(G, \mathbb{F}_p)$.
(b) If $G$ is $p$-perfect, i.e., $G_p^{ab} = 0$, $\tilde{G}$ is $p$-perfect too. Thus every $\tilde{G}$-module $M \in ob({}_{\tilde{G}}\mathbf{prf}_p)$, whose underlying abelian pro-$p$ group is isomorphic to $\mathbb{Z}_p$ and whose reduction mod $p$ $M/p.M$ is a trivial $\tilde{G}$-module, must be trivial. Hence in this case one can conclude that $\tilde{G}$ is indeed a orientable $p$-Poincaré duality group of dimension 2.
(c) In [17, Ex. 1.4] an example was given where for any maximal $p$-Frattini quotient $(\pi, \beta)$ of a morphism $\phi\colon \hat{G} \to PSl_2(7)$, the $p$-Frattini extension $\pi$ is of the type described in Theorem 4.5.
(d) One question which has been untouched completely is to describe all isomorphism types of extensions $\pi\colon \tilde{G} \to G$ satisfying (i) of Corollary 4.6. The construction we used does not give any evidence how one can achieve this goal.

## 5 Δ-Frattini extensions

Throughout this section we fix a prime number $p$. For a given finite group $G$ we denote by $\mathfrak{S}_p(G)$ the set of isomorphism types of irreducible (left) $\mathbb{F}_p[G]$-modules. For an irreducible $\mathbb{F}_p[G]$-module $S$ we use the symbol $[S] \in \mathfrak{S}_p(G)$ to denote its isomorphism type.

### 5.1 The Δ-head of an $\mathbb{F}_p[G]$-module

Let $\Delta \subseteq \mathfrak{S}_p(G)$ be a set of isomorphism types of irreducible $\mathbb{F}_p[G]$-modules. For short we call an $\mathbb{F}_p[G]$-module $M \in ob({}_G\,\mathrm{mod}_p)$ of finite $\mathbb{F}_p$-dimension a *Δ-module* if $M$ has a composition series $(M_k)_{0 \le k \le m}$, $0 = M_0 < M_1 < \cdots < M_m = M$, with each composition factor being contained in $\Delta$, i.e., $[M_k/M_{k-1}] \in \Delta$ for all $k = 1, \ldots, m$. We also assume that $0 \in ob({}_G\,\mathrm{mod}_p)$ is a Δ-module.

Let $M$ be an $\mathbb{F}_p[G]$-module of finite $\mathbb{F}_p$-dimension. We call an $\mathbb{F}_p[G]$-submodule $N \le M$ a *Δ-kernel*, if $M/N$ is a Δ-module. Obviously, the intersection of any set of $\Delta$-kernels $N_i \leq M$, $i \in I$, is again a $\Delta$-kernel. Hence there exists a minimal $\Delta$-kernel $M_\Delta \leq M$. For short we call

$$hd_\Delta(M) := M/M_\Delta \tag{5.1}$$

The $\Delta$-*head* of $M$.

### 5.2 The universal $\Delta$-Frattini extension

Let

$$1 \longrightarrow \Omega_2(G, \mathbb{F}_p) \stackrel{\iota}{\longrightarrow} \tilde{G}_{/p} \stackrel{\pi/p}{\longrightarrow} G \longrightarrow 1 \tag{5.2}$$

be the universal elementary $p$-abelian Frattini extension of $G$, where $\iota$ is considered to be given by inclusion. Factoring by the minimal $\Delta$-kernel $\Omega_2(G, \mathbb{F}_p)_\Delta$ of $\Omega_2(G, \mathbb{F}_p)$ yields a $\Delta$-Frattini extension

$$1 \longrightarrow hd(\Omega_2(G, \mathbb{F}_p)) \stackrel{\iota}{\longrightarrow} \tilde{G}_{/\Delta} \stackrel{\pi/\Delta}{\longrightarrow} G \longrightarrow 1 \tag{5.3}$$

which is easily seen to be universal with respect to all elementary $p$-abelian $\Delta$-Frattini extensions of $G$. Thus for $G_0 := G$, and $\pi_{i+1,i} \colon G_{i+1} \to G_i$ the universal elementary $p$-abelian $\Delta$-Frattini extension of $G_i$, we obtain an inverse system whose inverse limit

$$\tilde{G}_\Delta := \varprojlim_{i \in \mathbb{N}_0} G_i \tag{5.4}$$

together with the canonical map $\pi_\Delta \colon \tilde{G}_\Delta \to G$ is a $\Delta$-Frattini extension of $G$. The universality as well as the uniqueness up to isomorphism follows by the same arguments which were used to prove these statements for the universal $p$-Frattini extension (cf. [6]).

At this point we have to deal with the question how one characterize the universal $\Delta$-Frattini extension among all $\Delta$-Frattini extensions. This is the subject of the following proposition.

**Proposition 5.1** *Let $\pi \colon \tilde{G} \to G$ be a $\Delta$-Frattini extension of $G$, $\Delta \subseteq \mathfrak{S}_p(G)$. Then the following are equivalent:*

(i) *$\pi$ coincides with the universal $\Delta$-Frattini extension of $G$.*

(ii) *$H^2(\tilde{G}, S) = 0$ for all irreducible $\mathbb{F}_p[G]$-modules $S$, $[S] \in \Delta$.*

**Proof** Assume that $\pi \colon \tilde{G} \to G$ is the universal $\Delta$-Frattini extension of $G$, and that there exists an irreducible $\mathbb{F}_p[G]$-module $S$, $[S] \in \Delta$, with $H^2(\tilde{G}, S) \neq 0$. For $\eta \in H^2(\tilde{G}, S)$, $\eta \neq 0$, the associated extension of profinite groups

$$\mathbf{s}(\eta)\colon \quad 1 \longrightarrow S \longrightarrow X \stackrel{\tau}{\longrightarrow} \tilde{G} \longrightarrow 1 \tag{5.5}$$

is non-split and thus $\tau \circ \pi \colon X \to G$ is a $\Delta$-Frattini extension. The universality of $\pi$ implies that $\tau$ has a section $\sigma \colon \tilde{G} \to X$ contradicting the fact that $\mathbf{s}(\eta)$ is non-split. Thus (i) implies (ii).

Assume that $H^2(\tilde{G}, S) = 0$ for all $[S] \in \Delta$, and let $\pi_\Delta \colon \tilde{G}_\Delta \to G$ be the universa $\Delta$-Frattini extension of $G$. Then one has a surjective map $\beta \colon \tilde{G}_\Delta \to \tilde{G}$, and thus an isomorphism

$$\tilde{\beta}^{-1} \colon \tilde{G} \longrightarrow \tilde{G}_\Delta / ker(\beta). \tag{5.6}$$

Assume that $\ker(\beta) \neq 1$ is non-trivial, and let $U \leq \ker(\beta)$ be a maximal open subgroup of $\ker(\beta)$ which is normal in $\tilde{G}_\Delta$. Since $[\ker(\beta)/U] \in \Delta$, one has $H^2(\tilde{G}, \ker(\beta)/U) = 0$. Hence the embedding problem

$$\begin{array}{cccccccccc} & & & & & & \tilde{G} & & \\ & & & & & & \big\downarrow \tilde{\beta}^{-1} & & \\ \mathbf{s}\colon & 1 & \longrightarrow & \ker(\beta)/U & \longrightarrow & \tilde{G}_\Delta/U & \longrightarrow & \tilde{G}_\Delta/\ker(\beta) & \longrightarrow & 1 \end{array} \tag{5.7}$$

has a weak solution (cf. [17, Prop. 3.2]). This implies that $\mathbf{s}$ is split exact, which contradicts the fact that $\mathbf{s}$ is also a $p$-Frattini extension. Thus $\ker(\beta) = 1$, and this yields the claim. □

## 5.3 Chevalley groups over $\mathbb{Z}_p$

For a given Dynkin diagram $D$ let $X_D$ be the simple simply-connected $\mathbb{Z}$-Chevalley group scheme associated to $D$, i.e., if $D$ is of type $A_n$, one has $X_D = Sl_{n+1}$. It has been proved in [16, Thm. B] that

$$\pi_D \colon X_D(\mathbb{Z}_p) \longrightarrow X_D(\mathbb{F}_p) \tag{5.8}$$

is a $p$-Frattini extension apart from possibly 11 explicitly known values of $(D, p)$ It was also shown that in 8 of these 11 cases (5.8) fails to be a $p$-Frattini extension

In case $\pi_D$ is a $p$-Frattini extension, then it is also a $\Delta_D$-Frattini extension where $\Delta_D$ consists of all the $\mathbb{F}_p[X(F_p)]$-composition factors of the $\mathbb{F}_p$-Chevalley Lie algebra $\mathfrak{L}_D \otimes \mathbb{F}_p$ (cf. [16, (2.5)]). If one has additionally

$$\begin{aligned}(D,p) \notin \{\, &(A_n, p), p|(n+1), (B_n, 2), (C_n, 2), (D_n, 2), \ldots \\ &\ldots, (E_6, 3), (E_7, 2), (F_4, 2), (G_2, 2), (G_2, 3) \,\},\end{aligned} \tag{5.9}$$

then $\mathfrak{L}_D \otimes \mathbb{F}_p$ is an irreducible $\mathbb{F}_p[X_D(\mathbb{F}_p)]$-module (cf. [16, Lemma 2.10]), and thus $\Delta_D = \{[\mathfrak{L}_D \otimes \mathbb{F}_p]\}$.

The question raised in [6, Prob. 20.40] can now be restated in the following way

**Question 5.2** *Assume that $p$ is large with respect to the Coxeter number of $D$ Is it true that the $p$-Frattini extension $\pi_D \colon X(\mathbb{Z}_p) \to X(\mathbb{F}_p)$ coincides with the universal $\Delta_D$-Frattini extension?*

From Proposition 5.1 one concludes that the problem of Question 5.2 is equivalent to the following vanishing problem.

**Question 5.3** *Assume that $p$ is large with respect to the Coxeter number of $D$. Is it true that*

$$H^2(X_D(\mathbb{Z}_p), \mathfrak{L}_D \otimes \mathbb{F}_p) = 0? \tag{5.10}$$

As we see in the following theorem both questions have an affirmative answer for $X_D = Sl_2$.

**Theorem 5.4** *Let $p$ be a prime number different from 2, 3 or 5. Then*

$$\pi_{A_1} : Sl_2(\mathbb{Z}_p) \to Sl_2(\mathbb{F}_p) \tag{5.11}$$

*coincides with the universal $\Delta$-Frattini extension for all $\Delta \subseteq \mathfrak{S}_p(Sl_2(\mathbb{F}_p))$ satisfying $[M_2] \in \Delta$, $[M_{p-3}] \notin \Delta$, where $M_k$, $k = 0, \ldots, p-1$ denotes the irreducible $\mathbb{F}_p[Sl_2(\mathbb{F}_p)]$-module of heighest weight $k$ and $\mathbb{F}_p$-dimension $k+1$.*

**Proof** By the previously mentioned remark and Proposition 5.1 it suffices to show that $H^2(Sl_2(\mathbb{Z}_p), M_k) = 0$ for all $k \neq p-3$.

As $p \neq 2, 3$, $\tilde{G} := Sl_2(\mathbb{Z}_p)$ is $p$-torsionfree, and thus a $p$-Poincaré duality group of dimension $d$ (cf. [13, Prop. 4.4.1]). As we assumed $p \neq 2, 3$, $\tilde{G}$ is perfect (cf. [16, Prop. 3.2]). Thus its $p$-dualizing module $\mathbb{I}_{\tilde{G},p}$ is a trivial $\tilde{G}$-module. Hence by Poincaré duality and the Universal Coefficiant Theorem one has

$$H^2(Sl_2(\mathbb{Z}_p), M_k) \simeq H_1(Sl_2(\mathbb{Z}_p), M_k) \simeq H^1(Sl_2(Z_p), M_k)^*, \tag{5.12}$$

where $^*$ denotes the Pontryagin dual. Moreover, from [17, Prop. 3.1] and [18] one concludes that

$$H^1(Sl_2(\mathbb{Z}_p), M_k) \simeq H^1(Sl_2(\mathbb{F}_p), M_k) = 0 \tag{5.13}$$

for $k \neq p-3$. This yields the claim. □

**Remark 5.5** Theorem 5.4 does not hold for $p = 2$, 3 or 5, but in each case for a different reason.

For $p = 2$ or 3, $\pi_{A_1}$ is not a 2-Frattini extension (cf. [16, Thm. B]). For $p = 3$, $\pi_{A_1}$ is even a split extension, since in this case $\mathfrak{L}_{A_1} \otimes \mathbb{F}_3$ is isomorphic to the Steinberg module for $Sl_2(\mathbb{F}_3)$.

For $p = 5$, $\Omega_2(Sl_2(\mathbb{F}_5), \mathbb{F}_5)$ is a $\Delta_{A_1}$-module (cf. [18]). Hence the universal elementary $p$-abelian $\Delta_{A_1}$-extension coincides with the universal elementary $p$-abelian Frattini extension $\pi_{/p}$. However,

$$\dim_{\mathbb{F}_5}(\Omega_2(Sl_2(\mathbb{F}_5), \mathbb{F}_5) = 6, \quad \dim_{\mathbb{F}_5}(\ker(\pi_{A_1})^{ab}) = 3. \tag{5.14}$$

This phenomenon can also be explained by analyzing cohomology groups. Since $p-3 = 2$, Poincaré duality and [17, Prop. 3.1] implies that

$$H^2(Sl_2(\mathbb{Z}_5), \mathfrak{L}_{A_1} \otimes \mathbb{F}_5)^* \simeq H^1(Sl_2(\mathbb{Z}_5), \mathfrak{L}_{A_1} \otimes \mathbb{F}_5) \simeq H^1(Sl_2(\mathbb{F}_5), \mathfrak{L}_{A_1} \otimes \mathbb{F}_5) \simeq \mathbb{F}_5. \tag{5.15}$$

## References

[1] P. Bailey and M. D. Fried, Hurwitz monodromy, spin separation and higher levels of modular towers, *Proc. Sympos. Pure Math.* **70**, 79–200, Amer. Math. Soc., Providence, RI, 2000.

[2] A. Brumer, Pseudocompact algebras, profinite groups and class formations, *J. Algebra* **4** (1966), 442–470.
[3] J. Cossey, O. H. Kegel and L. G. Kovács, Maximal Frattini extensions, *Arch. Math. (Basel)* **35** (1980), no. 3, 210–217.
[4] A. Dress, *Contributions to the theory of induced representations*, Lecture Notes in Mathematics **342**, Springer-Verlag, Berlin, 1973.
[5] K. Erdmann, Principal blocks of groups with dihedral Sylow 2-subgroups, *Comm. Algebra* **5** (1977), 665–694.
[6] M. D. Fried and M. Jarden, *Field Arithmetic*, Ergebnisse der Mathematik und ihrer Grenzgebiete, 3.Folge, Band **11**, Springer-Verlag, New York, 1986.
[7] K. W. Gruenberg, Projective profinite groups, *J. London Math. Soc.* **47** (1967), 155–165.
[8] K. W. Gruenberg, *Relation modules for finite groups*, Conference Board of the Mathematical Sciences Regional Conference Series in Mathematics **25**, American Mathematical Society, Providence, R.I., 1976.
[9] S. Mac Lane, *Homology*, Classics in Mathematics, Springer-Verlag", Berlin, 1995.
[10] J. Neukirch, *Algebraische Zahlentheorie*, Springer-Verlag, Berlin, 1992.
[11] L. Ribes and P. Zalesskii, *Profinite Groups*, Ergebnisse der Mathematik und ihrer Grenzgebiete, 3.Folge **40**, Springer-Verlag, Berlin, 2000.
[12] J.-P. Serre, *Galois Cohomology*, cinquième édition, révisée et complétée, Springer-Verlag, Berlin, 1997.
[13] P. Symonds and Th. Weigel, Cohomology of p-adic analytic groups, in *New horizons in pro-p groups*, M. du Sautoy, D. Segal and A. Shalev (eds.), Progress in Mathematics **184**, 349–410, Birkhäuser, Boston, 2000.
[14] J. Tate, Relations between $K_2$ and Galois cohomology, *Invent. Math.* **36** (1976), 257–274.
[15] P. Webb, User guide to Mackey functors, in *Handbook of Algebra* **2**, M. Hazewinkel (ed.), Elsevier, 2000, 805–836.
[16] Th. Weigel, On the profinite completion of arithmetic groups of split type, in *Lois d'algèbres et variété algébrique*, M. Goze (ed.), Travaux en cours **50**, 79–101, Hermann Paris, 1996.
[17] Th. Weigel, Maximal $\ell$-Frattini quotients of $\ell$-Poincaré duality groups of dimension 2 *Arch. Math. (Basel)*, to appear.
[18] Th. Weigel, On the universal Frattini extension of a finite group, submitted.

# THE NILPOTENCY CLASS OF GROUPS WITH FIXED POINT FREE AUTOMORPHISMS OF PRIME ORDER

LAWRENCE WILSON

Department of Mathematics, University of Florida, 358 Little Hall, Gainesville, FL 32611-8105, U.S.A.
Email: larry@math.ufl.edu

## Abstract

In 1959, Thompson proved that a group with a fixed point free automorphism of prime order must be nilpotent. G. Higman had already asked whether one could say anything about their nilpotency class. We survey the literature on this question and give an improvement on the bound for the nilpotency class.

## 1 Introduction

Thompson proved, in his thesis of 1959, see [14], that groups with fixed point free automorphisms of prime order are nilpotent. Such groups are exactly the Frobenius kernels and Frobenius had conjectured that they must be nilpotent. This conjecture was informed by two fairly simple results; a group with a fixed point free automorphism of order 2 is abelian and a group with a fixed point free automorphism of order 3 is nilpotent. Proofs may be found in Chapter 10 of [3].

This last result can, via a simple argument, be strengthened to give that a group with a fixed point free automorphism of order 3 is nilpotent of class at most 2. This was first published by Neumann, see [10] and [11]. This led G. Higman to ask in [4] of 1959 whether a group with a fixed point free automorphism of order $p$ must be of nilpotency class at most some number, now called $h(p)$ in honor of Higman. He proved that in fact such a number does exist.

The intent of this paper is to look at the various attempts to understand this function. Certainly this question is of intrinsic interest but it is worth noting that information about $h(p)$ is relevant to other questions. For example, the following is Proposition 13.17 of [6]: if a $p$-group has an automorphism of order $p^n$ with $p$ fixed points, then it has a subgroup of index bounded by a function of $p$ and $n$ which has nilpotency class at most $h(p)$. This is relevant to, for example, $p$-groups of maximal class. If such a group has class at least $p$, then it has a unique maximal subgroup which is regular and an element outside this subgroup which acts on it by conjugation with exactly $p$ fixed points (the center of the group).

There has also been much research on fixed point free automorphisms of composite order. To keep this paper relatively short, we will not discuss this work.

In Section 2 we examine Higman's proof of the existence of $h(p)$. Then, in Section 3, we discuss the efforts to find $h(p)$ for small primes while, in Section 4, we look at the attempts to bound $h(p)$ for the larger primes. Finally, in Section 5, we improve the bound for the larger primes and end with a table tracing the decrease in the bound on $h(p)$ for $p \geq 13$.

## 2 The existence of $h(p)$

Here we wish to give the general structure of the proof that there exists a number $h(p)$ such that a group with a fixed point free automorphism of order $p$ has nilpotency class at most $h(p)$. Because Higman was working before Thompson's proof, he made the additional assumption that the groups in question were nilpotent. This assumption was widely shared during the late 1950's and, of course, became a known fact at the end of the decade.

Associated to every nilpotent group is a Lie ring (that is, a Lie algebra over $\mathbb{Z}$). For the full details, one may consult Chapter 6 of [6], a well-written reference. We will give a few details here to establish notation. The lower central series of the group $G$ is defined recursively as $\gamma_1(G) = G$ and $\gamma_{i+1}(G) = [\gamma_i(G), G]$, the subgroup generated by the elements $[x, y]$ with $x$ in $\gamma_i(G)$ and $y$ in $G$. The group is nilpotent if some $\gamma_k(G) = 1$ and the nilpotency class is the last $c$ such that $\gamma_c(G) \neq 1$. Another way of looking at this series is to define 'higher commutators' recursively by $[x_1, \ldots, x_m, x_{m+1}] = [[x_1, \ldots, x_m], x_{m+1}]$ and call this an $(m+1)$-long commutator. Then $\gamma_i(G)$ is the smallest normal subgroup containing all of the $i$-long commutators.

Each $\gamma_i(G)/\gamma_{i+1}(G)$ is contained in the center of $G/\gamma_{i+1}(G)$ (thus this is a 'central' series) and hence is abelian. Thus $\tilde{L}(G) = \bigoplus \gamma_i(G)/\gamma_{i+1}(G)$ forms an abelian group, which we write additively. We define a Lie bracket on $\tilde{L}(G)$ given by $[x\gamma_{i+1}(G), y\gamma_{j+1}(G)] = [x, y]\gamma_{i+j+1}(G)$; here the left hand side contains a Lie bracket and the right hand side a group commutator; all brackets from here on are Lie brackets. One can prove that this is well-defined and, when extended by linearity, satisfies anti-symmetry and the Jacobi Identity ($[a, b, c] + [b, c, a] + [c, a, b] = 0$). Thus $\tilde{L}(G)$ forms a Lie ring.

It is again simple to prove that if $\alpha$ acts fixed-point free on $G$ and $N$ is a normal, $\alpha$-invariant subgroup, then $\alpha$ acts fixed point free on $G/N$. This is Lemma 10.1.3 of [3] and Lemma 2.12 of [6]. Therefore, the fixed point free automorphism of $G$ also acts as a fixed point free automorphism of $\tilde{L}(G)$. It acts as a linear transformation of the $\mathbb{Z}$-module which is the additive group of $\tilde{L}(G)$ and, as it has order $p$, all its eigenvalues must be $p$-th roots of unity.

Higman's idea was to take the tensor product $L(G) = \tilde{L}(G) \otimes \mathbb{Z}[\omega]$, where $\omega$ is a $p$-th root of unity. We consider $L_i$, the $\omega^i$-eigenspace of the automorphism $\alpha$, which is a subgroup of the additive group of $L(G)$. Note that $L_0$ consists of fixed points of $\alpha$ and hence $L_0 = 0$. If $x$ is in $L_i$ and $y$ is in $L_j$, then

$$[x, y]^\alpha = [x^\alpha, y^\alpha] = [\omega^i x, \omega^j y] = \omega^{i+j}[x, y]$$

and hence $[L_i, L_j] \subseteq L_{i+j}$, where we take the addition mod $p$. Therefore $L_0 + L_1 + \cdots + L_{p-1}$ is a sub-algebra which we call $L$ from here on in. A Lie ring $L$ with 'homogeneous' components $L_0, L_1, \ldots, L_{p-1}$ satisfying $[L_i, L_j] \subseteq L_{i+j}$ with addition mod $p$ is said to be a $\mathbb{Z}_p$-graded Lie ring.

It turns out that $pL(G)$ is contained in $L$. As $\alpha$ acts fixed point freely on the abelian additive group of $L(G)$ we must have that $(p, |L(G)|) = 1$. From this, it follows that $L$ and $L(G)$ have the same nilpotency class (and derived length). Therefore it suffices to bound the nilpotency class of $L$. This reduces the problem to bounding the nilpotency class of a $\mathbb{Z}_p$-graded Lie ring $L$ with $L_0 = 0$.

At this point, it is useful to veer a bit from Higman's proof. One can consider the variety of such Lie rings and prove that all such are solvable, Theorem 7.1 of [6]. It was well-known at the time Higman wrote that solvable implies nilpotent for Frobenius kernels and a similar argument implies that these Lie rings are nilpotent. The variety contains a universal element and we call its nilpotency class $h(p)$. Every Lie ring in the variety is an image of the universal element and therefore has nilpotency class at most $h(p)$, proving the result.

We see that Higman's proof does not give a number for $h(p)$. In the next section we consider the efforts to find $h(p)$ for small $p$.

## 3 Finding $h(p)$ for small $p$

The goal is to find $h(p)$, the largest possible nilpotency class of a group with a fixed point free automorphism of order $p$. Obviously then, a lower bound for $h(p)$ can be achieved by producing examples. Essentially, Higman gives a Lie ring by generators and relations and proves that it has nilpotency class at least $(p^2-1)/4$. He does this by finding relations satisfied by the $(p^2 - 1)/4$ long commutators and noting that the same relations hold among certain linearly independent polynomials. Then, one gets a group example by using the Lazard correspondence and producing an $\ell$-group for any prime $\ell$ exceeding $(p^2 - 1)/4$.

It is perhaps worth noting that this leaves many open questions. For example, is there an $\ell$-group with a fixed point free automorphism of order $p$ and nilpotency class $h(p)$ for every $\ell \neq p$? Or, more specifically, if we set $h_2(p)$ to be the largest nilpotency class of a 2-group with a fixed point free automorphism of order $p$, is $h_2(p) = h(p)$ for all odd $p$? It is not difficult to find examples which prove that $h_2(3) = 2$. One can be accessed as `SmallGroup(64,245)` in GAP, [2]. This is the only prime for which we know that $h_2(p)$ is equal to $h(p)$.

Higman's examples remain the ones with the largest nilpotency class. Therefore, the best lower bound is $h(p) \geq (p^2 - 1)/4$ for any odd prime $p$. We have already noted that $h(3) = 2 = (9 - 1)/4$. Higman proved that $h(5) = 6 = (25 - 1)/4$ and we discuss two different proofs that $h(7) = 12 = (49 - 1)/4$. These are the only primes for which the exact value of $h(p)$ is known and it has lead to the conjecture that $h(p) = (p^2 - 1)/4$ for all odd $p$.

### 3.1 Higman's proof that $h(5) \leq 6$

In order to prove that $h(5) \leq 6$, it is necessary to prove that, in a $\mathbb{Z}_5$-graded Lie ring $L$ with $L_0 = 0$ we have that $[\ell_1, \ell_2, \ldots, \ell_7] = 0$ for all $\ell_i$ in $L$. The first thing to note is that it is sufficient to consider homogeneous elements for the $\ell_i$ since $L$ is generated by these elements and the bracket is linear. Then one notes that what matters is which homogeneous component the element is in, not which element we have. Therefore, one attempts to prove that $[L_{a_1}, L_{a_2}, \ldots, L_{a_7}] = 0$ with $1 \leq a_i \leq 4$.

Two commonly used elements in the proof are that $[L_a, L_{5-a}] = 0$ and that $[L_{a_1}, \ldots, L_{a_k}] = 0$ if $a_1 + \cdots + a_k \equiv 0 \bmod 5$. Thus, for example, Higman notes that $[L_{a_1}, \ldots, L_{a_5}] = 0$ if each $a_i$ is either 1 or 4. The same conclusion holds if each $a_i$ is either 2 or 3; certainly this can be argued in the same way. However, it is useful to note that we can think of regrading the ring, write it as $L = L_0 + L_2 + L_4 + L_1 + L_3$, for example. In this way, proving the result for 1 and 4 is also a proof for 2 and 3. This, in essence, lessens the amount of work to be done by a factor of $(p-1)$.

It is also useful to note that $[L_{a_1}, \ldots, L_{a_k}, L_{a_{k+1}}] = [L_{a_{k+1}}, [L_{a_1}, \ldots, L_{a_k}]]$ (as these are subgroups we do not need the minus sign that would occur if we had elements). The right hand side can then be expanded as a sum of standard higher commutators using the Jacobi Identity repeatedly. For example,

$$\begin{aligned}
[L_{a_1}, L_{a_2}, L_{a_3}, L_{a_4}] &= [L_{a_4}, [L_{a_1}, L_{a_2}, L_{a_3}]] \\
&= [L_{a_4}, [L_{a_1}, L_{a_2}], L_{a_3}] + [L_{a_4}, L_{a_3}, [L_{a_1}, L_{a_2}]] \\
&= [L_{a_4}, L_{a_1}, L_{a_2}, L_{a_3}] + [L_{a_4}, L_{a_2}, L_{a_1}, L_{a_3}] \\
&\quad + [L_{a_4}, L_{a_3}, L_{a_1}, L_{a_2}] + [L_{a_4}, L_{a_3}, L_{a_2}, L_{a_1}]
\end{aligned}$$

Higman uses this to argue that $[L_{a_1}, \ldots, L_{a_5}] = 0$ if exactly one of the $a_i$ is 2 and the others are either 1 or 4. By the above argument, it is enough to consider commutators of the form

$$[L_2, \overbrace{L_1, \ldots, L_1}^{r}, \overbrace{L_4, \ldots, L_4}^{s}] = [L_2, \overbrace{L_4, \ldots, L_4}^{s}, \overbrace{L_1, \ldots, L_1}^{r}]$$

where $r + s = 4$. The former is 0 if $r \geq 3$ (as the commutator $[L_2, L_1, L_1, L_1] = 0$) and so we are left with those where $r \leq 2$ and hence $s \geq 2$. But the latter is 0 in this case and we deduce that all such terms are 0. In this way, Higman comes to the proof that all 7-long commutators are 0 by eliminating all possible cases.

### 3.2 Scimemi's computer proof that $h(7) = 12$

For the small primes, no progress was made until the early 1970's. At that time, Scimemi began circulating the announcement, [12], of a computer verification that $h(7) = 12$, certainly one of the earliest computer proofs. While Scimemi's announcement is not readily available, the results are reproduced in the work [1] by Favaretto, a student of Scimemi's.

The computer argument works in the following manner. Assume I've got a list of $k$-long commutators $[L_{a_1}, \ldots, L_{a_k}]$ from which I've removed commutators that I know must always be 0. I then form a collection of $(k+1)$-long commutators by appending to each of these commutators all the $L_b$ with $1 \leq b \leq p-1$. I then prune this list down by eliminating commutators I know must be 0. If this determines that all $n$-long commutators are 0, then I know that $h(p) < n$.

There are several valid ways to prune the list. Obviously if $a_1 + \cdots + a_k + b \equiv 0 \bmod p$, then the commutator is 0. Also, if $[L_{a_1+a_2}, L_{a_3}, \ldots, L_b]$ is 0, then the new commutator is also 0. By anti-symmetry, if $[L_{a_2}, L_{a_1}, L_{a_3}, \ldots, L_{a_k}, L_b] = 0$, then also $[L_{a_1}, L_{a_2}, L_{a_3}, \ldots, L_{a_k}, L_b] = 0$. The Jacobi Identity also provides an easy check: if both $[L_{a_1}, \ldots, L_{a_{k-1}}, L_{a_k+b}] = 0$ and $[L_{a_1}, \ldots, L_{a_{k-1}}, L_b, L_{a_k}] = 0$, then the original commutator is also 0.

Just these simple methods suffice to prove that $h(7) < 14$. To get down to the better bound, one must throw in some other tests. One possibility is to bring one of the terms to the front (as in the last subsection) and expand your original term as a sum of a number of commutators which might all be 0, proving that the original commutator is 0. You can also take advantage of internal segments which are 0. For example, if $[L_{a_2}, L_{a_3}, L_{a_4}] = 0$, then

$$\begin{aligned} 0 &= [L_{a_1}, [L_{a_2}, L_{a_3}, L_{a_4}]] \\ &= [L_{a_1}, L_{a_2}, L_{a_3}, L_{a_4}] + [L_{a_1}, L_{a_3}, L_{a_2}, L_{a_4}] \\ &\qquad + [L_{a_1}, L_{a_4}, L_{a_2}, L_{a_3}] + [L_{a_1}, L_{a_4}, L_{a_3}, L_{a_2}] \end{aligned}$$

so that if three of these last are 0, then the fourth must be also.

Including these checks introduces performance issues. However, the case of 7 is small enough that even a somewhat inexperienced programmer could probably implement a verification that $h(7) \leq 12$. This can, for example, be done in GAP, [2], where the built-in list handling is quite helpful.

### 3.3 Hughes' computer-less proof

In 1985, Hughes announced, [5], a proof that $h(7) = 12$ which does not depend on computers. The argument again comes down to eliminating possible cases. The following is the last paragraph of the paper [5], the references therein are to parts of that paper. The reference to (A) is that one can regrade to make any specific term in the commutator any chosen number from 1 to $p - 1$. We should also note that he writes $[L_{a_1}, \ldots, L_{a_k}]$ as (not exactly) "$a_1 \cdots a_k$" and refers to this as a word.

> We now consider words of length 13 and show that each is zero. By (A) at the end of section 2 we need only consider words with middle digit 2. Let $w$ be a word of length 13 with middle digit 2 which is not zero. Then by 1(b), 2(b), 4(b) and 5(b) its middle three digits are restricted. We then consider all possible middle five digits, and using 1 to 7 above are left with eleven possible middle five digits for $w$. We conclude by considering each of these cases individually.

Hughes does give one result that holds for general $p$, stating that $[\overbrace{L_1, \ldots, L_1}^{p-3}]$ is contained in $Z_3(L)$, that is $[\overbrace{L_1, \ldots, L_1}^{p-3}, L, L, L] = 0$. While he does not provide the proof of this, it is fairly simple and we encourage the reader to give it a try. As a first step, note that $[L_1, \ldots, L_1, L_a]$ is a sum of commutators that are all equal to $[L_a, L_1, \ldots, L_1]$ and hence is 0 if $3 \leq a \leq p - 1$. He uses this to give a new proof that $h(5) < 7$; Favaretto also gave a proof of this using Hughes' result as a starting point and that proof is included below. For $p = 7$, Hughes states that $[L_1, L_1, L_1]$ is contained in $Z_5(L)$.

We noted earlier that "$a_1 \cdots a_k$" is not $[L_{a_1}, \ldots, L_{a_k}]$. Rather, it refers to $[L_{a_1}, L_{a_2-a_1}, L_{a_3-a_2}, \ldots, L_{a_k-a_{k-1}}]$. He calls the word the "accumulative weight sequence" of the commutator. Consider the Jacobi Identity for the word "$a_1a_2a_3$". The commutator is $[L_{a_1}, L_{a_2-a_1}, L_{a_3-a_2}]$ and by the Jacobi Identity this is contained in $[L_{a_1}, [L_{a_2-a_1+a_3-a_2}]] + [L_{a_1}, L_{a_3-a_2}, L_{a_2-a_1}]$. The words for these are "$a_1a_3$" and "$a_1(a_3 - a_2 + a_1)a_3$". Notice that if we had started instead with the word "$a_3a_2a_1$", then the words arising from the Jacobi Identity would similarly be reversed.

This leads Hughes to propose his "SYMMETRY PRINCIPLE". If we have proven that both the words "$a_1a_2 \cdots a_k$" and "$a_k \cdots a_2a_1$" are zero, then any consequence of the first which depends only on the Jacobi Identity immediately gives a consequence for the second. Hughes uses this principle repeatedly in his arguments. It is, however, somewhat difficult to tell whether or not this is appropriate.

As a specific example, proving the reverse of his result that $[\overbrace{L_1, \ldots, L_1}^{p-3}]$ is contained in $Z_3(L)$ seems to require proving that $[L_2, L_{p-1}, L_2, \overbrace{L_1, \ldots, L_1}^{p-4}]$ is zero and this seems difficult. It is unfortunate that Hughes never published a more complete version of his argument. Furthermore, in general one wants to use anti-symmetry and conclude that if $[L_{a_1}, L_{a_2}, \ldots, L_{a_k}] = 0$, then $[L_{a_2}, L_{a_1}, \ldots, L_{a_k}] = 0$. It is not immediate that the reversed conclusion can always be drawn. In particular, this is why I did not list this as a possible pruning method for a computer search.

### 3.4 Favaretto's investigation of $h(11)$

Favaretto, a student of Scimemi, in his Ph.D. thesis of 1998, [1], used a computer to investigate $h(11)$. We know that $h(11) \geq 30$ and hence storage of the not necessarily zero commutators becomes an issue. In addition to technical issues of compressing the data, Favaretto made two simplifications. First, he made use of the option to regrade by always assuming that the first term of the commutator was $L_1$. This leads to a 10-fold decrease in the number of sequences to be stored. Additionally, he only stores one of $[L_{a_1}, L_{a_2}, \ldots, L_{a_k}]$ and $[L_{a_2}, L_{a_1}, \ldots, L_{a_k}]$. This decreases the number of sequences by less than half but perhaps nearly half.

Both of these choices lead to an increase in performance time. Because of the latter, any time you want to check if a given commutator is in the list, you first have to figure out which of the two it would be stored as. Because of the former, whenever an $L_{a_k}$ is brought to the front, all of the terms need to scaled, introducing many multiplications. When you bring $L_{a_k}$ to the front and then expand, you get $2^{k-2}$ different commutators which you need to determine whether or not you know they are non-zero. A similar thing happens if you know $[L_{a_i}, L_{a_{i+1}}, \ldots, L_{a_{i+k}}]$ is zero and attempt to use this to determine that the commutator is zero.

In order to get the program to run in a reasonable amount of time, Favaretto makes choices for how far to keep pushing such things. As his choices lead to a quick computation that $h(7) = 12$, there is some evidence that he chose well. He implements the program using parallel computation and this allows him to make larger choices than otherwise. Still, he is only able to run his program up to 10-long commutators when he sets $p = 11$.

He is, however, able to use this to get a proof that $h(11) \leq 118$. Given that you've got pretty good information about 10-long commutators, you can at least say something about 11-long commutators by looking at $[L_{a_1+a_2}, L_{a_3}, \ldots, L_{a_{11}}]$. By limiting his program initially to extend only those commutators that begin $[L_1, L_1, L_1, L_1]$, he is able to conclude that all such 30-long commutators are zero. That is, he concludes that $[L_1, L_1, L_1, L_1]$ is contained in $Z_{26}(L)$.

Now he reruns the calculations for the 10-long commutators assuming that $[L_1, L_1, L_1, L_1] = 0$. This, in essence, now does calculations in $L/Z_{26}(L)$. He then limits the program to extend only commutators that begin $[L_1, L_1, L_1, L_2]$ and finds that all such 22-long commutators are zero. Therefore, $[L_1, L_1, L_1, L_2]$ is contained in $Z_{18}(L/Z_{26}(L))$ and hence in $Z_{44}(L)$. In a similar manner, he next finds that $[L_1, L_1, L_1]$ is in $Z_{68}(L)$ and then that $[L_1, L_1, L_2]$ is in $Z_{92}(L)$ and then it is possible to deduce that $L_1$ is in $Z_{118}(L)$. This leads to the conclusion that $h(11) \leq 118$. This is a far better bound then the theoretical bounds we will discuss in the next section.

### 3.5 Favaretto's proof that $h(5) = 6$

Favaretto combined his idea for bounding the nilpotency class with Hughes' result that $[\overbrace{L_1, \ldots, L_1}^{p-3}]$ is contained in $Z_3(L)$. For $p = 5$, we have that each $[L_a, L_a]$ is contained in $Z_3(L)$ and we will work in the quotient $\tilde{L} = L/Z_3(L)$. This is again a $\mathbb{Z}_p$-graded Lie ring with $\tilde{L}_0 = 0$ where each $\tilde{L}_a$ is the image of $L_a$. If we can prove that $\tilde{L}_1$ is contained in $Z_3(\tilde{L})$, then $h(5) \leq 6$.

We first note that $[\tilde{L}_1, \tilde{L}_a]$ is zero unless $a$ is 2 or 3. For $[\tilde{L}_1, \tilde{L}_2, \tilde{L}_b]$, this is zero if $b = 2$ or $b = 3$. Also,

$$[\tilde{L}_1, \tilde{L}_2, \tilde{L}_4] = [\tilde{L}_2, \tilde{L}_1, \tilde{L}_4] = [\tilde{L}_2, \tilde{L}_4, \tilde{L}_1] \subseteq [\tilde{L}_1, \tilde{L}_1] = 0$$

Finally, $[\tilde{L}_1, \tilde{L}_2, \tilde{L}_1, \tilde{L}_c]$ is 0 if $c = 1$ or $c = 4$. For $c = 2$, Jacobi Identity implies that this is contained in $[\tilde{L}_1, \tilde{L}_2, \tilde{L}_3] + [\tilde{L}_1, \tilde{L}_2, \tilde{L}_2, \tilde{L}_1]$. The latter is 0 as $[\tilde{L}_1, \tilde{L}_2, \tilde{L}_2] = 0$ and the former is 0 as $[\tilde{L}_1, \tilde{L}_2, \tilde{L}_3] \subseteq [\tilde{L}_3, \tilde{L}_3] = 0$. Finally, for $c = 3$, the Jacobi Identity gives that this is contained in $[\tilde{L}_1, \tilde{L}_2, \tilde{L}_4] + [\tilde{L}_1, \tilde{L}_2, \tilde{L}_3, \tilde{L}_1]$. The first we know is 0 and the second is as $[\tilde{L}_1, \tilde{L}_2, \tilde{L}_3] \subseteq [\tilde{L}_3, \tilde{L}_3] = 0$. Thus $[\tilde{L}_1, \tilde{L}_2]$ is in $Z_2(\tilde{L})$.

It only remains to prove that $[\tilde{L}_1, \tilde{L}_3, \tilde{L}_b, \tilde{L}_c] = 0$ for all choices of $b$ and $c$. This is equal to $[\tilde{L}_3, \tilde{L}_1, \tilde{L}_b, \tilde{L}_c]$ and, by regrading, this becomes $[\tilde{L}_1, \tilde{L}_2, \tilde{L}_{2b}, \tilde{L}_{2c}]$. We have just argued that all such commutators are 0 and hence we have the desired result and we conclude that $\tilde{L}_1$ is in $Z_3(\tilde{L})$, as desired.

We are led to ask if a proof along the lines introduced by Favaretto could produce a small upper bound for $h(p)$. We would need good bounds on $z(k)$ where $[\overbrace{\tilde{L}_1, \ldots, \tilde{L}_1}^{k}]$ is in $Z_{z(k)}(\tilde{L})$ and $\tilde{L}$ is a quotient of $L$ in which $[\overbrace{\tilde{L}_1, \ldots, \tilde{L}_1}^{k+1}] = 0$. The author has a proof that $z(p-4)$ is at most $2s+5$ where $s = \lfloor (p-2)/3 \rfloor$. Favaretto did find bounds on $z(1)$ for small primes. For the primes 11 and 13 he found that $z(1) \leq (p^2 - 1)/8$ and he has conjectured that this will hold for all primes at least 11. It would be very interesting to have a proof of this result.

## 4 Bounding $h(p)$ for larger $p$

Kreknin and Kostrikin, [7, 8], were the first to give an explicit upper bound for $h(p)$. Their proof proceeds in two parts; first, bound the derived length of $L$ and second, bound the nilpotency class assuming a fixed derived length. One notes immediately that this seems backwards; usually one finds a bound on derived length for a nilpotent group from a bound on the nilpotency class. This suggests that the bounds we will achieve are unnecessarily large. P. Hall first proved that nilpotency class could be bounded in terms of derived length.

It is well-known that $G/N$ and $N$ can both be nilpotent while $G$ is not. Less well-known is that if $G/N'$ and $N$ are nilpotent, then $G$ must be. And, in fact from the nilpotency class of each one can derive a bound on the nilpotency class of $G$. Assume we have a collection of groups which includes all quotients of its groups. Further assume we know that all such of derived length 2 have nilpotency class at most $c$. If $G$ has derived length 3, then $G/G''$ and $G'$ are nilpotent of class at most $c$ and we get a bound for the nilpotency class of $G$. Continuing in this way gives a bound on the nilpotency class for all groups in the collection based on their derived length. If you know that the derived lengths are bounded, then you can conclude that the nilpotency class is bounded.

Kreknin and Kostrikin find better bounds than arise from this method, but this does suggest that the method might succeed. We will now examine the two halves along with the subsequent arguments which give improved bounds.

### 4.1 Bounding the derived length

It turns out that the basic argument here does not depend on prime-ness;[1] a $\mathbb{Z}_n$-graded Lie ring with $L_0 = 0$ is solvable with a derived length bounded by a function of $n$. Let us consider $L'$. Every commutator can be expanded as a sum of commutators of homogeneous elements. These lie in a certain homogeneous component and hence $L'$ is also a sum of homogeneous components we can call $L'_1 + L'_2 + \cdots + L'_{n-1}$. Again, a commutator in $L'_1$ is a sum of commutators of homogeneous elements. We deduce that $L'_1 = [L_0, L_1] + [L_2, L_{n-1}] + [L_3, L_{n-2}] + \cdots$ and of course the first term is not necessary. The same argument now applies to $(L')'$ and so on and so we deduce that $L^{(k)}$ is the sum of homogeneous components $L_1^{(k)} + \cdots + L_{n-1}^{(k)}$ and that $L_a^{(k+1)} = \sum_{b+c\equiv a \bmod p}[L_b^{(k)}, L_c^{(k)}]$.

As any element of $L'_1$ can be written as a sum of commutators of homogeneous elements from $L_2, L_3, \ldots, L_{n-1}$ we deduce that $L'$ is contained in the subalgebra $\langle L_2, L_3, \ldots, L_{n-1}\rangle$. This is really the starting place for the bound derived

[1] It should be noted that this does not imply that groups with a fixed point free automorphism of any order must be solvable. It does imply this for Lie rings but there is no clear way to derive the result for groups from this. The result is, however, a consequence of the Classification of Finite Simple Groups if the order of the automorphism is relatively prime to the order of the group.

by Kreknin and Kostrikin. The goal now is to show that if $L^{(r)}$ is contained in $\langle L_k, L_{k+1}, \ldots, L_{n-1}\rangle$, then some $L^{(f(r))}$ is contained in $\langle L_{k+1}, L_{k+2}, \ldots, L_{n-1}\rangle$.

Under this assumption on $L^{(r)}$, they first show that $L_k^{(r+1)}$ is contained in $\langle L_{k+1}, L_{k+2}, \ldots, L_{n-1}\rangle$. By the hypothesis, any commutator in $[L^{(r)}, L^{(r)}]$ can be written as a linear combination of commutators like $[L_{a_1}, L_{a_2}, \ldots, L_{a_s}]$ where each $a_i$ has $k \le a_i \le n-1$. It suffices, then, to prove that such a commutator in which $a_1 + \cdots + a_s \equiv k \bmod n$ is contained in $\langle L_{k+1}, L_{k+2}, \ldots, L_{n-1}\rangle$. This is actually quite simple; first, if $a_s = k$, then $[L_{a_1}, \ldots, L_{a_{k-1}}] \subseteq L_0 = 0$. Second, if $k+1 \le a_s \le n-1$ and $a_1 + \cdots + a_{k-1} \equiv b \bmod n$ where $0 \le b \le n-1$, then, as $a + b \equiv k \bmod n$ we must have $a + b = k + n$ and $k+1 \le b \le n-1$ so that $[L_b, L_{a_s}]$ is contained in the desired subalgebra.

Now assume that we have proven that every $\mathbb{Z}_n$-graded Lie ring $L$ with $L_0 = 0$ has $L^{(r)}$ contained in $\langle L_k, L_{k+1}, \ldots, L_{n-1}\rangle$. As $L^{(r+1)}$ is such a Lie ring, we have that $L^{(2r+1)}$ is contained in $\langle L_k^{(r+1)}, L_{k+1}^{(r+1)}, \ldots, L_{n-1}^{(r+1)}\rangle$. We have just proven that $L_k^{(r+1)}$ is contained in $\langle L_{k+1}, L_{k+2}, \ldots, L_{n-1}\rangle$ and we deduce that $L^{(2r+1)}$ is contained in this subalgebra. In this way, Kreknin and Kostrikin determine that $L$ has derived length at most $2^{n-1} - 1$.

This remained the best published bound until the recent appearance of [13]. However, it is quite simple to prove that $L''$ is contained in $\langle L_3, L_4, \ldots, L_{n-1}\rangle$ (we encourage the reader to try to prove this). From this one can improve the bound on the derived length to $2^{n-2} + 2^{n-3} + 1$.

In [13], the argument begins by proving that $L^{(3)}$ is contained in the subalgebra $\langle L_1, L_5, L_6, \ldots, L_{n-1}\rangle$. Let us look at the proof that $L_2^{(3)}$ satisfies this. In $L_2^{(3)}$ there are the terms $[L_1^{(2)}, L_1^{(2)}]$, $[L_5^{(2)}, L_{n-3}^{(2)}]$, $[L_6^{(2)}, L_{n-4}^{(2)}]$, and so on. These are already in the desired subalgebra. We do have to work on the terms $[L_3^{(2)}, L_{n-1}^{(2)}]$ and $[L_4^{(2)}, L_{n-2}^{(2)}]$.

In looking at $L_3^{(2)}$, one term which arises is $[L_5', L_{n-2}']$ and clearly $[L_5, L_{n-2}, L_{n-1}]$ is in the subalgebra. However, the terms $[L_2', L_1, L_{n-1}]$ and $[L_4', L_{n-1}, L_{n-1}]$ are not written in generators of the subalgebra. The first is equal to $[L_2', L_{n-1}, L_1]$, which is contained in $[L_1, L_1]$. Similar arguments work if $L_4'$ is written as a sum of commutators. Consider $[L_2, L_2, L_{n-1}, L_{n-1}]$, which is contained in $[L_2, L_{n-1}, L_2, L_{n-1}] + [L_2, L_1, L_{n-1}]$ by the Jacobi Identity. The second is equal to $[L_2, L_{n-1}, L_1]$ and contained in $[L_1, L_1]$. The first is contained in $[L_2, L_{n-1}, L_{n-1}, L_2] + [L_2, L_{n-1}, L_1]$. The first of these is 0 and the second is contained in $[L_1, L_1]$.

One next must prove that if $L^{(r)}$ is contained in $\langle L_1, L_k, L_{k+1}, \ldots, L_{n-1}\rangle$, then $L_k^{(r+1)}$ is contained in $\langle L_1, L_{k+1}, L_{k+2}, \ldots, L_{n-1}\rangle$. This is somewhat more difficult than the corresponding result of Kreknin and Kostrikin, but not that difficult. One now concludes that if every $\mathbb{Z}_n$-graded Lie algebra $L$ with $L_0 = 0$ has $L^{(r)}$ contained in $\langle L_1, L_k, L_{k+1}, \ldots, L_{n-1}\rangle$, then they all have $L^{(2r+1)}$ contained in $\langle L_1, L_{k+1}, L_{k+2}, \ldots, L_{n-1}\rangle$.

Given the new starting place, one concludes that $L^{(2^{n-4}-1)}$ is contained in $\langle L_1, L_{n-1}\rangle$. As $L_1$ and $L_{n-1}$ commute, it is clear that this subalgebra has nilpotency class at most $n-1$ and hence derived length at most $\lfloor \log_2(n-1)\rfloor + 1$. Thus

$L$ has derived length at most $2^{n-4} + \lfloor \log_2(n-1) \rfloor$. In the case of a prime $p$, one can make a slightly more complex argument and conclude that $L$ has derived length at most $2^{p-5} + \lfloor \log_2(p-3) \rfloor + 2$.

Note that, if, as conjectured, the nilpotency class of $L$ is at most $(p^2-1)/4$, then the derived length is actually logarithmic in $p$ and not exponential.

## 4.2 Bounding nilpotency class given derived length

If one hopes to get a bound on nilpotency class assuming a given derived length, one must begin with derived length two. By the Jacobi Identity, the commutator $[L_{a_1}, L_{a_2}, L_{a_3}, L_{a_4}]$ is contained in $[L_{a_1}, L_{a_2}, L_{a_4}, L_{a_3}] + [L_{a_1}, L_{a_2}, [L_{a_3}, L_{a_4}]]$. If the derived length is two, then the latter is zero. Thus, under this assumption, we can transpose any two consecutive terms in a commutator past the second and therefore we can achieve an arbitrary permutation of these terms.

We wish to try to conclude that $[L_a, L_b, L_{c_1}, \ldots, L_{c_k}]$ is zero. One way we could do this is to find some collection of the $c_i$'s which add up to $-(a+b)$. It turns out that $p-1$ such $c$'s suffice to find such a collection. This follows from the following simple combinatorial result: $k$ not necessarily distinct non-zero elements of $\mathbb{Z}_p$ produce as sums of sub-collections at least either $k+1$ or $p$ distinct terms. We certainly cannot expect more than $p$ distinct terms and, given a collection of $k$ ones, the possible sums are 0 (for the empty collection) through $k$, for a total of $k+1$. Thus these bounds cannot be improved.

The proof of this combinatorial result is fairly simple. Given one non-zero term $a$ of $\mathbb{Z}_p$ the possible sums are 0 and $a$. Now assume the result for $k$ and look at $a_1, \ldots, a_{k+1}$ not necessarily distinct non-zero elements of $\mathbb{Z}_p$. As $a_1, \ldots, a_k$ produce all $p$ or at least $k+1$ terms, we are done unless they produce exactly $k+1$ and including $a_{k+1}$ produces no new sums. But then $a_{k+1}$ must be produced by $a_1, \ldots, a_k$ and hence $2a_{k+1}$ must be produced and hence $3a_{k+1}$ must be produced and so on until we deduce that all of $\mathbb{Z}_p$ must be produced, proving the result.

Thus if $L$ has derived length two, then it has nilpotency class at most $p$. Kreknin and Kostrikin give a slightly more complicated argument than Hall's argument outlined above and prove that if $L$ has derived length $s$, then it has nilpotency class at most $1 + (p-1) + (p-1)^2 + \cdots + (p-1)^{s-1} = ((p-1)^s - 1)/(p-2)$. Together with the earlier bound on $s$, this gave the first explicit bound on $h(p)$.

Meixner, in his Ph.D. thesis of 1979, see [9], was the first to improve on this bound. He proved that if $L$ has derived length 2, then its nilpotency class is at most $p-1$. From this, with some additional argument, he was able to conclude that if the derived length is $s$, then the nilpotency class is at most $(p-1)^{s-1}$, so that this takes the most significant term from Kreknin and Kostrikin's bound.

This part of the bound was also improved in [13]. The first step is to extend the combinatorial result given above to state that the example given is more or less the only one which produces exactly $k+1$ terms. This result is proven there: $k$ not necessarily distinct non-zero elements of $\mathbb{Z}_p$, call them $a_1, \ldots, a_k$, produce as sums of sub-collections exactly $k+1 < p$ distinct terms if and only if each $a_i = \pm a_1$. The 'if' part is clear; if there are $r$ terms equal to $a_1$ and $k-r$ terms equal to $-a_1$

then the sums that can be produced are the terms from 0 to $r$ times $a_1$ and from $p-k+r$ to $p-1$ times $a_1$, a total of $k+1$ terms.

For the necessity, we proceed by induction again. Two non-zero terms $a$ and $b$ produce 0, $a$, $b$, and $a+b$ and the only ways in which this can be three distinct terms is if $b=a$ or $a+b=0$. Now assume the result for $k$ and let us prove it for $k+1$. If $a_1,\ldots,a_k$ produce $k+2$ terms, then it must be that including $a_{k+1}$ adds no new sums and we've already seen that this implies that all of $\mathbb{Z}_p$ is produced. So $a_1,\ldots,a_k$ produce exactly $k+1$ terms and hence are all $\pm a_1$. The same argument, now including $a_{k+1}$ and not $a_k$ implies that also $a_{k+1}=\pm a_1$.

Thus, in considering $[L_a, L_b, L_{c_1},\ldots,L_{c_{p-2}}]$, the $c_i$'s produce $-(a+b)$ except possibly if each $c_i=\pm c_1$. However, in this case, it is possible to prove that $[L', L_{c_1},\ldots,L_{c_{p-2}}]=0$ also. This, then, is a new proof that if the derived length is two, then the nilpotency class is at most $p-1$.

In order to get an improved bound on the nilpotency class, it is necessary to consider homogeneous ideals. These are ideals $M$ such that $M=M_1+\cdots+M_{p-1}$ where each $M_i=M\cap L_i$. The result on derived length two generalizes to: if $M$ and $N$ are homogeneous ideals with $M\subseteq C_L([N,N])\cap L'$, then $[M,\overbrace{N,\ldots,N}^{p-2}]=0$. One then argues that if $M\subseteq L'$ and $M$ commutes with $L^{(s)}$, then $M\subseteq Z_{(p-2)^s}(L)$.

This leads to an inductive proof that if $L$ has derived length $s$, then $L$ has nilpotency class at most $1+(p-2)+\cdots+(p-2)^{s-1}=((p-2)^s-1)/(p-3)$. We know the result now for $s=1$ and $s=2$. Assume the result for $s$ and assume that $L$ has derived length $s+1$. Then $L/L^{(s)}$ has derived length $s$ and hence the given nilpotency class. Also, $[L^{(s)},L^{(s)}]=0$ and $L^{(s)}$ is a homogeneous ideal so $L^{(s)}$ is contained in $Z_{(p-2)^s}(L)$. This now proves the result for $s+1$.

In the next section we will give an improved bound on $h(p)$ by giving new bounds in this part of the argument.

## 5 An improved bound on $h(p)$

Our goal is to prove the following bound on $h(p)$.

**Theorem A** *If $L$ is a $\mathbb{Z}_p$-graded Lie ring with $L_0=0$, then the nilpotency class of $L$ is at most*

$$\frac{(p-3)^{2^{p-5}+\lfloor\log_2((p-3)/2)\rfloor+2}-1}{p-4}+(p-2)^{2^{(p-5)/2}+\lfloor\log_2((p-1)/2)\rfloor}.$$

Note that $s=2^{p-5}+\lfloor\log_2((p-3)/2)\rfloor+2$ is similar to the bound on the derived length from [13], but with a smaller term inside the logarithm. This slight improvement aside, the main improvement in this bound comes from replacing $((p-2)^s-1)/(p-3)$ by $((p-3)^s-1)/(p-4)$ in the bound on the nilpotency class. We see that this comes at the cost of adding something $(p-2)^{\sqrt{s}}$ which is a large number but far smaller than the other part.

Our proof will proceed very much along the lines outlined in Subsection 4.2. In particular, we will begin by looking at a combinatorial result on how many terms are needed to achieve so many different sums. Then, we will consider the case of derived length two (with an additional hypothesis here) and find a bound on the nilpotency class. We will then work with ideals to get a general bound for nilpotency class given derived length. Finally, we will deal with the additional hypothesis to prove the bound in Theorem A.

**Lemma 5.1** *Take $k \geq 3$ and let $a_1, \ldots, a_k$ be non-zero elements of $\mathbb{Z}_p$ which produce exactly $k+2 < p$ terms of $\mathbb{Z}_p$. Then there is a number $c$ in $\mathbb{Z}_p$ such that $k-1$ of the $a_j$ are equal to $c$ or $-c$ and the other term is $2c$ or $-2c$.*

**Proof** We proceed by induction on $k$ beginning with the case $k = 3$. We consider first the possibility that two of $a_1$, $a_2$, and $a_3$ are equal, say $a_1 = a_3 = c$. If $a_2 = \pm c$, then these produce exactly four terms so $a_2 \neq \pm c$. Therefore, 0, $c$, $2c$, and $c + a_2$ are four distinct terms produced by $a_1$, $a_2$, and $a_3$. These terms also produce $a_2$ and $2c + a_2$, making six terms, and hence two of them must be equal. If $a_2$ is equal to one of the other terms, it must be $2c$ as all other possibilities contradict our hypothesis. If $2c + a_2$ is equal to one of the other terms, it must be 0 as, again, all other possibilities contradict the hypothesis. (For example, $2c + a_2 = c$ requires $a_2 = -c$.) In this case, then $a_2 = -2c$ and the result holds.

We now assume that $a_1$, $a_2$, and $a_3$ are all distinct. Hence four distinct terms produced are 0, $a_1$, $a_2$ and $a_3$. If $a_1 + a_2$ is one of these four terms, then it must be either 0 or $a_3$. If $a_1 + a_2 = 0$, set $a_1 = c$ and $a_2 = -c$. The terms produced are now 0, $c$, $-c$, $a_3$, $a_3 + c$, and $a_3 - c$. If $a_3 = \pm c$, then only four terms are produced and so this cannot happen. One of the last two must equal one of the previous four, and the only possibilities are $a_3 + c = -c$ and $a_3 - c = c$. This gives $a_3 = \pm 2c$ as desired.

If $a_1 + a_2 = a_3$, then the terms produced are 0, $a_1$, $a_2$, $a_1 + a_2$, $2a_1 + a_2$, $a_1 + 2a_2$, and $2a_1 + 2a_2$. The first four are distinct. If $2a_1 + a_2$ is equal to one of the first four, it must be 0 (as $a_3 \neq 0$). This gives $a_2 = -2a_1$ and $a_3 = a_1 + a_2 = -a_1$ and hence the situation is as desired with $c = a_1$. Similarly, the result holds if $a_1 + 2a_2$ is equal to one of the first four. As we have that $2a_1 + a_2 \neq a_1 + 2a_2$, this concludes the proof for this case.

We have now reduced to the case that 0, $a_1$, $a_2$, $a_3$, and $a_1+a_2$ are the five distinct elements produced. The same argument as above handles the case of $a_1 + a_3$ equal to one of 0, $a_1$, $a_2$, and $a_3$. This leaves only $a_1 + a_3 = a_1 + a_2$, but this contradicts $a_2 \neq a_3$. This completes the proof of the base case, that $k = 3$. Assume the result for $k$ and let us now prove it for $k + 1$.

If $M$ is the set of elements produced by $a_1, \ldots, a_k$, then the set of elements produced by all $k + 1$ elements is $M \cup (M + a_{k+1})$. Therefore, the size of $M$ is at most $k+3$. However, if $M$ has size $k+3$, then $M$ contains the subgroup generated by $a_{k+1}$ and hence all of $\mathbb{Z}_p$, contrary to hypothesis.

Let us take now the case that $M$ has size $k + 1$, the minimum possible. Then all of $a_1, \ldots, a_k$ are equal to $\pm c$. Suppose that $\ell$ of them are equal to $c$ where $0 \leq \ell \leq k$. Then $M$ is the set of $rc$ with $\ell - k \leq r \leq \ell$. Hence $M + a_{k+1}$ is the set $rc + a_{k+1}$ with $\ell - k \leq r \leq \ell$. In order for these two sets to overlap in all but two places, we must have $a_{k+1}+(\ell-k)c=(\ell-k+2)c$ or $a_{k+1}+(\ell-k)c=(\ell-k-2)c$. These require $a_{k+1}=2c$ and $a_{k+1}=-2c$, respectively.

This leaves only the case that $M$ has size $k+2$ and hence meets the induction hypothesis. Choose $c$ so that the one term is $2c$ and all other terms are $\pm c$. Suppose that $\ell$ of them are $c$ where $0\le\ell\le k-1$. Here $M$ consists of the elements $rc$ with $\ell-(k-1)\le r\le \ell+2$. (Note that if $\ell=0$, then $(\ell+1)c$ is produced as $2c+(-c)$.) Hence $M+a_{k+1}$ is the set $rc+a_{k+1}$ with $\ell-(k-1)\le r\le\ell+2$. In order for these two sets to overlap in all but one place, we must have $a_{k+1}+(\ell-(k-1))c=(\ell-(k-1)-1)c$ or $a_{k+1}+(\ell-(k-1))c=(\ell-(k-1)+1)c$. These two cases require $a_{k+1}=-c$ or $a_{k+1}=c$ respectively and this completes the proof of the lemma. □

The next result implies that if $L$ has derived length two and each $[\overbrace{L_a,\dots,L_a}^{(p-1)/2}]=0$, then $L$ has nilpotency class at most $p-2$. This additional hypothesis will be achieved by taking a quotient by a certain term of the upper central series of $L$.

**Proposition 5.2** *If $L$ is a $\mathbb{Z}_p$-graded Lie ring with $L_0=0$ and $[\overbrace{L_a,\dots,L_a}^{(p-1)/2}]=0$ for all $1\le a\le p-1$ and $I$, $J$, and $K$ are homogeneous ideals of $L$ such that $[[I,J],[K,K]]=0$, then $[I,J,\overbrace{K,\dots,K}^{p-3}]=0$.*

**Proof** It suffices to prove that each $[I_a,J_b,K_{c_1},\dots,K_{c_{p-3}}]=0$ for all non-zero $a$, $b$, $c_1,\dots,c_{p-3}$ in $\mathbb{Z}_p$. As $[I,J]$ commutes with $[K,K]$, if $c_1,\dots,c_{p-3}$ produce $-(a+b)$, then the term is clearly equal to 0. We are left with the case that $c_1,\dots,c_{p-3}$ produce at most $p-1$ terms of $\mathbb{Z}_p$, not including $-(a+b)$.

Let us first deal with the case that they produce exactly $p-2$ terms of $\mathbb{Z}_p$. Then each $c_i$ is equal to $\pm c$ for some non-zero $c$ in $\mathbb{Z}_p$. Choose $c$ so that $\ell$ of the $c_i$ are equal to $c$ with $(p-3)/2\le\ell\le p-3$. The terms produced by the $c_i$ are $0,c,\dots,\ell c$ and $-c,-2c,\dots,(p-3-\ell)(-c)$.

We have that $-(a+b)$ is equal to either $(\ell+1)c$ or $(\ell+2)c$. Let us first take the case that $-(a+b)=(\ell+2)c$. Then $[I_a,J_b,\overbrace{K_{-c},\dots,K_{-c}}^{p-3-\ell},\overbrace{K_c,\dots,K_c}^{\ell}]$ is contained in $[L_c,\overbrace{L_c,\dots,L_c}^{\ell}]$ and by hypothesis this is 0 as $\ell+1\ge(p-1)/2$.

We now take $-(a+b)=(\ell+1)c$. Write $a=j(-c)$ and $b=k(-c)$. There are two possibilities, either $j$ and $k$ are both at least $\ell+2$ or they are both at most $\ell$.

We now assume $j$ and $k$ are at least $\ell+2$ and $j+k=p+\ell+1$. Repeated applications of the Jacobi Identity imply that

$$[I_a,J_b,\overbrace{K_{-c},\dots,K_{-c}}^{p-3-\ell},\overbrace{K_c,\dots,K_c}^{\ell}]$$
$$\subseteq\sum_{m=0}^{p-3-\ell}[I_a,\overbrace{K_{-c},\dots,K_{-c}}^{m},[J_b,\overbrace{K_{-c},\dots,K_{-c}}^{p-3-\ell-m}],\overbrace{K_c,\dots,K_c}^{\ell}]$$

Note that $[L_a, \overbrace{L_{-c}, \ldots, L_{-c}}^{p-j}] = 0$ so if $m \geq p - j$, then the $m$-th summand is 0. Similarly, the summand is 0 if $p - 3 - \ell - m \geq p - k$. Thus, the only terms we need consider are those in which $k - 3 - \ell < m < p - j$. Thus the term is zero unless $k - 3 - \ell < m < k - 1 - \ell$. Hence $m = k - 2 - \ell = p - j - 1$. Here the $m$-th summand is contained in

$$[L_{j(-c)}, \overbrace{L_{-c}, \ldots, L_{-c}}^{p-1-j}, [L_{k(-c)}, \overbrace{L_{-c}, \ldots, L_{-c}}^{p-1-k}], \overbrace{L_c, \ldots, L_c}^{\ell}] \subseteq [L_c, L_c, \overbrace{L_c, \ldots, L_c}^{\ell}] = 0$$

where the last equality follows by hypothesis as $\ell + 2 \geq (p+1)/2$.

Assume now that $j$ and $k$ are at most $\ell$. Again using the Jacobi Identity repeatedly, we conclude that

$$\begin{aligned}
&[I_a, J_b, \overbrace{K_c, \ldots, K_c}^{\ell}, \overbrace{K_{-c}, \ldots, K_{-c}}^{p-3-\ell}] \\
&\quad \subseteq \sum_{m=0}^{\ell} [I_a, \overbrace{K_c, \ldots, K_c}^{m}, [J_b, \overbrace{K_c, \ldots, K_c}^{\ell-m}], \overbrace{K_{-c}, \ldots, K_{-c}}^{p-3-\ell}]
\end{aligned}$$

Note that $[L_a, \overbrace{L_c, \ldots, L_c}^{j}] = 0$ and hence the $m$-th summand is 0 when $m \geq j$. Similarly, the $m$-th summand is 0 if $\ell - m \geq k$. Thus we consider $m$ such that $\ell - k < m < j$. As $j + k = \ell + 1$, this requires $j - 1 < m < j$. Hence, all summands are 0, completing the proof when $c_1, \ldots, c_{p-3}$ produce exactly $p - 2$ terms.

We now consider the case where the $c_1, \ldots, c_{p-3}$ produce exactly $p - 1$ terms. Using Lemma 5.1, choose $c$ so that the one additional term is $2c$. Let $\ell$ be the number of the $c_i$ that are equal to $c$, here $0 \leq \ell \leq p - 4$. The $c_i$ then produce all numbers except $(\ell + 3)c$ and hence we must have $-(a + b) = (\ell + 3)c$. Here now, $[I_a, J_b, K_{2c}, \overbrace{K_c, \ldots, K_c}^{\ell}, \overbrace{K_{-c}, \ldots, K_{-c}}^{p-4-\ell}]$ is contained in $[L_{-c}, \overbrace{L_{-c}, \ldots, L_{-c}}^{p-4-\ell}]$ and hence is 0 if $p - 3 - \ell \geq (p-1)/2$, or, equivalently, if $\ell \leq (p-5)/2$. Also, $[I_a, J_b, \overbrace{K_{-c}, \ldots, K_{-c}}^{p-4-\ell}, \overbrace{K_c, \ldots, K_c}^{\ell}, K_{2c}]$ is contained in $[L_c, \overbrace{L_c, \ldots, L_c}^{\ell}, L_{2c}]$ and hence is equal to 0 if $\ell + 1 \geq (p-1)/2$ or $\ell \geq (p-3)/2$. As this handles all cases of $\ell$, we see that the term must be 0. This completes the proof of the proposition. □

We now locate these ideals $[I, J]$ in the upper central series of $L$.

**Proposition 5.3** *If $L$ is a $\mathbb{Z}_p$-graded Lie ring with $L_0 = 0$ and $[\overbrace{L_a, \ldots, L_a}^{(p-1)/2}] = 0$ for all $1 \leq a \leq p - 1$ and $I$ and $J$ are homogeneous ideals of $L$ such that $[[I, J], L^{(s)}] = 0$, then $[I, J]$ is contained in $Z_{(p-3)^s}(L)$.*

**Proof** We proceed by induction on $s$. For $s = 0$, the result is immediate. For $s = 1$, this is a consequence of Proposition 5.2 using $L$ for the $K$ in the proposition.

We now assume the result for $s$ and prove it for $s+1$. Proposition 5.2 implies that $[I, J, \overbrace{L^{(s)}, \ldots, L^{(s)}}^{p-3}] = 0$. Therefore, $[[I, J, \overbrace{L^{(s)}, \ldots, L^{(s)}}^{p-5}], L^{(s)}, L^{(s)}] = 0$ and hence, by the induction hypothesis, $[I, J, \overbrace{L^{(s)}, \ldots, L^{(s)}}^{p-4}]$ is contained in the $(p-3)^s$-th term of the upper central series of $L$. Working in the quotient $L/Z_{(p-3)^s}(L)$, we have that $[[I, J, \overbrace{L^{(s)}, \ldots, L^{(s)}}^{p-6}], L^{(s)}, L^{(s)}] = 0$ and so $[I, J, \overbrace{L^{(s)}, \ldots, L^{(s)}}^{p-5}]$ is in the $(p-3)^s$-th center of this quotient ring by the induction hypothesis and hence in the $2(p-3)^s$-th center of $L$. Repeating the argument, we eventually deduce that $[I, J]$ is contained in the $(p-3)(p-3)^s$-th center of $L$. □

We now bound the nilpotency class for $L$ under this additional assumption.

**Proposition 5.4** *If $L$ is a $\mathbb{Z}_p$-graded Lie ring with $L_0 = 0$ and $[\overbrace{L_a, \ldots, L_a}^{(p-1)/2}] = 0$ for all $1 \leq a \leq p-1$ and $L$ has derived length $s$, then $L$ has nilpotency class at most*

$$\frac{(p-3)^s - 1}{p-4} = \sum_{i=0}^{s-1} (p-3)^i.$$

**Proof** We proceed by induction on $s$. For $s = 1$ the ring has nilpotency class 1, as desired. Now assume the result for $s$ and let us prove it for $s+1$.

Then $L/L^{(s)}$ has nilpotency class at most $\sum_{i=0}^{s-1}(p-3)^i$ and hence it suffices to prove that $L^{(s)}$ is contained in the $(p-3)^s$-th center of $L$. As $0 = L^{(s+1)} = [L^{(s-1)}, L^{(s-1)}, L^{(s)}]$, Proposition 5.3 implies that $L^{(s)} = [L^{(s-1)}, L^{(s-1)}]$ is contained in the $(p-3)^s$-th center of $L$, as desired. □

We are now in position to give the proof of Theorem A.

**Proof of Theorem A** Proposition 3.1 and Lemma 3.7 of [13] imply that the ideal $L^{(2^{(p-5)/2}-1)}$ is contained in the subalgebra $\langle L_1, L_{(p+1)/2}, L_{(p+3)/2}, \ldots, L_{p-1} \rangle$. One can check that $\langle L_{(p+1)/2}, L_{(p+3)/2}, \ldots, L_{p-1} \rangle$ is an ideal in this algebra, and so we can take the quotient. This quotient has nilpotency class at most $(p-1)/2$. Hence $L^{(2^{(p-5)/2}+\lfloor \log_2((p-1)/2) \rfloor)}$ is contained in $\langle L_{(p+1)/2}, L_{(p+3)/2}, \ldots, L_{p-1} \rangle$. Call $k = 2^{(p-5)/2} + \lfloor \log_2((p-1)/2) \rfloor$.

Let $j \geq (p+1)/2$ and consider $[\overbrace{L_1, \ldots, L_1}^{(p-1)/2}, L_j]$. As described in Subsection 3.2, this is contained in $[L_j, \overbrace{L_1, \ldots, L_1}^{p-j}, \overbrace{L_1, \ldots, L_1}^{((p-1)/2)-(p-j)}] = 0$. Therefore, $[\overbrace{L_1, \ldots, L_1}^{(p-1)/2}]$ commutes with $L^{(k)}$. By changing the grading, we can conclude that $[\overbrace{L_a, \ldots, L_a}^{(p-1)/2}]$ is commutes with $L^{(k)}$ for all $1 \leq a \leq p-1$. Call $I$ the ideal generated by $[\overbrace{L_a, \ldots, L_a}^{(p-1)/2}]$

for $1 \le a \le p-1$. This is a homogeneous ideal contained in $C_L(L^{(k)})$ and contained in $L'$. Hence, by Lemma 4.3 of [13], we conclude that $I$ is contained in $Z_{(p-2)^k}(L)$.

We now work in $\tilde{L} = L/Z_{(p-2)^k}(L)$. We know that $\tilde{L}^{(2^{p-5}-1)}$ is contained in $\langle \tilde{L}_1, \tilde{L}_{p-2}, \tilde{L}_{p-1}\rangle$. The subalgebra $\langle \tilde{L}_{p-2}, \tilde{L}_{p-1}\rangle$ is an ideal in here and the quotient has nilpotency class at most $(p-3)/2$ as $[\overbrace{\tilde{L}_1, \ldots, \tilde{L}_1}^{(p-1)/2}] = 0$. Continuing the argument as in [13], we conclude that $\tilde{L}$ has derived length at most $s = 2^{p-5} + \lfloor \log_2((p-3)/2)\rfloor + 2$. As each $[\overbrace{\tilde{L}_a, \ldots, \tilde{L}_a}^{(p-1)/2}] = 0$, we conclude that $\tilde{L}$ has nilpotency class at most $((p-3)^s-1)/(p-4)$. Therefore the nilpotency class of $L$ is at most $((p-3)^s-1)/(p-4) + (p-2)^k$, as desired. □

We end with a table giving the historical development of the bounds on $h(p)$ for $p \ge 13$. We abbreviate $\lfloor \log_2(x)\rfloor$ by $\lambda(x)$.

| | dl | nc (dl=$s$) | $p = 13$ |
|---|---|---|---|
| KK 1963 [7, 8] | $2^{p-1}-1$ | $\dfrac{(p-1)^s-1}{p-2}$ | $12^{4095}+12^{4094}+\cdots+1$ |
| Meixner 1980 [9] | | $(p-1)^{s-1}$ | $12^{2045}$ |
| STW 2005 [13] | $2^{p-5}+2+\lambda(p-3)$ | $\dfrac{(p-2)^s-1}{p-3}$ | $11^{261}+11^{260}+\cdots+1$ |
| Thm A | | $\dfrac{(p-3)^s-1+(p-4)(p-2)^k}{p-4}$ where $k = 2^{(p-5)/2}+\lambda(\frac{p-1}{2})$ and we can take $s = 2^{p-5}+\lambda(\frac{p-3}{2})+2$ | $10^{260}+10^{259}+\cdots+1+11^{18}$ |

We end by noting that the improvement on the bound for $h(13)$ in [13] exceeds the new bound by a factor of $6\times 10^{11}$.

## References


[1] M. Favaretto, *Contributi alla soluzione di una congettura relativa alle algebre di Lie nilpotenti*, Ph.D. Thesis, Università Ca' Foscari Venezia, 1998.

[2] The GAP Group, GAP – Groups, Algorithms, and Programming, Version 4.4; 2005. (http://www.gap-system.org)

[3] D. Gorenstein, *Finite groups*, Harper and Row, New York, 1968.

[4] G. Higman, Groups and Lie rings having automorphisms without non-trivial fixed points, *J. London Math. Soc.* **32** (1957), 321–334.

[5] I. Hughes, Groups with fixed-point-free automorphisms, *C. R. Math. Rep. Acad. Sci. Canada* **7** (1985), 61–66.

[6] E. I. Khukhro, *p-Automorphisms of Finite p-Groups* (Cambridge Univ. Press, 1998).

[7] V. A. Kreknin, Solvability of Lie algebras with a regular automorphism of finite period, *Soviet Math. Dokl.* **4** (1963), 683–685.

[8] V. A. Kreknin and A. I. Kostrikin, Lie algebras with regular automorphisms, *Soviet Math. Dokl.* **4** (1963), 355–358.
[9] T. Meixner, Metabelsche Gruppen mit einem fixpunktfreien Automorphismus von Primzahlordnung, *Arch. Math. (Basel)* **35** (1980), no. 6, 497–500.
[10] B. H. Neumann, On the commutativity of addition, *J. London Math. Soc.* **15** (1940), 203–208.
[11] B. H. Neumann, Groups with automorphisms that leave only the neutral element fixed, *Archiv der Math.* **7** (1956), 1–5.
[12] B. Scimemi, unpublished.
[13] P. Shumyatsky, A. Tamarozzi and L. Wilson, $\mathbb{Z}_n$-graded Lie rings, *J. Algebra* **283** (2005), 149–160.
[14] J. G. Thompson, Finite groups with fixed-point-free automorphisms of prime order, *Proc. Nat. Acad. Sci. USA* **45** (1959), 578–581.